INTRODUCTION TO SWITCHING THEORY AND LOGICAL DESIGN

INTRODUCTION TO
SWITCHING THEORY
AND LOGICAL DESIGN

THIRD EDITION

Frederick J. Hill

Gerald R. Peterson

University of Arizona

JOHN WILEY & SONS

New York • Chichester • Brisbane • Toronto • Singapore

Library of Congress Cataloging in Publication Data:

Hill, Frederick J.
    Introduction to switching theory and logical design.

    Includes bibliographies and index.
    1. Switching circuits.   2. Digital electronics.
3. Logic design.   I. Peterson, Gerald R., joint
author.   II. Title.

TK7868.S9H5 1981   621.3815′37   80-20333
ISBN 0-471-04273-0

Printed in the United States of America

10 9

PREFACE

In the field of digital computers, there is an area somewhere between circuit design and system formulation which is usually identified as switching theory. Although this topic must be mastered by the digital computer engineer, it does not answer all the questions concerning layout of an efficient digital computing system. The remaining questions, which are less susceptible to formalization, may be grouped under the heading "Digital System Theory."

Our approach is to follow the basic framework of switching theory. We hope, however, to motivate the student by presenting examples of the many problems which appear repeatedly in the design of digital systems. It is not our intent to provide the student with a detailed knowledge of such specialized system topics as computer arithmetic. Instead, we attempt to provide the framework through which the student might develop a sound design philosophy applicable to any digital design problem. Thus, we do not feel that the use of the term "logical design" in the title is unjustified.

The usefulness of switching theory is not restricted to engineers actively engaged in computer design. In fact, it is our contention that almost every design engineer in the field of electronics will have some opportunity to draw on this subject. Applications occur whenever information in communication, control, or instrumentation systems, for example, is handled in other than analog form.

The above three paragraphs, which began the preface of the first edition of this book, seem to us to be even more accurate today than they were 12 years ago. In these years, digital techniques have become virtually standard in many areas once considered the exclusive preserve of analog techniques. This "digitalization" of electrical engineering has become so pervasive that it is a rare electrical engineer who will not have some contact with digital design problems.

Since completion of the second edition of this book, the most important event in digital systems design has been the almost meteoric increase in both sophistication and application of microprocessors. At this writing 8- and 16-bit processors form the core of inexpensive microcomputer systems. These processors with associated memory are also used extensively in place of networks of hardwired combinational and sequential logic. Microprocessor realization will almost always result in fewer components and fewer wires than the hardwired networks they replace. Similarly the microprocessors have opened new applications for which the cost of hardwired networks would have been prohibitive.

v

Some might reason from the above that microprocessors have made the material of this book obsolete. This is not so! We now find ourselves in a situation in which an electrical engineer must master microprocessor programming as well as traditional logic design. In most curricula there are now two courses where previously there was one.

It is our opinion that microprocessor programming is a sufficiently comprehensive topic that it is best treated in a separate textbook. We have, therefore, not added programming to this third edition. The amount of engineering time devoted to conventional logic design has increased rather than decreased since the appearance of the microprocessor. Almost every microprocessor will be surrounded by a network of conventional parts. In addition, some applications cannot afford the speed penalty imposed by a software routine within a microprocessor.

Perhaps the major change in the third edition results from our admission that a hardware description language is an effective tool for the design of even relatively small sequential circuits. Formal-state table minimization will be employed only rarely in clock-mode design. We have therefore completely rewritten the chapter on the language AHPL and placed it immediately after Chapter 10, the first chapter to treat the synthesis of sequential circuits. Chapter 12 continues the development of AHPL and adapts this language for the specification of interconnections between MSI parts.

Chapter 5 has been rewritten to reflect recent innovations in digital circuit design. The important topic of programmed logic arrays (PLAs) has been added to Chapter 8. The chapter on level-mode analysis and synthesis has been improved and extended to include the edge-triggered D flip-flop. The chapter on threshold logic has been dropped with a small part of this material added to Chapter 16.

Chapters 1 and 2 were written with the goal of providing student motivation. We have endeavored to sustain this motivation throughout the book by treating with special care points that might otherwise become sources of confusion. It is our belief that if questions concerning justifications behind manipulative procedures go unanswered, the student will be discouraged. Our intention to avoid such pitfalls is exemplified in Chapter 4 by a reliance on Huntington's original postulates and careful distinction between the algebra of the Karnaugh map and the two-valued algebra of truth values. (This is essentially the same as the distinction between Boolean forms and switching functions, which is made with great care in some more advanced texts.) Our treatment of Boolean algebra is formal, to be sure, but we have found that it represents no conceptual difficulties to the student.

The introductory chapter should occupy no more than one day of class time, or it could be left as a reading assignment if class time is critical. The binary number system is taken up in Chapter 2 to provide a background for the binary notation of Boolean functions and maps in Chapters 4 and 6. The subject is allotted a separate chapter that is placed prior to the introduction of Boolean algebra to minimize the chance of confusing the two concepts. The additional material on binary arithmetic is essential only for examples and might be omitted, or referred back to, if desired.

Chapters 3 and 4 provide the basic logical and algebraic basis for switching theory and certainly have not diminished in importance in any way. We believe the benefits to the students will be great if the topics of these chapters are covered in order with no omissions. Coverage of Chapter 5 on components may be adjusted to suit the purposes of the instructor.

Chapter 6 is unchanged from the second edition. Chapter 7 has been expanded to provide a brief treatment of multilevel design. Rapid developments in components, particularly integrated circuits, have greatly changed the relative importance of various cost factors in digital design, but the basic correlation between simple algebraic forms and economical designs remains unchanged. And we believe that the Karnaugh map and Quine-McCluskey minimization remain unchallenged as the most natural, powerful, and widely applicable tools for simplifying switching functions. These two chapters should be covered in detail.

Chapter 9 has been expanded to provide more thorough coverage of various types of flip-flops. Chapter 10 is changed only in the inclusion of material contrasting Mealy and Moore model sequential circuits. The material on pulse-mode design has been updated and incorporated in Chapter 11 with an introduction to AHPL and controlled clocking.

In this book we provide only a brief introduction to AHPL. Our companion volume, *Digital Systems: Hardware Organization and Design*, 2nd Ed. (Wiley, 1978) provides a thorough treatment of the application of a register transfer language to the design of large-scale digital systems. This book applies AHPL to the design of circuits that are too large to be efficiently attacked by the logic design methods presented earlier but are not large enough to be considered full-scale digital systems in the usual sense. The reader who has used both texts will notice a difference in the primary hardware implementation of the control unit in the two books. In this book, we rely on the reader's familiarity with standard clock-mode sequential circuits, while in the other book we were primarily concerned with describing large digital hardware systems to students who may or may not have been expert in sequential circuits. The language used in Chapters 11 and 12 to describe hardware is completely consistent with the AHPL of the other book.

Since publication of the second edition, two programs have been developed to support AHPL. These include a function-level simulator and a hardware compiler. Both programs are available in versions for the DEC 10, the CDC CYBER, and the IBM 370.

The following diagram illustrates the prerequisite relationships among the various chapters, as an aid to instructors in planning courses based on this book. It can readily be seen from the chart that Chapters 2, 3, 4, 6, 9, 10, and 11 would

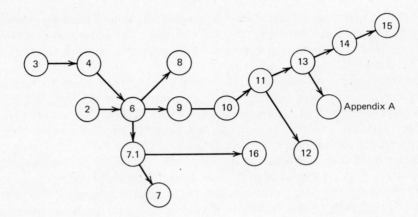

necessarily be included in almost any course structure. Beyond this, the instructor has considerable freedom in structuring the course. Of the remaining restrictions, the most critical is that the level-mode material of Chapters 14 and 15 has as a prerequisite the incompletely specified state table minimization of Chapter 13 which in turn requires Chapter 11. One perfectly satisfactory approach is to take up the chapters in numerical order. Where students already have a background in combinational logic, a course can be begun with any of Chapters 7, 8, or 9. There is more than enough material to constitute a one-semester course beginning with Chapter 9.

Obviously, many persons contribute to the development of any book, and there is no way we could acknowledge them all. Our primary thanks must go to our families, to whom it must seem as though we will never stop writing and revising books.

F. J. Hill
G. R. Peterson

CONTENTS

1

INTRODUCTION

2

NUMBER SYSTEMS

3

TRUTH FUNCTIONS

4

BOOLEAN ALGEBRA

5

SWITCHING DEVICES

6

MINIMIZATION OF BOOLEAN FUNCTIONS

7

TABULAR MINIMIZATION AND MULTIPLE-OUTPUT CIRCUITS

8

SPECIAL REALIZATIONS AND CODES

9

INTRODUCTION TO SEQUENTIAL CIRCUITS

10

SYNTHESIS OF CLOCK-MODE SEQUENTIAL CIRCUITS

11

CLOCK-CONTROL AND PULSE-MODE CIRCUITS

12

LOGICAL DESIGN USING MSI AND LSI PARTS

13

INCOMPLETELY SPECIFIED SEQUENTIAL CIRCUITS

14

LEVEL-MODE SEQUENTIAL CIRCUITS

15

A SECOND LOOK AT FLIP-FLOPS AND TIMING

16

COMBINATIONAL FUNCTIONS WITH SPECIAL PROPERTIES

APPENDIX **A**

SELECTION OF MINIMAL CLOSED COVERS

APPENDIX **B**

RELAY CIRCUITS

INTRODUCTION TO
SWITCHING THEORY
AND LOGICAL DESIGN

1

INTRODUCTION

1.1 Characteristics of Digital Systems

The purpose of this book is to present the basic mathematical tools used in the design of digital systems. This statement raises the question: What is a *digital* system? Broadly speaking, a *digital* system is any system for the transmission or processing of information in which the information is represented by physical quantities (signals) that are so constrained as to take on only discrete values. If the signals are constrained to only two discrete values, the system is *binary*. The idea of a digital system can be clarified by examining one problem that can be solved by incorporating such a system. As a very specific example, we choose the activation of the control surfaces of an aircraft.

The pilot controls the motion of the plane with various pedals and levers, such as the "stick." In older craft and even today in small planes, these pedals and levers move cables that directly move the control surfaces. In large aircraft, however, the control forces required are too large for this to be practical, and some sort of mechanical assistance must be provided. Figure 1.1 shows a commonly used system. Motion of the stick turns the shaft of a potentiometer, a variable resistor, which puts out a voltage corresponding to the position of the stick, indicating the desired position of the control surface. A second potentiometer, similarly mounted on the shaft of the control surface, puts out a voltage corresponding to the position of the surface. A device called a servo controller compares the two voltages and delivers a signal to control an actuator to move the control surface to the desired position.

This system is an example of a *continuous*, or analog, system, so called because the voltages from the potentiometers, and elsewhere in the system, vary continuously over some range. Such systems can work quite well and are widely used, but they do have some problems, such as those associated with externally induced disturbances (noise) that may cause the system to respond incorrectly. In terms of transmission of electrical signals the interior of an aircraft is a very "noisy" environment. Noise may be caused by resistance change due to temperature variation as well as natural and artificial electromagnetic disturbances in the plane and in the atmosphere. Since the control voltages vary continuously, even a small change in voltage because of noise may cause significant errors in the response of the system.

1

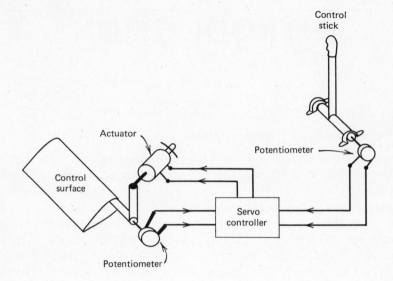

The susceptibility of the system to noise can be greatly reduced by applying *discrete*, or digital, techniques. Assume that the range of motion of the control surface is ± 30 degrees, and position must be specified to an accuracy of 1 degree. One possible approach is to replace the potentiometers with 60-position switches that will place a voltage on one of 60 wires, depending on the angle, virtually eliminating the possibility of error.

Although such a scheme may be theoretically attractive, it is almost certainly impractical to run sets of 60 wires from each switch to the controller, particularly on an airplane, where volume and weight are at a premium. To reduce the number of wires, we can use a more complicated switch, known as a *shaft position encoder*, to apply different combinations of voltage and no-voltage to 6 wires to indicate the position of the control stick or surface. Typically the 64 combinations of voltage or no voltage on the 6 output wires might be the binary coding of the integer measure of the angle between 0 and 63 degrees.

In addition to converting the position information to digital form, we can also digitize the control of the motor by using a *stepping motor*. A stepping motor is effectively a digital motor, with two control inputs. A pulse on one input will cause the motor to step forward a fixed amount, say 1 degree, and a pulse on the other line will cause it to step in the other direction. Regardless of pulse duration or magnitude above some minimum, the motor will move the same amount. Figure 1.2 shows a possible configuration using shaft position encoders

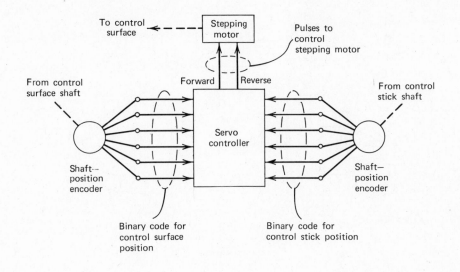

and a stepping motor. When the pilot moves the control stick, the servo controller senses the difference between the two position codes and issues pulses to the stepping motor to move the control surface until the positions coincide.

The servo controller could be a special-purpose digital system or it might equally well be a digital computer. Since digital computers are now readily available in single integrated circuit packages (*microprocessors*), the latter is probably more likely. The system, including a control computer, is depicted in Fig. 1.3. A single computer would probably be used for controlling all of the control surfaces on the aircraft. With a computer in the system, many other interesting possibilities open up. By providing the computer with inputs indicating course, elevation, and aircraft altitude, we have the basis of an autopilot. The pilot provides the desired course and altitude, and the computer flies the plane. It will also probably monitor engine performance, control thrust, etc.

Transmission of codes simultaneously over a set of lines is known as *parallel* data transmission. It works well but may be too expensive in some cases. Six lines is less than 60 but may still not be the most economical approach. Another possibility is to represent each position in terms of a *series* of pulses sent over a single line, in which case we have *serial* data transmission. There are many ways to encode information into serial form. One method, used in Morse code, is to represent each item of information, such as a letter of the alphabet or the position of a shaft, by a unique combination of long and short pulses. Serial transmission will be slower than parallel transmission, but it will usually be

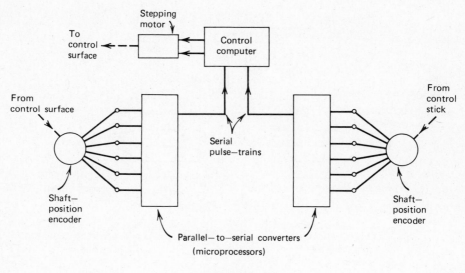

FIGURE 1.3

cheaper, since it requires only one data path instead of many. This "trade-off" between speed and cost almost always exists in the design of the digital systems, and one of the difficult tasks of the designer is to decide how much speed is needed and affordable.

It might be possible to develop a shaft-position encoder that would directly develop a serial output, but it is more likely that a parallel-to-serial converter would be used in conjunction with a conventional encoder. This is the approach depicted in Fig. 1.3. The converter might be a special-purpose unit or it might be a small microprocessor programmed to sample the parallel output of the encoder at appropriate intervals of, perhaps 1000 times a second, and convert it to a serial pulse train to be sent to the controller.

The advantages to be gained through the use of digital techniques go far beyond the increased immunity to noise discussed previously. In the servo controller of the analog system (Fig. 1.1) all the electronic components will have to operate in continuous mode. In the controllers and converters in the digital systems, all the electronic devices will operate in the discrete mode, often called the switching mode, because most of the devices act very much like switches, being either on or off. For a variety of technical reasons, devices that function in the switching mode are simpler and cheaper to construct than those that function in the continuous mode, and they function much more reliably. Using digital techniques, one can build a complete computer, capable of controlling the entire aircraft, for a fraction of the cost of a single analog servo controller such as that shown in Fig. 1.1, and this computer will be more reliable and require less service than the analog controller.

1.2   A Brief Historical Note

One of the earliest and the world's largest digital systems is the dial telephone system. When you turn the dial and then release it, the spinning dial activates a switch that puts out a number of pulses equal to the number dialed. These pulses are then counted and recorded by special switches at the central office. After all the numbers have been dialed and recorded, they control the operation of a number of switches that connect you to the desired party. A switch is a digital device since it has only two conditions, open or closed. The dial system is thus a digital system—in fact, the world's largest digital system.

The dial telephone was invented many years ago, and the dial system was the first digital system of any consequence. As a result, telephone scientists and engineers made the first contributions to a theory of digital systems. Because the telephone system was largely constructed of switches, this theory became known as *switching theory*, the name by which it is still known today.

Shortly before World War II, it was recognized that the theory and technology developed by telephone engineers could be applied to the design of digital computing machines. In 1939, Harvard University was given a contract to develop a machine to compute ballistic tables for the U.S. Navy. The result of this work was Harvard Mark 1, which went into operation in 1943. In the next few years, more machines were built in research laboratories around the world. The first commercially produced computer, Univac I, went into operation in 1951, and more and more large, digital computers were introduced in the next decade. These first computers all utilized vacuum tubes as the primary electronic element and were bulky, expensive, and consumed immense amounts of power.

The transistor, invented in 1948, first appeared in computers about 1958 and had totally supplanted vacuum tubes by 1960. The first "second-generation" computers were little more than transistor versions of their vacuum tube predecessors. By the middle of the 1960s, designers had found ways to exploit the many advantages of the transistor in areas such as reliability, size, and cost, leading to the first "third-generation" computer, the IBM System 360, and the first successful "minicomputer," the DEC PDP/8; the most significant development, however, was the integrated circuit, which made it possible to build complete digital building blocks, such as gates, flip-flops, counters, adders, etc., on single minute chips of silicon. As the technology developed, designers found ways to fit more and more complex systems onto integrated circuits, and the middle of the 1970s saw the introduction of the microprocessor, a complete computer on a single chip. Today, with continuing developments in microcircuits, we stand on the threshold of what some have christened the "digital age."

1.3 Organization of Digital Computers

Since digital computers form the most important single class of digital systems, it is appropriate to consider their basic organization. The basic block diagram of a typical general-purpose digital computer is shown in Fig. 1.4.

The input/output (I/O) section consists of such devices as card readers and punches, printers, magnetic tape units, and typewriters. All data and instructions are fed to the computer through the input section, and all results are returned through the output section.

All data and instructions passing through the input section go directly to *memory*, where they are stored. The memory consists of various devices, usually magnetic or electronic, by means of which information can be stored in such a manner as to be available to other sections of the computer at electronic speeds. All information fed to the computer is divided into basic units, called *words*, *bytes*, or *characters*, depending on the exact organization of the machine. The memory is in turn divided into individual locations, each capable of storing a single word, byte, or character. Each location has associated with it an identification number or *address* by means of which information can be directed to it or retrieved from it.

The information stored in memory consists of data and instructions. Within the machine there is no difference in form between the two; both are simply collections of numbers or characters. The difference lies in what is done with them. Data elements are sent to the *arithmetic-logic unit* (ALU), where they are processed. The ALU basically consists of a number of registers, electronic units

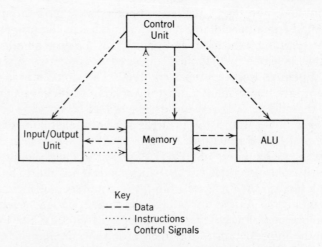

Key
--- Data
······· Instructions
—·—· Control Signals

FIGURE 1.4 Basic organization, typical digital computer.

for temporarily storing information, and certain logical circuits for performing operations on the numbers stored in the registers. There is often a special register, known as the *accumulator*, which is permanently connected to the *adder* circuits. The basic arithmetic operation of the computer consists of adding a number taken from memory to whatever number is already in the accumulator. The accumulator thus performs a function similar to that of the dials of a desk calculator, accumulating the running sum of numbers sent to it.

Instructions are sent to the control unit, where they are decoded or interpreted. The control unit decides what the instructions call for and sends electronic signals to the various other parts of the computer to cause the appropriate operations to take place. A typical command might be ADD4000, which tells the computer to take the number stored in location 4000 in memory and add it to whatever is already in the accumulator. The control unit will send out the proper signals to cause the contents of location 4000 to be read out and sent to the arithmetic unit, and then added to the accumulator. Other typical instructions might tell the computer to read a card and store the data in memory, or to print out certain data stored in memory, or to store the contents of the accumulator in memory.

A particularly important class of instructions is the class of decision, or *branch*, commands. The instructions making up a computer program are normally stored in order, in sequential locations in memory. The computer normally takes its instructions in sequence from succeeding locations. The branch command tells the computer to "branch" to a different location for its next instruction if certain conditions are met.

As an example, assume that a certain program calls for division. For obvious reasons, division by zero cannot be allowed, but, if the divisor is itself computed, we cannot be sure in advance that it will not be zero. To guard against this situation, we will first load the divisor into the accumulator. We will then use the *branch on zero* command, which tells the computer to inspect the number in the accumulator. If the number is not zero, the computer should take its next instruction from the next sequential location, in the normal manner. If the number is zero, the computer should branch to some other location, as specified, for its next command. At this branch location, the programmer will have stored a set of instructions for dealing with the contingency of a zero divisor.

1.4   Types of Digital Circuits

In this book, we will not be concerned with designing complete computers or other digital systems of comparable complexity. Such *systems design* is the subject of our other book. *Digital Systems: Hardware Organization and Design.* Here we will be concerned with the design of many types of circuits and subsystems that appear within computers and other types of digital systems. These circuits fall into two general categories, *combinational* and *sequential*. An example of a combinational circuit is one for making the branch decision described above. This circuit would have as its input the contents of the accumulator. Its output would take on one value if the accumulator contents were zero, another value if they were not. The name *combinational* conveys the idea that the output at any time is a function solely of the combination of all the inputs at that time. Another example of a combinational circuit in a computer is the adder; the inputs are the two numbers to be added, the output is the sum.

An example of a sequential circuit is the accumulator. Its output (i.e., its contents) at any given time will be a function not only of current inputs but also of some *sequence* of prior events, e.g., what numbers have been added into it since it was last cleared. Another sequential circuit in a computer is the program counter. Recall that the instructions making up a program are normally stored in consecutive (sequentially numbered) locations in memory and are executed in sequence unless a branch instruction is executed. The program counter is set to the address of the first instruction in the program and is incremented to succeeding values as instructions are executed sequentially, thus keeping track of progress through the program. It is a sequential circuit because its output (contents) is clearly a function of a series of prior events, i.e., its initial setting and how many times it has been incremented.

While digital computers certainly form the most prominent class of digital systems, digital techniques may be applied to many other areas, some quite unrelated to computers. Until fairly recently, digital techniques, were primarily restricted to computing applications, where their high cost could be justified. This situation was changed by the development of the integrated circuit, which has reduced the cost of individual logic elements from dollars to fractions of a cent. An even more dramatic development in recent years is the development of the microprocessor, which has made the power of computers available at a cost of tens or hundreds of dollars, instead of thousands or millions of dollars. As a result, digital techniques are now being applied to virtually every area of technology, and virtually every engineer will have occasion to use digital techniques no matter what the area of specialization.

One very important area for digital applications is instrumentation. For example, in a manufacturing plant, analog readings of various instruments could be connected to a multiplexer at a central point. The multiplexer, a sort of digital switch, would connect each line in turn to an analog-to-digital converter, which would convert each signal to digital form. Each digital signal would then be sent to an assembly register to be joined to an identifying signal generated by the multiplexer. The complete word, containing the identification of the signal and its value, could then be sent to a computer for processing or recorded for later processing. Such a system in which the data are recorded is often referred to as a *data-logging* system. Two more applications of digital systems are industrial process control and office word processing systems.

Digital traffic controllers measure traffic flow on cross streets and adjust the period of traffic signals to maximize flow. Digital clocks measure time simply by counting the cycles of line current. Digital antiskid controllers measure the speed of the wheels on trucks and adjust the braking force to prevent skids. Ignition controllers measure the operating parameters of automobile engines and adjust carburetion and timing to reduce emissions. Radios and other communications gear are tuned by digital frequency division and phase-locked loops. Digital meters, electronic calculators and digital watches have almost replaced their mechanical counterparts. Every day ingenious designers figure out new ways to apply digital techniques to their problems.

All these various types of systems vary widely in function and complexity, and many will have microprocessors as their central elements. Even where micro-processors are used, systems will also include interconnections of individual combinational and sequential circuits. The design of these individual circuits is the basic topic of this book.

2
NUMBER SYSTEMS

2.1 Introduction

A familiarity with the binary number system is necessary in this book for two primary reasons. First, binary numbers will appear frequently as a notational device. Second, digital computers generally work with binary numbers in some form, and many of the best examples of both combinational and sequential circuits are drawn from subsystems within computers. Some readers will already possess a working knowledge of the binary number system and, indeed, may have first encountered the topic in elementary school. Such readers may wish to skip this chapter and proceed directly to Chapter 3. However, this chapter should be covered by those readers who are not reasonably proficient in the binary number system. First, let us review what is involved in the definition of a number system. We write, for example:

$$7419$$

as a shortened notation for

$$7 \cdot 10^3 + 4 \cdot 10^2 + 1 \cdot 10^1 + 9 \cdot 10^0$$

Because the decimal number system is almost universal, the powers of 10 are implied without further notation. We say that 10 is the *radix* or *base* of the decimal number system. The fact that 10 was chosen by man as the base of the number system is commonly attributed to the 10 fingers on two hands. Most any other relatively small number (a large number of symbols would be awkward) would have done as well.

Consider the number

$$N = 110101$$

and assume that the base of N is 2. The number system with base-2 is called the *binary* number system. We may rewrite N as follows:

$$N = 1 \cdot 2^5 + 1 \cdot 2^4 + 0 \cdot 2^3 + 1 \cdot 2^2 + 0 \cdot 2^1 + 1 \cdot 2^0$$

$$= 32 + 16 + 0 + 4 + 0 + 1 = 53 \text{ (decimal)}$$

thus obtaining the decimal equivalent. Note that the binary number system requires only two symbols, 0 and 1, while the familiar 10 symbols are employed in the decimal system.

In contrast to men, digital computers do not have 10 fingers and are not biased toward the decimal system. They are constructed of physical devices that operate much more reliably when required to switch between only two stable operating points. Some examples are switches, relays, data transmission lines, transistors, etc. One of the stable operating points may be assigned to represent 1 and the other 0. Thus, base-2 must play some role within any digital system.

Many examples throughout the book will be problems in digital computer design. This, however, is not the only reason for introducing binary numbers at this point. We will find the binary number system indispensable as a notational device in the next few chapters.

The reader who is already familiar with arithmetic in other bases may skip rapidly over the next two sections.

2.2   Conversion Between Bases

Suppose that some number N is expressed in base-s. It may be converted to base-r by the following sequence of divisions carried out in base-s. The digits A_i are the remainders from each division, so that $A_i < r$.

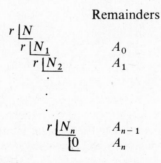

Alternatively, this division may be written as follows:

$$N = r \cdot N_1 + A_0$$
$$N_1 = r \cdot N_2 + A_1$$
$$\cdot$$
$$\cdot$$
$$\cdot$$
$$N_n = r \cdot 0 + A_n \tag{2.1}$$

or as

$$N = r(rN_2 + A_1) + A_0$$
$$= r^2 N_2 + rA_1 + A_0$$
$$= r^2(rN_3 + A_2) + rA_1 + A_0$$

$$= A_n r^n + A_{n-1} r^{n-1} + \cdots + A_1 r + A_0 \qquad (2.2)$$

Example 2.1

Convert 653_{10} (the subscript indicates base-10) to (a) base-2 and (b) base-5.

(a)

2	653	
2	326	1
2	163	0
2	81	1
2	40	1
2	20	0
2	10	0
2	5	0
2	2	1
2	1	0
	0	1

Check

$1010001101_2 = 512 + 128 + 8 + 4 + 1$
$= 653_{10}$

(b)

5	653	
5	130	3
5	26	0
5	5	1
5	1	0
	0	1

$10103_5 = 1 \times 5^4 + 1 \times 5^2 + 3 \times 5^0$
$= 1 \times 625 + 1 \times 25 + 3$
$= 653_{10}$

∎

Example 2.2

Convert the number 1606_{10} to base 12: The decimal numbers 10_{10} and 11_{10} become one-digit numbers in base 12, which may be represented by α and β, respectively.

12	1606	
12	133	$10 = \alpha$
12	11	1
	0	$11 = \beta$

Check

$\beta 1 \alpha_{12} = 11(144) + 1(12) + 10$
$= 1606_{10}$

∎

In each of the above examples, arithmetic was carried out in base-10. This is convenient because we are familiar with base-10 arithmetic. If a number is to be converted from base-r ($r \neq 10$) to some base-b, we have the choice of doing the arithmetic in base-r or else first converting to base-10 (as shown in Section 2.1) and then to base-b. Arithmetic in bases other than 10 is the subject of the next section.

The numbers we have discussed so far have been integers. Suppose it is desired to convert numbers with fractional parts from one number system to another. Consider a number

$$N = N_I + N_F = A_n r^n + \cdots + A_1 r + A_0 + A_{-1} r^{-1} + A_{-2} r^{-2} \cdots \quad (2.3)$$

where N_I and N_F are the integral and fractional parts, respectively. The fractional part in one base will always lead to the fractional part in any other base. Therefore, N_I may be converted as before and N_F converted separately as follows:

$$N_F = A_{-1} r^{-1} + A_{-2} r^{-2} + A_{-3} r^{-3} \cdots \quad (2.4)$$

To determine the coefficients A_{-1}, A_{-2}, A_{-3}, etc., for the base-r, we note that each of these coefficients is itself an integer. We first multiply by r.

$$r N_F = A_{-1} + A_{-2} r^{-1} + A_{-3} r^{-2} \quad (2.5)$$

Thus, the integral part of $r N_F$ is A_{-1}. We subtract A_{-1} and again multiply by r

$$r(r N_F - A_{-1}) = A_{-2} + A_{-3} r^{-1} + A_{-4} r^{-2} \quad (2.6)$$

thus determining A_{-2}. This process is continued until as many coefficients as desired are obtained. The process may not terminate.

Note that we have avoided the terms *decimal point* or *decimal places*. For base-2, for example, the appropriate terms are *binary point* and *binary places*.

Example 2.3

Convert the number 653.61_{10} to base-2.

$$\begin{array}{lll}
2 \cdot (0.61) = 1.22 & A_{-1} = 1 & \text{Check} \\
2 \cdot (0.22) = 0.44 & A_{-2} = 0 & 653.61_{10} = 1010001101.1001110 \cdots \\
2 \cdot (0.44) = 0.88 & A_{-3} = 0 & \qquad = 653 + \frac{1}{2} + \frac{1}{16} + \frac{1}{32} + \frac{1}{64} \cdots \\
2 \cdot (0.88) = 1.76 & A_{-4} = 1 & \qquad = 653 + 0.5 + 0.0625 + 0.03125 \\
2 \cdot (0.76) = 1.52 & A_{-5} = 1 & \qquad\quad + 0.015625 \\
2 \cdot (0.52) = 1.04 & A_{-6} = 1 & \qquad = 653.609375 \\
2 \cdot (0.04) = 0.08 & A_{-7} = 0 &
\end{array}$$

Base-8 (octal) and base-16 (hexadecimal) are of particular importance in the field of digital computers because of their close relationship to base-2. Since

Decimal Number	0	1	2	3	4	5	6	7	8	9	10	11	12	13	14	15
Hexadecimal Digit	0	1	2	3	4	5	6	7	8	9	*A*	*B*	*C*	*D*	*E*	*F*

FIGURE 2.1

$8 = 2^3$, each octal digit corresponds to three binary digits (bits). The procedure for converting a binary number into an octal number is to partition the binary number into groups of three bits, starting from the binary point, and assign the corresponding octal digit to each group.

$$111001010011.010110011_2 = 7123.263_8$$
$$7\quad 1\quad 2\quad 3\quad\; 2\quad 6\quad 3$$

The hexadecimal equivalent of a binary number is obtained in a similar fashion, dividing the binary number into groups of four bits. One complication is that we need 16 distinct symbols for a hexadecimal system. The most common convention is to use the first six letters of the alphabet plus the ten decimal digits, as shown in Fig. 2.1. With this specification of the hexadecimal system, the following conversion is straightforward:

$$110110111000 0110.10100011_2 = DB86.A3_{16} \qquad (2.8)$$
$$D\quad B\quad 8\quad 6\quad\; A\quad 3$$

2.3   Arithmetic with Bases Other than Ten

Our ability to perform arithmetic in base-10 depends on a set of basic addition and multiplication relations that are committed to memory. These operations become so familiar that we tend to forget the significance of these tables, particularly the addition table.   Doing arithmetic in some other base similarly requires a certain familiarization with the counterparts of these tables for that base. We see in Fig. 2.2 the addition and multiplication tables for base-2 and base-5.

All further rules of arithmetic, such as the carry and borrow operations, apply in precise analogy for bases other than 10. We are now ready to consider nondecimal arithmetic.

+	0	1
0	0	1
1	1	10

×	0	1
0	0	0
1	0	1

(a) Base-2

+	0	1	2	3	4
0	0	1	2	3	4
1	1	2	3	4	10
2	2	3	4	10	11
3	3	4	10	11	12
4	4	10	11	12	13

×	0	1	2	3	4
0	0	0	0	0	0
1	0	1	2	3	4
2	0	2	4	11	13
3	0	3	11	14	22
4	0	4	13	22	31

(b) Base-5

(*Note:* Where two digits are included in a square, the most
significant may be regarded as a carry.)

FIGURE 2.2   *Addition and multiplication tables.*

Example 2.4

Add 321 to 314 and multiply 142 × 32 in base-5.

$$
\begin{array}{rrrl}
1 & \;1 & & \leftarrow \text{Carry} \\
3 & 1 & 4 & \\
+3 & 2 & 1 & \\
\hline
1\;1 & 1 & 0 & \\
\end{array}
\qquad
\begin{array}{rrrl}
 & 1 & 4 & 2 \\
 & \times\; 3 & 2 \\
\hline
 & & & 1 \quad \leftarrow \text{Carry} \\
 & 2 & 3 & 4 \\
 & 1 & 2 & 1 \quad \leftarrow \text{Carry} \\
 & 3 & 2 & 1 \\
\hline
1 & 1 & 1 & 4\;4 \\
\end{array}
$$

Example 2.5 Using base-5 arithmetic, convert 431_5 to base-2

$$
\begin{array}{r|l}
2 & 431 \\
2 & 213 \quad 0 \\
2 & 104 \quad 0 \\
2 & 24 \quad 1 \\
2 & 12 \quad 0 \\
2 & 3 \quad 1 \\
2 & 1 \quad 1 \\
 & 0 \quad 1 \\
\end{array}
$$

$N_2 = 1110100$

Check:

$N = 1 \cdot 2^{11} + 1 \cdot 2^{10} + 1 \cdot 2^4 + 0 \cdot 2^3$
$\quad\quad + 1 \cdot 2^2 + 0 \cdot 2 + 0 \cdot 1$
$\quad = (2^3)^2 + 2^2 \cdot 2^3 + 2^2 \cdot 2^2 + 2^2$
$\quad = (13)^2 + 4(13) + 4 \cdot 4 + 4$
$\quad = 224 + 112 + 31 + 4 = 431$

It is not difficult to acquire a facility for base-2 arithmetic because of the simplicity of the base-2 addition and multiplication tables. We will see in succeeding chapters that, for this same reason, physical implementation of base-2 arithmetic is economical.

In the decimal number system, the multiplication of a number by a power of 10 requires only a shifting operation. The same thing is true for powers of 2 in base-2. Consider, for example,

$$25_{10} \times 2_{10}^4 = (11001 \times 10000)_2 = 110010000$$

Example 2.6

Carry out the following arithmetic operations in base-2. Convert the answers to base-10 as a check.

$$24_{10} + 11_{10} \qquad 24_{10} + 31.5_{10} \qquad 24_{10} - 11_{10} \qquad 24_{10} \times 11_{10}$$

2 | 24			2 | 31			2 | 11	
2 | 12	0		2 | 15	1		2 | 5	1
2 | 6	0		2 | 7	1		2 | 2	1
2 | 3	0		2 | 3	1		2 | 1	0
2 | 1	1		2 | 1	1		0	1
0	1		0	1			

$$24_{10} = 11000_2 \qquad 31.5_{10} = 11111.1_2 \qquad 11_{10} = 1011_2$$

(24 + 11)   (24 + 31.5)

	Check
11000	32 + 2 + 1 = 35
+ 1011	
100011	

	Check
11000	32 + 16 + 4 + 2 + 1 + ½
+ 11111.1	= 55.5
110111.1	

(24 − 11)   (24 × 11)

1111 ← borrows
11000
− 1011
1101 = 8 + 4 + 1 = 13_{10}

11000
× 1011
11000
11000
11000
11000
100001000 = 256 + 8 = 264_{10} ∎

2.4 Negative Numbers

Negative numbers in base-2 may be treated in much the same way as in base-10. For example, in adding a positive number to a negative number, the number smaller in magnitude is subtracted from the larger, and the sign of the larger is assigned to the result. Thus, the mind must carry on a sequence of logical manipulations independent of the actual arithmetic. We shall see that these logical operations may be implemented in the circuitry of an electronic digital computer.

If a slightly different approach is used, the amount of such circuitry required may be greatly reduced, with the additional benefit that the computer need not distinguish between addition and subtraction. That is, the circuitry employed in the operation of addition may also be used for subtraction.

Most computers deal with numbers in the form

$$A \cdot 2^x$$

where A is a number less than 1. Machines equipped with a *floating point* feature provide separate circuitry to accomplish shifting and to keep track of the exponent α. In some machines, the task of assuring that $A < 1$ is left to the programmer.

Suppose x is a negative number expressed in base-2 such that $|x| < 1$. We define the 2s *complement* of x to be the quantity

$$x_C = 2 - x \tag{2.9}$$

Example 2.7

Determine the 2s complement of $x = 0.110101100_2$

$$
\begin{array}{r}
11.111111 \quad \leftarrow \text{borrows} \\
10.000000000 \\
-\ \ 0.110101100 \\
\hline
01.001010100
\end{array}
$$

Note that the three least significant digits are the same for the number x as for its 2s complement. The remaining digits, however, are complemented. Any 0s at the right end of a fraction are nonsignificant and will have no effect on the complement, since they are subtracted from zero. The least significant 1 when subtracted from zero yields the un-derlined 1 in the difference. From there on, due to the repeated borrows, each digit is subtracted from a 1. Therefore, each of these digits is complemented in the answer. ■

From the above example, we may state the following rule for obtaining the 2s complement of a number less than 1.

Rule 2.1 The 2s complement of a number $x < 1$ may be obtained (1) by writing a 1 for the digit corresponding to the least significant 1 of x together with 0s for all less significant digits, and (2) by complementing (changing 0s to 1s and 1s to 0s) the remaining digits.

Implementation of Rule 2.1 may be accomplished within a computer without restoring to a separate addition (subtraction) cycle.

Note, from Example 2.7, that negative numbers in 2s complement form will always be distinguished by a 1 immediately to the left of the binary point. This digit will be zero for properly shifted positive numbers. Thus, the information regarding the sign of a number is not lost in the complementing operation.

The subtraction of a positive number x from a positive number y may be expressed as the addition of the 2s complement of x as follows:

$$y + x_c = y + (2 - x) = (y - x) + 2 \tag{2.10}$$

Suppose that $x < y$, in which case $y + x_c$ will exceed by two the proper difference $y - x$. This excess 2 will take the form of a carry to the second digit left of the binary point. If, as is actually done in a computer, we merely ignore all but the first digit left of the binary point, the result will be correct. This point of view will yield consistent results for any combination of signs and magnitudes of x and y.

If $x > y$, we may write

$$y + x_c = 2 - (x - y) \tag{2.11}$$

Now the result is a negative number expressed in 2s complement notation. This case will always be distinguished from the former by the 1 which will appear left of the binary point. Such a 1 would never appear if $x < y$ because the result must necessarily be a positive number that is less than 1.

If two numbers of the same sign, each of magnitude less than 1, are added, the magnitude of the result might be greater than 1. Unless special steps are taken, a computer could not distinguish between the resulting 1 to the left of the binary and the 2s complement indicator just mentioned. When the magnitude of a result exceeds 1, we have what is termed an *overflow* condition. Most computers provide for the detection of overflow and the generation of an overflow signal. This signal may, at the option of the programmer, initiate some corrective action, such as a right shift and adjustment of the exponent.

Example 2.8

Accomplish the following operations in 2s complement arithmetic.

$$\text{(a) } 0.1011 - 0.1001 \qquad \text{(b) } 0.1011 - 0.1101$$

(a) $(-0.1001)_c$ is found by inspection to be 1.0111

$$
\begin{array}{ll}
1.111 \leftarrow \text{Carry} & \\
0.1011 & \text{Check:} \\
+\,1.0111 & 0.1011 - 0.1001 = \left(\frac{11}{16} - \frac{9}{16}\right)_{10} = \frac{1}{8}{}_{10} \\
\hline
0.0010 = \frac{1}{8}{}_{10} &
\end{array}
$$

(b) $(-0.1101)_c = (1.0011)$

$$
\begin{array}{ll}
\quad 11 & (0.1110)_c = (0.0010) \qquad \text{The 1 to the left of the binary point indicates} \\
0.1011 & \hspace{4.5cm} \text{that this result is negative.} \\
+\,1.0011 & \\
\hline
1.1110 & \text{Check:} \\
& 0.1011 - 0.1101 = \left(\frac{11}{16} - \frac{13}{16}\right)_{10} = -\frac{1}{8}{}_{10} = -0.0010 \qquad\blacksquare
\end{array}
$$

2.5 Binary-Coded Decimal Numbers

Although virtually all digital systems are binary in the sense that all signals within the systems may take on only two values, some nevertheless perform arithmetic in the decimal system. Indeed, some digital computers are referred to as "decimal machines." In such systems, the identity of decimal numbers is retained to the extent that each decimal digit is individually represented by a binary code. There are ten decimal digits, so four binary bits are required in each code element. The most obvious choice is to let each decimal digit be represented by the corresponding 4-bit binary number, as shown in Fig. 2.3. This form of representation is known as the BCD (binary-coded decimal) representation or code.

In computer systems, most input or output is in the decimal system because this is the system most convenient for human users. On input, the decimal numbers are converted into some binary form for processing, and this process is reversed on output. In a "straight binary" computer, a decimal number, such as 75 would be converted into the binary equivalent, 1001011. In a computer using the BCD system, 75 would be converted into 0111 0101. If the BCD form is used, it must be preserved in arithmetic processing. For example, addition in a BCD machine would be carried out as shown below.

$$
\begin{array}{rr}
37 & 0011\ 0111 \\
+24 & +0010\ 0100 \\
\hline
61 & 0110\ 0001 \\
\end{array}
$$

(Decimal)   (BCD)

Decimal Digit	Binary Representation $X_3\ X_2\ X_1\ X_0$
0	0000
1	0001
2	0010
3	0011
4	0100
5	0101
6	0110
7	0111
8	1000
9	1001

FIGURE 2.3   Binary-coded decimal digits.

The principal advantage of the BCD system is the simplicity of input/output conversion; the principal disadvantage is the complexity of arithmetic processing. The choice (between binary and BCD) depends on the type of problems the system will be handling. Computers are built both ways, and the proper choice is a subject of some debate.

The BCD code is not the only possible coding for decimal digits. Any set of at least 10 distinct combinations of at least four binary bits could theoretically be used. Indeed, some writers use the term *BCD* to refer to any set of 10 binary codes used to represent the decimal digits, but we will use BCD only to refer to the code of Fig. 2.3. Some other commonly used codes will be discussed in Chapter 8.

Problems

2.1. Convert the following base-10 numbers to base-2 and check.
   (a) 13   (b) 94   (c) 356

2.2. Repeat Problem 2.1, converting to base-5.

2.3. Repeat Problem 2.1 for base-8 and base-16.

2.4. Convert the following base-10 numbers to base-2.
   (a) 0.00625   (b) 43.32   (c) 0.51

2.5. Find the base-8 equivalent of the following base-2 numbers.
   (a) 10111100101   (b) 1101.101   (c) 1.0111

2.6. Express (a) 734_8 and (b) 41.5_8 in base-2.

2.7. Express the base-2 numbers of Problem 2.5 in base-4 without employing division.

2.8. Construct addition and multiplication tables for base-4.

2.9. Perform the indicated arithmetic operations in base-5.

   (a) 143_5 (b) 124.22_5 (c) 413_5 (d) 404_5
   $+221_5$ $+\ 21.34_5$ $\times\ 32_5$ -213_5

2.10. Convert the base-5 number $4{,}433{,}214_5$ directly to base-12.

2.11. Perform the indicated arithmetic operations after converting the operands to base-2.

   (a) 695_{10} (b) 695_{10} (c) 272_{10} (d) $\dfrac{272_{10}}{23_{10}}$
   $+272_{10}$ -272_{10} $\times\ 23_{10}$

2.12. Repeat Problem 2.11 for the following.

   (a) 97_{10} (b) 131_{10} (c) 97_{10} (d) $\dfrac{586_{10}}{39_{10}}$
   $+83_{10}$ -219_{10} $\times 43_{10}$

2.13. Carry out the following computations after first converting each negative operand to 2s complement notation. Check your results.

(a)   0.11101   (b)   0.10001   (c)   −0.10101
    −0.10111   −0.11011   −0.11010

2.14. Carry out the operations of Problem 2.11 in BCD; that is, convert the operands to BCD and express the results in BCD.

2.15. Repeat Problem 2.14 for the operations of Problem 2.12.

2.16. The method of subtracting by adding complements may also be employed in the decimal number system.
(a) Determine the 10s complement of 373.
(b) Accomplish the subtraction 614 − 373 by adding the 10s complement.

2.17. One of the most common uses of the octal system is as a "shorthand" notation for binary numbers. Shown in Fig. P2.17 is a table listing a stan-

A	011000	N	000110
B	010011	O	000011
C	001110	P	001101
D	010010	Q	011101
E	010000	R	001010
F	010110	S	010100
G	001011	T	000001
H	000101	U	011100
I	001100	V	001111
J	011010	W	011001
K	011110	X	010111
L	001001	Y	010101
M	000111	Z	010001

FIGURE P2.17

dard binary code for the letters of the alphabet. Produce a new table with the binary codes replaced by their octal equivalents. Note the comparative simplicity of the new table.

Bibliography

1. Ware, W. H. *Digital Computer Technology and Design*, Vol. I, Wiley, New York, 1963.
2. Flores, I. *The Logic of Computer Arithmetic*, Prentice-Hall, Englewood Cliffs, N.J., 1962.
3. Kline, R. M. *Digital Computer Design*, Prentice-Hall, Englewood Cliffs, N.J., 1977.
4. Hwang, Kai. *Computer Arithmetic*, Wiley, New York, 1979.

3

TRUTH FUNCTIONS

3.1 Introduction

The power of many digital systems, particularly computers, is largely dependent on the capability of their components to (1) make decisions and (2) store information, including the results of decisions. We will defer consideration of the storage capability until Chapter 9 and first explore the problem of physical implementation of logical decision processes.

Digital computers are not the only physical systems that utilize electronic components to make logical decisions although in digital computers this practice has reached its highest degree of sophistication. In order to explore the relation of electrical systems to logical decisions, let us consider a relatively simple design problem.

A logic network is to be designed to implement a seat-belt alarm system. A set of sensor switches is available to supply the inputs to the network. One switch will be turned on if the gearshift is engaged (not in neutral). A switch is placed under each front seat, and each will turn on if someone sits in the corresponding seat. Finally, a switch is attached to each front seat that will turn on if and only if that seat belt is fastened. *An alarm buzzer is to sound when the ignition is turned on and the gearshift is engaged provided that either of the front seats is occupied and the corresponding seat belt is not fastened.*

These conditions are sufficiently simple that the reader could most likely work out a circuit design by trial and error. We will take another approach requiring a formal analysis of the statements in the previous paragraph. In this way we will lead up to formal procedures that can be applied to much more complicated systems.

The statement in the design specification may be listed as follows:

Alarm will sound:	*A*
Ignition is on:	*I*
Gearshift is engaged:	*G*
Left front seat is occupied:	*L*

Right front seat is occupied: R

Left seat belt is fastened: B_L

Right seat belt is fastened: B_R

To each statement, we have assigned a *statement variable*, which will represent the statement and which may take on a *truth value*, **T** or **F**, according to whether the corresponding statement is true or false. In general, we define a statement as any declarative sentence that may be classified as true or false.

The mathematics of manipulating statement variables and assigning truth values is known as *truth-functional calculus*. The application of truth-functional calculus is not restricted to simple positive statements such as those above. First, for any positive statement, there is a corresponding negative statement. For example,

"The left seat belt is *not* fastened": $\bar{B}_L$

Since the statement $\bar{B}_L$ is true whenever B_L is false and is false whenever B_L is true, $\bar{B}_L$ is called the *negation* of B_L. (Other commonly used symbols for the negation of a statement A are $\sim A$ and A'.)

Consider next this statement:

"The left seat belt is not fastened
and the left front seat is occupied": $\bar{B}_L \wedge L$

This is a *truth-functional compound*, a compound statement, the truth value of which may be determined from the truth values of the component statements. The exact relationship between the truth of the component statements and the truth of the compound statements is determined by the *connective* relating the components. In this case, the connective is AND, indicating that the statement $\bar{B}_L \wedge L$ is true if and only if both the component statements, $\bar{B}_L$ AND L, are true. Since AND is a common relationship between statements, we will assign it a standard symbol, $\wedge$, so that a compound of two statements, A and B, related by AND may be represented by $A \wedge B$. There are only four possible combinations of truth values of the two component statements A and B. We may therefore completely define $A \wedge B$ by listing its truth values in a four-row table such as Fig. 3.1. A table of this form is called a *truth table*.

This table also defines $A \vee B$, which symbolizes a statement that is true if and only if either statement A is true, or statement B is true, or both of these statements are true. Common usage of the word *or* is not always in agreement with the definition of $A \vee B$. Consider the sentence

"Joe will be driving either a blue sedan or a green convertible."

A	B	$A \wedge B$	$A \vee B$	$A \oplus B$
F	F	F	F	F
F	T	F	T	T
T	F	F	T	T
T	T	T	T	F

FIGURE 3.1

Clearly, it does not mean that Joe might be driving both cars. The latter usage of "or" is called the *exclusive-or* and is symbolized by $A \oplus B$. The truth values of $A \oplus B$ are also tabulated in Fig. 3.1. The connective $\vee$ is known by contrast as *inclusive-or*. Since this connective turns out to be more common in logical design problems than the exclusive-or, it is standard to refer to it simply as OR, leaving the "inclusive" understood.

One more connective, $\equiv$, must be defined before truth-functional calculus can be applied to the seat-belt problem. By $A \equiv B$ it is meant that A and B always have the same truth value, i.e., A is true if and only if B is true. This connective is tabulated in Fig. 3.2.

Let us now attempt to represent the information contained in the verbal specification of the alarm system by a truth-functional equation. Rephrasing the specifications in terms of the simple statements, we see that *the alarm will sound* (A) if and only if *the ignition is on* (I) and *the gearshift is engaged* (G) and *the left front seat is occupied* (L) and *the left seat belt is not fastened* ($\bar{B}_L$), or *the right front seat is occupied* (R) and *the right seat belt is not fastened* ($\bar{B}_R$). As a truth-functional equation, all this is simply written as

$$A \equiv I \wedge [G \wedge ((L \wedge \bar{B}_L) \vee (R \wedge \bar{B}_R))] \tag{3.1}$$

The design process is certainly not complete, but we have seen that the truth-functional calculus enables us to put the specifications in a form that is compact, unambiguous, and suitable for mathematical manipulation. It should be noted

A	B	$A \equiv B$
F	F	T
F	T	F
T	F	F
T	T	T

FIGURE 3.2

that the designer must be able to express the specifications in the form of statements related by clearly defined connectives. Not all declarative sentences are subject to such expression. Consider the sentence

"The alarm sounded *because* the right seat belt was not fastened."

This is not a truth-functional compound, since the truth of the statement cannot be determined solely from the truth of the component statements. For example, even if both components are true, the compound statement may not be true, since the right seat may have been unoccupied, in which case the latter half of the statement is true, but the alarm was set off by the driver not fastening the seat belt. Such sentences cannot be handled by truth-functional calculus. In the next four sections, it will become clear that the basic usefulness of electrical circuits in making logical decisions is limited to situations in which truth-functional calculus is applicable. Thus, the designer must avoid formulating a design statement in such a way that it does not constitute a truth-functional compound.

3.2 Binary Connectives

The connectives defined by the truth tables of Figs. 3.1 and 3.2 are *binary*, since they relate only two component statements. Each binary connective corresponds to a unique assignment of truth values to the four rows of the truth table, corresponding to the four possible combinations of truth values of two component statements. There are $2^4 = 16$ ways of arranging **T**s and **F**s on four rows, so there are 16 possible binary connectives, as tabulated in Fig. 3.3. Each connective is numbered, and symbols are indicated for those already defined and for others of particular interest.

Note particularly connective 13, which may be written

$$A \supset B$$

A	B		$\wedge$		A		B	$\oplus$	$\vee$	$\downarrow$	$\equiv$	$\bar{B}$		$\bar{A}$	$\supset$	$\uparrow$
		0	1	2	3	4	5	6	7	8	9	10	11	12	13	14	15
F	F	F	F	F	F	F	F	F	F	T	T	T	T	T	T	T	T
F	T	F	F	F	F	T	T	T	T	F	F	F	F	T	T	T	T
T	F	F	F	T	T	F	F	T	T	F	F	T	T	F	F	T	T
T	T	F	T	F	T	F	T	F	T	F	T	F	T	F	T	F	T

FIGURE 3.3

and which represents the compound sentence:

"If A is true, then B is true."

This is similar to the *if and only if* statement discussed previously, except that B is not necessarily false if A is false. For example, consider the statement

$$(2 > X > 1) \supset (X > 0) \tag{3.2}$$

If $X = 0.5$, for example, the statement $X > 0$ is true; but the statement $2 > X > 1$ is false. There is no contradiction to Equation 3.2, and it is true as defined in Fig. 3.3.

The truth values for the *if-then* statement may seem unnatural at first, but this concept is necessary in many mathematical arguments. The distinction between $\supset$ and $\equiv$ is very important. One is required to prove both

$$A \supset B \qquad \text{and} \qquad B \supset A$$

in order to assert that

$$B \equiv A \tag{3.3}$$

We will leave it as a problem for the reader to prove that $(A \supset B) \wedge (B \supset A)$ means the same as $B \equiv A$. That is,

$$(B \equiv A) \equiv [(A \supset B) \wedge (B \supset A)] \tag{3.4}$$

Since the logic designer is often concerned with converting natural language specifications into precise mathematical form, the designer should keep in mind that, in conventional usage, *if-then* may be thoughtlessly used when *if and only if* is intended. For example, a carelessly written specification might say:

"If the switch is closed, then the circuit output will be 5 volts,"

although what is almost certainly meant is:

"The circuit output will be 5 volts if and only if the switch is closed."

The designer thus has a dual responsibility—to be precise in writing specifications and to be careful in interpreting specifications written by others.

Also of special interest are connectives 8 and 14,* written

$$A \uparrow B \equiv \overline{A \wedge B} \tag{3.5}$$

and

$$A \downarrow B \equiv \overline{A \vee B} \tag{3.6}$$

---

* Connective 14 in Fig. 3.3 is often designated by a simple stroke, $A|B$, and called *Sheffer's stroke* in honor of H. M. Sheffer. In 1913, Sheffer presented, using "|", a set of postulates defining both $\uparrow$ and $\downarrow$.

$A \uparrow B$ is the negation of the AND relation, i.e., its truth values are precisely opposite those of $A \wedge B$. It is commonly known as the NAND (Not AND) relation. $A \downarrow B$ is similarly the negation of $A \vee B$ and is known as the NOR relation. The usefulness of these two connectives will become clear in later sections.

3.3 Evaluation of Truth Functions

We are now ready to consider compounds related by more than one binary connective. Suppose, for example, we are interested in the truth value of

$$\bar{A} \vee (A \supset B) \tag{3.7}$$

for the case where both A and B are false. We replace the statements A and B in the expression by the truth value **F**, yielding

$$\bar{\mathbf{F}} \vee (\mathbf{F} \supset \mathbf{F}) \equiv \mathbf{T} \vee \mathbf{T} \equiv \mathbf{T} \tag{3.8}$$

As $A \supset B$ is defined in Fig. 3.3 to be true where A and B are both false, $\mathbf{F} \supset \mathbf{F}$ may be replaced by the value **T**. Similarly, $\bar{\mathbf{F}}$ may be replaced by **T**. Furthermore, two true statements related by OR also produce a true statement, completing the evaluation of expression 3.8. In each case, successive steps of the evaluation are connected by $\equiv$, indicating that the truth value is the same for the expressions on either side of the symbol. Expression 3.7 is evaluated for the three remaining combinations of values of A and B in Fig. 3.4.

Note that the resulting truth value in each row is the same as that defined in Fig. 3.3 for the statement $A \supset B$. We may therefore write

$$\bar{A} \vee (A \supset B) \equiv (A \supset B) \tag{3.9}$$

Note also the parentheses in expression 3.7; without them, this expression might be interpreted as

$$(\bar{A} \vee A) \supset B \tag{3.10}$$

A	B	$\bar{A}$	$A \supset B$	$\bar{A} \vee (A \supset B)$	$\bar{A} \vee B$
F	F	T	T	T	T
F	T	T	T	T	T
T	F	F	F	F	F
T	T	F	T	T	T

FIGURE 3.4

A	B	$\bar{A}$	$\bar{A} \vee A$	$(\bar{A} \vee A) \supset B$	B
F	F	T	T	F	F
F	T	T	T	T	T
T	F	F	T	F	F
T	T	F	T	T	T

FIGURE 3.5

The truth values of this expression as compiled in Fig. 3.5 are not the same as for expression 3.7. We note instead that

$$(\bar{A} \vee A) \supset B \equiv B \tag{3.11}$$

We may conclude, then, that the placement of parentheses is essential to the meaning of these expressions.

Any expression in terms of only two statements, A and B, related by any number of binary connectives with parentheses designating a unique order of operation can be reduced to a list of four truth values corresponding to each combination of values of A and B.

3.4 Many-Statement Compounds

Consider a statement Z, which can be symbolized by

$$Z \equiv (H \wedge R) \supset D \tag{3.12}$$

A truth-table analysis of this sort of function is distinguished from those in the previous section only in that the truth tables have more than four rows. In this case, an eight-row truth table is required to provide for all possible combinations of truth or falsity of H, R, and D. This table, with a derivation of truth values for expression 3.12, appears in Fig. 3.6. By reference to Fig. 3.3, the reader can verify that

$$A \supset B \equiv \bar{A} \vee B \tag{3.13}$$

It then follows immediately that

$$Z \equiv (H \wedge \bar{R}) \supset D \equiv (\overline{H \wedge \bar{R}}) \vee D \tag{3.14}$$

The validity of expression 3.14 is also demonstrated in Fig. 3.6. Note that the right side of 3.14 employs only $\wedge$, $\vee$, and NOT. We will see that any truth-

H	R	D	$\bar{R}$	$H \wedge \bar{R}$	$(H \wedge \bar{R}) \supset D$	$\overline{H \wedge \bar{R}}$	$\overline{(H \wedge \bar{R})} \vee D$
F	F	F	T	F	T	T	T
F	F	T	T	F	T	T	T
F	T	F	F	F	T	T	T
F	T	T	F	F	T	T	T
T	F	F	T	T	F	F	F
T	F	T	T	T	T	F	T
T	T	F	F	F	T	T	T
T	T	T	F	F	T	T	T

FIGURE 3.6

functional compound, of any number of statements, may be expressed in terms of these three binary connectives. A proof of this assertion must await methods to be developed in Chapter 6.

Following the same procedure, an abbreviated truth table for the seat-belt problem is determined in Fig. 3.7. Only the final 16 of the 64 rows of a six-variable truth table are shown. On the other 48 rows, I or G are false so that $A \equiv \mathbf{F}$.

I	G	L	B_L	R	B_R	$\bar{B}_R$	$(R \wedge \bar{B}_R)$	$(L \wedge \bar{B}_L) \vee (R \wedge \bar{B}_R)$	A
T	T	F	F	F	F	T	F	F	F
T	T	F	F	F	T	F	F	F	F
T	T	F	F	T	F	T	T	T	T
T	T	F	F	T	T	F	F	F	F
T	T	F	T	F	F	T	F	F	F
T	T	F	T	F	T	F	F	F	F
T	T	F	T	T	F	T	T	T	T
T	T	F	T	T	T	F	F	F	F
T	T	T	F	F	F	T	F	T	T
T	T	T	F	F	T	F	F	T	T
T	T	T	F	T	F	T	T	T	T
T	T	T	F	T	T	F	F	T	T
T	T	T	T	F	F	T	F	F	F
T	T	T	T	F	T	F	F	F	F
T	T	T	T	T	F	T	T	T	T
T	T	T	T	T	T	F	F	F	F

FIGURE 3.7

In some problems, it may be less confusing to identify the statement variables and compile the truth table directly rather than trying first to formulate a truth-functional compound. This can be accomplished by first listing the 2^n combinations of truth values of the n variables. Then the truth value of the compound can be deduced from the problem statement for each row of the truth table.

3.5 Physical Realizations

At this point, a brief consideration of the physical realization of binary connectives may help clarify the relationship between the truth-functional calculus and logic circuits. Although any of the connectives of Fig. 3.3 might theoretically be realized physically, in practice only AND, OR, NAND, NOR, exclusive-OR, and negation are implemented in logic circuits. Circuits realizing the first five of these connectives are known as *gates*, and circuits realizing negation are known as *inverters*. A logic gate is an electrical or electronic device* with one output lead and an arbitrary number of input leads. The voltage potential, with respect to ground, of any input or output lead may take on one of only two distinct values. One of the voltages will represent true throughout the system, and the other will represent false.

In an AND gate, the output will be at the voltage representing true if all inputs are at that voltage and will be at the voltage representing false if any of the inputs are at that voltage. For an OR gate, the output will be true if any inputs are true and will be false only if all inputs are false. For other types of gates, the input and output voltages must correspond in the same manner to the truth values given in Fig. 3.3 for the corresponding connective. An inverter has one input and one output, with the output being false when the input is true, and vice versa. A discussion of the circuit aspects of these logic elements will be deferred until Chapter 5.

The most common pictorial representation of logic circuits is the *block diagram*, in which the logic elements are represented by standard symbols. The standard logic symbols specified by IEEE Standard No. 91 (ANSI Y 32.14, 1973) are shown in Figs. 3.8a and 3.8b. Although the symbols are shown with two inputs for the gates, AND, OR, NAND, and NOR gates may have any number of inputs. Exclusive-OR, however, is defined only for two inputs.

The reader will note that there are two types of symbols, the *uniform-shape* symbols and the *distinctive-shape* symbols. The uniform-shape symbols are those established by the International Electrotechnical Commission (IEC Publication

---

*Although electronic gates are by far the most common, there are other possible forms of physical realization (e.g., hydraulic and pneumatic).

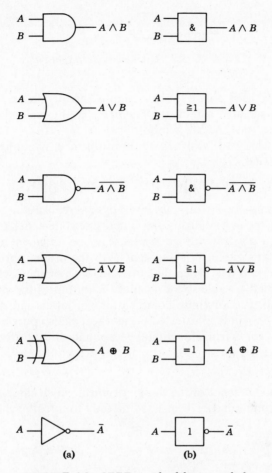

FIGURE 3.8   IEEE standard logic symbols.

117-15) and are widely used in Europe. The IEEE has included these symbols in its standard, but the distinctive-shape symbols remain the standard of preference in the United States and have found wide acceptance in other parts of the world. The distinctive-shape symbols will be used in this volume.

There are two ways of indicating inversion or negation. One is the inverter symbol as shown in Fig. 3.8. The second is a small circle on the signal line where it enters or leaves a logic symbol, sometimes referred to as *dot negation*. This system is used for NAND and NOR symbols, as shown in Fig. 3.8, and Fig. 3.9 gives an example of a more complex use of this notation. Generally, the dot

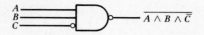

FIGURE 3.9   *Example of dot negation.*

negation should be used only if the inversion is actually accomplished within the circuitry of the associated gate. If the inverter is a physically distinct element, it is preferable to use the separate symbol.

Let us now return to the seat-belt alarm problem as symbolized by expression 3.1, which is repeated as

$$A \equiv I \wedge [G \wedge ((L \wedge \bar{B}_L) \vee (R \wedge \bar{B}_R))] \tag{3.15}$$

As in the determination of truth tables, a physical realization of expression 3.15 may be obtained by first realizing the innermost statements and then working outward. Letting a positive voltage represent a true statement and 0 volts a false statement, we obtain Fig. 3.10. Consistent with the original problem statement, a statement variable is true when the corresponding switch is closed.

The reader should be able to verify that the output of the circuit of Fig. 3.10 will correspond to true for those combinations of inputs indicated in the truth table of Fig. 3.7. As will be illustrated presently, it is also possible to determine a logical expression for a truth-functional compound directly from its truth table. With this in mind, we may phrase the design procedure for decision circuits as follows.

Step 1   Obtain a statement of the design problem which can be symbolized as a truth-functional compound or translated directly to a truth table.

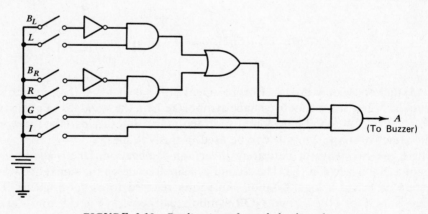

FIGURE 3.10   *Realization of seat belt alarm logic.*

Step 2 Obtain an expression for the output of the problem in terms of the connectives AND, OR, and NOT.

Step 3* Obtain the logical expression, equivalent to that obtained in Step 2, that will result in the most economical physical realization.

Step 4 Construct the physical realization corresponding to Step 3. This procedure will be illustrated in the following example.

Example 3.1

The flight of a certain satellite may be controlled in any one of three ways: by ground control, manually by the astronaut, or automatically by a computer aboard the satellite. The ground station, which has ultimate authority, will cause a voltage source, S, aboard the satellite to assume a true level when

"Control is aboard the satellite": S

is true. The astronaut then chooses between manual and automatic control by means of a switch M corresponding to

"Control is manual": M

is false. A check signal C generated within the computer will remain at the true level as long as the computer is functioning properly and is able to assume or remain in control.

There are panel lights at both the ground station and aboard the satellite that should go out, sounding an alarm, if the computer fails to function properly at a time when it should be in control. Letting the variable L correspond to the lights being lit, the above conditions may be represented by

$$\bar{L} = (S \wedge \bar{M}) \wedge \bar{C} \tag{3.16}$$

Alternatively, we may state that the lights should be lit if the computer is functioning or if it is not supposed to be in control, i.e.,

$$L = (\overline{S \wedge \bar{M}}) \vee C \tag{3.17}$$

A physical implementation of the right side of Equation 3.17 is shown in Fig. 3.11. The reader should verify that the lamp will light for the proper combinations of input truth values. ∎

3.6 Functionally Complete Sets of Connectives

We saw in the previous section that certain of the binary connectives in Fig. 3.3 could be expressed in terms of various other connectives. The question arises: Can a smaller set of connectives be found in terms of which all 16 functions can be expressed? Let us search for such a smaller set.

* This step is the topic of Chapters 6 and 7.

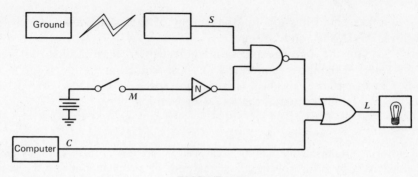

FIGURE 3.11

For example, consider the set of connectives containing only $\land$ and $\lor$. The case where A and B are both false is critical. Clearly, $A \land B$ and $A \lor B$ must both be false. By way of induction, visualize a set of statements, arrived at by relating false statements A and B by $\land$ and $\lor$ in a fashion of arbitrary complexity, but such that they are all false. Now relate any two of these statements, or relate one of them to A or B, utilizing $\land$ or $\lor$. By definition, the result must be another false statement. Because the set initially contained only false statements A and B, induction tells us that the set will always include only false statements. In Fig. 3.3, however, note that functions 8 through 15 must be true when A and B are both false. Therefore, these eight functions cannot be realized by any expression written in terms of only the connectives $\land$ and $\lor$.

Suppose the NOT operation is added to our set of connectives. The reader who intuitively expects that all functions can be written in terms of these three connectives is quite correct. That each of the functions in Fig. 3.3 can be expressed in terms of $\land$, $\lor$, and " $^-$ " may be verified by determining truth values for the corresponding expression listed in Table 3.1. This task is left to the reader.

A subset of the connectives sufficient for realizing all the connectives and, by extension, any function expressed in terms of any of the connectives, is said to be

Table 3.1

$f_0 \equiv A \land \bar{A}$	$f_8 \equiv \overline{A \lor B}$
$f_1 \equiv A \land B$	$f_9 \equiv (\bar{A} \land \bar{B}) \lor (A \land B)$
$f_2 \equiv A \land \bar{B}$	$f_{10} \equiv \bar{B}$
$f_3 \equiv A$	$f_{11} \equiv A \lor \bar{B}$
$f_4 \equiv B \land \bar{A}$	$f_{12} \equiv \bar{A}$
$f_5 \equiv B$	$f_{13} \equiv \bar{A} \lor B$
$f_6 \equiv (\bar{A} \land B) \lor (A \land \bar{B})$	$f_{14} \equiv \overline{A \land B}$
$f_7 \equiv A \lor \bar{B}$	$f_{15} \equiv A \lor \bar{A}$

functionally complete. The reader may wonder whether any list of fewer than three connectives might be functionally complete. The answer is yes. Either AND and NOT or OR and NOT is a functionally complete set. As will be apparent in Chapter 5, statements can usually be expressed most conveniently by using all three.

At this point, we pause briefly in our discussion of functionally complete sets of connectives to verify, for truth-functional calculus, two very important theorems called *DeMorgan's theorems*. We will prove these theorems again, algebraically, in Chapter 4. Because they facilitate our discussion, we introduce them here.

THEOREM 3.1

(1) $A \downarrow B \equiv \overline{A \vee B} \equiv \bar{A} \wedge \bar{B}$
(2) $A \uparrow B \equiv \overline{A \wedge B} \equiv \bar{A} \vee \bar{B}$

Proof As elsewhere in the chapter, the proof consists of a truth table analysis as given in Fig. 3.12.

A	B	$A \vee B$	$\overline{A \vee B}$	$\bar{A}$	$\bar{B}$	$\bar{A} \wedge \bar{B}$	$A \wedge B$	$\overline{A \wedge B}$	$\bar{A} \vee \bar{B}$
F	F	F	T	T	T	T	F	T	T
F	T	T	F	T	F	F	F	T	T
T	F	T	F	F	T	F	F	T	T
T	T	T	F	F	F	F	T	F	F

FIGURE 3.12   Truth functional proof of DeMorgan's theorems.

Using DeMorgan's theorems, we see that $\wedge$, $\vee$, and NOT can be expressed in terms of $\uparrow$ and in terms of $\downarrow$. Clearly $X \wedge X \equiv X$, so we have an expression for $\bar{X}$ in Equation 3.18. Using the obvious

$$\bar{X} \equiv \overline{X \wedge X} \equiv X \uparrow X \qquad (3.18)$$

theorem $\bar{\bar{A}} \equiv A$ and DeMorgan's theorems, we have expressions for $X \wedge Y$ and $X \vee Y$ in Equations 3.19 and 3.20.

$$X \wedge Y \equiv \overline{\overline{X \uparrow Y}} \equiv (X \uparrow Y) \uparrow (X \uparrow Y) \qquad (3.19)$$

$$X \vee Y \equiv \overline{\overline{X \vee Y}} \equiv \overline{\overline{X} \wedge \overline{Y}} \equiv \bar{X} \uparrow \bar{Y} \equiv (X \uparrow X) \uparrow (Y \uparrow Y) \qquad (3.20)$$

We similarly obtain expressions 3.21, 3.22, and 3.23 in terms of $\downarrow$.

$$\bar{X} \equiv \overline{X \vee X} \equiv X \downarrow X \qquad (3.21)$$

$$X \wedge Y \equiv \overline{\overline{X \wedge Y}} \equiv \overline{\overline{X} \vee \overline{Y}} \equiv \bar{X} \downarrow \bar{Y} \equiv (X \downarrow X) \downarrow (Y \downarrow Y) \qquad (3.22)$$

$$X \vee Y \equiv \overline{X \downarrow Y} \equiv (X \downarrow Y) \downarrow (X \downarrow Y) \qquad (3.23)$$

We now conclude that any truth-functional compound may be expressed in terms of a single connective, which may be either $\downarrow$ or $\uparrow$. An $\wedge$, $\vee$, " $^-$ " expression may be converted to a $\uparrow$ expression by replacing the former connectives and the statements to which they apply by the appropriate expression out of Equations 3.18, 3.19, and 3.20. For instance, f_{13} in Fig. 3.3 becomes

$$\bar{A} \vee B \equiv (A \uparrow A) \vee B \equiv [(A \uparrow A) \uparrow (A \uparrow A)] \uparrow [B \uparrow B] \qquad (3.24)$$

If the "obvious" theorem $\bar{\bar{A}} \equiv A$ (easily verified by truth table analysis) is applied, a much simpler expression will result. That is,

$$[(A \uparrow A) \uparrow (A \uparrow A)] \uparrow (B \uparrow B) \equiv (\bar{A} \uparrow \bar{A}) \uparrow (B \uparrow B)$$

$$\equiv \bar{\bar{A}} \uparrow (B \uparrow B)$$

$$\equiv A \uparrow (B \uparrow B) \qquad (3.25)$$

Certainly, other theorems might be used to simplify complex logical expressions. A systematic exposition of such theorems in the form of Boolean algebra is the topic of Chapter 4.

The observation that all logical expressions can be expressed in terms of either $\downarrow$ or $\uparrow$ previews an important notion that will be developed in Chapter 5. That is, all logic circuits can be constructed using only NOR gates or only NAND gates. More formally, we may say that either NAND or NOR is functionally complete.

3.7 A Digital Computer Example

Before leaving this chapter, let us formulate one final example. For this, we have chosen a complex statement that must be realized physically in the design of most digital computers.

On most computers, there is a provision for the detection of *overflow*. Overflow occurs when an arithmetic operation produces a result that is too large to be handled correctly by the machine. When this occurs, a signal must be generated that may, typically, cause the computer to switch from the regular program to a special routine for dealing with this contingency. For this example, we will assume that the operands and results are all placed in registers, or temporary storage locations, of the same fixed capacity of N digits. Then the addition of two N-digit numbers may produce an $N + 1$-digit number too large to be stored, in which case we have *additive overflow*. Multiplication and division can also cause overflow, but we will restrict our attention in this example to additive overflow.

If only addition of the positive numbers is considered, overflow is always indicated by a carry from the most significant digit; but if subtraction or addition of negative numbers is also considered, the situation becomes more complicated. As discussed in Chapter 2, subtraction is often accomplished in computers by addition of complements, and negative numbers are often represented by their complements. When complement arithmetic is used, the relationship between overflow and carry from the most significant digit depends on the operation and the signs of operands. For details, we refer the interested reader to books on computer arithmetic [1, 2]. For this example, we ask the reader to accept the following statements as being true for a machine using complement arithmetic.

1. When the operation is addition and both addend and augend are positive, overflow is indicated by a carry from the most significant digit.
2. When the operation is addition and both addend and augend are negative, overflow is indicated by the absence of carry from the most significant digit.
3. When the operation is subtraction and the minuend is positive and the subtrahend negative, overflow is indicated by a carry from the most significant digit.
4. When the operation is subtraction and the minuend is negative and the subtrahend positive, overflow is indicated by absence of a carry from the most significant digit.

In order to symbolize the above statements as truth-functional compounds, let us assign variables to the various component statements as follows.

Statement	Truth variable
Operation is addition	A
Operation is subtraction	S
Sign of the augend is negative	W
Sign of the addend is negative	X
Sign of the minuend is negative	Y
Sign of the subtrahend is negative	Z
Carry from the most significant digit	C
Additive overflow error has occurred	E

Note that we have not included statements about the signs of the operands being positive. The reason for this is that all numbers in a computer are either positive or negative. Therefore, the statement "the augend is positive" is properly represented by $\bar{W}$. It would be equally valid to assign its truth variables to statements that the signs of the operands are positive and let the conditions of negative signs be represented by the negative truth variables.

To simplify discussion, let us designate the overflow in each of the above cases as E_1, E_2, E_3, and E_4, respectively. Now let us consider the first statement about overflow. This statement says that overflow exists—that is, E_1 is true when

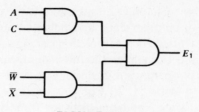

FIGURE 3.13

statements A and C are true and $\bar{W}$ and $\bar{X}$ are true. Thus, the statement is represented by the truth-functional equation

$$E_1 = (A \wedge C) \wedge (\bar{W} \wedge \bar{X}) \tag{3.26}$$

If the two AND statements enclosed by parentheses in Equation 3.26 are generated by physical AND circuits whose outputs are the inputs to a third AND circuit, E_1 will be the output of this third circuit. This is illustrated in Fig. 3.13. The following expressions for E_2, E_3, and E_4 are similarly developed:

$$E_2 = (A \wedge \bar{C}) \wedge (W \wedge X) \tag{3.27}$$

$$E_3 = (S \wedge C) \wedge (\bar{Y} \wedge Z) \tag{3.28}$$

$$E_4 = (S \wedge \bar{C}) \wedge (Y \wedge \bar{Z}) \tag{3.29}$$

The physical implementations of these expressions are similar to the configuration in Fig. 3.13.

The statement E is true; that is, overflow occurs in the machine whenever E_1 or E_2 or E_3 or E_4 is true. Relating Equations 3.26 to 3.29 in this manner results in

$$E = [[(A \wedge C) \wedge (\bar{W} \wedge \bar{X})] \vee [(A \wedge \bar{C}) \wedge (W \wedge X)]] \vee$$
$$[[(S \wedge C) \wedge (\bar{Y} \wedge Z)] \vee [(S \wedge \bar{C}) \wedge (Y \wedge \bar{Z})]] \tag{3.30}$$

Implementing the three OR connectives together with the circuits for E_1, E_2, E_3, and E_4 results in the circuit in Fig. 3.14. The output of this circuit will be the voltage representing true whenever overflow occurs in the machine.

The reader may have noted that the procedure followed here is an example of the design philosophy described in the introduction. A mathematical model, consisting of the truth-functional Equation 3.30 and the corresponding physical realization in Fig. 3.14, was developed for a problem, which was first specified as a set of natural language statements. It remains to simplify this expression and

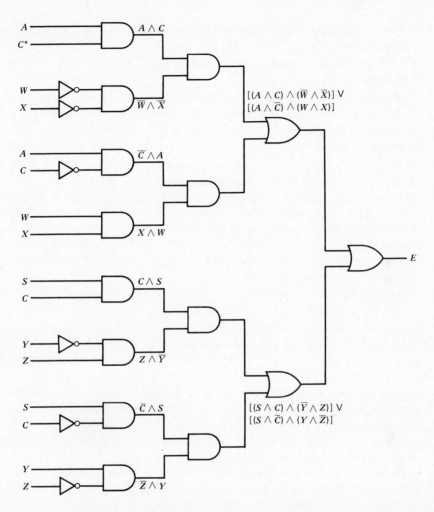

FIGURE 3.14 *Initial block diagram of overflow detector. *All lines labeled with the same switching variable (e.g., the four lines labeled C) are assumed to be electrically common; to draw in all the interconnections would complicate the diagram unnecessarily.*

thereby the physical realization, utilizing an algebra of logic to be defined in the next chapter.

Problems

3.1. Symbolize the truth-functional compound describing the following design problem. Determine the truth values of the compound.

The flow of water into the brine solution to be used in a chemical process will be turned off if and only if (1) the tank is full, or (2) the output is shut off, the salt concentration does not exceed 2.5%, and the water level is not below a designated minimum level.

3.2. Symbolize the following statement and determine its truth values:

If it is hot in Arizona and it is raining outside or demonstrators are in the streets, then it is hot in Arizona, demonstrators are in the streets, and it is snowing in Argentina.

3.3. A burglar alarm system for a bank is to be operative only if a master switch at the police station has been turned on. Subject to this condition, the alarm will ring if the vault door is disturbed in any way, or if the door to the bank is opened unless a special switch is first operated by the security guard's key. The vault door will be equipped with a vibration sensor that will cause a switch to close if the vault door is disturbed, and a switch will be mounted on the bank door in such a way that it will close whenever the bank door is opened.

Symbolize the above system as a truth-functional compound and construct the corresponding logic block diagram.

3.4. In many cars, the seat-belt alarm buzzer is also used to warn against leaving the key in the ignition or leaving the lights on when the car is unoccupied. The following statement describes how such a system might operate:

The alarm is to sound if the key is in the ignition when the door is open and the motor is not running, or if the lights are on when the key is not in the ignition, or if the driver seat belt is not fastened when the motor is running, or if the passenger seat is occupied and the passenger seat belt is not fastened when the motor is running.

Symbolize the above system as a truth-functional compound and construct the corresponding logic block diagram. You may assume that switch contacts are available to indicate the occurrence of the various individual conditions.

3.5. Using a truth table, verify that

$$(A \supset B) \supset [\overline{A \vee (\overline{A \wedge B})}] \equiv A \wedge \bar{B}$$

3.6. Express $X \oplus Y$ in terms of only the connective $\downarrow$.

3.7. Prove that no connective in Fig. 3.3 other than $\downarrow$ or $\uparrow$ will serve as the only binary connective in expressions for every function in that figure.

3.8. Tabulate the truth values of:
(a) Example 3.1.
(b) The example of Section 3.7.

3.9. Verify Expression 3.3 by a tabulation of truth values.

3.10. Suppose three one-digit binary numbers, a, b, and c, are to be added together to form a two-digit number whose digits are denoted by $s_1 s_2$. For each of the above five binary digits, define the corresponding capital letter to be a statement variable that is true whenever the small letter digit is 1 and false otherwise. Determine a truth-functional expression for S_2 such that S_2 will be true whenever $s_2 = 1$. Determine a similar expression for S_1. Construct the diagram of logic circuits realizing S_2 and S_1.

3.11. An argument consists of several compound statements regarded as *premises* with one additional compound statement called the conclusion, which is asserted to follow logically from the premises. We say that the argument is valid if the conclusion is true whenever the conjuction of the premises is true. Determine by using truth tables the validity of each of the following arguments:

(a) If I work I earn money, and if I don't work I enjoy myself. Therefore, if I don't earn money I enjoy myself.

(b) The governor of California hails from either Los Angeles or San Francisco. Jones doesn't come from San Francisco. Therefore, if Jones is not from Los Angeles, Jones is not governor of California.

(c) If Harvard wins the Ivy League football championship, then Princeton will be second; and if Princeton is second, then Dartmouth will place in second division. Harvard will win or Dartmouth will not finish among the top four teams. Therefore, Princeton will not take second place.

Bibliography

1. Flores, I. *Logic of Computer Arithmetic*, Prentice-Hall, Englewood Cliffs, N.J., 1963.
2. Hill, F. J. and G. R. Peterson. *Digital Systems: Hardware Organization and Design*, 2nd ed., Wiley, New York, 1978.
3. Copi, I. M. *Symbolic Logic*, Macmillan, New York, 1954.
4. Quine, W. V. *Mathematical Logic*, Harvard University Press, Cambridge, 1955.

4

BOOLEAN ALGEBRA

4.1 Introduction

Associated with the physical realization of any logic element is a cost. Similarly, a cost may be assigned to any logic circuit equal to the total cost of all the logic elements in the circuit. For example, the cost of the circuit in Fig. 3.14 is the cost of 3 OR gates plus the cost of 12 AND gates plus the cost of 6 inverters.

Where it is necessary to construct a physical realization of any truth function, it is highly desirable that this be done at minimum cost. It is possible to verify that a given circuit realizes the proper truth function by checking its output against the truth table for each combination of inputs. Similarly, two functions may be shown equivalent by this method. Truth-functional calculus, however, offers no convenient way for finding the most economical realization of a given function.

For this purpose, we introduce Boolean algebra. Perhaps the most convenient set of postulates for a Boolean algebra is the one set forth by Huntington in 1904 [1]. The first postulate can be thought of as establishing the system under study.

I. *There exists a set* K *of objects or elements, subject to an equivalence* relation, denoted " =," which satisfies the principle of substitution.*

By substitution, it is meant: if $a = b$, a may be substituted for b in any expression involving b without affecting the validity of the expression.

The remainder of Huntington's postulates are:

IIa. *A rule of combination " + " is defined such that* a + b *is in* K *whenever both* a *and* b *are in* K.

IIb. *A rule of combination "·" is defined such that* a · b *(abbreviated* ab*) is in* K *whenever both* a *and* b *are in* K.

IIIa. *There exists an element 0 in* K *such that, for every* a *in* K, $a + 0 = a$.

IIIb. *There exists an element 1 in* K *such that, for every* a *in* K, $a \cdot 1 = a$.

---

* A formal definition of an equivalence relation may be found in Chapter 10. In the present chapter, the reader will encounter no difficulty in the use of the standard notion of equality from ordinary algebra.

42

IVa. $a + b = b + a$ ⎫
IVb. $a \cdot b = b \cdot a$ ⎭ (commutative laws).

Va. $a + (b \cdot c) = (a + b) \cdot (a + c)$ ⎫
Vb. $a \cdot (b + c) = (a \cdot b) + (a \cdot c)$ ⎭ (distributive laws)

VI.   *For every element* a *in* K *there exists an element* $\bar{a}$ *such that*

$$a \cdot \bar{a} = 0$$

and

$$a + \bar{a} = 1.$$

VII.   *There are at least two elements* x *and* y *in* K *such that* $x \neq y$.

Note that nothing has been said that would fix the number or type of elements that make up *K*. As a matter of fact, many systems satisfy these postulates. A few will be illustrated later.

The reader may have observed a similarity between these postulates and those of ordinary algebra. Note, however, that the first distributive law—that is, distribution over addition—does not hold for ordinary algebra. Also, there is no $\bar{a}$ defined in ordinary algebra.

If a set of postulates is to be of any use, it must be consistent. That is, none of the postulates in the set may contradict any other postulate of the set. To verify consistency, we might attempt a postulate-by-postulate examination to ascertain that no postulate contradicted any possible group of postulates. This is rather awkward; in fact, a much easier way is available. It is merely necessary to find an example of a Boolean algebra that is known independently to be consistent. If such a consistent system satisfies all of Huntington's postulates, then the postulates themselves must be consistent.*

The simplest example of a Boolean algebra consists of only two elements, 1 and 0, defined to satisfy

$$\bar{1} = 0 \qquad \bar{0} = 1$$
$$1 \cdot 1 = 1 + 1 = 1 + 0 = 0 + 1 = 1$$

and

$$0 + 0 = 0 \cdot 0 = 1 \cdot 0 = 0 \cdot 1 = 0$$

We see that Postulates I, II, III, and VII are satisfied by definition. Consider, for example, IIIb. If $a = 1$, we have

$$a \cdot 1 = 1 \cdot 1 = 1$$

If $a = 0$, we have

$$a \cdot 1 = 0 \cdot 1 = 0$$

---

* This matter of the consistency of a set of postulates is a subtle mathematical concept. The interested reader should consult Huntington's paper [1] for a full discussion.

The satisfaction of both of the above equations results from our definition of the
"·" function for all possible combinations of arguments. Satisfaction of the commutative laws is evident, and verification of the distributive laws requires only a
truth-table listing of the values of each side of both equations for all combination of values of a, b, and c. This task is left to the reader.

We observe also the satisfaction of Postulate VI upon letting a take on the
values 1 and 0. That is, letting $a = 1$,

$$a \cdot \bar{a} = 1 \cdot \bar{1} = 1 \cdot 0 = 0$$
$$a + \bar{a} = 1 + \bar{1} = 1 + 0 = 1 \tag{4.1}$$

and letting $a = 0$,

$$a \cdot \bar{a} = 0 \cdot \bar{0} = 0 \cdot 1 = 0$$

$$a + \bar{a} = 0 + \bar{0} = 0 + 1 = 1 \tag{4.2}$$

In addition to consistency, the question of *independence* of the postulates has
been of some interest. By independence, it is meant that none of the postulates
can be proved from the others. The postulates presented here are, in fact,
independent. However, a demonstration of this fact would be lengthy and is not
essential to our discussion.

It is not necessary that we begin with an independent set of postulates. In fact,
some authors save effort by including some of the theorems that we will develop
later as postulates. It is our opinion, however, that the proof of these theorems
constitutes the best possible introduction to the manipulation of Boolean
algebra.

4.2   Truth-Functional Calculus as a Boolean Algebra

There is a one-to-one correspondence between the truth-functional calculus of
Chapter 3 and the two-element algebra introduced in the previous section. Table
4.1 interprets each truth value or logical connective as an element or a rule of
combination, respectively, in a Boolean algebra. These interpretations imply the
truth-table correspondence shown in Fig. 4.1.

Table 4.1

Truth Calculus	Boolean Algebra
$\wedge \longleftrightarrow$	$\cdot$
$\vee \longleftrightarrow$	$+$
F $\longleftrightarrow$	0
T $\longleftrightarrow$	1
$\bar{A} \longleftrightarrow$	$\bar{A}$

AB	ab	$A \wedge B \quad a \cdot b$	$A \vee B \quad a + b$
FF $\longrightarrow$ 00		F $\longrightarrow$ 0	F $\longrightarrow$ 0
FT $\longrightarrow$ 01		F $\longrightarrow$ 0	T $\longrightarrow$ 1
TF $\longrightarrow$ 10		F $\longrightarrow$ 0	T $\longrightarrow$ 1
TT $\longrightarrow$ 11		T $\longrightarrow$ 1	T $\longrightarrow$ 1

FIGURE 4.1

A precise distinction between the two columns of Table 4.1 is not always made. Generally, "·" is identified in the literature as AND while " + " is called OR. We follow this convention throughout the book.

In the remainder of this chapter, Boolean algebra will be developed into a convenient manipulative tool. As this is accomplished, the results will be immediately applicable to truth functions and therefore to logic circuits.

4.3 Duality

Notice that Huntington's postulates are presented in pairs. A closer look reveals that in each case one postulate in a pair can be obtained from the other by interchanging 0 and 1 along with + and "·" symbols. For example,

$$a + 0 = a$$
$$\downarrow \; \downarrow$$
$$a \cdot 1 = a \tag{4.3}$$

and

$$a + (b \cdot c) = (a + b) \cdot (a + c)$$
$$\downarrow \quad \downarrow \qquad \downarrow \quad \downarrow \quad \downarrow$$
$$a \cdot (b + c) = (a \cdot b) + (a \cdot c) \tag{4.4}$$

Every theorem that can be proved for Boolean algebra has a dual that is also true. That is, every step of a proof of a theorem may be replaced by its dual, yielding a proof of the dual of the theorem. In a sense, this doubles our capacity for proving theorems.

4.4 Fundamental Theorems of Boolean Algebra

In this section, we prove the theorems necessary for the convenient manipulation of Boolean algebra. Some of these are labeled *theorems*, while others are labeled *lemmas*. Our criterion for making this distinction is that only equalities that are valuable tools for working problems are designated as theorems. Intermediate results necessary in the proof of theorems and results, which are included only for the sake of logical completeness, are called lemmas.

LEMMA 4.1   The 0 and 1 elements are unique.

Proof   By way of contradiction, assume that there are two zero elements, 0_1 and 0_2. For any elements a_1 and a_2 in K we have

$$a_1 + 0_1 = a_1 \quad \text{and} \quad a_2 + 0_2 = a_2$$

$$\text{(Post. IIIa)}$$

Now let $a_1 = 0_2$ and $a_2 = 0_1$. Therefore,

$$0_2 + 0_1 = 0_2 \quad \text{and} \quad 0_1 + 0_2 = 0_1$$

Thus, using the first commutative law and the transitive property of equality, we have

$$0_1 = 0_2$$

As an example of duality, let us make the following change of symbols in our proof:

$$
\begin{array}{cc}
a_1 + 0_1 = a_1 & a_2 + 0_2 = a_2 \\
\downarrow\downarrow & \downarrow\downarrow \\
a_1 \cdot 1_1 = a_1 & a_2 \cdot 1_2 = a_2
\end{array}
$$

and

$$
\begin{array}{cc}
0_2 + 0_1 = 0_2 & 0_1 + 0_2 = 0_1 \\
\downarrow\;\downarrow\downarrow\quad\downarrow & \downarrow\;\downarrow\downarrow\quad\downarrow \\
1_2 \cdot 1_1 = 1_2 & 1_1 \cdot 1_2 = 1_1
\end{array}
$$

and

$$
\begin{array}{c}
0_1 = 0_2 \\
\downarrow\quad\downarrow \\
1_1 = 1_2
\end{array}
$$

Because we were aware that the dual of each postulate exists, it was unnecessary to cite postulates as the dual of each step was compiled. In fact, we could have merely stated that the dual of the theorem was true by the principle of duality. This approach will be followed in the remaining theorems of this section.

LEMMA 4.2   For every element a in K, $a + a = a$ and $a \cdot a = a$.

Proof

$$
\begin{array}{ll}
a + a = (a + a) \cdot 1 & \text{(Post. IIIb)} \\
a + a = (a + a)(a + \bar{a}) & \text{(Post. VI)} \\
a + a = a + a\bar{a} & \text{(Post. Va)} \\
a + a = a + 0 & \text{(Post. VI)} \\
a + a = a & \text{(Post. IIIa)} \\
a \cdot a = a & \text{(Duality)}
\end{array}
$$

LEMMA 4.3   For every a in K, $a + 1 = 1$ and $a \cdot 0 = 0$.

Proof

$a + 1 = 1 \cdot (a + 1)$	(Post. IIIb)
$a + 1 = (a + \bar{a})(a + 1)$	(Post. VI)
$a + 1 = a + \bar{a} \cdot 1$	(Post. Va)
$a + 1 = a + \bar{a}$	(Post. IIIb)
$a + 1 = 1$	(Post. VI)
$a \cdot 0 = 0$	(Duality)

LEMMA 4.4   The elements 1 and 0 are distinct and $\bar{1} = 0$.

Proof   Let a be any element in K.

$a \cdot 1 = a$	(Post. IIIb)
$a \cdot 0 = 0$	(Lemma 4.3)

Now assume that $1 = 0$. In this case, the above expressions are satisfied only if $a = 0$. Postulate VII, however, tells us that there are at least two elements in K. The contradiction can be resolved only by concluding that $1 \neq 0$. To prove the second assertion, we need only write

$\bar{1} = \bar{1} \cdot 1$	(Post. IIIb)
$= 0$	(Post. VI)

LEMMA 4.5   For every pair of elements a and b in K, $a + ab = a$ and $a(a + b) = a$.

Proof

$a + ab = a \cdot 1 + ab$	(Post. IIIb)
$= a(1 + b)$	(Post. Vb)
$= a \cdot 1$	(Lemma 4.3)
$= a$	(Post. IIIb)

and

$a(a + b) = a$	(Duality)

LEMMA 4.6   The $\bar{a}$ defined by Postulate VI for every a in K is unique.

Proof   By contradiction. Assume that there are two distinct elements, $\bar{a}_1$ and $\bar{a}_2$, that satisfy Postulate VI, i.e., assume that

$$a + \bar{a}_1 = 1, \qquad a + \bar{a}_2 = 1, \qquad a\bar{a}_1 = 0, \qquad a\bar{a}_2 = 0$$

$\bar{a}_2 = 1 \cdot \bar{a}_2$	(Post. IIIb)
$= (a + \bar{a}_1)\bar{a}_2$	Assumption
$= a\bar{a}_2 + \bar{a}_1\bar{a}_2$	(Post. Vb)
$= 0 + \bar{a}_1\bar{a}_2$	Assumption
$= a\bar{a}_1 + \bar{a}_1\bar{a}_2$	Assumption
$= (a + \bar{a}_2)\bar{a}_1$	(Post. Vb)
$= 1 \cdot \bar{a}_1$	Assumption
$= \bar{a}_1$	(Post. IIIb)

LEMMA 4.7   For every element a in K, $a = \bar{\bar{a}}$.

Proof   Let $\bar{\bar{a}} = x$. Therefore,

$$\bar{a}x = 0 \qquad \text{and} \qquad \bar{a} + x = 1 \qquad\qquad \text{(Post. VI)}$$

but

$$a\bar{a} = 0 \qquad \text{and} \qquad \bar{a} + a = 1 \qquad\qquad \text{(Post. VI)}$$

Thus, both x and a satisfy Postulate VI as the complement of $\bar{a}$. Therefore, by Lemma 4.6,

$$x = a$$

LEMMA 4.8 $a[(a + b) + c] = [(a + b) + c]a = a$.

Proof
$$
\begin{aligned}
a[(a + b) + c] &= a(a + b) + ac & \text{(Post. Vb)} \\
&= a + ac & \text{(Lemma 4.5)} \\
&= a = [(a + b) + c]a & \text{(Post. IVb, Lemma 4.5)}
\end{aligned}
$$

THEOREM 4.9   For any three elements a, b, and c in K, $a + (b + c) = (a + b) + c$ and $a \cdot (bc) = (ab) \cdot c$.

The reader will recognize Theorem 4.9 as the associative laws from ordinary algebra. Some authors include these laws among the postulates, although, as we will see, this is unnecessary.

Proof   Let

$$
\begin{aligned}
Z &= [(a + b) + c] \cdot [a + (b + c)] \\
&= [(a + b) + c]a + [(a + b) + c] \cdot (b + c) \\
&= a + [(a + b) + c] \cdot (b + c) & \text{(Lemma 4.8)} \\
&= a + \{[(a + b) + c]b + [(a + b) + c] \cdot c\} & \text{(Post. Vb)} \\
&= a + \{b + [(a + b) + c] \cdot c\} & \text{(Lemma 4.8, Post. IVb)} \\
&= a + (b + c) & \text{(Lemma 4.5)}
\end{aligned}
$$

However, we may also write

$$
\begin{aligned}
Z &= (a + b)[a + (b + c)] + c[a + (b + c)] & \text{(Post. Vb)} \\
&= (a + b)[a + (b + c)] + c & \text{(Lemma 4.8)} \\
&= \{a[a + (b + c)] + b[a + (b + c)]\} + c & \text{(Post. Vb)} \\
&= \{a[a + (b + c)] + b\} + c & \text{(Lemma 4.8)} \\
&= (a + b) + c & \text{(Lemma 4.5)}
\end{aligned}
$$

Therefore, by transitivity

$$a + (b + c) = (a + b) + c$$

and

$$(a \cdot b)c = a(b \cdot c) \qquad\qquad \text{(Duality)}$$

Now that we have established the associative laws, certain expressions may be simplified by omitting parentheses as follows:

$$(a + b) + c = a + b + c \tag{4.5}$$

and

$$(a \cdot b) \cdot c = abc \tag{4.6}$$

This format may be extended to Boolean sums and products of any number of variables.

THEOREM 4.10 For any pair of elements a and b in K, $a + \bar{a}b = a + b$; $a(\bar{a} + b) = ab$.

Proof

$$a + \bar{a}b = (a + \bar{a})(a + b) \qquad \text{(Post. Va)}$$
$$a + \bar{a}b = a + b \qquad \text{(Post. VI, Post. IIIb)}$$
$$a \cdot (\bar{a} + b) = a \cdot b \qquad \text{(Duality)}$$

THEOREM 4.11 For every pair of elements a and b in K, $\overline{a + b} = \bar{a} \cdot \bar{b}$ and $\overline{a \cdot b} = \bar{a} + \bar{b}$.

The expressions in Theorem 4.11 are the two forms of the very important DeMorgan's law, which we introduced in Chapter 3. The second form is the dual of the first, but, for the sake of instruction, we will arrive at the second form in another way.

$$(a + b) + \bar{a} \cdot \bar{b} = [(a + b) + \bar{a}] \cdot [(a + b) + \bar{b}] \qquad \text{(Post. Va)}$$
$$(a + b) + \bar{a} \cdot \bar{b} = [\bar{a} + (a + b)] \cdot [\bar{b} + (b + a)] \qquad \text{(Post. IVa)}$$
$$(a + b) + \bar{a} \cdot \bar{b} = 1 \cdot 1 = 1 \qquad \text{(Theorem 4.9, Lemma 4.3)}$$
$$(a + b) \cdot (\bar{a} \cdot \bar{b}) = a(\bar{a} \cdot \bar{b}) + b(\bar{b} \cdot \bar{a}) \qquad \text{(Post. Vb, Post. IVb)}$$
$$(a + b) \cdot (\bar{a} \cdot \bar{b}) = 0 + 0 = 0 \qquad \text{(Theorem 4.9, Lemma 4.3)}$$

Both requirements of Postulate VI have been satisfied, so $a + b$ is the unique complement of $\bar{a} \cdot \bar{b}$. Therefore, we may write

$$a + b = \overline{\bar{a} \cdot \bar{b}} \qquad \text{or} \qquad \overline{a + b} = \bar{a} \cdot \bar{b}$$

The above holds equally well for $\bar{a}$ and $\bar{b}$ in place of a and b, so we can write

$$\overline{\bar{a} + \bar{b}} = \bar{\bar{a}} \cdot \bar{\bar{b}} = a \cdot b$$

or

$$\overline{\bar{a} + \bar{b}} = \overline{a \cdot b} \qquad \text{(Lemma 4.6, 4.7)}.$$

The classic development of Boolean algebra based on Huntington's postulates as given above has the advantage of illustrating the same techniques required in the application of the algebra. An alternative development derives Boolean algebra as a distributive and complemented lattice. See, for example, Stone [6].

4.5   Set Theory as an Example of Boolean Algebra

We have already discussed truth-functional calculus as one example of Boolean algebra. Since this example satisfies all of Huntington's postulates, it must also satisfy each of the theorems and lemmas proved so far.

Set theory is a second example of a Boolean algebra. A *set* may be regarded as any collection of objects. In order to talk of an algebra of sets, we must limit the objects that make up these sets with some meaningful criteria. We define a *universal set* to include every object that satisfies these criteria. For a reason that will soon be apparent, the universal set must necessarily contain at least one object. The universal set may, in fact, contain a finite or an infinite number of objects. The collection of points in a plane or in a finite region of a plane make interesting examples of a universal set. When we say that R is a subset of S, we mean that every object in R is also found in S. This is symbolized by $R \subset S$. Only those sets that are subsets of the universal set are material to the development of an algebra of sets. A second set of importance is the *null set*, or the set containing no objects. The universal set and the null set will be symbolized by S_U and S_Z, respectively. The *union* of two sets, R and S, is the set that contains all objects contained in either R or S. This is symbolized by $R \cup S$. The *intersection* of two sets, R and S, is that set containing these elements found in both set R and set S. The intersection is designated by $R \cap S$. The *complement* of a set S, designated $C(S)$, contains those objects found in the universal set but not in set S.

A helpful illustration of a universal set is the collection of points in a rectangle such as shown in Fig. 4.2. The two sets R and S contain those points of the universal set found within the respective circles.* A representation of this type is known as a *Venn diagram*.

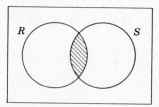

FIGURE 4.2 $R \cap S$.

The darkened area in Fig. 4.2 is the intersection of sets R and S. Similarly, $R \cup S$ and $C(S)$ are darkened in Figs. 4.3 and 4.4, respectively.

---

* The student of advanced calculus may be concerned about the status of points on the boundaries of the circles, a problem not critical to this development. One possible resolution of the conflict is to imagine the universal set to include only a large but finite number of points in the rectangle. In this way, boundaries can be deliberately arranged to avoid points in the universal set.

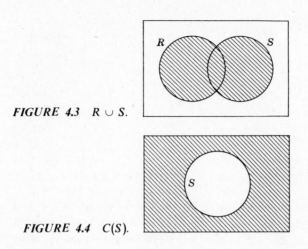

FIGURE 4.3 $R \cup S$.

FIGURE 4.4 $C(S)$.

Recalling the postulates in Sec. 4.1, let us choose a set of K objects to be those sets containing no points other than points found in the universal set. Such sets, one of which is the null set, are subsets of the universal set. If equality is taken to relate only identical sets, then Postulate I is satisfied. We will proceed to show that, in light of the correspondences in Table 4.2, the remaining postulates will also be satisfied.

The sets $A \cap B$ and $A \cup B$ are subsets of S_U, so Postulate II is satisfied. Because $R \cup S_Z$ and $R \cap S_U$ contain precisely the same objects as R, Postulate III is satisfied. Satisfaction of Postulate IV, the commutative laws, is evident. Under the set theory interpretation, the distributive laws become

$$A \cup (B \cap C) = (A \cup B) \cap (A \cup C) \tag{4.7}$$

and

$$A \cap (B \cup C) = (A \cap B) \cup (A \cap C) \tag{4.8}$$

The fact that Equations 4.7 and 4.8 are satisfied is best illustrated by using the Venn diagram. Note in Fig. 4.5a that set $B \cap C$ is shaded with vertical lines while set A is shaded by horizontal lines. The set $A \cup (B \cap C)$ consists of the

Table 4.2

$\cap$	$\longleftrightarrow$	$\cdot$
$\cup$	$\longleftrightarrow$	$+$
S_Z	$\longleftrightarrow$	0
S_U	$\longleftrightarrow$	1
$C(S)$	$\longleftrightarrow$	$\bar{S}$

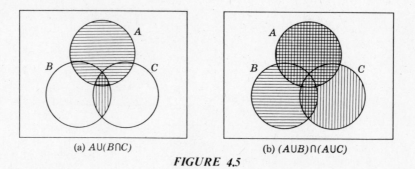

(a) $A \cup (B \cap C)$ (b) $(A \cup B) \cap (A \cup C)$

FIGURE 4.5

areas darkened in either manner. In Fig. 4.5b, the set $A \cup B$ is indicated by horizontal lines, and $A \cup C$ is shaded by vertical lines. The intersections of these two sets, $(A \cup B) \cap (A \cup C)$ is therefore given by the crisscrossed area in Fig. 4.5b. Note that the areas representing both sides of Equation 4.7 are identical. The validity of Equation 4.8 can be similarly demonstrated by using the Venn diagram. By the definition of the complement, it follows immediately that

$$R \cap \bar{R} = S_Z \tag{4.9}$$

and

$$R \cup \bar{R} = S_U \tag{4.10}$$

Thus, Postulate VI is satisfied. The fact that S_U and S_Z are always defined and distinct guarantees the satisfaction of Postulate VII.

4.6 Examples of Boolean Simplification

We are now ready to utilize Boolean algebra in the simplification of some meaningful examples. The first two examples are a problem in set theory and a needlessly complex logic circuit.

Example 4.1

Scholars A, B, and C collect old manuscripts. Mr. A is a collector of English political works and foreign-language novels. Mr. B is a collector of all political works, except English novels, and English works that are not novels. Mr. C collects nonfictional items that are English works or foreign-language political works. Determine those books for which there is competition—that is, those desired by two or more collectors.

Solution   Let us first define the various sets involved in the problem.

> A set of books collected by Mr. A
> B set of books collected by Mr. B
> C set of books collected by Mr. C
> E set of all English language books
> N set of all novels
> P set of all political works

The set of books collected by two or more persons can be symbolized by

$$Z = (A \cap B) \cup (A \cap C) \cup (B \cap C) \tag{4.11}$$

Translating the problem statements into set theoretic expressions gives

$$A = (E \cap P) \cup (\bar{E} \cap N)$$
$$B = (P \cap \overline{E \cap N}) \cup (E \cap \bar{N})$$
$$C = [E \cup (\bar{E} \cap P)] \cap \bar{N}$$

Note that it was more convenient to use the Boolean algebra symbol than the set theoretic symbol for the complement. In order to utilize Boolean algebra in obtaining a solution, the symbols for intersection and union will also be replaced by the corresponding "." and " + " symbols. Before substituting into Equation 4.11, let us write the expression for A, B, and C in Boolean form and make whatever simplifications are possible.

$$A = EP + \bar{E}N$$
$$B = P(\overline{EN}) + E\bar{N} = P(\bar{E} + \bar{N}) + E\bar{N} \qquad \text{(DeMorgan's theorem)}$$
$$= P\bar{E} + P\bar{N}(E + \bar{E}) + E\bar{N}$$
$$= P\bar{E} + P\bar{N}\bar{E} + PE\bar{N} + E\bar{N}$$
$$= P\bar{E}(1 + \bar{N}) + (P + 1)E\bar{N} = P\bar{E} + E\bar{N}$$
$$C = (E + \bar{E}P)\bar{N} = (E + P)\bar{N} \qquad \text{(Theorem 4.10)}$$

Now writing Equation 4.11 as a Boolean expression and substituting we get

$$Z = (EP + \bar{E}N)(P\bar{E} + E\bar{N}) + (EP + \bar{E}N)(E\bar{N} + P\bar{N})$$
$$+ (P\bar{E} + E\bar{N})(E\bar{N} + P\bar{N})$$

or

$$Z = (EP + \bar{E}N)(P\bar{E} + E\bar{N} + E\bar{N} + P\bar{N}) + (P\bar{E} + E\bar{N})(E\bar{N} + P\bar{N}) \quad \text{(Distributive laws)}$$
$$= (EP + \bar{E}N)(P\bar{E} + E\bar{N} + P\bar{N}) + (P\bar{E} + E\bar{N})(E\bar{N} + P\bar{N})$$
$$= (EP + \bar{E}N)(P\bar{E} + E\bar{N} + P\bar{E}\bar{N} + PE\bar{N}) + (P\bar{E} + E\bar{N})(E\bar{N} + P\bar{N})$$
$$= (EP + \bar{E}N)(P\bar{E}(1 + \bar{N}) + (P + 1)E\bar{N}) + (P\bar{E} + E\bar{N})(E\bar{N} + P\bar{N})$$
$$= (EP + \bar{E}N)(P\bar{E} + E\bar{N}) + (P\bar{E} + E\bar{N})(E\bar{N} + P\bar{N})$$
$$= (P\bar{E} + E\bar{N})(EP + \bar{E}N + E\bar{N} + P\bar{N}) \qquad \text{(Distributive law)}$$
$$= E\bar{N} + P\bar{E}(PE + \bar{E}N + P\bar{N}) \qquad \text{(Distributive law)}$$
$$= E\bar{N} + P\bar{E}E + P\bar{E}N + P\bar{E}\bar{N}$$
$$= E\bar{N} + P\bar{E}(N + \bar{N}) = E\bar{N} + P\bar{E}$$

Thus, we conclude that more than one person is interested in English nonfiction and foreign-language political works.

An alternative solution of this problem constitutes an interesting application of the Venn diagram. Note that sets A, B, and C are represented by the darkened areas in Figs. 4.6a, 4.6b, and 4.6c, respectively.

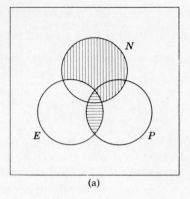

(a)

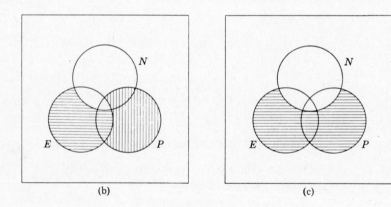

(b)   (c)

FIGURE 4.6

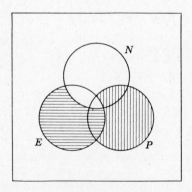

FIGURE 4.7

By inspection, we see that the area darkened in at least two of these three diagrams is that shown in Fig. 4.7. This can be expressed as

$$(E \cap \bar{N}) \cup (P \cap \bar{E})$$

which is the same result obtained by the algebraic method.   ■

The reader may have noticed that expressions similar to

$$xy + \bar{x}z + yz \qquad (4.12)$$

appeared several times in Example 4.1. Each time, a simplification involving several lines of algebraic manipulation was made. The following theorem makes it possible to shortcut this manipulation in future examples.

THEOREM 4.12   For any three elements a, b, and c in K,

$$ab + \bar{a}c + bc = ab + \bar{a}c$$

and

$$(a + b)(\bar{a} + c)(b + c) = (a + b)(\bar{a} + c)$$

Proof
$$\begin{aligned}
ab + \bar{a}c + bc &= ab + \bar{a}c + bc(a + \bar{a}) \\
&= ab + abc + \bar{a}c + \bar{a}cb \\
&= ab(1 + c) + \bar{a}c(1 + b) \\
&= ab + \bar{a}c
\end{aligned}$$
$$(a + b)(\bar{a} + c)(b + c) = (a + b)(\bar{a} + c) \qquad \text{(Duality)}$$

Theorem 4.12 is illustrated by the Venn diagram in Fig. 4.8. Note that the outlined area bc is partially covered by ab and partially by $\bar{a}c$. The theorem may be applied whenever two products jointly cover a third in this manner.

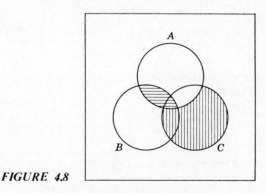

FIGURE 4.8

Example 4.2

The logic circuit shown in Fig. 4.9 was designed by trial and error. To simplify the figure, some of the actual connections to the inputs and to the outputs of the inverters are not

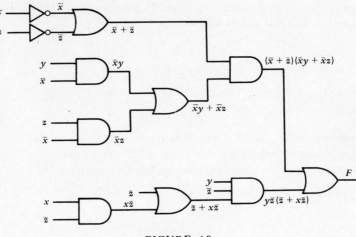

FIGURE 4.9

shown. Let us utilize Boolean algebra to design an equivalent circuit free from unnecessary hardware.

Solution It is first necessary to determine the function realized by this circuit. To this end, the output functions for the various logic circuits are shown in Fig. 4.9. The output function is given by

$$F(x, y, z) = y\bar{z}(\bar{z} + \bar{z}x) + (\bar{x} + \bar{z})(\bar{x}y + \bar{x}z) \tag{4.13}$$
$$= y\bar{z} + (\bar{x} + \bar{z})(\bar{x}y + \bar{x}z) \tag{Lemma 4.5}$$
$$= y\bar{z} + (\bar{x} + \bar{z})\bar{x}(y + z) \tag{Lemma 4.5}$$
$$= \bar{z}y + \bar{x}z + \bar{x}y \tag{Lemma 4.5}$$

Now if $\bar{z}$ is identified as a, y as b, and $\bar{x}$ as c in Theorem 4.12, we have

$$F(x, y, z) = \bar{z}y + \bar{x}z \tag{Theorem 4.12}$$

We see that the considerably simplified and therefore cheaper circuit in Fig. 4.10 realizes the same function as the one in Fig. 4.9.   ■

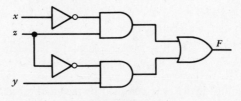

FIGURE 4.10

Example 4.3

Simplify the following Boolean expression, which represents the output of a logical decision circuit.

$$F(w, x, y, z) = x + xyz + yz\bar{x} + wx + \bar{w}x + \bar{x}y$$
$$= x + yz(x + \bar{x}) + \bar{x}y + x(w + \bar{w})$$
$$= (x + \bar{x}y) + yz + x$$
$$= x + y + x + yz$$
$$= x + (y + yz)$$
$$= x + y$$

Another approach to the problem involves the use of Lemma 4.7. That is,

$$\overline{F(w, x, y, z)} = \overline{x + xyz + yz\bar{x} + \bar{x}y + wx + \bar{w}x}$$
$$= \bar{x}(\overline{xyz})(\overline{\bar{x}yz})(\overline{\bar{x}y})(\overline{wx})(\overline{\bar{w}x})$$
$$= [\bar{x}(\bar{x} + \bar{y} + \bar{z})](x + \bar{y} + \bar{z})(x + \bar{y})(\bar{w} + \bar{x})(w + \bar{x})$$
$$= [\bar{x} \cdot (\bar{w} + \bar{x})(w + \bar{x})](x + \bar{y})(\bar{y} + \bar{z} + x)$$
$$= [\bar{x}](x + \bar{y})$$
$$\overline{F(w, x, y, z)} = \bar{x} \cdot \bar{y}$$

Therefore, by Lemma 4.7 we have

$$F(w, x, y, z) = \overline{\overline{F(w, x, y, z)}} = \overline{\bar{x} \cdot \bar{y}} = x + y. \qquad \blacksquare$$

4.7   Remarks on Switching Functions

The reader may have wondered why the Venn diagram that was so helpful in Example 4.1 was not utilized in the two subsequent examples. It should be recalled that this device has so far been introduced only as an illustration of set theory. The Boolean algebra of sets has many elements, not just two, as is the case with truth values and logic voltage levels. The Venn diagram can, however, be made useful in the simplification of switching circuits and truth functions if the following definitions and interpretations are kept in mind.

First, it is convenient to distinguish Boolean functions of only two elements by calling them *switching* functions. We then define a *switching variable* as a letter (usually chosen from the beginning or end of the alphabet, excluding F) that may take on either of the element values 0 or 1. In the case of logic circuits, a fixed number of variables, say n, serve as inputs to a circuit under consideration. The situation is similar in the case of truth functions.

There are 2^n possible ways of assigning values to n variables. The three-variable case is illustrated in Fig. 4.11. If the eight question marks are replaced

x_1	x_2	x_3	F
0	0	0	?
0	0	1	?
0	1	0	?
0	1	1	?
1	0	0	?
1	0	1	?
1	1	0	?
1	1	1	?

FIGURE 4.11

by any combination of zeros and ones, a specific *function* of x_1, x_2, and x_3 is defined. There are 2^8, or 2^{2^3}, ways of replacing the eight question marks by zeros and ones, so there are 2^{2^3} *switching functions* of three variables. The value of F for a particular row of the truth table is known as the functional value for the corresponding combination of input values. Formally, we have the following definition.

Definition 4.1 A *switching function* of n variables is any one particular assignment of functional values (1s or 0s) for all 2^n possible combinations of values of the n variables.

In general, then, there are 2^{2^n} distinct switching functions of n variables.

We have seen that there are many *expressions* for a given switching function of truth-table assignment. In the two-element algebra, many distinct expressions are equal to each of the two elements 1 and 0. For example,

$$1 \cdot (1 + 0) = 1$$

Boolean algebra provides a means of determining whether such expressions are equal to 1 or 0.

Let us consider now the possibility of an extended Boolean algebra whose *elements* are *all possible switching functions* of n variables. Such an algebra will provide us with a means of determining which *function* is represented by a given *expression* in the n variables. For example, we can determine that

$$x + xy + y + yz$$
$$x + \bar{x}y + x(w + \bar{w})$$

, and

$$x + y$$

x	y	F
0	0	0
0	1	1
1	0	1
1	1	1

FIGURE 4.12

are all expressions for the same function, specifically that function defined by the truth table of Fig. 4.12.

The theorems we have proved so far, when specialized for a two-valued algebra, express equality between switching functions or, if you prefer, between elements of the extended algebra. Thus, as in ordinary algebra, two switching functions are identically equal if their evaluations are equal for every combination of variable values. This notion of equality satisfies Postulate I. It remains to be shown that the proposed algebra of 2^{2^n} elements satisfies the remainder of the postulates and is indeed a Boolean algebra. Functional expressions may certainly be related by the connectives AND, OR, and NOT in the same way variables are related. Therefore, Postulates II, IV, V are satisfied. The zero element is the function that is identically 0, and the unit element is the function that is identically 1. The definition of the negation of F must certainly be $\bar{F}$. If a function h is to satisfy the expression

$$h(x_1, x_2, \ldots, x_n) = \bar{F}(x_1, x_2, \ldots, x_n) \tag{4.14}$$

then h must be 0 whenever F is evaluated as 1 and vice versa. Satisfaction of Postulate VII is immediate.

We will now see that a Venn diagram provides a convenient illustration of the algebra whose elements are the 2^8 functions of three variables. In Fig. 4.13, areas

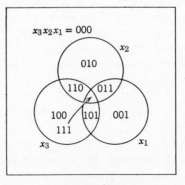

FIGURE 4.13

are assigned to each of the eight combinations of values. A function can then be indicated by darkening those areas corresponding to combinations of values for which the function is 1. Every possible area configuration stands for a distinct function. For example, the function

$$F(x, y, z) = \bar{z}y + \bar{x}z + \bar{x}y = \bar{z}y + \bar{x}z \qquad (4.15)$$

resulting from Example 4.2 is depicted in Fig. 4.14. The differently lined regions indicate the simplification resulting from the application of Theorem 4.12.

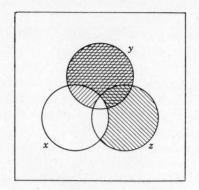

FIGURE 4.14

Almost every problem in Boolean algebra will be found to be some variation of the following statement:

Given one of the 2^{2^n} functions of n variables, determine from the large number of equivalent expressions of this function one which satisfies some criteria for simplicity.

If the problem concerns an electronic switching circuit, the simplicity criteria will result from the desirability of using the cheapest possible circuit that provides the desired performance. The Venn diagram is a means of solving this type of problem where three or fewer variables are involved.

For more than three variables, the basic illustrative form of the Venn diagram is inadequate. Extensions are possible, however, the most convenient of which is the Karnaugh map, to be discussed in Chapter 6.

4.8 Summary

In this chapter, we have developed a formal algebra for the representation and manipulation of switching functions. We have also laid the necessary foundation for the development in later chapters of systematic procedures for simplification of algebraic expressions and the corresponding circuits.

Following is a list of the important postulates, lemmas, and theorems for ready reference in problems and proofs.

PIIIa $a + 0 = a$	PIIIb $a \cdot 1 = a$
PIVa $a + b = b + a$	PIVb $ab = ba$
PVa $a + bc = (a + b)(a + c)$	PVb $a(b + c) = ab + ac$
PVIa $a \cdot \bar{a} = 0$	PVIb $a + \bar{a} = 1$
L 2a $a + \acute{a} = a$	L 2b $a \cdot a = a$
L 3a $a + 1 = 1$	L 3b $a \cdot 0 = 0$
L 4 $1 = 0$	
L 5a $a + ab = a$	L 5b $a(a + b) = a$
L 7 $\bar{\bar{a}} = a$	
T 9a $a + (b + c) = (a + b) + c$ $= a + b + c$	T 9b $a(bc) = (ab)c = abc$
T10a $a + \bar{a}b = a + b$	T10b $a(\bar{a} + b) = a \cdot b$
T11a $\overline{a + b} = \bar{a} \cdot \bar{b}$	T11b $\overline{a \cdot b} = \bar{a} + \bar{b}$
T12a $ab + \bar{a}c + bc = ab + \bar{a}c$	T12b $(a + b)(\bar{a} + c)(b + c)$ $= (a + b)(\bar{a} + c)$

Problems

4.1. (a) Verify the distributive law

$$a(b + c) = ab + ac$$

for the two-valued Boolean algebra.

(b) Repeat (a) for $a + bc = (a + b)(a + c)$.

4.2. Write out a proof of the second associative law.

$$(a \cdot b) \cdot c = a(b \cdot c)$$

4.3. Verify by a truth-table analysis (using 1 and 0 instead of **T** and **F**) that the following theorems of Boolean algebra hold for the two-valued truth-functional calculus.

(a) Lemma 4.2
(b) Lemma 4.5
(c) Lemma 4.7
(d) Theorem 4.9
(e) Theorem 4.10
(f) Theorem 4.11

4.4 Write in its simplest possible form the Boolean function realized by the logic network of Fig. P4.4.

4.5. Use the Venn diagram to illustrate the validity of Equation 4.8.

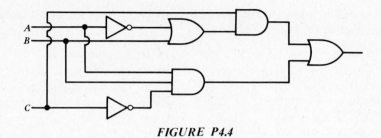

FIGURE P4.4

4.6. Verify the following by Boolean algebraic manipulation. Justify each step with a reference to a postulate or theorem.

(a) $(X + \bar{Y} + XY)(X + \bar{Y})\bar{X}Y = 0$

(b) $(X + \bar{Y} + X\bar{Y})(XY + \bar{X}Z + YZ) = XY + \bar{X}\bar{Y}Z$

(c) $(AB + C + D)(\bar{C} + D)(\bar{C} + D + E) = AB\bar{C} + D$

4.7. Using the postulates and theorems of Boolean algebra, simplify the following to a form with as few occurrences of each variable as possible.

(a) $(X + \bar{Y})[XYZ + \bar{Y}(Z + X)] + XY\bar{Z}(X + \bar{X}Y)$

(b) $(X + \bar{Y}\bar{X})[XZ + X\bar{Z}(Y + \bar{Y})]$

4.8. Simplify the following as far as possible by manipulation of Boolean algebra.

(a) $X\bar{Z}Y + (X\bar{Z}Y + Z\bar{X})[Y(Z + X) + \bar{Y}Z + \bar{Y}X\bar{Z}]$

(b) $[(a \downarrow b)] \uparrow [(a + (\bar{a} + xz)\bar{a})(\bar{z} + \bar{a})]$

4.9. Repeat Problem 4.8, utilizing a Venn diagram.

4.10. Prove that $(a + b)(\bar{a} + c)(b + c) = (a + b)(\bar{a} + c)$.

4.11. By manipulation of Boolean algebra verify that

$$[\bar{X}_1\bar{X}_2(X_3X_1 + \bar{X}_2)] + (X_1 + X_2)\overline{(X_1\bar{X}_2\bar{X}_3)} \downarrow (\bar{X}_1X_2X_3)$$

$$= \bar{X}_2\bar{X}_3 + \bar{X}_1X_3$$

4.12. Let the dual of a Boolean function, f, be designated by f^D. Prove the theorem $f^D(X_1, X_2, \ldots, X_n) = \bar{f}(\bar{X}_1, \bar{X}_2, \ldots, \bar{X}_n)$ for all functions of 1, 2, or 3 variables. Assume that the variables may take on only the values 1 and 0.

4.13. A Boolean function, F, is called self-dual if the dual of the function is equal to the function itself $(F = F^D)$. Show that there are 2^{2n-1} self-dual functions of n variables.

4.14. Prove that

$$f(X_1, X_2, \ldots, X_n) = X_1 f(1, X_2, \ldots, X_n) + \bar{X}_1 f(0, X_2, \ldots, X_n)$$

where the variables may take on only the values 1 and 0.

4.15. State and prove the dual of the expression in Problem 4.14.

4.16. Consider the possibility of a Boolean algebra of three elements, say, 0, 1, and 2. Show that a Boolean algebra cannot be defined for three elements.

4.17. Honest John, the used-car dealer, has some two-toned, 8-cylinder cars. He also has in stock some air-conditioned cars and some cars with fewer than 8 cylinders, all of which are overpriced. Another dealer, Insane Charlie, has on his lot some cars without air-conditioning or 8 cylinders that are not overpriced. He also has some air-conditioned, 8-cylinder models and some overpriced solid color cars. Write the simplest possible Boolean expression for the categories of cars currently stocked by both dealers.

Bibliography

1. Huntington, E. V. "Sets of Independent Postulates for the Algebra of Logic," *Trans. American Math. Soc.*, **5,** 288–309 (1904).
2. Sheffer, H. M. "A Set of Five Independent Postulates for Boolean Algebras, with Applications to Logical Constants," *Trans. American Math. Soc.*, **14,** 481–488 (1913).
3. Boole, George. *An Investigation of the Laws of Thought*, Dover Publications, New York, 1854.
4. Shannon, C. E. "Symbolic Analysis of Relay and Switching Circuits," *Transactions AIEE*, **57,** 713–723 (1938).
5. Carroll, Lewis. *The Complete Works.* The Nonesuch Press, London, 1939.
6. Stone, H. S. *Discrete Mathematical Structures and Their Applications*, SRA Inc., Palo Alto, 1973.
7. Tremblay, J. P. and R. Manochar. *Discrete Mathematical Structures and Applications to Computer Science*, McGraw-Hill, N.Y. 1975.

5

SWITCHING DEVICES

5.1 Introduction

The theory discussed in the last two chapters is not dependent on specific hardware for its validity. In the following chapters, we will generally assume ideal realizations of the various logical operations, so we can theoretically proceed with no knowledge whatever of the actual hardware involved. However, the less-than-ideal characteristics of practical devices impose significant limitations on the designer. It is thus important for the designer to have some knowledge of the characteristics of practical switching devices.

5.2 Switches and Relays

The simplest switching device is the switch itself. A switch is any mechanical device by means of which two (or more) electrical conductors may be conveniently connected or disconnected. The simplest form of switch consists of two strips of spring metal on which are mounted electrical contacts. A lever or push-button controls whether the switch is *open* (contacts separated) or *closed* (contacts touching). The manner in which switches may be used to implement logic functions is illustrated in Fig. 5.1. In the circuit of Fig. 5.1*a*, the light bulb will be lit only if both switch *A* AND switch *B* are closed. In Fig. 5.1*b*, the light bulb will be lit if either switch *A* OR switch *B* is closed.

A relay is a switch operated by an electromagnet. When an appropriate current flows through the coil, a magnetic force displaces the armature, in turn

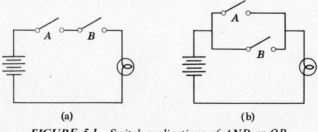

(a)   (b)

FIGURE 5.1 *Switch realizations of AND or OR.*

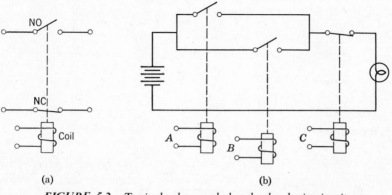

FIGURE 5.2   *Typical relay symbol and relay logic circuit.*

causing the switch contacts to open or close. The position of the switch contacts when the coil is not energized is known as the *normal* position. Thus, a relay may have both *normally open* (NO) and *normally closed* (NC) contacts (Fig. 5.2*a*). In Fig. 5.2*b*, we have a simple relay circuit for controlling a light bulb. The light bulb will be lit if relay *A* OR relay *B* is energized AND relay *C* is NOT energized. Thus, if we let *T* represent the statement, "The bulb is lit," the circuit of Fig. 5.2*b* could be conveniently described by the equation

$$T = (A + B)\bar{C} \tag{5.1}$$

While relays are far too slow for use in the "main frame" of a digital computer, they are still useful for various auxiliary functions in peripheral equipment. The bilateral nature of relay circuits gives rise to some special problems. These are treated in Appendix B.

5.3   Logic Circuits

Switches and relays are of great historic importance, but today the vast majority of digital systems are constructed of electronic devices, principally semiconductor diodes and transistors. The schematic symbol for a diode is shown in Fig. 5.3*a*. Ideally, a diode offers zero resistance to the flow of current from anode to cathode (forward direction) and infinite resistance to current flow from cathode to anode (reverse direction). The ideal current-voltage characteristic would thus be that shown in Fig. 5.3*b*. A typical characteristic for an actual semiconductor diode is shown in Fig. 5.3*c*. When the diode is conducting in the forward direction, there is a small, nearly constant positive voltage across the diode. When the voltage across the diode goes negative, a very small reverse current flows. When the voltage goes sufficiently negative, a phenomenon known as the

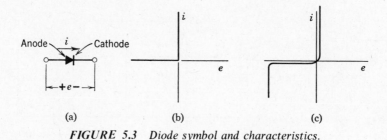

FIGURE 5.3 *Diode symbol and characteristics.*

Zener* breakdown occurs, and the reverse current increases sharply. In normal switching circuit applications, diodes are never operated in the Zener range.

The circuits for the two basic forms of diode gates are shown in Fig. 5.4. In analyzing them, we first assume that the diodes are ideal and then consider the effects of the nonideal characteristics. In these gate circuits, there are two possible input voltages, V_- and V_+, and two supply voltages, V_L and V_H, such that $V_H > V_+ > V_- > V_L$. In the circuit of Fig. 5.4a, $e_0 = V_+$ only if both inputs are at V_+. If one or more inputs are at V_-, then $e_0 = V_-$. This behavior is summarized in the table of Fig. 5.4b. In the circuit of Fig. 5.4e, $e_0 = V_-$ only if both inputs are at V_-. If one or more inputs are at V_+, then $e_0 = V_+$. This behavior is summarized in the table of Fig. 5.4f.

In order to classify these gates as logic devices, we must specify which voltage levels shall represent the switching values 0 and 1. The specification $V_+ = 1$ and $V_- = 0$ is denoted as *positive logic*. In this case, the voltage tables of Figs. 5.4b and 5.4f become the truth tables of Figs. 5.4c and 5.4g. We see that the circuits of Figs. 5.4a and 5.4e are AND gates and OR gates, respectively, for positive logic. The designation $V_- = 1$ and $V_+ = 0$, denoted as *negative logic*, yields the truth tables of Figs. 5.4d and 5.4h. We see that the circuits of Figs. 5.4a and 5.4e are OR gates and AND gates, respectively, for negative logic.

This duality of function of the diode gates may be a source of some confusion to beginners. In the logical design process, the designer works in terms of the switching values, 0 and 1, not the actual voltage levels, so it does not matter whether positive or negative logic is being used. After the logic design is complete, the designer will choose one form of gate for AND and the other for OR, in accordance with the type of logic desired. The choice of positive or negative logic may be governed by compatibility requirements with connected systems, power supply limitations, or any of a number of factors beyond the scope of this book.

---

* There are actually two types of breakdown, *Zener* and *avalanche*, that are due to quite different physical phenomena. From an external point of view, however, there is no important difference.

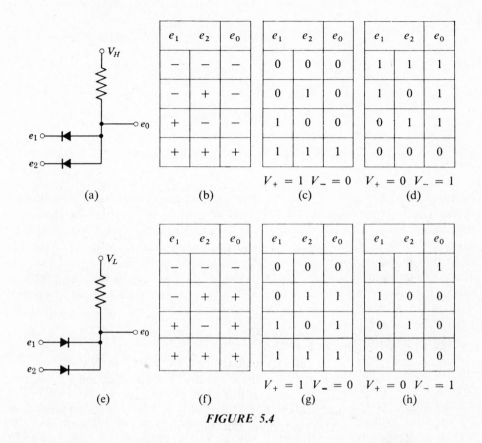

FIGURE 5.4

If the diodes were ideal, there would be no limit on the number of inputs on a single gate (fan-in), or on the number of other gates a single gate could drive (fan-out), or on the number of gates that could be connected in series one after another (levels of gating). However, the voltage across conducting diodes and the finite reverse current of nonconducting diodes result in continual degradation signal level as the circuit complexity increases. For example, even in a single gate, the output voltage will always be slightly different from the input voltages because of the drops across the conducting diodes.

The signal degradation problem associated with diode logic may be alleviated by introducing amplification at appropriate points, most frequently through the use of transistors. The symbols for the two basic types of transistors, *PNP* and *NPN*, are shown in Fig. 5.5.

The transistors are used as switches in logic circuits. In the *PNP* transistor, if the base is driven slightly positive with respect to the emitter, the transistor will

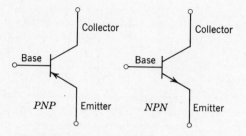

FIGURE 5.5   *PNP and NPN transistors.*

be *cut off*, thus acting as an open circuit to current flow from emitter to collector. If the base is driven slightly negative with respect to the emitter, the transistor will be *saturated*, thus acting essentially as a short-circuit flow from emitter to collector. There will be a small voltage drop from emitter to collector in saturation, but this drop will be essentially independent of the magnitude of the emitter-to-collector current. The *NPN* transistor behaves in the same fashion except that the polarities are reversed. A negative voltage from base to emitter will cut off an *NPN* transistor while a positive voltage will saturate it, and the current flow will be from collector to emitter.

Transistors exhibit both voltage and current gain. Figure 5.6 shows an *NPN* transistor connected in the standard "grounded emitter" configuration. When the transistor is cut off, the collector will be at the collector supply level, V_+. When it is saturated, the collector will be typically about 0.1 volt above the emitter supply level, V_-. This collector voltage swing will be in the range of 5 to 50 volts for typical transistors. The base voltage swing necessary to take the transistor from cutoff to saturation will be in the range of 0.1 to 0.5 volt. In

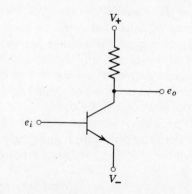

FIGURE 5.6   *Transistor in grounded emitter circuit.*

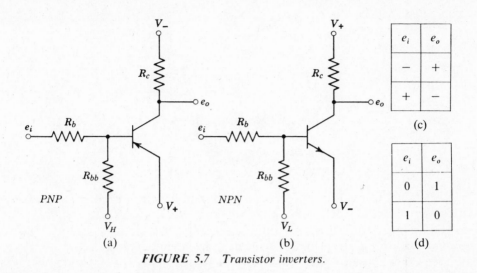

FIGURE 5.7   *Transistor inverters.*

addition, the current flowing in the base circuit will typically be on the order of one-hundredth of the current in the collector and emitter circuits.

The basic circuits for *PNP* and *NPN* inverters are shown in Fig. 5.7. As before $V_H > V_+ > V_- > V_L$. Consider the *NPN* inverter, Fig. 5.7b. When $e_i = V_-$, the voltage at the base will be negative with respect to V_-, so the transistor will be cut off, and $e_0 = V_+$. The values of R_b and R_{bb} are so chosen that when $e_i = V_+$, the voltage at the base will be positive with respect to V_-. The transistor will thus be conducting so that $e_0 = V_-$, neglecting the small drop across the transistor in saturation. This behavior is summarized in the table of Fig. 5.7c. For positive logic, this will result in the truth table of Fig. 5.7d. For negative logic, the truth table will be the same, except the position of the rows will be reversed. We leave it to the reader to satisfy himself that the *PNP* inverter will function in the same general fashion.

In order to correct the loading deficiencies of the diode gates, we may simply let the output of each gate drive an inverter, as shown in Figs. 5.8 and 5.9. Consider the circuit of Fig. 5.8a, which consists of a first-form diode gate driving an *NPN* inverter. If one or more inputs is at V_-, e_g will also be at V_-, and the voltage at the base will be negative with respect to the emitter. The transistor will be cut off, so $e_0 = V_+$. If both inputs are at V_+, the voltage at e_g, which will depend on the circuit parameters in the base circuit, will be such as to drive the base positive with respect to the emitter. The transistor will conduct, and $e_0 = V_-$. This behavior is summarized in the table of Fig. 5.8b. For positive logic, this results in the truth table of Fig. 5.8c, which is seen to be the table for NOT-AND, or NAND. For negative logic, we obtain the truth table of Fig. 5.8d,

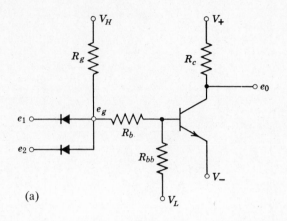

e_1	e_2	e_0
−	−	+
−	+	+
+	−	+
+	+	−

e_1	e_2	e_0
0	0	1
0	1	1
1	0	1
1	1	0

e_1	e_2	e_0
1	1	0
1	0	0
0	1	0
0	0	1

$V_+ = 1 \quad V_- = 0$ $\qquad$ $V_+ = 0 \quad V_- = 1$

(b) $\qquad\qquad\qquad$ (c) $\qquad\qquad\qquad$ (d)

FIGURE 5.8 Diode-transistor gate, first form.

the table for NOT-OR, or NOR. By a similar argument, we find that the circuit of Fig. 5.9*a* implements NOR for positive logic and NAND for negative logic.

The addition of the transistor inverters considerably alleviates the loading problems associated with diode logic, but it does not entirely eliminate them. Consider the situation shown in Fig. 5.10, where one NAND gate drives another. When the output of the first gate is at the low level (transistor in saturation), a load current, I_L, will flow as shown. If other gates are connected in parallel, each will contribute a similar I_L. However, there is a limit, set by base current and current gain, to the amount of current that can flow into the transistor if it is to remain in saturation. If so many gates are connected that this limit is exceeded, the transistor will come out of saturation and the output voltage will rise above the low logic level. I_L is of the same order of magnitude as the base current, so the gain of the transistor nevertheless considerably reduces fan-out restrictions.

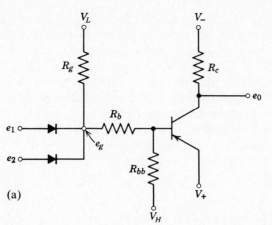

(a)

e_1	e_2	e_0
+	+	−
−	+	−
+	−	−
−	−	+

(b)

e_1	e_2	e_0
1	1	0
0	1	0
1	0	0
0	0	1

$V_+ = 1 \quad V_- = 0$

(c)

e_1	e_2	e_0
0	0	1
1	0	1
0	1	1
1	1	0

$V_+ = 0 \quad V_- = 1$

(d)

FIGURE 5.9 *Diode-transistor gate, second form.*

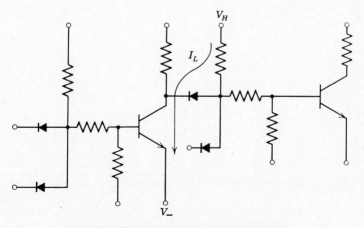

FIGURE 5.10 *Loading on NAND gate.*

The fan-in can also be considerably increased over that possible with diode logic. The finite back resistance of the diodes will still cause deterioration in signal level at the gate output, but now the gate voltage, e_g, need only have enough swing to provide reliable control of the transistor. As long as we observe the fan-in and fan-out restrictions, there will be no deterioration of signal level through successive levels of diode-transistor gating. Because of the small voltage across the transistor in saturation and minor loading effects in cutoff, the signal levels will differ slightly from the supply voltages, V_+ and V_-. However, the power gain of the transistors will eliminate any progressive degradation of signals.

5.4 Speed and Delay in Logic Circuits

Faster operation generally means greater computing power, so there is a continual search for faster and faster logic circuits. The development of faster circuits is the province of the electronic circuit engineer, not the logic designer, and the actual speed of the electronics is of only indirect interest to the logic designer. However, the qualitative nature of the transitions and delays in logic circuits plays an important role in the logical theory of sequential circuits.

Figure 5.11 shows a grounded-emitter inverter and the response to a positive pulse at the base, taking the transistor from cutoff to saturation and back. A number of complex physical factors enter into the determination of this response, of which we can consider only a few. First, from the time of the start of the base pulse, there is a delay time, t_d, before the collector current starts to rise. This

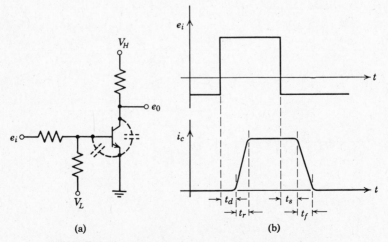

(a)   (b)

FIGURE 5.11 Transient response of transistor inverter.

delay is primarily caused by the effective base-to-emitter capacitance of the transistor (shown dotted in Fig. 5.11a). This capacitance must be charged to a positive level before the transistor starts to turn on. After the transistor starts to conduct, there is a finite rise time, t_r, primarily controlled by the collector capacitance. On the trailing edge of the base pulse, there is again a delay, t_s, before the collector current starts to drop. This delay is due to the base-to-emitter capacitance as well as the storage of charge in the base region during saturation. The delay due to the storage effect t_s is longer than t_d. Finally, there is a finite fall time, t_f, again due to the collector capacitance.

In terms of specifying delay times through gates, the times t_d and t_s shown in Fig. 5.11b are meaningful only if the input rise and fall times are negligible compared to those at the output. This is rarely the case, since logic circuits are usually driven by other logic circuits of the same type, so that rise and fall times are comparable at input and output, and are generally of the same order of magnitude as the delay times. It is thus necessary to specify just what points on the input and output waveforms are to be used in specifying delay times.

In this regard it is important to note that the transistors in logic circuits are operating in the switching mode. Even though the input may vary over the full logic range (0 to 3.6 for transistor-transistor logic—TTL), only a very small change in input voltage is required to cause switching. For TTL the output switches fully from one logic level to the other as the input passes through a narrow range of about 0.1 volt, nominally centered at 1.5 volts. Unless the input change is so slow that it takes longer than the nominal rise time of the circuit to pass through this 0.1-volt range, which is most unlikely, the output rise and fall times will be totally determined by the internal parameters of the logic circuit.

On the basis of this switching at a nominal voltage of 1.5 volts, delay time is normally measured from the time the input waveform passes through 1.5 volts until the corresponding output change passes through 1.5 volts. A typical test setup for measuring delay time is shown in Fig. 5.12, along with the corresponding

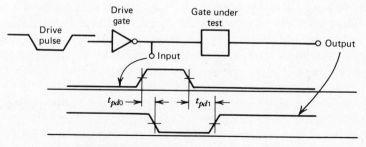

FIGURE 5.12   Typical test for gate delay.

waveforms. As shown, the gate under test is driven by another gate of the same type, which is in turn driven by a drive pulse having rise and fall times of the same order of magnitude of those of the gates.

As shown, the delay times are measured from the 1.5-volt points of the input transitions to the 1.5-volt points of the corresponding output transitions. Because one transition involves the transistor turning on, the other the transistor turning off, the delays, as noted above, are not equal. The notation for these delays is unfortunately not standardized among manufacturers. The most common notations are t_{pd0} or t_{pd-} for the negative-going output transition and t_{pd1} or t_{pd+} for the positive-going output transition. The average gate delay, t_{pd}, is the arithmetic mean of these two delays. The transition voltages will be different for other logic families. The user should consult the manufacturer's data sheet for details.

5.5   Integrated Circuit Logic

In Section 5.3, we discussed some elementary implementations of logic elements that could be realized in terms of discrete diodes, transistors, and resistors, interconnected by printed circuits or wires. Such discrete logic was dominant through most of the 1960s, but today it has been almost entirely supplanted by integrated-circuit (IC) logic. In integrated circuits, entire gate circuits made up of diodes, transistors, and resistors are formed on a single monolithic chip of silicon by chemical and metallurgical techniques.

Integrated circuits offer significant advantages over discrete circuits in three primary areas: size, power, and cost. A number of complete gate circuits can be deposited on a single chip about the size of the head of a common pin. Practical problems of connecting these minute circuits may require mounting them in much larger packages, but the size advantage relative to discrete circuits is nevertheless substantial.

Closely related to the size advantage is the reduction in power dissipation. Whatever the size of the individual circuit packages, the density with which large numbers of packages can be mounted is largely determined by their power dissipation because of the problems associated with dissipating the heat generated by large numbers of closely packed circuits. The reduced power requirements also reduce power supply requirements, providing for still further reductions in equipment size and cost. Some families of IC logic require so little power that battery-powered operation of even very complex devices becomes practical, a real advantage where portability is important.

Perhaps most important are the reductions in cost made possible by integrated circuits. The fabrication processes of integrated circuits are basically similar to those of transistors and diodes, and the cost of a complete integrated circuit is generally comparable to the cost of a single transistor. Further, all the costs of interconnecting discrete components are eliminated. As a result, the cost of logic has been reduced dramatically, to the point where the logic cost is a minor part of the cost of most digital systems.

Integrated circuits are commonly classified in terms of the number of gates or active devices on a single chip. *Small-scale integration* (SSI) denotes integrated circuits with fewer than 12 gates per chip; *medium-scale integration* (MSI) indicates between 12 and 100 gates per chip; *large-scale integrating* (LSI) corresponds to more than 100 gates per chip. A newer category, *very large-scale integration* (VLSI), has attracted much interest recently. The exact location of the boundary between LSI and VLSI is somewhat uncertain, but it is probably in the range of 10,000 to 20,000 active devices per chip. In this book (except for Chapter 12) we will be concerned mostly with SSI and MSI. Within each of these various size categories, several technologies are available, each offering a different balance of advantages and disadvantages in such areas as speed, power, cost, and noise immunity. In SSI and MSI, three technologies are commonly available: TTL, emitter-coupled logic (ECL), and complementary metal oxide semiconductor (CMOS). In LSI three types are also commonly available: integrated injection logic (I^2L), N-channel metal oxide semiconductor (NMOS), and CMOS. A complete treatment of all of these would properly be the subject of several books, but we will briefly consider some of the more significant characteristics from the point of view of the system designer.

The most popular family of logic at this time is TTL (transistor-transistor logic). In order to gain an understanding of this type of logic, it is useful first to consider the IC form of diode-transistor logic (DTL). This type of logic is now obsolete, but TTL may be viewed in terms of an evolution from DTL. The basic circuit of the standard DTL gate (Fig. 5.13) is similar to the discrete NAND gate shown in Fig. 5.8. The extra transistor provides additional gain to increase fan-out, and the extra diode provides biasing for the output transistor without a separate negative supply (V_L).

The first change from DTL to TTL is based on the fact that, in integrated circuits, it is easier to fabricate transistors than diodes. Where diodes are required, it is usual to use the base-emitter junction of a transistor, with the base serving as anode, the emitter as cathode, and the collector tied to the base. In the circuit of Fig. 5.13 we note that the anodes of the input diodes are common, so that these diodes can be realized in the form of a multiple-emitter transistor, as shown in Fig. 5.14a.

The remaining changes in the evolution from DTL to TTL are made to achieve increased speed. When the circuit of Fig. 5.14a is to switch from the low-output condition to the high-output condition, transistor $Q1$ must go from saturation to cutoff. As noted earlier, this requires removal of charge from the base of $Q1$. This switching action is brought about by transistor $Q2$ going from

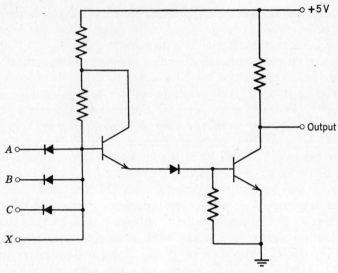

FIGURE 5.13   Standard DTL gate.

saturation to cutoff, so that the only path for discharging the base of $Q1$ is through resistor $R1$, and the speed of this discharge is limited by the time-constant of this circuit. To obtain faster switching we take advantage of the fact that the input diodes are realized in terms of a transistor and change the circuit to that shown in Fig. 5.14b. Now the charge on the base of $Q1$ is removed through transistor $Q3$, resulting in a considerable reduction in the storage delay time, t_s.

Another factor limiting the transition speed is the collector capacitance, which must be charged as the output voltage switches from the low value to the high value. In the circuit of Fig. 5.14b the only path for charging this capacitance is via the collector resistance $R2$. The charging speed can be increased by reducing $R2$, but this will result in increased power dissipation. An output circuit of this form is known as a *passive pullup* circuit, since the output capacitance is "pulled up" via the passive element $R2$. An alternative that provides faster charging without increased power dissipation is the *active pullup* circuit shown in Fig. 5.14c. In the low-output condition, the *phase-splitter* transistor $Q3$ is in saturation and the saturation current through $R1$ results in a sufficiently positive voltage to saturate output transistor $Q1$. Because of the diode $D1$ the emitter of output transistor $Q2$ is more positive than the collector of $Q3$, so that $Q2$ is cut off. When the gate goes to the high-output condition, transistor $Q3$ cuts off, so that the base of $Q1$ goes to 0, the base of $Q2$ goes sharply positive, $Q1$ cuts off, and $Q2$ saturates. Thus the output circuit essentially acts as a 2-pole switch,

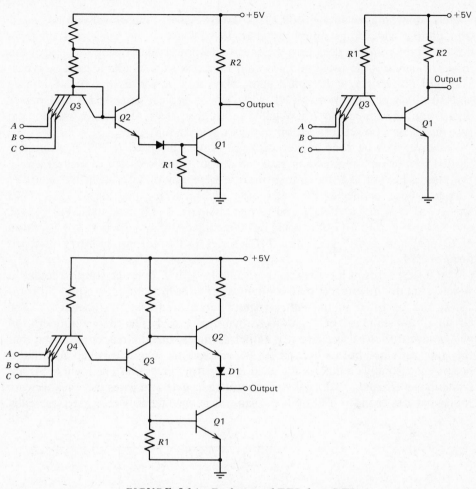

FIGURE 5.14 *Evolution of TTL from DTL.*

switching the output between ground and the supply voltage. Because of voltage drops in the output circuit, the actual output voltages, with $V_{CC} = 5v$, are 0.2 (low) and 3.4 volts (high). In terms of speed, the result of these changes in circuitry is that the typical TTL gate is approximately three times as fast as the corresponding DTL gate.

TTL gates are available in AND, OR, NAND, and NOR, with 2, 3, 4, or 8 inputs per gate, one to four gates per chip. Also available are chips with six inverters. Fan-out is typically 8 to 10, expressed as the number of similar gate inputs that can be driven.

One important advantage of TTL is that the speed and power can be varied over quite a wide range simply by changing the size of the resistors. Increasing the size of the resistors decreases the speed and power; decreasing the size of the resistors increases the speed and power. TTL circuits are available in three types: Standard (74 series), with typical gate delays and power dissipation of 10 nsec and 10 mW/gate; low power (74L series), with typical gate delays and power dissipation of 30 nsec and 1 mW/gate; and high speed (74H series), with typical gate delays and power dissipation of 6 nsec and 25 mW/gate.

Whatever may be done to the resistance values, the speed of TTL is ultimately limited by the time required to pull the output transistors out of saturation. In *Schottky* TTL (STTL), the transistors are kept out of saturation by Schottky barrier diodes connected between base and collector. Standard STTL (74S series) has typical gate delays and power dissipation of 3 nsec and 20 mW/gate, and low-power STTL (74LS series) has typical gate delays and power dissipation of 10 nsec and 2 mW/gate. The 74LS series is the current industry-standard form of TTL.

The fastest form of logic currently available is ECL (emitter-coupled logic) in which a totally different circuit configuration is utilized to ensure that all transistors are operated in the *active* region between saturation and cut off. The circuit of the standard ECL gate is shown in Fig. 5.15. The theory of operation will not be discussed, because it is rather complex, but the reader will note that the gate provides both OR and NOR outputs. This is a distinct advantage, because it virtually eliminates the need for separate inverters. The logic levels are typically -0.75 and -1.75 volts. The inputs per gate and gates per packages are somewhat less than for TTL, but the fan-out is considerably increased, typically

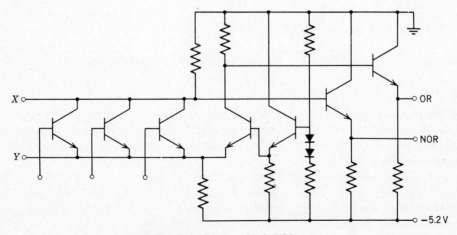

FIGURE 5.15   Standard ECL gate.

20 to 25. Expansion, to a maximum fan-in of 20, is possible by connecting additional input transistors between points X and Y. Several types of ECL circuits are available, but the two most commonly used provide average gate delays of 2 nsec and 0.7 nsec, with power dissipations of 25 mW/gate and 50 mW/gate, respectively.

The transistors used in TTL and ECL are *bipolar* transistors, and these types of logic are known as bipolar logic. A third type of bipolar logic is integrated injection logic (I^2L). As was the case with TTL, it is useful to consider this form of logic in terms of its evolution from an earlier form of logic, direct-coupled transistor logic (DCTL). Figure 5.16*a* shows a DCTL NOR gate. If either transistor is on, the output will be low; only if both inputs go high, cutting off both transistors, will the output be high, so the function is NOR. In use, the input variables, A and B, would normally be supplied by other gates of a similar form, and the output would drive other gates of similar form. In Fig. 5.16*b* we show a NOR gate, G1, together with the output sections of two driving gates, G2 and G3, and the input gate of a driven gate, G4. In Fig. 5.16*c*, we have the same circuit except that the resistors that the resistors have been moved from the collectors of the driving gates to the bases of the driven gates. Since each collector drives a base this makes no difference in the logical behavior of the circuits, but makes a big difference in the way the gates are realized in IC form. Note that the circuit of Fig. 5.16*c* utilizes passive pullup. In Fig. 5.16*d* we have changed to active pullup, replacing the resistors with transistors. This change is important for IC realization because transistors are simpler to realize than resistors and occupy less area on an integrated circuit chip.

Elimination of the resistors provides significant economies in fabrication, and the ingenious interconnections of the transistors provide further benefits. In Fig. 5.17*a* the circuit of the I^2L NOR gate is repeated with the *PNP* and *NPN* sections of each transistor labeled accordingly. Note first that the N-emitters and the N-bases of the four transistors are electrically common. Further the P-collector and P-base of $Q3$ and $Q1$ are common, as is the case with $Q4$ and $Q2$. Finally, the P-emitters of $Q3$ and $Q4$ are common. A cross section of the resultant structure in IC form is shown in Fig. 5.17*b*. Because of the sharing of areas on the chip between the four transistors, the complete gate can be realized in little more area than would be required for a single transistor. The *injector*, the common emitter of $Q3$ and $Q4$, serves to inject the pullup currents—hence the name *integrated injection logic*. As a result of this ingenious structure, the complete gate requires only about one-tenth the area (on an IC chip) that is required for a corresponding TTL or ECL gate. In terms of speed and power, I^2L is comparable to Schottky TTL, so that its chief advantage is its economical utilization of chip "real estate," which makes it the only bipolar logic family really practical for LSI, where a large number of gates must be packed into a small area.

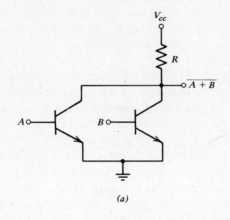

(a)

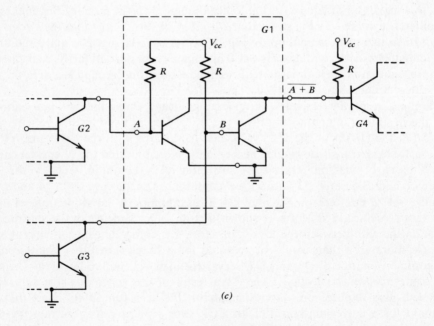

(c)

FIGURE 5.16   *Evolution of I²L from DCTL.*

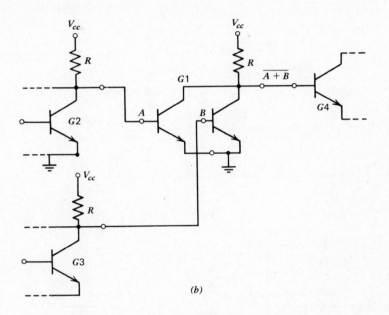

(b)

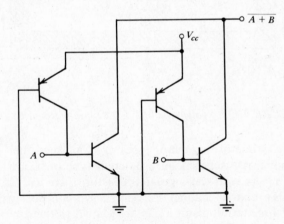

(d)

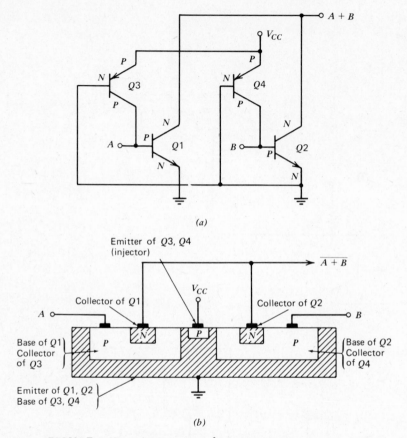

FIGURE 5.17 *Realization of I^2L gate in integrated circuit form.*

Bipolar logic is one of two major types of logic. The other type, MOS (metal oxide semiconductor), is based on a different type of transistor, the field effect transistor (FET). As with the bipolar transistor, there are two types, known as N-channel FETs and P-channel FETs, the symbols for which are shown in Fig. 5.18. The three terminals of FETs are known as the drain, source, and gate.

In the N-channel FET, if the gate is negative with respect to the source, the FET is an open circuit from drain to source. If the gate is positive with respect to the source, the FET is a short circuit from drain to source. The operation is the same for the P-channel FET except that polarities are reversed. The operation of the FET may appear to be the same as that of the bipolar, but there are important differences. In the bipolar, although we characterize the operation in terms of the base-to-emitter voltage, the base current is the controlling factor.

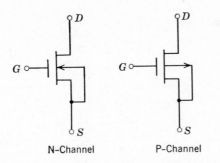

FIGURE 5.18 *N-channel and P-channel FETs.*

For example, in an *NPN* transistor, when the base goes positive with respect to the emitter, current flows from base to emitter; this base current turns the transistor on. In addition, the transistor goes into saturation, i.e., the collector-to-emitter voltage drops to a negligible value, only when a sufficient collector current flows. In the FET, by contrast, the gate-to-source voltage controls the drain-to-source impedance by electrostatic field action (hence the name *field effect*) and the gate current is essentially unmeasurable. Further, when the FET is turned on, the drain-to-source voltage is essentially zero even if no current flows in the drain circuit. As we will see, these special characteristics of the FET make possible logic circuits with extremely low power consumption.

There are three basic types of MOS logic circuits: *N*-channel (NMOS), *P*-channel (PMOS), and complementary (CMOS). PMOS is rarely used today. NMOS is slower than any of the bipolar logic families, but has very low power drain and high packing density, with the result that it is the most popular form of LSI. CMOS has even lower power drain than NMOS, but the packing density is not as high. Of the three types of MOS logic, only CMOS is used in SSI and MSI, because its very low power requirements may outweigh its low speed in portable applications, where low current drain is essential for long battery life.

The circuit of the standard CMOS inverter is shown in Fig. 5.19a. The circuit consists of an *N*-channel FET and a *P*-channel FET in series (hence the name *complementary*), normally connected between ground and a positive supply voltage. The signal levels are the same as the supply levels, i.e., 0 and V_+. When the input is at 0 volts, the NFET is cut off and the PFET is turned on. As a result, there is a low impedance path from the output to V_+ and an open circuit to ground, resulting in an output voltage equal to V_+. Similarly, when the input is at V_+ volts, the NFET becomes a low impedance while the PFET becomes an open circuit, resulting in 0 volts at the output.

Note that, in a static condition, one FET or the other is always cut off, and there are no resistive paths. Further, if the circuit drives another CMOS device,

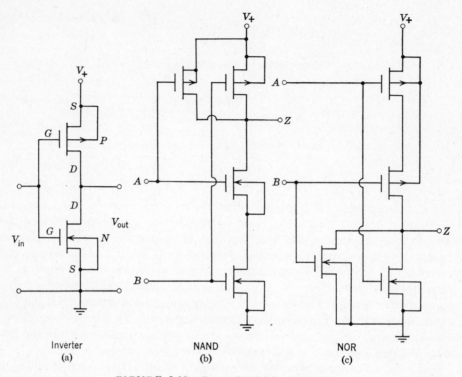

FIGURE 5.19 *Basic CMOS logic circuits.*

there is no output current, since the gate of an FET draws no current. As a result, the only quiescent power dissipation is that due to leakage in the cutoff FET. This quiescent dissipation is typically less than 1 μW/gate and in some circuits is in the nanowatt range.

When the inverter switches from one level to another, both gates will be on momentarily, and the change in voltage levels requires the charging of parasitic capacitances within the inverter and the interconnections between circuits. As a result, current will flow during switching and the average power dissipation is directly dependent on the switching rate. At an average switching rate of 1 MHz (one logic change per microsecond), the typical dissipation is 1 mW/gate.

NAND and NOR gates are constructed by using one inverter for each input, with the upper (P) transistors in parallel and the lower (N) transistors in series for NAND, and vice versa for NOR, as shown in Figs. 5.19b and 5.19c. In the NAND gate (Fig. 5.19b), if any of the inputs are at the 0 level, the corresponding upper transistors are biased on and the corresponding lower transistors are biased off, so that the output is at the V_+ level. If all the inputs are at the V_+

level, all the upper transistors are off and all the lower transistors are on, so that the output is at the 0 level. The operation of the NOR gate may be analyzed in an analogous fashion.

Standard CMOS circuits will run properly at supply voltages of 3 to 15 volts, a characteristic that makes it possible to use unregulated power supplies. Special CMOS circuits can be operated at 1.3 volts, making it possible to use single dry cells as the power source. There is no standard gate type in CMOS, with NAND and NOR being equally available. Maximum fan-in is 4. Since FET gates draw no current, fan-out is primarily determined by parasitic wiring capacitances and is, for all practical purposes, unlimited. Average gate delay is dependent on the power supply voltage, being typically 35 nsec at 5 volts, 25 nsec at 10 volts.

5.6 Noise Margin

The term *noise* refers to spurious signals generated in a system by any of a variety of environmental causes. In digital systems most noise is due to inductive or capacitive coupling between signal lines. For example, if the current in one line changes rapidly, the resultant rapid change in the surrounding magnetic field may generate a voltage in adjacent lines. Whatever the source of the noise, if it is large enough it may change the signal enough to cause it to take on the opposite logical value. The *noise margin* is a measure of how much noise can be added to the logic inputs of a circuit without the circuit responding improperly.

Figure 5.20 shows the *transfer characteristic* of a TTL inverter, i.e., a plot of output voltage, V_O, as a function of the input voltage, V_I. Similar curves can be drawn for any family of logic. Because the use of actual voltages will make the basic concepts easier to understand, we will use the specific example of TTL in the discussion to follow, but the ideas involved apply to any family of logic. With a low-level voltage applied at the input, the output will be at the high level,

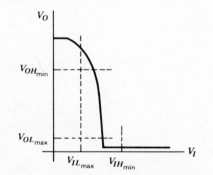

FIGURE 5.20 *Transfer characteristic of TTL inverter.*

nominally 3.4 volts for TTL. The exact voltage will vary from one gate to another, depending on statistical variations in the characteristics of gates. To place bounds on this variation, the manufacturer specifies a minimum high-level output voltage, $V_{OH_{min}} = 2.4$ volts. With this specification, the manufacturer guarantees that, if you apply a low-level voltage at the input, the output will never be less than 2.4 volts. Similarly, the manufacturer specifies a maximum low-level output voltage, $V_{OL_{max}} = 0.4$ volts. If a high-level input is applied, the output is guaranteed not to exceed 0.4 volts.

These two values, $V_{OH_{min}}$ and $V_{OL_{max}}$, place bounds on the output voltages, but what about the inputs? We need similar bounds to define the input levels. To define the low-level input the manufacturer specifies $V_{IL_{max}} = 0.8$ volts. This means that the circuit is guaranteed to interpret an input as low level and produce the corresponding output (high for an inverter) if the input voltage does not exceed 0.8 volts. Similarly, an input voltage is guaranteed to be interpreted as a high-level input as long as the voltage is not less than $V_{IH_{min}} = 2.0$ volts.

The manner in which these specifications determine the noise margin can be considered in terms of Fig. 5.21a, showing one TTL gate driving another, together with a graphic interpretation of the relevant levels in Fig. 5.21b. Assume that the inputs to $G1$ are such that the output is at the low level. This voltage is guaranteed not to exceed $V_{OL_{max}} = 0.4$ volts. The gate $G2$ will interpret an input signal as low level as long as it does not exceed $V_{IL_{max}} = 0.8$ volts. Thus, additive

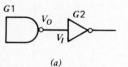

(a)

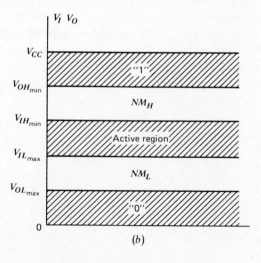

(b)

FIGURE 5.21   *Noise margin in logic circuits.*

noise up to 0.4 volts can appear on the line connecting these two gates without causing gate $G2$ to respond improperly. We thus define the *low noise margin* as

$$NM_L = V_{IL_{max}} - V_{OL_{max}}$$

Similarly, if the output of $G1$ is at the high level, the voltage will be no less than $V_{OH_{min}} = 2.4$ volts. Gate $G2$ will interpret the input as high level as long as it is not less than $V_{IH_{min}} = 2.0$ volts, so that noise can reduce the voltage from $G1$ by as much as 0.4 volts without $G2$ responding improperly. We define the *high noise margin* as

$$NM_H = V_{OH_{min}} - V_{IH_{min}}$$

The overall noise margin is defined as the smaller of the two noise margins. For TTL they are the same and $NM = 0.4$ volts.

The noise margin provides information as to how a logic family responds to noise. Equally important in considering the overall noise performance is the amount of noise generated by the logic. The noise margin of TTL is good, but TTL also generates a lot of noise, so that its overall noise performance is considered only fair. Of the three types of logic used in SSI and MSI, CMOS has the best noise performance and ECL the worst, with TTL in the middle. In LSI, CMOS again has the best noise performance, followed by NMOS and I^2L, which are comparable.

5.7   Logical Interconnection of NAND and NOR Gates

We saw in several examples in Chapters 3 and 4 that many problems seem most naturally formulated in AND-OR terms, i.e., in expressions of the general form

$$f = AB + CD + \bar{A}D$$

We will see in the next chapter that *any* Boolean function can be expressed in AND-OR form, or in the OR-AND form, such as

$$f = (A + B) \cdot (B + C) \cdot (\bar{C} + D)$$

We will see that powerful and completely general design techniques are available that lead to the AND-OR or OR-AND forms. Because of these factors, most designers prefer to design for these forms. However, in terms of actual realization, the use of AND-gates and OR-gates may not be desirable, for two reasons. First, AND and OR are more complex in terms of realization in most types of logic. As we saw in DTL, the single stage of amplification inherently provides inversion, leading to NAND or NOR; AND or OR requires another level of inversion. Second, reducing the number of different types of gates can provide

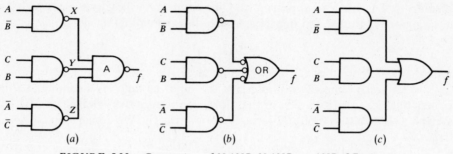

FIGURE 5.22 *Conversion of NAND-NAND to AND-OR circuit.*

significant economies through volume purchasing and simplification of inventory control. We saw in Chapter 3 that the NAND and NOR functions are functionally complete, so that it is possible to realize any function using nothing but NAND or NOR gates.

While it may be theoretically possible to realize designs utilizing only one gate type, we must consider the practical problems of finding economical realizations by simple design procedures. Fortunately, it turns out that the AND-OR and OR-AND forms, for which such procedures exist, can be easily converted into all-NAND or all-NOR form. Consider the simple logical circuit of Fig. 5.22a, which consists of three NAND-gates driving another NAND gate. From DeMorgan's law,

$$\overline{XYZ} = \bar{X} + \bar{Y} + \bar{Z} \qquad \text{(Theorem 4.11)}$$

we see that the final NAND-gate can be replaced by an OR-gate with inversion on the inputs (Fig. 5.22b). From Lemma 4.7,

$$\bar{\bar{X}} = X \qquad \text{(Lemma 4.7)}$$

so that the two successive inversions on the lines between the input and output gates cancel, giving the circuit of Fig. 5.22c. Algebraically, the same result is obtained as follows:

$$f = \overline{(\overline{A\bar{B}}) \cdot (\overline{BC}) \cdot (\overline{\bar{A}\bar{C}})}$$
$$= (\overline{\overline{A\bar{B}}}) + (\overline{\overline{BC}}) + (\overline{\overline{\bar{A}\bar{C}}}) \qquad \text{(Theorem 4.11)}$$
$$= A\bar{B} + BC + \bar{A}\bar{C} \qquad \text{(Lemma 4.7)}$$

Thus, we see that a two-level NAND circuit is equivalent to a two-level AND-OR circuit. In a similar fashion, we can show that a two-level NOR circuit is equivalent to a two-level OR-AND circuit (Fig. 5.23). The reader should note carefully that these simple relationships hold only for two-level AND-OR or

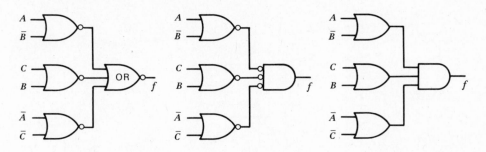

FIGURE 5.23   *Conversion of NOR-NOR to OR-AND circuit.*

OR-AND forms. Where there are more levels of gating, the procedures for finding all-NAND or all-NOR designs are considerably more complex. This topic will be discussed in the next chapter.

5.8   Wired Logic and AND-OR-INVERT Gates

The use of NAND or NOR gates in place of AND or OR gates not only provides economy through standardization, but also provides some improvement in speed. Because of the extra level of inversion, AND and OR gates are generally somewhat slower than NAND or NOR gates of the same family. Nevertheless, the speed is still limited by the fact that we have two levels of gating, so that the total delay is twice that of a single gate. In some families of logic it is possible to eliminate the output gate simply by wiring the outputs of the input gates together. The resultant circuit is called a *wired-OR* or *wired-AND*, depending on the type of logic. Wired-logic is widely used with ECL. Referring to Fig. 5.15, it is seen that the two outputs, OR and NOR, are taken from the emitters of the output transistors, so that the effective circuit when the outputs are wired together is as shown in Fig. 5.24a. If either transistor saturates, the output is effectively shorted to the supply voltage, producing a high output. The output will be low only if both transistors are cut off. Thus, in terms of positive logic, the resultant function is the logical OR of the functions realized by the gates. The OR and NOR outputs may be wired together in any combination, all ORs, all NORs, or any combination. Figure 5.24b shows the block diagram representation of two possible interconnections. For the OR outputs wired together we have

$$f_1 = (A + B) + (C + D) = A + B + C + D$$

i.e., input expansion. For the NOR outputs wired together, we have

$$f_2 = \overline{(A + B)} + \overline{(C + D)} = \overline{(A + B)} \cdot \overline{(C + D)}$$

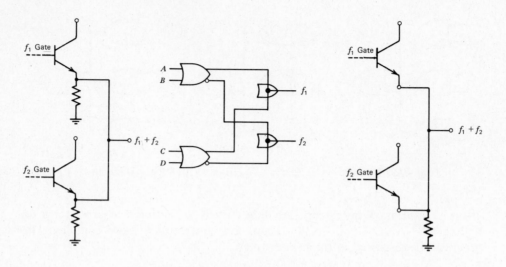

FIGURE 5.24   *Wired logic with ECL.*

the complement of the function realized by NOR-NOR connection of the same gates. ECL gates may have the emitter resistors in the individual gates, as shown in Fig. 5.24*a*, or may have open emitters, which are wired together with a common external emitter resistor, as shown in Fig. 5.24*c*.

Wired-logic cannot be used with regular TTL gates with totem-pole outputs. If two gates with totem-pole outputs were wired together and one was in the high state and the other in the low state, the power supply would be shorted to ground through the output transistors, resulting in one or both gates burning out. Special TTL gates, known as *open-collector gates*, can be wired together. Figure 5.25*a* shows two open-collector NAND gates wired together. Note that the output sections of these gates consist solely of the lower transistor of the usual totem-pole pair. The collectors are tied together and connected to V_{CC} through an external common pullup resistor. Note that the common output will be low if either gate is in the low state and will go high only if both gates go high. The result of the connection is to AND the outputs of the gates, and the connection is known as a wired-AND, symbolized as shown in Fig. 5.25*b*. The overall function realized depends on the functions realized by the gates wired together. In this case, with NAND gates, we have

$$f = (\overline{AB}) \cdot (\overline{CD}) = \overline{AB + CD}$$

Note that the function realized is the complement of the function that would be realized if the same gate were connected to a NAND output gate in the standard

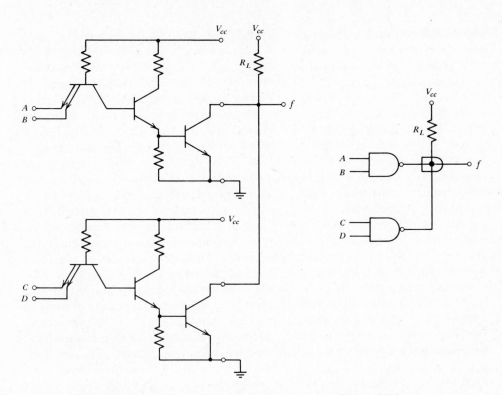

FIGURE 5.25 *Wired-AND connection of open-collector TTL gates.*

NAND-NAND connection. AND and NOR are also available in open-collector form in TTL, for which cases we have

$$f = (AB) \cdot (CD) = ABCD$$

and

$$f = (\overline{A + B}) \cdot (\overline{C + D}) = \overline{A + B + C + D}$$

so that the wired-AND simply provides input expansion for these types of gates.

Wired-logic connections offer obvious advantages in economy and, in the case of ECL, increased speed. For TTL the wired-logic connection is actually slower, since the common-collector output is basically the passive pullup of DTL, which is only about one-third as fast as the totem-pole output. Wired logic also has a significant drawback in ease of servicing. Suppose that we have a circuit of the form of Fig. 5.22a, and that one of the input NAND gates goes bad. We can determine which gate is defective simply by checking the outputs at points X, Y, and Z. Now suppose we have a wired-AND connection of the form of Fig. 5.25*b* and one of the gates fails. Since their outputs are wired together we cannot tell which one is bad without disconnecting them, which may be very difficult to do.

One application where the use of wired-logic is often a necessity is connections to a bus. A *bus* in a digital system is simply a common data path for connecting a number of devices. A bus may be implemented in several ways. When the devices are physically separate—for example, input/output devices in a computer system—a bus may be simply a set of wires to which the devices may be connected. This amounts to wired logic since the outputs of all the devices are connected together by the bus. Open collector gates can be used for busing, but will introduce some speed degradation and may cause loading problems. Every additional gate tied onto the bus results in an additional load on gates driving the bus. Further, the optimum size of common collector resistor depends on the number of gates connected, so that the resistors may have to be adjusted as devices are added to or removed from the bus. These problems can be alleviated through the use of gates with *three-state outputs*. The circuit of a typical TTL three-state output stage is shown in Fig. 5.26. When the control input, C, is at the high level, neither input diode conducts and the circuit functions normally as a totem-pole output state. When C goes low, however, both $Q2$ and $Q3$ will be cut off, putting the output into the *high impedance state*, in which the output will have no effect on anything to which it is connected, such as a bus. The net result is that we can totally disconnect a device from a bus. Figure 5.26*b* shows standard symbols for three-state gates. The downward-pointing triangle at the output

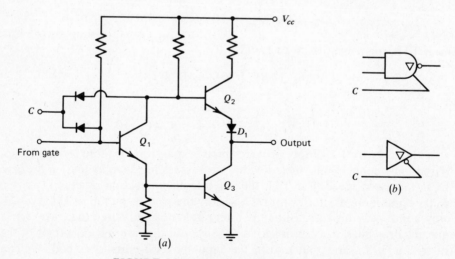

FIGURE 5.26 *Gates with three-state output.*

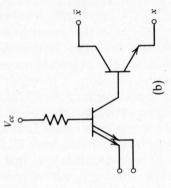

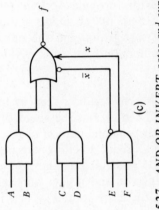

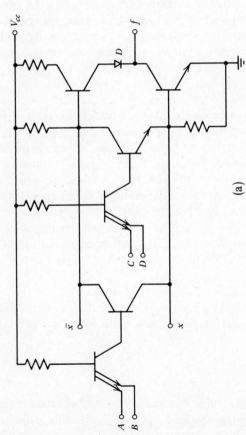

FIGURE 5.27 AND-OR-INVERT gates and expanders.

is the symbol for three-state output, and the input at the side is the control, or enable, input. The upper symbol is for a three-state NAND with active-high enable, i.e., the gate is enabled when C is high, as in the circuit of Fig. 5.26a. The lower symbol is for a buffer amplifier with active-low enable. When C is high the output is in the high impedance state; when C is low the output follows the input.

While wired logic does not provide any speed advantage for TTL, a special form of TTL gate, the AND-OR-INVERT gate (AOI), does provide the equivalent of two levels of gating with the delay of a single level. Figure 5.27a shows the circuit of a basic AOI gate. Comparing this circuit with the basic TTL NAND gate of Fig. 5.24c, we see that it is of the same basic form, but that we have added a second multiemitter input transistor driving a second phase-splitter transistor, and that the two phase splitters are connected in parallel. The input transistors provide ANDing of the signals at their emitter inputs, the parallel connection of the phase splitters provides the ORing of the signals from the input transistors, and the output section provides inversion, so that the complete function realized is AND-OR-INVERT (AOI); i.e.,

$$f = \overline{AB + CD}$$

The logical result is seen to be the same as for the wired-AND connection of NAND gates, but the speed of the totem-pole output section has been retained and the implementation of the OR function by parallel transistors has provided the additional function with no additional delay.

AOI gates are available in all five types of TTL, with various numbers of AND gates and inputs, and some AOI gates are expandable. In an expandable gate, the lines x and $\bar{x}$ in Fig. 5.27a are brought out for connection to the corresponding lines on *expander gates*, which have the basic configuration shown in Fig. 5.27b. Figure 5.27c shows the block diagram symbols for an AOI gate and expander, connected to realize the function

$$f = \overline{AB + CD + EF}$$

AOI gates are not widely used, probably because they are less flexible than separate gates. Nevertheless, they should be considered when speed is a primary concern.

5.9   Conclusion

It may seem to the reader that the wide variety of gate types and characteristics will pose a serious problem to the logic designer, possibly requiring different design techniques for each type of logic. Fortunately, this is not the case. In the next chapter, we will develop powerful design methods that can be adapted readily to any type of logic. There have been some attempts to develop special design methods tailored to some specific type of logic. We feel that this is most unwise, because it will tend to limit the designer to a specific type of logic and prevent him from exploiting the full power of all the many types of logic available. The logic types discussed in this chapter represent only a portion of those currently available, and new types are constantly being developed. An understanding of the basic characteristics and limitations of logic devices will enable the resourceful designer to adapt powerful general design techniques to whatever physical realizations may best suit his purpose.

Bibliography

1. Garrett, L. S. "Integrated-Circuit Digital Logic Families," *IEEE Spectrum*, 46–58 (Oct. 1970), 63–72 (Nov. 1970), 30–42 (Dec. 1970).
2. Torero, E. A. "Focus on CMOS," *Electronic Design*, **20**: 8, 54–61 (Apr. 13, 1972).
3. Torero, E. A. "Focus on Fast Logic," *Electronic Design*, **20**: 12, 50–57 (June 8, 1972).
4. Femling, Don. "Enhancement of Modular Design Capability by Use of Tri-State Logic," *Computer Design*, **10**: 6, 59–64 (June 1971).
5. Sheets, John. "Three State Switching Brings Wired-OR to TTL," *Electronics*, **43**: 19, 78–84 (Sept. 14, 1970).
6. Carr, W. N. and J. P. Mize. *MOS/LSI Design and Application*, McGraw-Hill, New York, 1972.
7. Kohonen, T. *Digital Devices and Circuits*, Prentice-Hall, Englewood Cliffs, N.J., 1972.
8. Barna, A. and D. I. Porat. *Integrated Circuits in Digital Electronics*, Wiley, New York, 1973.
9. Casasent, D. *Digital Electronics*, Quantum, New York, 1974.
10. Deem, B. R., K. Muchow, and A. Zeppa. *Digital Computer Circuits and Concepts*, Reston, Reston, Va., 1977.
11. Williams, G. E. *Digital Technology*, SRA, Chicago, 1977.
12. Taub, H. and D. Schilling. *Digital Integrated Electronics*, McGraw-Hill, New York, 1977.
13. Sandige, R. S., *Digital Concepts Using Standard Integrated Circuits*, McGraw-Hill, New York, 1978.

6

MINIMIZATION OF BOOLEAN FUNCTIONS

6.1 Introduction

In Chapter 3, we showed how the concept of truth-function analysis could be used in setting up a mathematical model of a logical (or switching) system. In Chapter 4, we stated the formal rules of an algebra governing the manipulation of the mathematical model. In Example 4.2, utilization of the mathematical model resulted in an obvious simplification of the switching circuit. In general, our basic purpose is design; the reason for using a mathematical model is that it provides a convenient form for exploring possible designs. Further, the *best* design is generally the *simplest* design that will get the job done, from which it follows that our objective in manipulating the mathematical model will usually be simplification or minimization. Because of the myriad possible forms of Boolean expressions, no completely general criteria for the simplest expression have been developed. As we will see, however, it is possible to define a simplest form of the two-level, or minimum-delay-time, circuit as discussed in Chapter 5.

We have seen that the rules of the algebra can be applied formally to manipulation and simplification of Boolean expressions, but such methods are far from easy to apply. The algebraic manipulation of Boolean functions is often quite involved, and finding the right line of attack requires considerable ingenuity and, sometimes, just plain luck. It is obvious that standard, systematic methods of minimization would be very useful. In this chapter and the next, two such methods are set forth—one graphical, the Karnaugh map, and one tabular, the Quine-McCluskey method.

6.2 Standard Forms of Boolean Functions

A standard, or general, method of analysis would seem to imply a standard starting point and a standard objective. In this section, we will develop two standard forms that may be used as starting points for simplification. The

96

Table 6.1

Term	Definition	Synonym
Literal	Variable or its complement. $(A, \bar{A}, B, \bar{B},$ etc.)	
Product term	Series of literals related by AND, e.g., $A\bar{B}D$, $AC\bar{D}E$, etc.	Conjunction
Sum term	Series of literals related by OR, e.g., $A + C + \bar{D}, A + B + \bar{D} + E$, etc.	Disjunction
Normal term	Product or sum term in which no variable appears more than once	

definitions of *literal, product term, sum term,* and *normal term,* which are given in Table 6.1, will be useful in discussing these standard forms. In a sense, the normal form is the "best" form. Since

$$A \cdot A = A = A + A$$

and   (Lemma 4.2)

$$A + \bar{A} = 1 \qquad \text{and} \qquad A\bar{A} = 0$$

a multiple occurrence of a variable in a sum or product term is always redundant or results in a trivial function.

As a first step, we will consider the form known as the *sum of products.* The nature of this form and manner of developing it are illustrated in the following examples.

Example 6.1

Find the sum-of-products (SOP) form of the function

$$f(A, B, C, D) = (AC + B)(CD + \bar{D})$$

First, let $AC = x$ and $CD = y$. Then

$$
\begin{aligned}
f(A, B, C, D) &= (x + B)(y + \bar{D}) \\
&= (x + B)y + (x + B)\bar{D} && \text{(Distributive law)} \\
&= xy + By + x\bar{D} + B\bar{D} && \text{(Distributive law)} \\
&= ACD + BCD + AC\bar{D} + B\bar{D} && \text{(6.1)}
\end{aligned}
$$

■

Example 6.2

Repeat Example 6.1 for the function

$$f(A, B, C, D, E) = (\overline{AC} + \bar{D})(\overline{B + CE})$$
$$= [(\bar{A} + \bar{C}) + \bar{D}](\bar{B} \cdot \overline{CE}) \qquad \text{(DeMorgan's law)}$$
$$= [(\bar{A} + \bar{C})\bar{B} + \bar{B}\bar{D}](\bar{C} + \bar{E}) \qquad \text{(DeMorgan's law)}$$
$$\qquad\qquad\qquad\qquad\qquad\qquad\qquad \text{(Distributive law)}$$
$$= (\bar{A}\bar{B} + \bar{B}\bar{C} + \bar{B}\bar{D})(\bar{C} + \bar{E}) \qquad \text{(Distributive law)}$$
$$= \bar{A}\bar{B}\bar{C} + \bar{B}\bar{C} + \bar{B}\bar{C}\bar{D} + \bar{A}\bar{B}\bar{E} + \bar{B}\bar{C}\bar{E} + \bar{B}\bar{D}\bar{E} \qquad (6.2)$$

∎

From these examples, the general method to be followed in converting any Boolean function to the SOP form should be evident. If no terms other than single variables are negated, only repeated application of the second distributive law is required. Where terms other than single variables are negated, DeMorgan's law must also be applied. Recalling the definition of a normal term, we see that Equation 6.1 or 6.2 could be referred to as a *sum of normal products*, but the shorter notation is preferred.

Continuing with the sum of products, Equation 6.1, we can write

$$f(A, B, C, D) = ACD + BCD + AC\bar{D} + B\bar{D} = ACD(B + \bar{B})$$
$$+ BCD(A + \bar{A}) + AC\bar{D}(B + \bar{B}) + B\bar{D}(A + \bar{A})$$
$$(a + \bar{a} = 1, a \cdot 1 = a)$$
$$= ABCD + A\bar{B}CD + ABCD + \bar{A}BCD + ABC\bar{D}$$
$$+ A\bar{B}C\bar{D} + B\bar{D}A + B\bar{D}\bar{A} \qquad \text{(Distributive law)}$$
$$= ABCD + A\bar{B}CD + \bar{A}BCD + ABC\bar{D} + A\bar{B}C\bar{D}$$
$$+ B\bar{D}A(C + \bar{C}) + B\bar{D}\bar{A}(C + \bar{C})$$
$$(a + a = a, a + \bar{a} = 1, a \cdot 1 = a)$$
$$= ABCD + A\bar{B}CD + \bar{A}BCD + ABC\bar{D} + A\bar{B}C\bar{D}$$
$$+ ABC\bar{D} + A\bar{B}C\bar{D} + \bar{A}BC\bar{D} + \bar{A}B\bar{C}\bar{D}$$
$$\text{(Distributive law)}$$
$$f(A, B, C, D) = ABCD + A\bar{B}CD + \bar{A}BCD + ABC\bar{D} + A\bar{B}C\bar{D}$$
$$+ AB\bar{C}\bar{D} + \bar{A}BC\bar{D} + \bar{A}B\bar{C}\bar{D} \qquad (a + a = a)$$
$$(6.3)$$

Note that Equation 6.3 is a sum of normal products with every product containing as many literals as there are variables in the function. Such products are called *canonic products, standard products,* or *minterms.* The terms *standard* or *canonic sum of products* and *full disjunctive normal form* have been used for expressions of the form of Equation 6.3. *Standard sum of products* will be preferred here.

We are led to the following theorem, the validity of which may already be evident to the reader. A proof of this theorem will be presented in the next section.

THEOREM 6.1   Any switching function of *n* variables $f(x_1, x_2, \ldots, x_n)$ may be expressed as a standard sum of products.

In view of the principle of duality, we expect that a *standard product of sums, full conjunctive normal form,* or *product of maxterms* will also exist.

Example 6.3

Express $f(A, B, C, D) = A + C + \bar{B}\bar{D}$ as a standard product of sums.

$$A + (C + \bar{B}\bar{D}) = A + (C + \bar{B})(C + \bar{D}) \qquad \text{(Distributive law)}$$
$$= (A + C + \bar{B})(A + C + \bar{D})$$

Then, since

$$A \cdot \bar{A} = 0 \qquad \text{(Postulate VI)}$$
$$A + 0 = A \qquad \text{(Postulate IIIa)}$$

and

$$(a + b\bar{b}) = (a + b)(a + \bar{b}) \qquad \text{(Distributive law)}$$

we can write

$$f(A, B, C, D) = (A + \bar{B} + C + D\bar{D})(A + C + \bar{D} + B\bar{B})$$
$$= (A + \bar{B} + C + D)(A + \bar{B} + C + \bar{D})(A + B + C + \bar{D})$$
$$\cdot (A + \bar{B} + C + \bar{D})$$
$$= (A + \bar{B} + C + D)(A + \bar{B} + C + \bar{D})(A + B + C + \bar{D}) \qquad (6.4)$$

∎

Equation 6.4 is seen to consist of a product of normal sums, with each sum containing as many literals as there are variables in the function. Such sums are known as *canonic sums,* or *standard sums,* or *maxterms.**

Generalizing, we have the dual of Theorem 6.1.

THEOREM 6.2   Any switching function of *n* variables $f(x_1, x_2, \ldots, x_n)$ may be expressed as a standard product of sums.

It may seem to the reader that we are moving in the wrong direction, since the standard forms found in the above examples were more complicated than the

---

* The reasons for the names *minterm* and *maxterm* will become evident in a later section.

expressions we started with. It is often the case that the standard sum or product will be a very complex form, but our basic purpose in using them is to provide a common form for starting simplification procedures.

6.3 Minterm and Maxterm Designation of Functions

It may have occurred to the reader, in following the examples of the preceding section, that writing out all the minterms or maxterms of a given function may be laborious. A shorthand notation for switching functions would certainly be useful. A switching function is defined by its truth table, a listing of the function values for all possible input combinations. Therefore, a simple and precise means of designating functions is obtained by numbering the rows of the truth table for a function and then listing the rows (input combinations) for which the function has the value 1, or, alternately, the rows for which it has the value 0.

Figure 6.1 shows the truth table of a particular three-variable function, with the rows assigned identifying numbers. The row numbers are simply the decimal equivalents of the input combinations on each row interpreted as binary numbers. For example, the input combination $A = 0$, $B = 1$, $C = 1$, interpreted as a binary number, gives us $011_2 = 3_{10}$, so that row is called row 3. Now we can specify the function by listing the rows for which it has the value 1

$$f(A, B, C) = \sum m(0, 4, 5, 7) \tag{6.5}$$

or the rows for which it has the value 0

$$f(A, B, C) = \prod M(1, 2, 3, 6) \tag{6.6}$$

The symbols "$\sum m$" and "$\prod M$" in the above lists are not just arbitrarily chosen but rather indicate a direct correspondence between these lists and the standard forms. To show this, we write out the standard sum-of-products form

Row. No.	A	B	C	f
0	0	0	0	1
1	0	0	1	0
2	0	1	0	0
3	0	1	1	0
4	1	0	0	1
5	1	0	1	1
6	1	1	0	0
7	1	1	1	1

FIGURE 6.1   Truth table with row numbers assigned.

(Equation 6.5a) and the standard product-of-sums form (Equation 6.6a) for the function of Fig. 6.1.

$$f(A, B, C) = \bar{A}\bar{B}\bar{C} + A\bar{B}\bar{C} + A\bar{B}C + ABC \qquad (6.5a)$$

$$f(A, B, C) = (A + B + \bar{C})(A + \bar{B} + C)$$
$$\cdot (A + \bar{B} + \bar{C})(\bar{A} + \bar{B} + C) \qquad (6.6a)$$

Consider first the sum-of-products form. From Lemma 4.3 $(a + 1 = 1)$, we see that the function will take on the value 1 whenever any one (or more) of the products takes on the value 1. Since these standard products, or minterms, contain all three variables, there is only one combination of inputs for which a given minterm will equal 1. Since each minterm is unique, no two minterms will equal 1 for the same input combination. For example, for input $A = 1$, $B = 0$, $C = 0$ (row 4), the second minterm in Equation 6.5a equals 1, since

$$A\bar{B}\bar{C} = 1 \cdot 1 \cdot 1 = 1$$

It should be obvious that this is the only input combination for which $A\bar{B}\bar{C}$ equals 1. Thus, there is a one-to-one correspondence between $A\bar{B}\bar{C}$ and the 1 on row 4 of the truth table. We indicate this by saying that this minterm "produces" the 1 on row 4, and we designate it as minterm 4, denoted m_4. Similarly, the other minterms in Equation 6.5a are m_0, m_5, and m_7, and Equation 6.5 may then be interpreted as a list of the minterms in the standard sum-of-products form.

Consider next the standard product of sums, Equation 6.6a. From the dual version of Lemma 4.3 $(a \cdot 0 = 0)$, we see that this function will be 0 whenever one (or more) of the products is 0. By the dual of the above argument, a given maxterm can be 0 for only one input combination, and vice versa. For example, for input $A = 0$, $B = 1$, $C = 0$ (row 2),

$$A + \bar{B} + C = 0 + \bar{1} + 0 = 0 + 0 + 0 = 0$$

Thus, we may say that this maxterm "produces" the 0 in row 2 and designate it as maxterm 2, denoted M_2. Similarly, the other maxterms in Equation 6.5 are M_1, M_3, and M_6, and Equation 6.6 may be regarded as a list of maxterms in the standard product-of-sums form.

In summary, we associate each minterm with the input combination for which it would produce a 1 in the function and each maxterm with the combination for which it would produce a 0. Figure 6.2 shows the minterms and maxterms associated with each row of a 3-variable truth table. The extension to more variables should be obvious. From this table, we can derive a rule for determining the actual product or sum, given the row number or vice versa. For the minterms, each uncomplemented variable is associated with a 1 in the corre-

Row No.	A	B	C	Minterms	Maxterms
0	0	0	0	$\bar{A}\bar{B}\bar{C} = m_0$	$A + B + C = M_0$
1	0	0	1	$\bar{A}\bar{B}C = m_1$	$A + B + \bar{C} = M_1$
2	0	1	0	$\bar{A}B\bar{C} = m_2$	$A + \bar{B} + C = M_2$
3	0	1	1	$\bar{A}BC = m_3$	$A + \bar{B} + \bar{C} = M_3$
4	1	0	0	$A\bar{B}\bar{C} = m_4$	$\bar{A} + B + C = M_4$
5	1	0	1	$A\bar{B}C = m_5$	$\bar{A} + B + \bar{C} = M_5$
6	1	1	0	$AB\bar{C} = m_6$	$\bar{A} + \bar{B} + C = M_6$
7	1	1	1	$ABC = m_7$	$\bar{A} + \bar{B} + \bar{C} = M_7$

FIGURE 6.2

sponding position in the binary row number, and each complemented variable is associated with a 0. For the maxterms, the rule is just the opposite.

Example 6.4

Convert Eq. 6.3 to the minterm list form.

$f(A, B, C, D) =$

$ABCD + A\bar{B}CD + \bar{A}BCD + ABC\bar{D} + A\bar{B}C\bar{D} + AB\bar{C}\bar{D} + \bar{A}BC\bar{D} + \bar{A}B\bar{C}\bar{D}$

1111 1011 0111 1110 1010 1100 0110 0100

15   11   7   14   10   12   6   4

$f(A, B, C, D) = \sum m(4, 6, 7, 10, 11, 12, 14, 15)$

Example 6.5

Convert the following equation to the maxterm list form.

$f(A, B, C, D) =$

$(A + \bar{B} + C + \bar{D})(\bar{A} + B + C + D)(A + B + \bar{C} + \bar{D})(\bar{A} + B + C + \bar{D})$

0 1 0 1 1 0 0 0 0 0 1 1 1 0 0 1

5   8   3   9

$f(A, B, C, D) = \prod M(3, 5, 8, 9)$ ■

Since the truth table provides the complete specification of any switching function and we have now demonstrated a procedure for converting any truth table to a sum of minterms or product of maxterms, Theorems 6.1 and 6.2 have now been proved.

6.4 Karnaugh Map Representation of Boolean Functions

The Karnaugh map [1] is one of the most powerful tools in the repertoire of the logic designer. The power of the Karnaugh map does not lie in its application of

A	B	$A \cdot B$
0	0	0
0	1	0
1	1	1
1	0	0

A	B	$A + B$
0	0	0
0	1	1
1	1	1
1	0	1

FIGURE 6.3   Truth tables for AND and OR.

any marvelous new theorems but instead in its utilization of the remarkable ability of the human mind to perceive patterns in pictorial representations of data. This is not a new idea. Anytime we use a graph instead of a table of numeric data, we are utilizing the human ability to recognize complex patterns and relationships in a graphical representation far more rapidly and surely than in a tabular representation.

A Karnaugh map may be regarded either as a pictorial form of a truth table or as an extension of the Venn diagram. First consider a truth table for two variables. We list all four possible input combinations and the corresponding function values, e.g., the truth tables for AND and OR (Fig. 6.3).

As an alternative approach, let us set up a diagram consisting of four small boxes, one for each combination of variables. Place a 1 in any box representing a combination of variables for which the function has the value 1. There is no logical objection to putting 0s in the other boxes, but they are usually omitted for clarity. Figure 6.4 shows two forms of Karnaugh maps for AB and $A + B$.

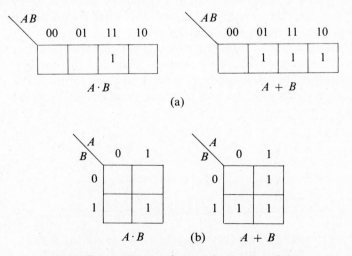

FIGURE 6.4   Karnaugh maps for AND and OR.

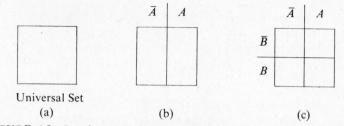

Universal Set
(a)   (b)   (c)

FIGURE 6.5   *Development of Karnaugh maps by Venn diagram approach.*

The diagrams of Fig. 6.4a are perfectly valid Karnaugh maps, but it is more common to arrange the four boxes in a square, as shown in Fig. 6.4b.

As an alternative approach, recall the interpretation of the Venn diagram discussed in Section 4.7. We interpret the universal set as the set of all 2^n combinations of values of n variables, divide this set into 2^n equal areas, and then darken the areas corresponding to those combinations for which the function has the value 1. We start with the universal set represented by a square (Fig. 6.5a) and divide it in half, corresponding to input combinations in which $A = 1$ and combinations in which $\bar{A} = 1$ (Fig. 6.5b). We then divide it in half again, corresponding to $B = 1$ and $\bar{B} = 1$ (Fig. 6.5c).

With this notation, the interpretations of AND as the intersection of sets and OR as the union of sets make it particularly simple to determine which squares should be darkened (Fig. 6.6).

We note that the shaded areas in Fig. 6.6 correspond to the squares containing 1s in Fig. 6.4. Thus, both interpretations lead to the same result. We might say that the Karnaugh map is essentially a diagrammatic form of truth table and that the Venn-diagram concepts of union and intersection of areas aid us in setting up or interpreting a Karnaugh map.

Since there must be one square for each input combination, there must be 2^n squares in a Karnaugh map for n-variables. Whatever the number of variables, we may interpret the map in terms of a graphical form of the truth table (Fig. 6.7a) or in terms of union and intersection of areas (Fig. 6.7b).

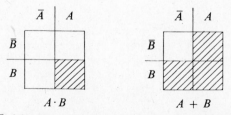

$A \cdot B$ $A + B$

FIGURE 6.6   *Karnaugh maps of AND and OR (Venn form).*

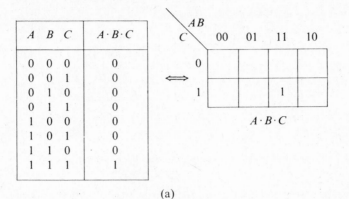

A	B	C	$A \cdot B \cdot C$
0	0	0	0
0	0	1	0
0	1	0	0
0	1	1	0
1	0	0	0
1	0	1	0
1	1	0	0
1	1	1	1

(a)

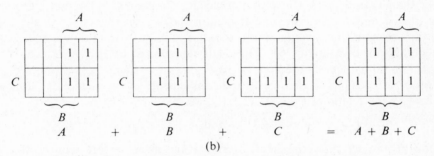

(b)

FIGURE 6.7   Karnaugh maps for 3-variable AND and OR.

In Fig. 6.7b, we have changed the labeling slightly and placed 1s in the squares instead of darkening them. The two types of map (Figs. 6.7a and b) are thus the same, except for labeling, and both types will be called Karnaugh maps* or, for short, K-maps. The K-maps for some other 3-variable functions are shown in Fig. 6.8 to clarify the concepts involved.

Note particularly the functions mapped in Figs. 6.7a and 6.8b. These are both minterms, m_7 and m_4, respectively. Each is represented by one square; obviously, each one of the eight squares corresponds to one of the eight minterms of three variables. This is the origin of the name *minterm*. A minterm is the form of Boolean function corresponding to the minimum possible area, other than 0, on a Karnaugh map. A *maxterm*, on the other hand, is the form of Boolean function corresponding to the maximum possible area, other than 1, on a Karnaugh map. Figures 6.7b and 6.8c are maps of maxterms M_0 and M_3; we see that each map covers the maximum possible area—all the squares but one.

* Both forms appear in Karnaugh's original paper [1].

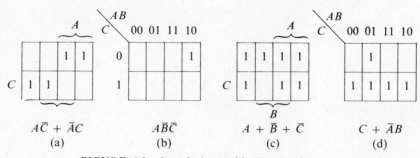

FIGURE 6.8   *Sample 3-variable Karnaugh maps.*

We have just referred to the smallest area other than 0 and the largest area other than 1. In Section 4.7, we remarked that, in the Venn-diagram interpretation of switching algebra, the unit element is the *function* that is identically 1 and the zero element, the *function* that is identically 0. In truth-table terms, we have the situation shown in Fig. 6.9. Thus, 1 is represented by the *entire* area of the K-map and 0 by *none* of the area of the K-map.

Next, we recall that the negative (complement) of an element is uniquely defined by the relationships

$$A + \bar{A} = 1 \quad \text{and} \quad A \cdot \bar{A} = 0$$

From this, we see that consistent interpretation of the algebra requires that the negation of a function be represented by the squares not covered on the map of the function. For example, consider DeMorgan's laws

$$\overline{A \cdot B} = \bar{A} + \bar{B} \quad \text{and} \quad \overline{A + B} = \bar{A} \cdot \bar{B}$$

On K-maps, we have the interpretations of Fig. 6.10.

Since each square on a K-map corresponds to a row in a truth table, it is appropriate to number the squares just as we numbered the rows. These standard K-maps (as we shall call them) are shown in Fig. 6.11 for 2 and 3 variables.

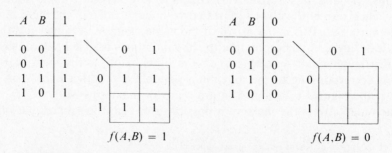

FIGURE 6.9   *Truth tables and K-maps of unit and zero elements.*

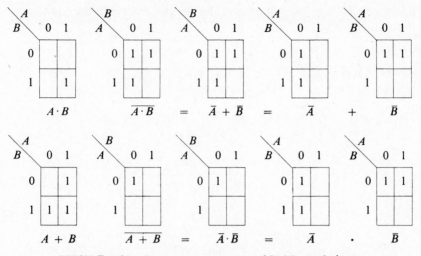

FIGURE 6.10   *K-map interpretation of DeMorgan's laws.*

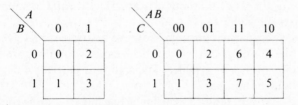

FIGURE 6.11   *Standard K-maps for two and three variables.*

If a function is stated in the form of the minterm list, all we need to do is enter 1 in the corresponding squares to produce the K-map.

Example 6.6

Develop the K-map of $f(A, B, C) = \sum m(0, 2, 3, 7)$.

Solution

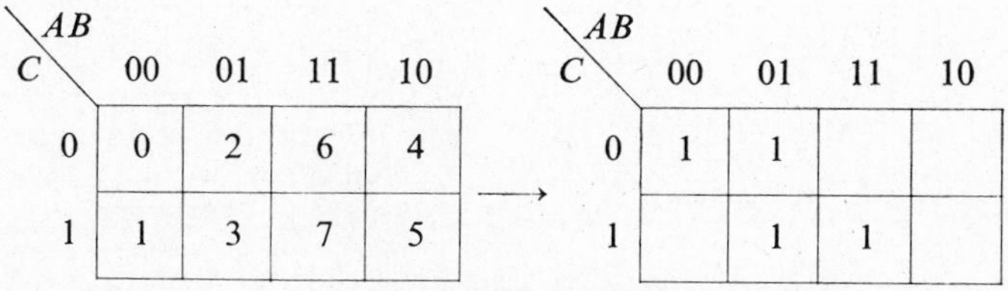

FIGURE 6.12

If a function is stated as a maxterm list, we can enter 0 in the squares listed or 1 in those not listed.

Example 6.7

Develop the K-map of $f(A, B, C) = \prod M(0, 1, 5, 6)$.

Solution

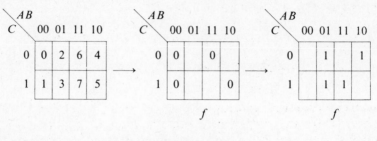

FIGURE 6.13

A map showing the 0s of a function is a perfectly valid K-map, although it is more common to show the 1s.

In developing these basic concepts, we have restricted ourselves to the simple 2- and 3-variable K-maps, but in practical cases we will more often be using maps for functions of more variables. The standard map for 4 variables is shown in Fig. 6.14 in both notations.

We have seen that one requirement for a Karnaugh map is that there must be a square corresponding to each input combination; the maps of Fig. 6.14 satisfy

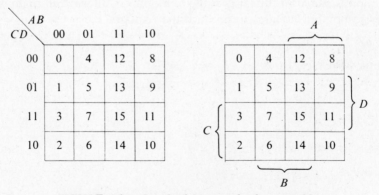

FIGURE 6.14   *Standard K-maps for four variables.*

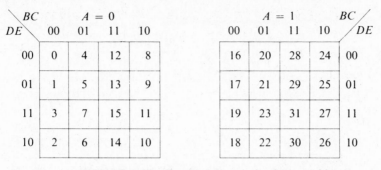

FIGURE 6.15   Standard K-map for five variables.

this requirement. Another requirement is that the squares must be so arranged that any pair of squares immediately adjacent to each other (horizontally or vertically) must correspond to a pair of input conditions that are *logically adjacent*, i.e., differ in only one variable. For example, squares 5 and 13 on the maps in Fig. 6.14 correspond to input combinations $\bar{A}B\bar{C}D$ and $AB\bar{C}D$, identical except in A. Note that squares at the ends of columns or rows are also logically adjacent.

The standard K-map for 5 variables is shown in Fig. 6.15. Here we have two 4-variable maps placed side by side. They are identical in $BCDE$, but one corresponds to $A = 1$, the other to $A = 0$. The standard 4-variable adjacencies apply in each map. In addition, squares in the same relative position on the two maps, e.g., 4 and 20, are also logically adjacent. Any maps that satisfy the requirements of 2^n squares in proper adjacency can be considered K-maps. In Figs. 6.16 and 6.17, we give the most common forms of K-maps, for 2 to 6 variables, in the two alternative notations.

There is no particular preference between the two types of notation. The notation of Fig. 6.16 makes it simple to determine the number of a square, since the binary equivalent is directly available. The alternative notation emphasizes the areas associated with each variable and may be preferable when starting from forms other than minterm or maxterm lists. We suggest that the reader try working with both forms and then use whichever seems most convenient.

Example 6.8

Find K-maps for the following functions:

(a) $f(V, W, X, Y, Z) = \sum m(9, 20, 21, 29, 30, 31)$.

(b) $f(A, B, C, D, E) = AB + \bar{C}D + DE$.

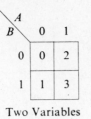

Two Variables

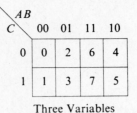

Three Variables

Four Variables

Five Variables

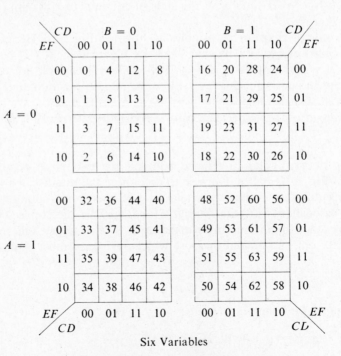

Six Variables

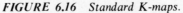

FIGURE 6.16 *Standard K-maps.*

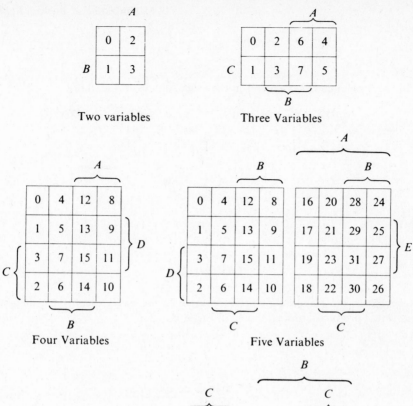

Two variables

Three Variables

Four Variables

Five Variables

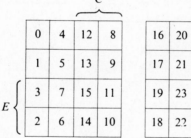

Six Variables

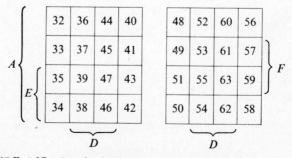

FIGURE 6.17 Standard K-maps—alternate notation.

111

Solution

(a) For the first function, we enter 1 in the listed squares in Fig. 6.18.

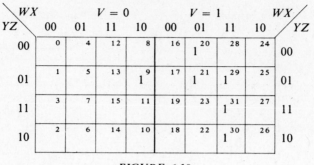

WX	V = 0				V = 1				WX
YZ	00	01	11	10	00	01	11	10	YZ
00	0	4	12	8	16	20 1	28	24	00
01	1	5	13	9 1	17	21 1	29 1	25	01
11	3	7	15	11	19	23	31 1	27	11
10	2	6	14	10	18	22	30 1	26	10

FIGURE 6.18

(b) Referring to the standard map for 5 variables, we easily identify the maps of the individual product terms, as shown in Fig. 6.19.

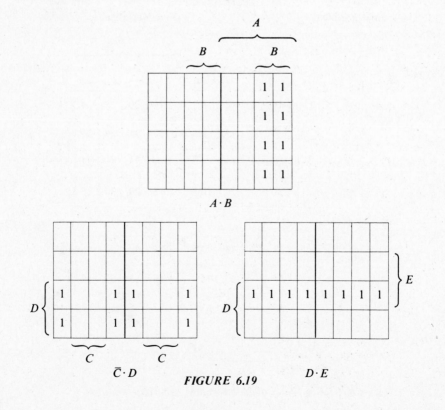

FIGURE 6.19

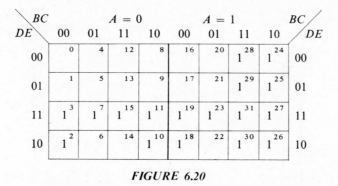

FIGURE 6.20

We then take the union of these to form the final K-map (Fig. 6.20).

The minterm list of this function may now be read directly from the map.

$$f(A, B, C, D, E) = AB + \bar{C}D + DE$$
$$= \sum m(2, 3, 7, 10, 11, 15, 18, 19, 23, 24, 25, 26, 27, 28, 29, 30, 31) \quad \blacksquare$$

In part (a) of Example 6.8, note that the variables do not have to be A, B, C, etc. Obviously, we can call the variables anything we want. The only precaution to be observed is that the variables must appear on the K-map in the proper manner corresponding to the order of their listing in the function statement.

6.5 Simplification of Functions on Karnaugh Maps

It has taken us a while to establish the necessary tools, but we are now ready to use them in minimizing functions. As mentioned earlier, we must have some circuit format in mind before a criterion for the simplest circuit can be defined. Let us consider direct realizations of three different forms of a 4-variable function as shown in Fig. 6.21.

$$f(A, B, C, D) = (A + B)(C\bar{D} + \bar{C}D)$$
$$= AC\bar{D} + A\bar{C}D + BC\bar{D} + B\bar{C}D$$
$$= \sum m(5, 6, 9, 10, 13, 14) \qquad (6.7)$$

Now, which of these forms is the simplest? Obviously, circuit (c), the realization of the standard sum of products, is the most complicated of the three circuits and would be the most expensive to build. Comparing circuits (a) and (b), the reader will probably feel that (a) is the simpler of the two. It is simpler, in the sense of having fewer gates, with fewer inputs; but it has one drawback relative to circuit (b). Note that the signals A and B, entering the left gate, pass through three gates, or three *levels of gating*, before reaching the output. By

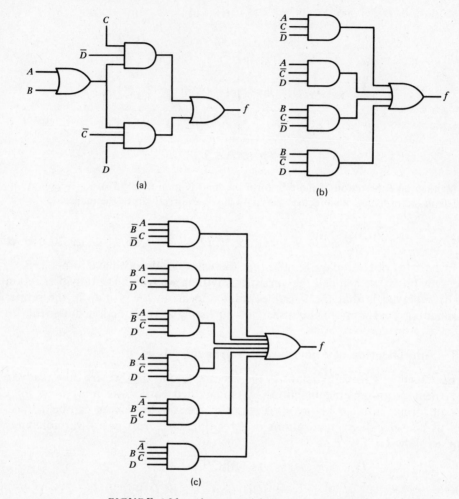

(a)

(b)

(c)

FIGURE 6.21   *Alternate realizations of a function.*

comparison all of the signals into circuit (b) pass through only two levels of gating.

In Chapter 5, it was pointed out that each level of logic adds to the delay in the development of a signal at the circuit output. In high-speed digital systems, it is desirable that this delay be as small as possible.

The choice between a simpler circuit and a faster circuit is generally a matter of engineering judgment. More speed almost invariably costs more money, and one of the most difficult jobs of the design engineer is to decide just how much speed is affordable. For the present, we will assume that speed is the dominant factor and we will design for the fastest possible circuit. Any sum-of-products or product-of-sums expression can be realized by two levels of gating. Whether these be AND-OR, OR-AND, NAND-NAND, or NOR-NOR configurations, they will be referred to as *second-order* circuits.

The other reason for concentrating on second-order circuits is that straightforward and completely general procedures for the minimization of this type of circuit configuration are available. There are no general procedures for finding designs like that of Fig. 6.21a. Circuits of this type, of higher order than second, are often referred to as *factored* circuits. This name is used because the process of getting the expression implemented by this type of circuit from a second-order form is very similar to the process of factoring a polynomial in conventional algebra. This sort of factoring is based largely on trial and error and the ability to remember what the products of certain types of factors look like.

In a sum-of-products (SOP) form, each product corresponds to a gate and each literal to a gate input. The same holds for each sum in a product-of-sums (POS) form. The exact ratio between the cost of a gate and the cost of a gate input will depend on the type of gate, but in practically every case the cost of an additional gate will be several times that of an additional input on an already-existing gate. On this basis, the elimination of gates will be the primary objective of any minimization process, leading to the following definition of a minimal expression.

Definition 6.1   A second-order sum-of-products expression will be regarded as a *minimal* expression if there exists (1) no other equivalent expression involving fewer products, and (2) no other equivalent expression involving the same number of products but a smaller number of literals. The minimal product of sums is the same with the word *products* replaced by the word *sum*, and vice versa.

Note that *a* minimal rather than *the* minimal expression is characterized by Definition 6.1. As we will see, there may very well be several distinct but equivalent expressions satisfying this definition and having the same number of both products and literals.

For the remainder of this section, only the SOP form will be considered, for the reason that the K-map as we have defined it represents each minterm by a square. In the next section, adaption of the K-map to the minimization of POS expressions will be discussed.

Simplification of functions on the K-map is based on the fact that sets of minterms that can be combined into simpler product terms either will be adjacent or will appear in symmetric patterns on the K-map.

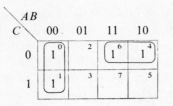

FIGURE 6.22   K-map for Equation 6.8.

Consider this function

$$f(A, B, C) = \sum m(0, 1, 4, 6)$$
$$= \bar{A}\bar{B}\bar{C} + \bar{A}\bar{B}C + A\bar{B}\bar{C} + AB\bar{C} \tag{6.8}$$

By algebraic manipulation, we can simplify this function as follows:

$$f(A, B, C) = \bar{A}\bar{B}(C + \bar{C}) + A\bar{C}(B + \bar{B})$$
$$= \bar{A}\bar{B} + A\bar{C}$$

On the map, we have the pattern shown in Fig. 6.22. Note that the minterms that have combined into a simpler term are adjacent on the K-map. This is a general principle. *Any pair of n-variable minterms that are adjacent on a K-map may be combined into a single product term of n − 1 literals.* As we noted earlier, K-maps are so arranged that minterms in adjacent squares are identical, except in one variable. This variable appears in the true form in one minterm and in the complemented (false) form in the other. Thus, the value of the function will be independent of the value of this variable. For example, in the function given by Equation 6.8, if $A = 0$ and $B = 0$, then $f = 1$, regardless of the value of C. Similarly, if $A = 1$ and $C = 0$, then $f = 1$, regardless of the value of B.

Now consider

$$f(A, B, C, D) = \sum m(0, 8, 12, 14, 5, 7)$$
$$= \bar{A}\bar{B}\bar{C}\bar{D} + A\bar{B}\bar{C}\bar{D} + AB\bar{C}\bar{D} + ABC\bar{D} + \bar{A}\bar{B}C D + \bar{A}BCD$$
$$= \bar{B}\bar{C}\bar{D}(A + \bar{A}) + AB\bar{D}(C + \bar{C}) + \bar{A}BD(C + \bar{C})$$
$$= \bar{B}\bar{C}\bar{D} + AB\bar{D} + \bar{A}BD \tag{6.9}$$

which is mapped in Fig. 6.23. Note that here the pairing includes 0 and 8, and 12 and 14, which do not appear to be adjacent. The variables are arranged in "ring" pattern of symmetry, so that these squares would be adjacent if the map were inscribed on a torus (a doughnut-shaped form). If you have difficulty visualizing the map on a torus, just remember that squares in the same row or column, but on opposite edges of the map, may be paired.

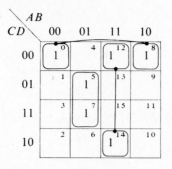

FIGURE 6.23   *Map of Equation 6.9.*

Now consider the function

$$f(A, B, C, D) = \sum m(5, 7, 10, 13, 15)$$

$$= BD + A\bar{B}C\bar{D} \qquad (6.10)$$

which is displayed on the map in Fig. 6.24. Here we see that a set of four adjacent minterms combines into a single term, with the elimination of two literals. In other words, if $B = 1$ and $D = 1$, then $f = 1$, regardless of the values A and C. Also note that m_{10} does *not* combine with any other minterm of the function. The adjacencies must be row or column adjacencies. For example, the diagonally adjacent m_{10} and m_{15} cannot be combined.

Other possible sets of four are shown in Fig. 6.25. Note the adjacency on the edges and corners.

Since sets of two minterms combine to eliminate one variable, and sets of four combine to eliminate two, it is to be expected that sets of eight will combine to eliminate three variables. Figure 6.26 illustrates some of these sets.

The same general principles apply as we go to the 5- and 6-variable maps, but we also have logical adjacency between squares, or sets of squares, in the same position on different sections of the map.

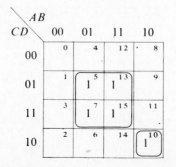

FIGURE 6.24   *Map of Equation 6.10.*

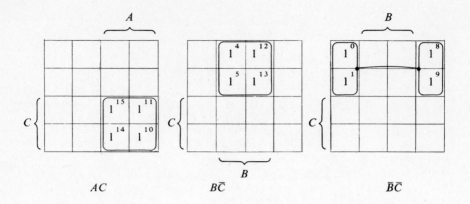

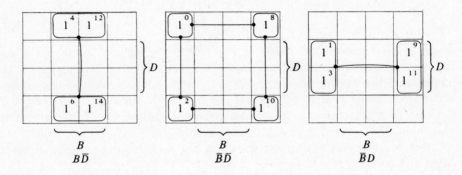

FIGURE 6.25   *Sets of four on the K-map.*

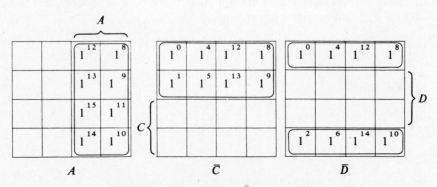

FIGURE 6.26   *Sets of eight on the K-map.*

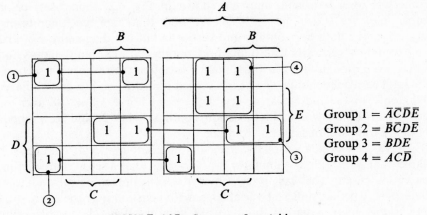

$$\begin{aligned} \text{Group 1} &= \bar{A}\bar{C}D\bar{E} \\ \text{Group 2} &= \bar{B}\bar{C}D\bar{E} \\ \text{Group 3} &= BDE \\ \text{Group 4} &= AC\bar{D} \end{aligned}$$

FIGURE 6.27 *Sets on a 5-variable map.*

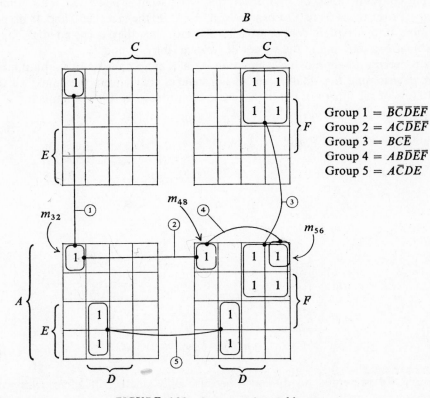

$$\begin{aligned} \text{Group 1} &= \bar{B}\bar{C}\bar{D}E\bar{F} \\ \text{Group 2} &= A\bar{C}\bar{D}E\bar{F} \\ \text{Group 3} &= BC\bar{E} \\ \text{Group 4} &= AB\bar{D}E\bar{F} \\ \text{Group 5} &= A\bar{C}DE \end{aligned}$$

FIGURE 6.28 *Sets on a 6-variable map.*

Some sets of a 5-variable map are shown in Fig. 6.27. Some sets on a 6-variable map are shown in Fig. 6.28. We note that here the "map-to-map" adjacency works both horizontally and vertically but not diagonally. For example, 32 combines vertically with 0 and horizontally with 48; but 0 and 48 do *not* combine, even though they are in the same position in their respective sections of the map. The reader should also keep in mind that the "end-to-end" adjacency applies only to the individual 4-variable section. For example, 48 and 56, at opposite ends of a row in a single section, combine, while 32 and 56, at opposite ends of separate sections, do not combine.

We have seen that a set of two *logically* adjacent minterms eliminates one variable, that a set of four eliminates two variables, that a set of eight eliminates three variables, etc. The way to test whether a set is in fact logically adjacent is to determine whether sufficient variables remain constant over the entire set. On an n-variable map, a pair of minterms is adjacent if $n - 1$ variables remain constant over the pair; a set of four minterms is adjacent if $n - 2$ variables remain constant; a set of eight minterms is adjacent if $n - 3$ variables remain constant over the set; etc. For example, in Fig. 6.28, the fact that 0 and 48 do not combine into a pair is consistent with the fact that they correspond to input combinations differing in the values of two variables, A and B.

The process of simplifying a function on a K-map consists of nothing more than determining the smallest set of adjacencies that covers (contains) all the minterms of the function. Let us illustrate the process by a few examples.

Example 6.9

Simplify $f(A, B, C, D) = \sum m(0, 1, 2, 3, 13, 15)$.

Solution

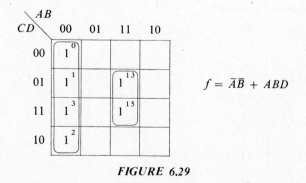

$$f = \bar{A}\bar{B} + ABD$$

FIGURE 6.29   ∎

Example 6.9 presented no difficulty because only one set of adjacencies was possible.

Example 6.10

Simplify $f(A, B, C, D) = \sum m(0, 2, 10, 11, 12, 14)$.

Solution

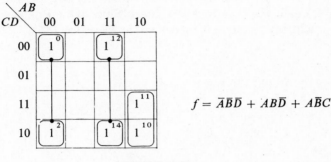

$$f = \overline{A}\overline{B}\overline{D} + AB\overline{D} + A\overline{B}C$$

FIGURE 6.30

Here there are choices. We could also combine m_2 and m_{10}, or m_{10} and m_{14}. However, there is no reason to use these combinations since m_2, m_{10}, and m_{14} are already covered by (contained in) the necessary pairings with m_0, m_{11}, m_{12}, respectively, for which there is no choice. *Thus, a rule in using Karnaugh maps is to start by combining those terms for which there is only one possibility.*

Example 6.11

Simplify $f(A, B, C, D) = \sum m(0, 2, 8, 12, 13)$.

Solution

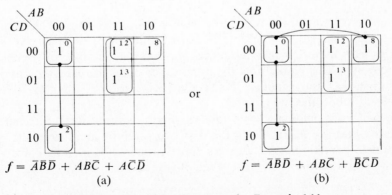

$f = \overline{A}\overline{B}\overline{D} + AB\overline{C} + A\overline{C}\overline{D}$ or $f = \overline{A}\overline{B}\overline{D} + AB\overline{C} + \overline{B}\overline{C}\overline{D}$

(a)   (b)

FIGURE 6.31 *Alternative maps for Example 6.11*

In this case, there are two equally valid choices.

Note that, in both realizations of Example 6.11, there is some redundancy. In the first, m_{12} is covered by the terms $AB\bar{C}$ and $A\bar{C}D$; in the second, m_0 is covered by $\bar{A}\bar{B}\bar{D}$ and $\bar{B}\bar{C}\bar{D}$. This procedure of covering a minterm more than once causes no trouble. In terms of AND-OR realization, it simply means that, when the variable values are such that the particular minterm is 1, the output of more than one AND gate will take on the value 1. Since the OR is inclusive, it does not matter how many of the AND gates are at the value 1.

Example 6.12

Simplify $f(A, B, C, D) = \sum m(1, 5, 6, 7, 11, 12, 13, 15)$

Solution

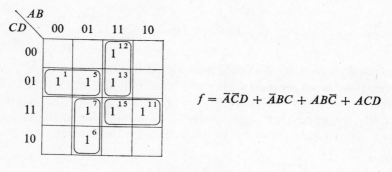

$$f = \bar{A}\bar{C}D + \bar{A}BC + AB\bar{C} + ACD$$

FIGURE 6.32

■

Example 6.12 illustrates a possible hazard in using K-maps. The temptation is great to use the set of four in the center; but when we make the *necessary* pairings with the other four minterms, we find that the four in the center have been covered. This emphasizes the importance of determining the essential products first, i.e., the products containing at least one minterm that can be combined in no other way.

We will now present a few more examples, without further comment. It is suggested that the reader study the first two carefully and then try to work the others before looking at the answers. Once mastered, Karnaugh maps will seem almost second nature, but mastery requires practice.

Example 6.13

$$f(A, B, C, D, E) = \sum m(0, 1, 4, 5, 6, 11, 12, 14, 16, 20, 22, 28, 30, 31)$$

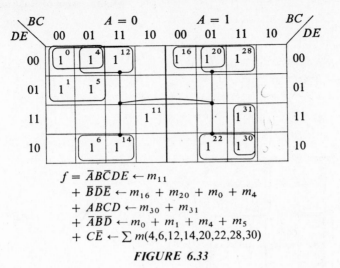

$$f = \bar{A}B\bar{C}DE \leftarrow m_{11}$$
$$+ \bar{B}\bar{D}E \leftarrow m_{16} + m_{20} + m_0 + m_4$$
$$+ ABCD \leftarrow m_{30} + m_{31}$$
$$+ \bar{A}\bar{B}\bar{D} \leftarrow m_0 + m_1 + m_4 + m_5$$
$$+ C\bar{E} \leftarrow \sum m(4,6,12,14,20,22,28,30)$$

FIGURE 6.33

■

Example 6.14

$$f(A, B, C, D, E, F) = \sum m(2, 3, 6, 7, 10, 14, 18, 19, 22, 23, 27, 37, 42, 43, 45, 46)$$

Solution

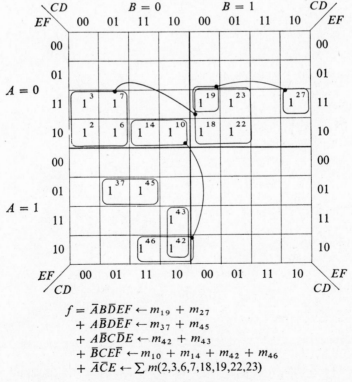

$$f = \bar{A}B\bar{D}EF \leftarrow m_{19} + m_{27}$$
$$+ A\bar{B}DEF \leftarrow m_{37} + m_{45}$$
$$+ A\bar{B}C\bar{D}E \leftarrow m_{42} + m_{43}$$
$$+ \bar{B}CE\bar{F} \leftarrow m_{10} + m_{14} + m_{42} + m_{46}$$
$$+ \bar{A}\bar{C}E \leftarrow \sum m(2,3,6,7,18,19,22,23)$$

FIGURE 6.34

■

Example 6.15

$f(A, B, C, D) = \prod M(7, 9, 13)$

Solution

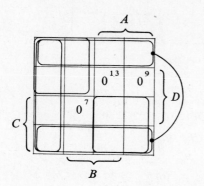

$$f = \bar{D} + \bar{A}\bar{C} + AC + \bar{A}\bar{B}$$

FIGURE 6.35

■

Example 6.16

$f(A, B, C, D) = \sum m(0, 1, 2, 4, 5, 8, 10)$

Solution

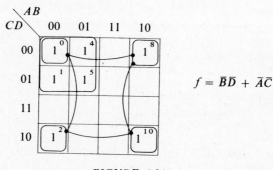

$$f = \bar{B}\bar{D} + \bar{A}\bar{C}$$

FIGURE 6.36

■

Example 6.17

$f(A, B, C, D, E) = \sum m(0, 1, 3, 4, 5, 7, 8, 9, 10, 12, 13, 21, 24, 25, 26, 28, 29)$

Solution

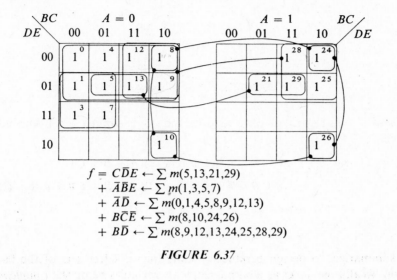

$$f = C\bar{D}E \leftarrow \sum m(5,13,21,29)$$
$$+ \ \bar{A}\bar{B}E \leftarrow \sum m(1,3,5,7)$$
$$+ \ \bar{A}\bar{D} \leftarrow \sum m(0,1,4,5,8,9,12,13)$$
$$+ \ B\bar{C}\bar{E} \leftarrow \sum m(8,10,24,26)$$
$$+ \ B\bar{D} \leftarrow \sum m(8,9,12,13,24,25,28,29)$$

FIGURE 6.37

6.6 Map Minimizations of Product-of-Sums Expressions

With only minor changes, the procedure described in Section 6.5 can be adapted
to product-of-sums design. We have seen that each 1 of a function is produced
by a single minterm and that the process of simplification consists of combining
minterms into products that have fewer literals and produce more than a single
1. We have also seen that each 0 of a function is produced by a single maxterm.
It would seem reasonable, then, to expect that maxterms might combine in a
similar fashion.

Consider the function

$$f(A, B, C) = \prod M(3, 6, 7)$$
$$= (A + \bar{B} + \bar{C})(\bar{A} + \bar{B} + C)(\bar{A} + \bar{B} + \bar{C}) \qquad (6.11)$$

which is shown in Fig. 6.38. Maxterm 6 produces a 0 when $A = 1$, $B = 1$, $C = 0$,
and maxterm 7 produces a 0 when $A = 1$, $B = 1$, $C = 1$. Taking them together,
they produce 0 whenever $A = 1$ and $B = 1$, regardless of C. Furthermore, the
single sum $(\bar{A} + \bar{B})$ will produce both 0s, since it will be 0 whenever $A = 1$ and
$B = 1$, regardless of C. We also note that M_3 and M_7 together produce 0s for

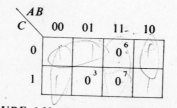

FIGURE 6.38   K-map of Equation 6.11.

$B = 1$ and $C = 1$, regardless of A, as does the single sum $(\bar{B} + \bar{C})$. Algebraically, we have

$$f(A, B, C) = (A + \bar{B} + \bar{C})(\bar{A} + \bar{B} + C)(\bar{A} + \bar{B} + \bar{C})$$
$$= (A + \bar{B} + \bar{C})(\bar{A} + \bar{B} + \bar{C})(\bar{A} + \bar{B} + C)(\bar{A} + \bar{B} + \bar{C})$$
$$= (A\bar{A} + \bar{B} + \bar{C})(\bar{A} + \bar{B} + C\bar{C})$$
$$= (\bar{B} + \bar{C})(\bar{A} + \bar{B})$$

To summarize, to design product-of-sums forms, select sets of the 0s of the function. Realize each set as a sum term, with variables being the *complements* of those that would be used if this same set were being realized as a product to produce 1s. Let us consider some examples.

Example 6.18

Obtain a minimal POS realization of $f(A, B, C, D) = \sum m(0, 2, 10, 11, 12, 14) = \prod M(1, 3, 4, 5, 6, 7, 8, 9, 13, 15)$

Solution

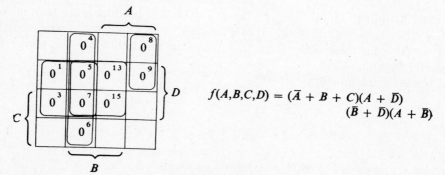

$$f(A,B,C,D) = (\bar{A} + B + C)(A + \bar{D})$$
$$(\bar{B} + \bar{D})(A + \bar{B})$$

FIGURE 6.39   ∎

Example 6.19

$f(A, B, C, D) = \sum m(0, 2, 8, 12, 13) = \prod M(1, 3, 4, 5, 6, 7, 9, 10, 11, 14, 15)$

Solution

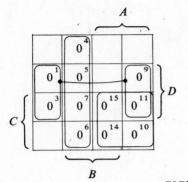

$$f(A,B,C,D) = (\bar{A} + \bar{C})(A + \bar{B})(B + \bar{D})$$

FIGURE 6.40

∎

If we compare Examples 6.18 and 6.10, we see that the sum-of-products form is preferable to the product-of-sums form for this function, since there are three products in the SOP form and four sums in the POS form. Comparing Examples 6.19 and 6.11, we see that the POS form is preferable. The number of products and sums is the same, but there are fewer literals in the sums. It would be useful to have some method of determining in advance which form (SOP or POS) will be best for a particular function. Unfortunately, no such method exists. Assuming there is no hardware preference, the designer should try both forms.

An important variation of the K-map method of POS design is design for the wired-AND connection of NAND gates or for AND-OR-INVERT (AOI) gates. The reader will recall from Chapter 5 that in both of these forms the function realized is the complement of the function that would be realized by connecting the same gates in a NAND-NAND configuration. From our interpretation of negation on the K-map, we recognize that the 0s of a function are the 1s of the complement of that function. Figure 6.41 shows the map of the 0s of the function of Example 6.19 and the map of the complement function. From the complement map we easily read off the function,

$$\overline{f(A, B, C, D)} = AC + \bar{A}B + \bar{B}D$$

from which we have an alternate form of the original function,

$$f(A, B, C, D) = \overline{AC + \bar{A}B + \bar{B}D}$$

This form of the function may be realized in wired-AND form as shown in Fig. 6.42a, or in AOI form as shown in Fig. 6.42b. Note that the groupings are the same on the two maps of Fig. 6.41, so that it does not matter which map you

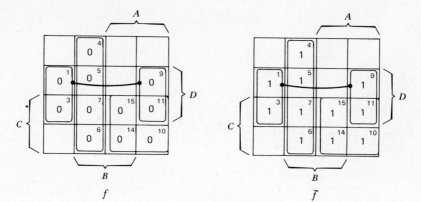

$$f \qquad\qquad \bar{f}$$

FIGURE 6.41

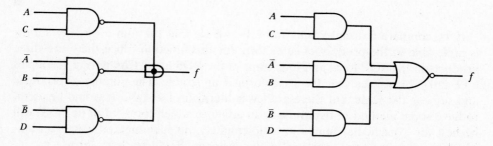

FIGURE 6.42 *Wired-AND and AOI realizations of Example 6.19.*

use—the 0s of the original function or the 1s of the complement function. In either case, the groupings are realized as products and their sum is complemented by use of the wired-AND or AOI configuration.

6.7 Incompletely Specified Functions

Recall that the basic specification of a switching function is the truth table, i.e., a listing of the values of the function for the 2^n possible combinations of n variables. Our design process basically consists of translating a (generally) verbal description of a logical job to be done into a truth table and then finding a specific function that realizes this truth table and satisfies some criterion of minimum cost. So far, we have assumed that the truth values were strictly specified for all of the 2^n possible input combinations. This is not always the case.

Sometimes the circuit we are designing is a part of a larger system in which certain inputs occur only under circumstances such that the output of the circuit will not influence the overall system. Whenever the output has no effect, we obviously *don't care* whether the output is a 0 or a 1. Another possibility is that certain input combinations never occur due to various external constraints. Note that this does not mean that the circuit would not develop some output if this forbidden input occurred. Any switching circuit will respond in some way to any input. However, since the input will never occur, we don't care whether the final circuit responds with a 0 or a 1 output to this forbidden-input combination.

Where such situations occur, we say that the output is *unspecified*. This is indicated on the truth table by entering an X as the functional value instead of 0 or 1.* Such conditions are commonly referred to as *don't-cares*, and functions including don't-cares are said to be *incompletely specified*. A realization of an incompletely specified function is any circuit that produces the same outputs for *all input combinations for which output is specified.*

Example 6.20

A digital computer has five output modes as follows:

Mode No.	Description	Code
1	Punched card	001
2	Mag tape	010
3	Typewriter	011
4	Punched tape, numeric	100
5	Punched tape, alphabetic	101

The selection of an output device is controlled by signals on three parallel lines, denoted x, y, z, according to the code indicated above. Since both modes 4 and 5 utilize the tape punch, a decoding circuit is required in the tape punch that will have 1 as its output if either 100 or 101 is received. The truth table for this decoding circuit is shown in Fig. 6.43a.

Since the codes 000, 110, and 111 are not used, these are don't-care conditions. A realization of this truth table is *any* circuit that produces the 1s and 0s specified in the truth table, regardless of its output for the three don't-care conditions.

The first step in minimization is to convert the truth table to a K-map. For a sum-of-products design, the 1s and don't-cares of the function map are as shown in Fig. 6.44a. Since the output is optional for the don't-cares, we assign them to be 0s or 1s in whatever manner will result in the simplest realization. On the K-map, this means that we group

---

* The symbol for the unspecified output condition is unfortunately not standard. Other symbols used include ϕ, d, and $-$.

xyz	$f(x, y, z)$
000	X
001	0
010	0
011	0
100	1
101	1
110	X
111	X

(a)

xyz	$f = x$
000	0
001	0
010	0
011	0
100	1
101	1
110	1
111	1

(b)

FIGURE 6.43

the Xs with the 1s whenever this results in a larger set (product with fewer literals), and we ignore the Xs if there is no advantage to be gained from using them. In this case, we include the Xs in 110 and 111, i.e., we assign them as 1s, and we ignore the X in 000 (Fig. 6.44b). Without the don't-cares, the function would be realized by $f(x, y, z) = x\bar{y}$, but with them it is $f(x, y, z) = x$. When the truth table of $f(x, y, z) = x$ (Fig. 6.43b) is compared with that of the original incompletely specified function (Fig. 6.43a), we see that it is a correct realization. The output is correct for all input conditions for which the function is specified. ∎

It would be convenient to have some compact algebraic way of stating an incompletely specified function. Since each row in the truth table corresponds to an input combination, we simply add a list of the rows for which the output is

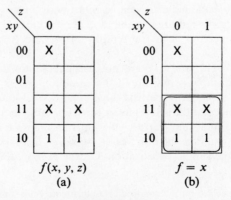

$f(x, y, z)$
(a)

$f = x$
(b)

FIGURE 6.44

unspecified. Thus, the minterm list form for the function of Example 6.20 would be

$$f(x, y, z) = \sum m(4, 5) + d(0, 6, 7)$$

Example 6.21

Obtain a minimal sum-of-products representation for

$$f(w, x, y, z) = \sum m(0, 7, 8, 10, 12) + d(2, 6, 11)$$

Solution

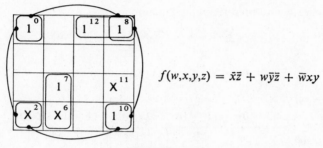

$$f(w,x,y,z) = \bar{x}\bar{z} + w\bar{y}\bar{z} + \bar{w}xy$$

FIGURE 6.45

We use the don't-cares in squares 2 and 6 to obtain larger sets than would otherwise be possible, but we ignore the X in 11, since the only possible combination is with 10, which is already covered.

Example 6.22

Obtain a minimal product-of-sums realization for the function of Example 6.21.

Solution

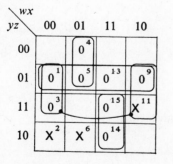

$$f(w,x,y,z) = (y + \bar{z})(x + \bar{z})(w + \bar{x} + y)$$
$$(\bar{w} + \bar{x} + \bar{y})$$

FIGURE 6.46

Since we want the POS form in Example 6.22, we design in the 0s, again using the don't-cares if they improve the combination, ignoring them otherwise. Again we start with necessary sets. For example, 4 combines only with 5, and 14 only with 15. We could combine 4 with the don't-care in 6, and also 14 with 6, but this would violate our rule of using don't-cares only if they make *larger* sets possible. The don't-care at 11 is used because it does place 3 in a larger set $(x + \bar{z})$ than could otherwise be obtained.

Example 6.23

Obtain a minimal sum-of-products representation for

$$f(A, B, C, D, E) = \sum m(1, 4, 6, 10, 20, 22, 24, 26) + d(0, 11, 16, 27).$$

Solution

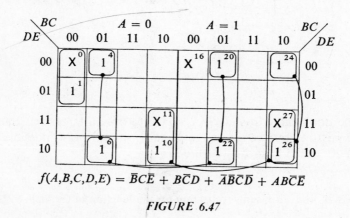

$$f(A,B,C,D,E) = \bar{B}C\bar{E} + B\bar{C}D + \bar{A}\bar{B}\bar{C}\bar{D} + AB\bar{C}\bar{E}$$

FIGURE 6.47

■

Here the don't-care in 16 could be used in several ways, but it would not increase the size of any set.

Problems

6.1. Convert the following Boolean forms to minterm lists.

(a) $f(w, x, y, z) = wy + x(w + y\bar{z})$

(b) $f(U, V, W, X, Y) = \bar{V}(\bar{W} + \bar{U})(X + \bar{Y}) + \bar{U}\bar{W}\bar{Y}$

(c) $f(V, W, X, Y, Z) = (X + \bar{Z})(\overline{Z + W\bar{Y}}) + (VZ + W\bar{X})(\bar{Y} + Z)$

6.2. Convert the forms of Problem 6.1 to maxterm lists.

6.3. By using the Karnaugh map, determine minimal sum-of-products realizations of the following functions.

(a) $f(A, B, C, D) = \sum m(0, 4, 6, 10, 11, 13)$

(b) $f(w, x, y, z) = \sum m(3, 4, 5, 7, 11, 12, 14, 15)$

(c) $f(a, b, c, d) = \prod M(3, 5, 7, 11, 13, 15)$

(d) $f(V, W, X, Y, Z) = \sum m(0, 2, 3, 4, 5, 11, 18, 19, 20, 23, 24, 28, 29, 31)$

6.4. A logic circuit is to be designed having 4 inputs y_1, y_0, x_1, and x_0. The pairs of bits $y_1 y_0$ and $x_1 x_0$ represent 2-bit binary numbers with y_1 and x_1 as the most significant bits. The only circuit output, z, is to be 1 if and only if the binary number $x_1 x_0$ is greater than or equal to the binary number $y_1 y_0$. Determine a minimal sum of products expression for z.

6.5. Determine a minimal sum-of-products expression equivalent to each of the following Boolean expressions.

(a) $f(A, B, C, D, E) = (\bar{C}\bar{E} + CE)(\bar{A} + B)D + (\overline{\bar{A} + B})D\bar{C}E$

(b) $f(w, x, y, z) = \overline{(\bar{w} + x)} + \overline{(\bar{x} + z)} + \overline{(\bar{y} + \bar{z})}$

6.6. By using Karnaugh maps, determine minimal product-of-sums realizations for the functions of Problems 6.3b and 6.3c.

6.7. (a) Determine the minimal product-of-sums realization for the function of Example 6.15, and compare the cost to that of the minimal sum-of-products realization.

(b) Repeat (a) for Example 6.16.

6.8. A prime number is a number that is only divisible by itself and 1. Suppose the numbers between 0 and 31 are represented in binary in the form of the five bits

$$x_4 x_3 x_2 x_1 x_0$$

where x_4 is the most significant bit. Design a prime detector. That is, design a combinational logic circuit whose output, Z, will be 1 if and only if the 5 input bits represent a prime number. Do not count 0 as a prime. Base your design on obtaining a minimal two-level expression for Z.

6.9. A 2-output, 4-input logic circuit is to be designed that will carry out addition modulo-4. The addition table for modulo-4 addition is given in Fig. P6.9. For example, $(3 + 3) \bmod\text{-}4 = 2$. Therefore, a 2 is entered in row 3 column 3 of the table, and so on. The input numbers are to be coded in straight binary, with one input number given by $x_2 x_1$ and the other by $y_2 y_1$. The output is also to be coded as the binary number $z_2 z_1$. That is, $z_2 z_1 = 00$ if the sum is 0, 01 if the sum is 1, 10 if the sum is 2, and 11 if the sum is 3.

(a) Determine a two-level Boolean expression for z_1.

(b) Determine a two-level Boolean expression for z_2.

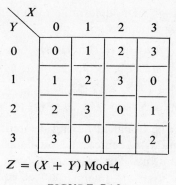

$$Z = (X + Y)\,\text{Mod-4}$$

FIGURE P6.9

6.10. Determine minimal sum-of-products realizations for the following incompletely specified functions.

(a) $f(A, B, C, D) = \sum m(1, 3, 5, 8, 9, 11, 15) + d(2, 13)$
(b) $f(W, X, Y, Z) = \sum m(4, 5, 7, 12, 14, 15) + d(3, 8, 10)$
(c) $f(A, B, C, D, E) = \sum m(1, 2, 3, 4, 5, 11, 18, 19, 20, 21,$
$\qquad\qquad\qquad\qquad 23, 28, 31) + d(0, 12, 15, 27, 30)$
(d) $f(a, b, c, d, e) = \sum m(7, 8, 9, 12, 13, 14, 19, 23, 24, 27, 29, 30)$
$\qquad\qquad\qquad\qquad + d(1, 10, 17, 26, 28, 31)$
(e) $f(u, v, w, x, y, z) = \sum m(0, 2, 14, 18, 21, 27, 32, 41, 49, 53, 62)$
$\qquad\qquad\qquad\qquad + d(6, 9, 25, 34, 55, 57, 61)$

6.11. Determine minimal realizations of the functions of Problem 6.10 in terms of wired-AND connections of NAND gates, and AOI gates.

6.12. The four lines into the combinational logic circuit depicted in Fig. P6.12 carry one binary-coded decimal digit. That is, the binary equivalents of the decimal digits 0–9 may appear on the lines $x_0 x_1 x_2 x_3$. The most significant bit is x_0. The combination of values corresponding to binary equivalents of the decimal numbers 10–15 will never appear on the lines. The single output Z of the circuit is to be 1 if and only if the inputs represent a number that is either 0 or a power of 2. Construct the logic block diagram of a minimal two-level realization of the circuit.

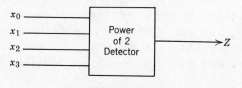

FIGURE P6.12

6.13. A shaft-position encoder provides a 4-bit signal indicating the position of a shaft in steps of 30 degrees, using a reflected (Gray) code as listed in Fig. P6.13. It may be assumed that the four possible combinations of four

Shaft Position	Encoder Output $E_3E_2E_1E_0$			
0–30°	0	0	1	1
30–60°	0	0	1	0
60–90°	0	1	1	0
90–120°	0	1	1	1
120–150°	0	1	0	1
150–180°	0	1	0	0
180–210°	1	1	0	0
210–240°	1	1	0	1
240–270°	1	1	1	1
270–300°	1	1	1	0
300–330°	1	0	1	0
330–360°	1	0	1	1

FIGURE P6.13

bits not used above will never occur. Design a minimal SOP realization of a circuit to produce an output whenever the shaft is in the first quadrant (0–90 degrees).

6.14. A circuit receives two 3-bit binary numbers, $A = A_2 A_1 A_0$, and $B = B_2 B_1 B_0$. Design a minimal SOP circuit to produce an output whenever A is greater than B.

6.15 In digital computers, letters of the alphabet are coded in the form of unique combinations of five or more bits. One of the most common codes is the 6-bit ASCII code. This code is given in Fig. P6.15 in terms of the octal equivalents. For example,

$$C = 03|_8 = \underbrace{000}_{0} \ \underbrace{011}_{3} \quad \text{(See Problem 2.17)}$$

Design a minimal SOP circuit that will receive this code as an input and produce an output whenever the letter is a vowel. The alphabet uses only 26 of the possible codes and the others are used for numerals and punctuation marks. However, you may assume that the data are pure alphabetic, so that only the codes listed will occur.

A	01	N	16
B	02	O	17
C	03	P	20
D	04	Q	21
E	05	R	22
F	06	S	23
G	07	T	24
H	10	U	25
I	11	V	26
J	12	W	27
K	13	X	30
L	14	Y	31
M	15	Z	32

FIGURE P6.15

6.16. Shown in Fig. P6.16 are the ASCII codes for the ten decimal numerals. Assume that a device receives alphanumeric data in this code, i.e., either these codes for numerals or the codes for letters as listed in Problem 6.15.

0	40
1	41
2	42
3	43
4	44
5	45
6	46
7	47
8	48
9	49

FIGURE P6.16

Design a minimal circuit that will develop a 1 out when the data are numeric, a 0 out when they are alphabetic. You may assume that any possible 6-bit codes not used for either letters or numerals will not occur.

6.17. Referring again to the ASCII code listed in Problems 6.15 and 6.16, design an "error circuit," i.e., a circuit that will put out a signal if a code other than one of the 36 "legal" alphanumeric codes is received.

6.18. In a certain computer, three separate sections of the computer proceed independently through four phases of operation. For purposes of control, it is necessary to know when two of three sections are in the same phase at the same time. Each section puts on a 2-bit signal (00, 01, 10, 11) in parallel on two lines. Design a circuit to put out a signal whenever it receives the same phase signal from any two or all three sections.

6.19. Prove again the result in Problem 4.14, assuming that the variables may take on values other than 0 and 1. *Hint:* Use the sum of products form:

$$f(x_1, x_2, \ldots, x_n) = x_1 f_1(x_2, x_3, \ldots, x_n) + \bar{x}_1 f_2(x_2, x_3, \ldots, x_n)$$
$$+ f_3(x_2, x_3, \ldots, x_n)$$

Bibliography

1. Karnaugh, M. "The Map Method for Synthesis of Combinational Logic Circuits," *Trans. AIEE*, **72**, Pt. I, 593–598 (1953).
2. Veitch, E. W. "A Chart Method for Simplifying Truth Functions," *Proc. ACM*, Pittsburgh, Pa., 127–133 (May 2, 3, 1952).
3. Tanana, E. J. "The Map Method," in E. J. McCluskey and T. G. Bartee (eds.). *A Survey of Switching Circuit Theory*, McGraw-Hill, New York, 1962.
4. Caldwell, S. H. *Switching Circuits and Logical Design*, Wiley, New York, 1958.
5. McCluskey, E. J. *Introduction to the Theory of Switching Circuits*, McGraw-Hill, New York, 1965.
6. Sloan, M. E. *Computer Hardware and Organization*, SRA Publishers, Chicago, 1976.
7. Dietmeyer, D. L. *Logic Design of Digital Systems*, 2nd Ed., Allyn and Bacon, Boston, 1978.
8. Roth, C. H., Jr. *Fundamentals of Logic Design*, 2nd Ed., West Publishing, St. Paul, 1979.
9. Nagle, H. T., Jr., B. D. Carroll, and J. D. Irwin. *An Introduction to Computer Logic*, Prentice-Hall, Englewood Cliffs, N.J. 1975.
10. Mange, D. *Analyse et Synthèse des Systemes Logiques*, Editions Georgi, St-Saphorin (Switzerland), 1978.

7
TABULAR MINIMIZATION AND MULTIPLE-OUTPUT CIRCUITS

7.1 Introduction

The Karnaugh map is a very powerful design tool, but it does have certain drawbacks. First, it is essentially a trial-and-error method that does not offer any guarantee of producing the best realization. Second, its dependence on the somewhat "intuitive" human ability to recognize patterns makes it unsuitable for any form of mechanization, such as programming for a digital computer. For functions of 6 or more variables, it is difficult for the designer to be sure that he has selected the smallest possible set of products from a Karnaugh map. You will recall that no specific straightforward procedure has been developed by which this proposition might be checked.

To some readers in this day of inexpensive integrated circuits, the need for optimization of combinational logic beyond what can be provided by the Karnaugh map may not seem evident. These readers may skip this chapter without fear that necessary background material will be missed. The subject of this chapter is the Quine–McCluskey [3] algorithm, a formal minimization procedure for two-level Boolean expressions. This algorithm will be introduced as a special case of a very general procedure for optimization of a set of objects subject to some algebraic criteria. Other forms of this procedure can be applied to various problems in digital system design even in an integrated-circuit environment. In Section 7.2 we digress to develop a new representation of Boolean functions that will permit a more compact development of the Quine–McCluskey algorithm.

7.2 Cubical Representation of Boolean Functions

At this point, it is convenient to introduce a new representation for Boolean functions that will provide convenient terminology for further work. We are all familiar with the notion of a geometric representation of a continuous variable as a distance along a straight line (Fig. 7.1a). In a similar fashion, a switching

138

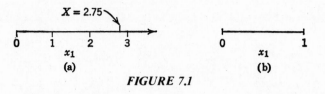

FIGURE 7.1

variable, which can take on only two values, can be represented by two points at the ends of a single line (Fig. 7.1b). We extend this to represent two variables by points in a plane (Fig. 7.2a). Similarly, the four possible values of two switching variables can be represented by the four vertices of a square (Fig. 7.2b). The extension to three variables, as shown in Fig. 7.3, should be obvious. The extension to more than three variables, requiring figures of more than three dimensions, is geometrically difficult but conceptually simple enough.

In general, we say that we represent the various possible combinations of n variables as points in n-space, and the collection of all 2^n possible points will be said to form the *vertices of an n-cube*, or a *Boolean hypercube*.

To represent functions on the n-cube, we set up a one-to-one correspondence between the minterms of n-variables and the vertices of the n-cube. Thus, in the 3-cube, the vertex (000) corresponds to m_0, the vertex (001) to m_1, etc., as shown in Fig. 7.3c. The cubical representation of a function of n-variables then consists of the set of vertices of the n-cube corresponding to the minterms of the function. For example, the function

$$f(A, B, C) = \sum m(0, 2, 3, 7) \tag{7.1}$$

would be represented on the 3-cube as shown in Fig. 7.4a, where the vertices corresponding to m_0, m_2, m_3, and m_7 are indicated by heavy back dots. These vertices, corresponding to the minterms, may also be referred to as the *0-cubes* of the function.*

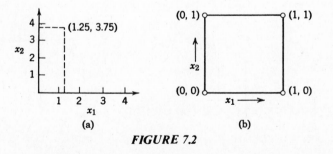

FIGURE 7.2

---

* The 0-cubes, 1-cubes, 2-cubes, etc., are more formally known as *k-dimensional product-subcubes* ($k = 0$, 1, 2, etc.) but this lengthy expression is generally shortened to *k-cube*. See Reference [1], pp. 100–101.

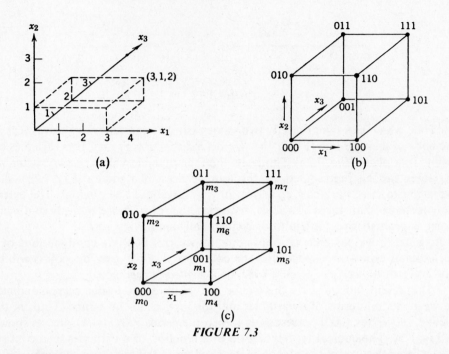

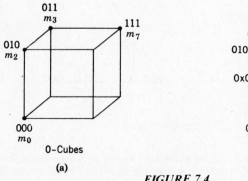

FIGURE 7.3

Two 0-cubes of a function are said to form a 1-cube if they differ in only one coordinate. In the function of Fig. 7.4, we have three 1-cubes, consisting of the pairs of 0-cubes 000 and 010, 010 and 011, and 011 and 111. The 1-cubes may be denoted by placing an x in the coordinate having different values and darkening the line between the pair of 0-cubes (Fig. 7.4b). In a similar fashion, a set of four 0-cubes, whose coordinate values are the same in all but two variables, are said to form a 2-cube of the function. Pictorially, a 2-cube may be represented as a

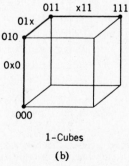

FIGURE 7.4

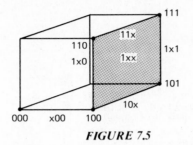

FIGURE 7.5

shaded plane. Figure 7.5 shows the cubic representation of a function exhibiting five 0-cubes, five 1-cubes, and one 2-cube.

When all the vertices (0-cubes) of a k-cube are in the set of vertices making up a larger k-cube, we will say that the smaller cube is *contained in*, or *covered by*, the larger cube. In Fig. 7.5, for example, the 0-cube 100 is contained in the 1-cubes x00, 10x, and 1x0 and in the 2-cube 1xx. Similarly, the 1-cubes 1x0, 10x, 11x, and 1x1 are all contained in the 2-cube 1xx.

The correspondences between the cubical representation and the Karnaugh map should be clear to the reader. The 0-cubes correspond to the squares on the K-map, the 1-cubes to pairs of adjacent squares, etc. The cubical *nomenclature* can be applied directly to the K-map without reference to the cubical *representation*, but the cubical representation makes the origin and significance of these terms much more obvious. The cubic representation will also be very helpful in discussing coding in Chapter 8.

7.3 Determination of Prime Implicants

We will soon see that the goal of the Quine–McCluskey algorithm is a special case of the following.

OBJECT: Select the smallest possible subset, D, *of a set of objects,* A, *so that some criterion,* f, *is satisfied.*

In the context of finding a minimal sum of products realization, set A is all possible Boolean cubes of any dimension that fall within a given Boolean function, f. The set D will consist of all products in the minimal sum of products realization as defined in Definition 6.1. The algorithm for computing D is given in Fig. 7.6. This method is based principally on the work of W. V. Quine [2] and E. J. McCluskey. In this chapter we will develop the method in tabular form. The conversion to a computer program is straightforward.

The first step, represented by block I of Fig. 7.6, is to find a subset, B, consisting of all cubes that are actual candidates for inclusion in the minimal realization. The members of B will be called *prime implicants*. The formal definition of

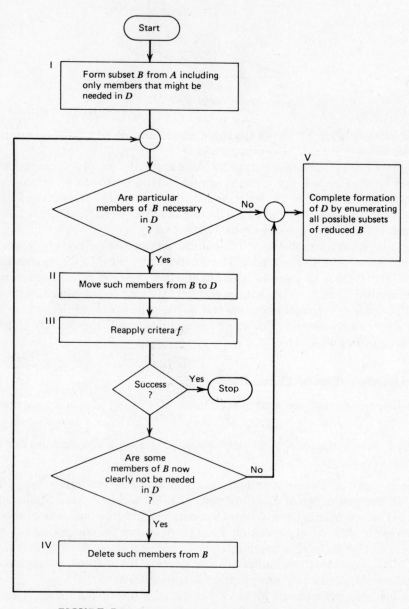

FIGURE 7.6   Generalized Quine–McCluskey algorithm.

prime implicant will be given later in this section, which develops the process implementing block I.

The starting point for block I of Fig. 7.6 is the minterm list of the function. If the function is not in this form, it must be converted to this form by the methods discussed earlier. Let us assume the following function:

$$f(A, B, C, D) = \sum m(0, 2, 3, 6, 7, 8, 9, 10, 13) \tag{7.2}$$

The cubical representation and K-map for this function are shown in Fig. 7.7.

The first step is to find all possible 1-cubes of the function. By definition, a 1-cube is found by combining two 0-cubes, which are identical except in one variable. First, convert the minterms to binary form, and then count (and list) the number of 1s in the binary representation (Fig. 7.8). Now reorder the minterm list in accordance with the number of 1s in the binary representations (Fig. 7.9a). Separate the minterms into groups with the same number of 1s by horizontal lines. This grouping of minterms is made to reduce the number of comparisons that must be made in determining the 1-cubes. If two minterms are to combine, their binary representation must be identical except in one position, in which one minterm will have a 0, the other a 1. Thus, the latter has one more 1 than the former, so that the two minterms must be in adjacent groups in the list of Fig. 7.9a.

With the minterms grouped, the procedure for finding the 1-cubes is quite simple. Compare each minterm in the top group with each minterm in the next

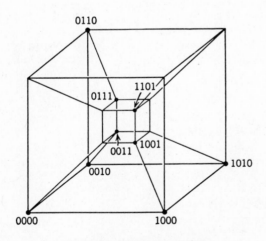

(a) Cubical Representation

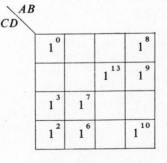

(b) K-map

FIGURE 7.7

Minterm	Binary Form	No. of 1's
m_0	0000	0
m_2	0010	1
m_3	0011	2
m_6	0110	2
m_7	0111	3
m_8	1000	1
m_9	1001	2
m_{10}	1010	2
m_{13}	1101	3

FIGURE 7.8   *Conversion of minterm list to binary form.*

lower group. If two minterms are the same in every position but one, place a check ($\sqrt{}$) to the right of both minterms to show that they have been covered and enter the 1-cube in the next column (Fig. 7.9b). List the cubical form of the 1-cube with an x in the position that does not agree, and also list the decimal numbers of the combining minterms.

In this case, m_0 (0000) combines with m_2 (0010) to form (00x0). The combination of two vertices into a 1-cube represents an application of the distributive

1's		Minterms
0	m_0	0000 ✓
1	m_2	0010 ✓
	m_8	1000 ✓
2	m_3	0011 ✓
	m_6	0110 ✓
	m_9	1001 ✓
	m_{10}	1010 ✓
3	m_7	0111 ✓
	m_{13}	1101 ✓

(a)

	1-Cubes
0,2	00x0 ✓
0,8	x000 ✓
2,3	001x ✓
2,6	0x10 ✓
2,10	x010 ✓
*8,9	100x
8,10	10x0 ✓
3,7	0x11 ✓
6,7	011x ✓
*9,13	1x01

(b)

2-Cubes	
*0,2,8,10	x0x0
*2,3,6,7	0x1x

(c)

FIGURE 7.9   *Determination of prime implicants by Quine–McCluskey.*

law just as does combining two squares on a Karnaugh map. For example, this combination of m_0 and m_2 is equivalent to the algebraic operation

$$\bar{A}\bar{B}\bar{C}\bar{D} + \bar{A}\bar{B}C\bar{D} = \bar{A}\bar{B}\bar{D}(C + \bar{C}) = \bar{A}\bar{B}\bar{D} \tag{7.3}$$

Minterm m_0 also combines with m_8 to form (x000). This completes the comparison between minterms in the first two groups, so we draw a line beneath the resultant 1-cubes. Next, the second and third groups of minterms are compared in the same manner. This comparison results in five more 1-cubes, formed from m_2 and m_3, m_8 and m_9, etc. A line is drawn beneath these five 1-cubes to indicate completion of comparisons between the second and third groups. Note that each minterm in a group must be compared with *every* minterm in the other group, even if either or both has already been checked ($\sqrt{}$) as having formed a 1-cube. *Every* 1-cube must be found, although there is no need to check off a minterm more than once.

This comparison process is repeated between successive groups until the minterm list is exhausted. In this case, all the minterms have been checked, indicating that all combine at least into 1-cubes. Thus, none of the minterms will appear explicitly in the final sum-of-products form.

The next step is a search of Fig. 7.9b for possible combinations of pairs of 1-cubes into 2-cubes. Again, cubes in each group need to be compared only with cubes in the next group down. In addition, cubes need be compared only if they have the same digit replaced by an x. In this case, 1-cube 0,2 (00x0) need be compared only with 8,10 (10x0). They differ in only one of the specified (non-x) positions and therefore combine to form the 2-cube 0,2,8,10 (x0x0), which is entered in the next column (Fig. 7.9c). The two 1-cubes are also checked off to show that they have combined into a 2-cube. Next, 1-cube 0,8 (x000) is found to combine with 2,10 (x010) to form 0,8,2,10 (x0x0), which is the same 2-cube as that already formed. Therefore, the two 1-cubes are checked off; but no new entry is made in the 2-cube column.

To clarify the above process, consider the algebraic interpretation. First, we have

$$m_0 + m_2 + m_8 + m_{10} = \bar{A}\bar{B}\bar{C}\bar{D} + \bar{A}\bar{B}C\bar{D} + A\bar{B}\bar{C}\bar{D} + A\bar{B}C\bar{D}$$

$$= \bar{A}\bar{B}\bar{D}(C + \bar{C}) + A\bar{B}\bar{D}(C + \bar{C}) \tag{7.4}$$

$$= \bar{A}\bar{B}\bar{D} + A\bar{B}\bar{D}$$

$$= \bar{B}\bar{D}(A + \bar{A}) \tag{7.5}$$

$$= \bar{B}\bar{D}$$

Here we see the reason for requiring the x to be in the same position in both 1-cubes. The x represents the variable eliminated. If the same variable (C) had not been eliminated from both 1-cubes in Equation 7.4, it is obvious that the second elimination (of A) could not have been made in Equation 7.5. The second combination of 1-cubes into the same 2-cubes is seen in the following.

$$m_0 + m_8 + m_2 + m_{10} = \bar{A}\bar{B}\bar{C}\bar{D} + A\bar{B}\bar{C}\bar{D} + \bar{A}\bar{B}C\bar{D} + A\bar{B}C\bar{D}$$

$$= \bar{B}\bar{C}\bar{D}(A + \bar{A}) + \bar{B}C\bar{D}(A + \bar{A}) \tag{7.6}$$

$$= \bar{B}\bar{C}\bar{D} + \bar{B}C\bar{D}$$

$$= \bar{B}\bar{D}(C + \bar{C}) \tag{7.7}$$

$$= \bar{B}\bar{D}$$

Equations 7.6 and 7.7 represent the same elimination of variables as Equations 7.4 and 7.5, except in the reverse order.

No further combinations can be made between the first and second groups of Fig. 9.7b, so we draw a line beneath the 2-cube formed. The second and third groups are then compared in the same fashion, resulting in another 2-cube. This completes the comparison of 1-cubes, and any unchecked entries—8,9 and 9,13 in this case—are marked with an asterisk (*), to indicate that they are *prime implicants*.

Definition 7.1 A *prime implicant* is any cube of a function that is not totally contained in some larger cube of the function.

If any minterms had failed to combine in the first step, they would also have been marked as prime implicants. The importance of prime implicants will become evident shortly.

Finally, the 2-cubes are checked for possible combination into 3-cubes. Since the x's are in different positions, these cubes do not combine and are thus prime implicants. For this example, the determination of prime implicants is now complete. In the general case, the same procedure continues as long as larger cubes can be formed.

This procedure of comparing binary representations is quite suitable for computer mechanization, particularly for computers that have some sort of "binary compare" command. For manual processing, this method has the drawback that errors will inevitably occur in trying to compare 0s and 1s over long lists of cubes. This difficulty is considerably reduced by the following alternative method, which utilizes only decimal notation. We will use the same example as above.

Again, the first step is to list the minterms, grouped according to the number of 1s in their binary representation. However, only the decimal minterm numbers are listed (Fig. 7.10a). To make the comparison, we draw on the fact that

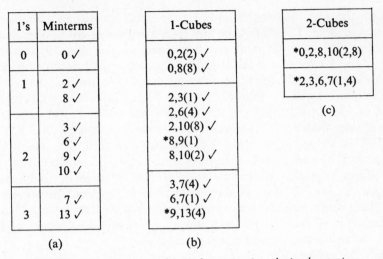

FIGURE 7.10   *Prime implicant determination, decimal notation.*

the numbers of minterms that combine must differ by a power of two. Consider the combination of m_{13} (1101) and m_9 (1001):

$$13 = 1 \times 2^3 + \boxed{1 \times 2^2} +0 \times 2^1 + 1 \times 2^0$$
$$9 = 1 \times 2^3 + \boxed{0 \times 2^2} +0 \times 2^1 + 1 \times 2^0$$

Each 1 in a binary number has the numeric weight of a power of two. The two combining minterms have the same 0s and 1s, except in one position. Thus, the number of the minterm with the extra 1 must be larger than the number of the other minterm by a power of two. The procedure, then, is as follows: compare each minterm number with each *larger* minterm number in the next group down. If they differ by a power of two, they combine to form a 1-cube.

To represent the 1-cube, list the two minterm numbers, followed by their numeric difference in parentheses. For the above example, the listing would be 9,13(4). The number in parentheses indicates the position in which the x would appear in the binary representation. Since the binary digit with the weight of 4 is the third from the right, the (4) in parentheses corresponds to the x in the third position in the binary notation 9,13 (1x01). Except for this change in notation, the process of comparing minterms and listing 1-cubes is the same as before. The resultant 1-cube column is shown in Fig. 7.10b.

In the binary-character method, it was seen that 1-cubes could combine only if the x's were in the same position. Since the number in parentheses indicates the

position of the x, it follows that only 1-cubes with the same number in parentheses can combine. Therefore, compare each 1-cube with all 1-cubes in the next
lower group that have the same number in parentheses. If the lowest minterm
number of the 1-cube in the lower group is greater by a power of two than the
corresponding number in the other cube, they combine. List the 2-cube by listing
all four minterm numbers, followed by both powers of two, in parentheses. For
example, in Fig. 7.10b, 0,2(2) and 8,10(2) have the same number in parentheses
and 8 is greater than 0 by 8. The resultant 2-cube is thus 0,2,8,10(2,8).

Figure 7.10c shows the complete 2-cube column in decimal notation. To compare 2-cubes, the same procedure applies. Just as both x's, in both 2-cubes, have
to be in the same position for combination, so *both* numbers in parentheses must
be the same for both 2-cubes. A comparison of Figs. 7.9 and 7.10 will show that
the resultant charts are identical. Furthermore, the procedures are identical in
principle, differing only in the substitution of mental subtraction of decimal
numbers for visual comparison of binary numbers.

7.4   Selection of an Optimum Set of Prime Implicants

In this section we turn our attention to obtaining the realization, *D*, from the set
of prime implicants, *B*. That this is possible is guaranteed by Theorem 7.1, first
proved by Quine [2]. All of the prime implicants determined for the example of
the previous section are indicated on the 4-cube and on the map of Fig. 7.11.
Interpreting the *k*-cubes as sets of 1s on the K-map, we see that each *k*-cube will
be represented in a sum-of-products realization by a product term with $n - k$

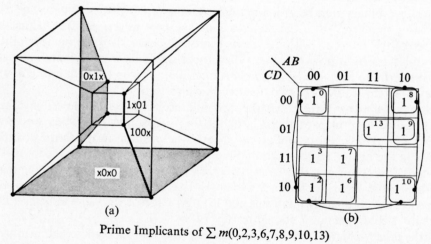

(a)   (b)

Prime Implicants of $\sum m(0,2,3,6,7,8,9,10,13)$

FIGURE 7.11

literals. Thus, any set of k-cubes containing all the 0-cubes (minterms) of the function will provide an SOP realization.

THEOREM 7.1   Any sum-of-products realization that is minimal in the sense of Definition 6.1 must consist of a sum of products representing prime implicants.

Proof   Consider any set of k-cubes containing all the 0-cubes of the function. Any cube that is not a prime implicant is, by definition, contained in some prime implicant. Therefore, we can replace this cube in the set by any prime implicant containing it. Since the prime implicant is of higher dimension, its product representation will have fewer literals than that of the cube it replaces. The replacement will therefore lower the cost of the SOP realization.

Theorem 7.1 tells us that we can restrict our attention to the prime implicants in selecting an SOP realization. (In the future we will use the term *prime implicant* to mean either the cubes themselves or the corresponding product.) However, the theorem does not say that *any* sum of prime implicants covering all the minterms is minimal. To find a minimal sum, we construct a *prime implicant table* (Fig. 7.12).

Each column corresponds to a minterm. At the left of each row are listed the prime implicants of the function, arranged in groups according to cost (number of literals in the product). Also included are an extra column on the left and an extra row at the bottom, for reasons that will be explained shortly. In each row (except the bottom), checks are placed in the columns corresponding to the minterms contained in the prime implicant listed on that row. For example, the first prime implicant listed in Fig. 7.12 contains the 0-cubes 0,2,8, and 10, so checks are placed in the first row in columns 0, 2, 8, and 10.

The completed prime implicant table is then inspected for columns containing only a single check mark. In this example, we find only one check in the first

		0	2	3	6	7	8	9	10	13
0,2,8,10(2,8)		✓	✓				✓		✓	
2,3,6,7(1,4)			✓	✓	✓	✓				
8,9(1)							✓	✓		
9,13(4)								✓		✓

FIGURE 7.12   Prime implicant table—first step.

		0	2	3	6	7	8	9	10	13
*	0,2,8,10(2,8)	✓	✓				✓		✓	
	2,3,6,7(1,4)		✓	✓	✓	✓				
	8,9(1)						✓	✓		
	9,13(4)							✓		✓
		✓	✓				✓		✓	

FIGURE 7.13 *Prime implicant table—second step.*

column, on the first row. This indicates that the corresponding prime implicant, 0,2,8,10(2,8), is the only one containing m_0 and must be included in the SOP realization. We therefore say that it is an *essential prime implicant* and mark it with an asterisk in the leftmost column. We also place checks in the bottom row in columns 0, 2, 8, and 10 to indicate that these are included in an already-selected prime implicant (Fig. 7.13).

Continuing the search, we find a single check in column 3, indicating that the prime implicant 2,3,6,7(1,4) is the only one covering m_3. We mark this prime implicant as essential and check off the minterms it covers, which were not covered by the first one selected. Finally, we find that 9,13(4) is also essential (Fig. 7.14). The determination of essential prime implicants corresponds to block II of Fig. 7.6. When the search for essential prime implicants is complete we inspect the bottom row to see if all columns have been checked. This constitutes

		0	2	3	6	7	8	9	10	13
*	0,2,8,10(2,8)	✓	✓				✓		✓	
*	2,3,6,7(1,4)		✓	✓	✓	✓				
	8,9(1)						✓	✓		
*	9,13(4)							✓		✓
		✓	✓	✓	✓	✓	✓	✓	✓	✓

FIGURE 7.14 *Prime implicant table—final.*

block III of Fig. 7.6. If so, then all the 0-cubes are contained in the essential prime implicants and the sum of the essential prime implicants is the minimal SOP realization. This is the case for our example.

The final step is to determine the actual products. To do this, we recall that difference numbers for each prime implicant indicate the variables that have been eliminated. Write out the binary equivalent of any one of the minterms in the prime implicant, cross out the positions having the binary weights in parentheses, and convert the remaining digits to the appropriate literals.

For the above example, we have

$$0,2,8,10(2,8) = 0\cancel{0}0\cancel{0} \rightarrow \bar{B}\bar{D}$$

$$2,3,6,7(1,4) \;\; = 00\cancel{1}\cancel{0} \rightarrow \bar{A}C$$

$$9,13(4) \qquad = 1\cancel{0}01 \rightarrow A\bar{C}D$$

Thus, the minimal sum of products is

$$f(A, B, C, D) = \sum m(0, 2, 3, 6, 7, 8, 9, 10, 13) = \bar{B}\bar{D} + \bar{A}C + A\bar{C}D \quad (7.8)$$

This very simple example was chosen to clarify the exact significance of the various steps rather than as a practical illustration of the value of the technique. Example 7.1 illustrates the application of the method to a problem of more practical scope.

Example 7.1

Determine the minimal SOP form for

$$f(A, B, C, D, E) = \sum m(0, 6, 8, 10, 12, 14, 17, 19, 20, 22, 25, 27, 28, 30)$$

for which the prime implicant chart and table are shown in Fig. 7.15.

Solution   Note that we have assigned a letter to represent each prime implicant, in order to simplify the prime implicant table. Thus,

$$\mathbf{a} = 8, 10, 12, 14(2, 4) = 01\cancel{0}0\cancel{0} = \bar{A}B\bar{E}, \text{ etc.}$$

Translating all essential prime implicants in this fashion, we obtain the desired minimal form

$$f(A, B, C, D, E) = \bar{A}B\bar{E} + CD\bar{E} + A\bar{C}E + AC\bar{E} + \bar{A}\bar{C}\bar{D}\bar{E} \quad (7.9)$$

∎

In both of the previous examples, the essential prime implicants covered all the minterms. This is not always the case. Figure 7.16a shows the prime implicant table for the function

$$f(A, B, C, D, E) = \sum m(1, 2, 3, 5, 9, 10, 11, 18, 19, 20, 21, 23, 25, 26, 27)$$

0 ✓	*0,8(8) f	*8,10,12,14(2,4) a
8 ✓	8,10(2) ✓ 8,12(4) ✓	*6,14,22,30(8,16) b *12,14,28,30(2,16) c
6 ✓ 10 ✓ 12 ✓ 17 ✓ 20 ✓	6,14(8) ✓ 6,22(16) ✓ 10,14(4) ✓ 12,14(2) ✓ 12,28(16) ✓	*17,19,25,27(2,8) d *20,22,28,30(2,8) e
14 ✓ 19 ✓ 22 ✓ 25 ✓ 28 ✓	17,19(2) ✓ 17,25(8) ✓ 20,22(2) ✓ 20,28(8) ✓	
27 ✓ 30 ✓	14,30(16) ✓ 19,27(8) ✓ 22,30(8) ✓ 25,27(2) ✓ 28,30(2) ✓	

	0	6	8	10	12	14	17	19	20	22	25	27	28	30
*a			✓	✓	✓	✓								
*b		✓				✓				✓				✓
c					✓	✓							✓	✓
*d							✓	✓			✓	✓		
*e									✓	✓			✓	✓
*f	✓		✓											
	✓	✓	✓	✓	✓	✓	✓	✓	✓	✓	✓	✓	✓	✓

FIGURE 7.15

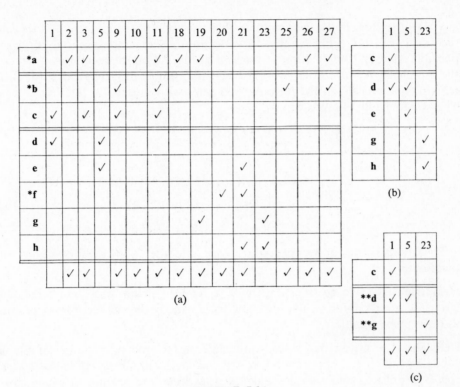

FIGURE 7.16

When we select the essential prime implicants on this table, we find that they do not cover all the minterms of the function. We must now apply some cost criteria in making a selection from the remaining nonessential terms.

To make it easier to find the appropriate terms, a *reduced* prime implicant table is set up, listing only the 0-cubes not contained in the essential prime implicants as columns and those nonessential prime implicants containing any of the uncovered 0-cubes as rows (Fig. 7.16b). In the reduced table, we note that there could be no possible advantage to using **e**, since it covers only m_5, while **d**, which has the same cost, covers both m_5 and m_1. Thus, removing **e** from the table cannot prevent us from finding a minimal sum. We next note that **g** and **h** cover the same minterm at the same cost. Thus, removing **h** from the table cannot prevent our finding a minimal sum. Since rows **e** and **h** are not needed in Fig. 7.16b, the answer to the lower question in the flowchart of Fig. 7.6 is yes. Thus, these rows are removed as called for by block IV of the flowchart to form a still further reduced table, as shown in Fig. 7.16c. On this table, it may be seen that **d** is the only prime implicant left to cover m_5 and **g** is the only left to cover

m_{23}. We therefore mark them with double asterisks (**) to indicate that they are *secondary essential* prime implicants. Selection of secondary essential prime implicants is another application of block II of Fig. 7.6 on the second pass through the flowchart. We also note that these essential prime implicants cover all of the remaining minterms, so the process terminates with success on this second pass. The minimal sum is given by

$$f(A, B, C, D, E) = \underbrace{a + b + f}_{\text{essential}} + \underbrace{d + g}_{\substack{\text{secondary} \\ \text{essential}}} \tag{7.10}$$

$$= (2, 3, 10, 11, 18, 19, 26, 27) + (9, 11, 25, 27)$$

$$+ (20, 21) + (1, 5) + (19, 23) \tag{7.11}$$

$$= \bar{C}D + B\bar{C}E + A\bar{B}C\bar{D} + \bar{A}\bar{B}\bar{D}E + A\bar{B}DE \tag{7.12}$$

To formalize the above procedure, we state the following definitions and theorem.

Definition 7.2   Two rows, **a** and **b**, of a reduced prime implicant table, that cover the same minterms, i.e., have checks in exactly the same columns, are said to be interchangeable.

Definition 7.3   Given two rows, **a** and **b**, in a reduced prime implicant table, row **a** is said to *dominate* row **b** if row **a** has checks in all the columns in which row **b** has checks and also has a check in at least one column in which row **b** does not have a check.

THEOREM 7.2   Let **a** and **b** be rows of a reduced prime implicant table, such that the cost of **a** is less than or equal to the cost of **b**. Then if **a** dominates **b** or **a** and **b** are interchangeable, there exists a minimal sum of products that does not include **b**.

Theorem 7.2 is applied as necessary as part of the lower question or of block IV of the flowchart of Fig. 7.6.

Example 7.2

Given the function
$$f(A, B, C, D, E, F) = \sum m(1, 2, 3, 4, 5, 8, 9, 10, 17, 20, 21, 24, 25, 27, 32,$$
$$33, 34, 36, 37, 40, 41, 42, 43, 44, 45, 46, 47, 48,$$
$$56, 59, 62)$$

By the procedures already covered, the essential prime implicants have been determined. The reduced prime implicant table left after this removal is shown in Fig. 7.17. Complete the selection of prime implicants.

Solution   We first see that prime implicant **n** covers only m_{34} while **g** covers m_{34}, m_{10}, and m_2. So **g** dominates **n**, and **n** can be removed. Similarly, **e** dominates **c**, **a** dominates **i**,

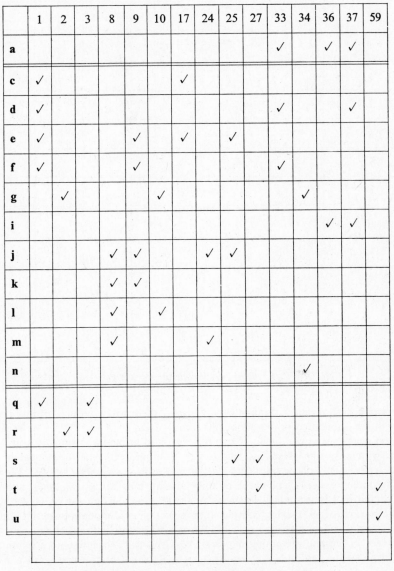

	1	2	3	8	9	10	17	24	25	27	33	34	36	37	59
a											✓		✓	✓	
c	✓						✓								
d	✓										✓			✓	
e	✓				✓		✓		✓						
f	✓				✓						✓				
g		✓				✓						✓			
i													✓	✓	
j				✓	✓			✓	✓						
k				✓	✓										
l				✓		✓									
m				✓				✓							
n												✓			
q	✓		✓												
r		✓	✓												
s									✓	✓					
t										✓					✓
u															✓

FIGURE 7.17 Reduced prime implicant table.

	1	2	3	8	9	10	17	24	25	27	33	34	36	37	59
**a											✓		✓	✓	
d	✓										✓			✓	
**e	✓				✓		✓		✓						
f	✓				✓				✓						
**g		✓				✓						✓			
**j				✓	✓			✓	✓						
l				✓		✓									
q	✓		✓												
r		✓	✓												
s								✓	✓						
**t										✓					✓
	✓	✓		✓	✓	✓	✓	✓	✓	✓	✓	✓	✓	✓	✓

FIGURE 7.18   Prime implicant table—second reduction.

j dominates **k** and **m**, and **t** dominates **u**. Figure 7.18 is a new reduced prime implicant table, with the dominated prime implicants removed.

We next check the new table for secondary essential prime implicants by searching for columns with only one check, as before. For example, with the dominated terms eliminated, **e** is the only prime implicant left that covers m_{17}, so **e** must be used. We mark it (******) as secondary essential and check off the minterms it covers, m_1, m_9, m_{17}, and m_{25}, as before. Similarly, **j** is secondary essential by virtue of m_{24}, **g** because of 34, **a** because of 36, and **t** because of 59. Having completed the selection of secondary essential implicants, we find that only one minterm, m_3, is left. We can see by inspection that **q** and **r** are equal with respect to the one remaining minterm. We arbitrarily select **q**, and a minimal form will be given by the sum of **a + e + g + j + t + q +** the essential prime implicants.   ■

In Example 7.2, one application of row dominance was sufficient to complete the design. In some cases, the first selection of secondary essential prime implicants may leave a large group of minterms still uncovered. In this case, simply repeat the process, i.e., reduce the table by removing the secondary essential rows and apply row dominance again. This process may be repeated as many times as necessary. If at any point no secondary essential prime implicants are found or Theorem 7.2 cannot be applied, a "no" answer to a question in the flowchart results, leading to block V.

One implementation of block V, called Petrick's method [5], will be used on the reduced prime implicant table shown in Fig. 7.19. This table cannot be further reduced, since no column has a single check and no row dominates any other row.

A list of all possible nonredundant sets of prime implicants that will cover the remaining minterms may be obtained in the following manner. The letters representing the remaining prime implicants are interpreted as Boolean variables, which will take on the value 1 if that prime implicant is selected, the value 0 if it is not. From the first column, either **a** or **f** must be selected to cover m_6, so we write

$$(\mathbf{a} + \mathbf{f}) = 1 \tag{7.13}$$

Similarly, either **a** or **c** must be selected to cover m_7; so

$$(\mathbf{a} + \mathbf{f})(\mathbf{a} + \mathbf{c}) = 1 \tag{7.14}$$

In this fashion, we build up a product of sums, each sum listing the prime implicants that contain a particular minterm. In this case, the complete expression is

$$(\mathbf{a} + \mathbf{f})(\mathbf{a} + \mathbf{c})(\mathbf{b} + \mathbf{c})(\mathbf{e} + \mathbf{f})(\mathbf{d} + \mathbf{e})(\mathbf{b} + \mathbf{d}) = 1 \tag{7.15}$$

	6	7	15	38	46	47
a	✓	✓				
b			✓			✓
c		✓	✓			
d					✓	✓
e				✓	✓	
f	✓			✓		

FIGURE 7.19   Reduced prime implicant table.

Applying the distributive laws (multiplying out), we obtain

$$\textbf{abe + abdf + acde + bcef + cdf} = 1 \tag{7.16}$$

To satisfy Equation 7.16, at least one of the five products must be 1. Of these, **abe** and **cdf** represent a covering of the table by three prime implicants. The total cost is least for **a**, **b**, and **e** (smallest total number of inputs), so these three are included with any essential prime implicants in a minimal sum of products.

The Quine–McCluskey method can be modified to handle don't-care terms very simply. In determining the prime implicants, include the don't-cares in the minterm list. In this manner, you insure that you have included groupings that can be formed by using the don't-cares. Then, in setting up the prime implicant table, list only the minterms across the top. This insures that only prime implicants necessary to cover minterms will be selected.

Example 7.3

Find the minimal SOP realization of

$$f(A, B, C, D, E) = \sum m(1, 4, 7, 14, 17, 20, 21, 22, 23) + d(0, 3, 6, 19, 30)$$

On the chart in Fig. 7.20a, we list both minterms and don't-cares in column I.

Solution   On the prime implicant table (Fig. 7.20b), we list only the minterms. From the table, we have

$$f(A, B, C, D, E) = \underset{\underset{\text{essential}}{\uparrow}}{\textbf{e}} \quad + \underset{\underset{\substack{\text{secondary}\\\text{essential}}}{\uparrow}}{\underbrace{\textbf{a} + \textbf{b} + \textbf{d}}} + \underset{\underset{\text{optional}}{\uparrow}}{\begin{cases} \text{f} \\ \text{or } \text{g} \end{cases}}$$

or

$$f(A, B, C, D, E) = CD\bar{E} + \bar{B}\bar{C}E + \bar{B}C\bar{E} + \bar{B}CD + \begin{cases} A\bar{B}E \\ \text{or } A\bar{B}C \end{cases} \tag{7.17}$$

■

Like the K-map, the Quine–McCluskey method can be adapted to POS design by designing on the 0s instead of the 1s of the function. Only two changes are necessary. In column I of the Quine–McCluskey chart, list the 0s (maxterms) of the function plus the don't-cares (if any). Then proceed exactly as before until the last step, the conversion of the prime implicants, at which point you convert to sums instead of products, using complemented variables, just as with the K-map method.

It should be evident that the Quine–McCluskey method does not qualify as a labor-saving technique. The map method is always quicker, but for more than 5 variables, the reliability in determining a minimal realization is questionable.

I	II	III
0 ✓	*0,1(1) h *0,4(4) i	*1,3,17,19(2,16) a *4,6,20,22(2,16) b
1 ✓		
4 ✓	1,3(2) ✓ 1,17(16) ✓	*3,7,19,23(4,16) c *6,7,22,23(1,16) d
3 ✓ 6 ✓ 17 ✓ 20 ✓	4,6(2) ✓ 4,20(16) ✓	*6,14,22,30(8,16) e *17,19,21,23(2,4) f *20,21,22,23(1,2) g
	3,7(4) ✓ 3,19(16) ✓	
7 ✓ 14 ✓ 19 ✓ 21 ✓ 22 ✓	6.7(1) ✓ 6,14(8) ✓ 6,22(16) ✓ 17,19(2) ✓ 17,21(4) ✓ 20,21(1) ✓	(a)
23 ✓ 30 ✓	20,22(2) ✓	
	7,23(16) ✓ 14,30(16) ✓ 19,23(4) ✓ 21,23(2) ✓ 22,23(1) ✓ 22,30(8) ✓	

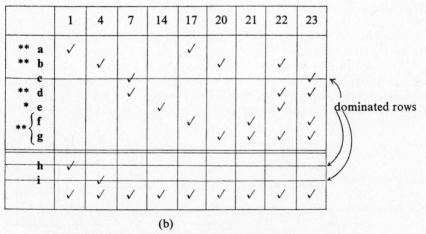

	1	4	7	14	17	20	21	22	23	
** a	✓				✓					
** b		✓				✓		✓		
c			✓						✓	
** d			✓					✓	✓	
* e				✓				✓		dominated rows
** { f					✓		✓	✓	✓	
g						✓	✓	✓	✓	
h	✓									
i		✓								
	✓	✓	✓	✓	✓	✓	✓	✓	✓	

(b)

FIGURE 7.20 *Chart and prime implicant table, Example 7.3.*

The map method can generally be relied on to come fairly close to a minimal design. If you are designing a circuit that will only be built once, the savings resulting from eliminating one or two gates or literals will not justify the extra time required for the tabular method. On the other hand, if you are designing a circuit to be produced in large quantity, the elimination of a single literal may warrant many hours of engineering time.

The real importance of the Quine–McCluskey method lies in the fact that it is a formal procedure, in no way dependent on human intuition. It is therefore suitable for computer mechanization, and a number of programs based on this method have been written [6, 7]. A thorough understanding of the basic principles of this method is thus vitally important to logic designers even though they may seldom use it personally.

7.5   Multiple-Output Circuits

So far we have considered only the implementation of a single function of a given set of variables. In the design of complete systems, we frequently wish to implement a number of different functions of the same set of variables. We can implement each function completely independently by the techniques already developed, but considerable savings can often be achieved by the sharing of hardware between various functions.

Example 7.4

The circuit **BB** is to serve as an interface between the two computers of Fig. 7.21. The first four letters of the alphabet must intermittently be transmitted from computer 1 to computer 2. In computer 1, these letters are coded on three lines, x_1, x_2, x_3, as shown in Fig. 7.22a. In computer 2, they are coded on two lines, y_2 and y_1, as shown in Fig. 7.22b.

The translating functions $y_1(x_1, x_2, x_3)$ and $y_2(x_1, x_2, x_3)$ are easily compiled directly on Karnaugh maps, as in Fig. 7.23. To accomplish this, we determine from Fig. 7.22b which letters of the alphabet require 1 for either y_1 or y_2. We then determine from Fig. 7.22a which input combinations correspond to these letters. For example, y_1 is to be 1 for c or d (Fig. 7.22b). From Fig. 7.22a, we see that c or d is represented by $x_1 x_2 x_3 = 000,001,101$, so we enter 1 in the corresponding squares of Fig. 7.23a.

From these maps, the minimal realizations of y_1 and y_2, when considered individually,

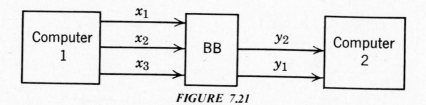

FIGURE 7.21

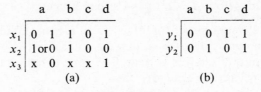

	a	b	c	d
x_1	0	1	1	0	1
x_2	1or0	1	0	0
x_3	x	0	x	x	1

(a)

	a	b	c	d
y_1	0	0	1	1
y_2	0	1	0	1

(b)

FIGURE 7.22

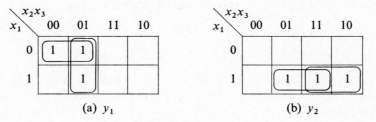

(a) y_1 (b) y_2

FIGURE 7.23

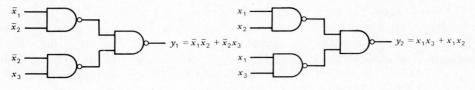

$y_1 = \bar{x}_1\bar{x}_2 + \bar{x}_2 x_3$ $y_2 = x_1 x_3 + x_1 x_2$

FIGURE 7.24

are easily seen to be those of Fig. 7.24. Alternatively, we note that y_1 and y_2 could be expressed as

$$y_1 = \bar{x}_1\bar{x}_2 + x_1\bar{x}_2 x_3 \qquad \text{and} \qquad y_2 = x_1 x_2 + x_1\bar{x}_2 x_3 \qquad (7.18)$$

Taking advantage of the common term, $x_1\bar{x}_2 x_3$, permits the implementation found in Fig. 7.25. Notice that this configuration requires only five NAND gates as opposed to 6 in Fig. 7.24.

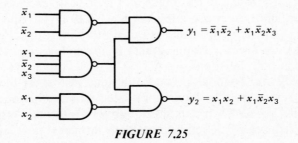

$y_1 = \bar{x}_1\bar{x}_2 + x_1\bar{x}_2 x_3$

$y_2 = x_1 x_2 + x_1\bar{x}_2 x_3$

FIGURE 7.25   ∎

7.6   Map Minimization of Multiple-Output Circuits

The value of a systematic procedure that would identify common products and guarantee an optimum realization (minimum total number of gates for all outputs) of multiple-output circuits is clear. The general convenience of Karnaugh maps lead us to explore this approach first.

Let us assume that we wish to find a minimal SOP realization for the three functions

$$f_\alpha(A, B, C, D) = \sum m(2, 4, 10, 11, 12, 13)$$

$$f_\beta(A, B, C, D) = \sum m(4, 5, 10, 11, 13)$$

$$f_\gamma(A, B, C, D) = \sum m(1, 2, 3, 10, 11, 12)$$

We first draw the Karnaugh maps of all three functions (Fig. 7.26). We also draw maps of the *intersections*, or products, of the three functions, by pairs and all together. Recall that, in terms of K-maps, the intersection of two functions consists of their common squares.

We now locate the prime implicants on the maps, starting with the product of all the functions. The only prime implicant of $f_\alpha \cdot f_\beta \cdot f_\gamma$ is 10,11(1). We now look for prime implicants of the function pairs, but we do not mark a grouping as a prime implicant if it is a prime implicant of a higher-order function product. For example, we do not mark 10,11(1) on $f_\beta \cdot f_\gamma$ since it is a prime implicant of $f_\alpha \cdot f_\beta \cdot f_\gamma$. On $f_\alpha \cdot f_\gamma$, we mark only 2,10(8) and 12, since the other group is also in $f_\alpha \cdot f_\beta \cdot f_\gamma$. Similarly, we mark only 4 and 13 on $f_\alpha \cdot f_\beta$. Now we go to the maps of the functions themselves, marking only those prime implicants that do not also appear in the function products.

The next task is to select the sets of prime implicants leading to the realization with the lowest total cost. For a problem of this size, the selection could be made directly from the maps. A more systematic approach uses a modified version of the prime implicant table (Fig. 7.27). The columns correspond to all the minterms for each function. The rows correspond to the prime implicants grouped by cost. Also indicated are the function, or function products, in which they appear. On each row, we check the minterm columns *for the function(s) in which the prime implicant appears*. For example, in Fig. 7.27 we check columns 2, 3, 10, 11 under f_γ *only*, since the prime implicant **a** = 2,3,10,11(1,8) appears only in f_γ, although some of the minterms also appear in other functions. For prime implicant **g** = 2,10(8), we check columns 2 and 10 under both f_α and f_γ, since this 1-cube appears in both functions.

We next check for essential prime implicants, just as in the single-output Quine–McCluskey method, by looking for columns with only one check. In this case, the essential prime implicants do not cover all the minterms of all three functions, so a reduced prime implicant table is formed (Fig. 7.28).

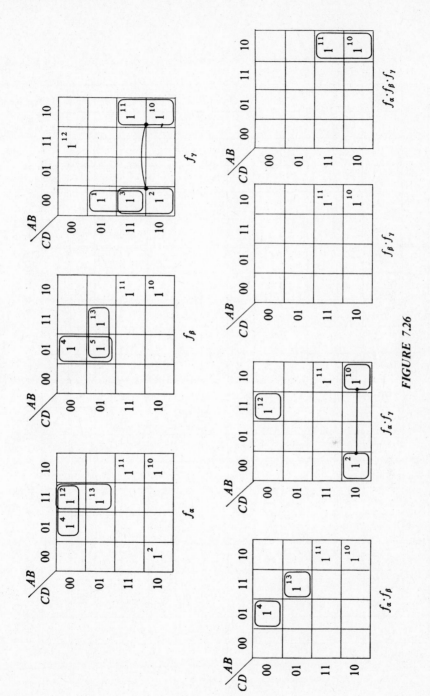

FIGURE 7.26

Fcn.	Pr. Imp.	f_α 2	4	10	11	12	13	f_β 4	5	10	11	13	f_γ 1	2	3	10	11	12
γ	a													√	√	√	√	
α	b		√			√												
α	c					√	√											
β	d							√	√									
β	e								√			√						
γ *	f												√		√			
$\alpha\gamma$ *	g	√		√										√		√		
$\alpha\beta\gamma$ *	h			√	√					√	√					√	√	
$\alpha\beta$	i		√					√										
$\alpha\gamma$	j					√												√
$\alpha\beta$ *	k						√					√						

$a = 2,3,10,11(1,8)$ $b = 4,12(8)$ $c = 12,13(1)$ $d = 4,5(1)$ $e = 5,13(8)$ $f = 1,3(2)$ $g = 2,10(8)$
$h = 10,11(1)$ $i = 4$ $j = 12$ $k = 13$

FIGURE 7.27

Fcn.	P. Imp.	f_α		f_β		
		4	13	4	5	13
α	**b**	✓				
α	**c**		✓			
β	**d**			✓	✓	
β	**e**				✓	✓
$\alpha\beta$	**i**	✓		✓		
$\alpha\beta$	**k**		✓			✓

FIGURE 7.28   Reduced multiple-output prime implicant table.

The next step is to apply dominance to eliminate rows in exactly the same manner as for the single-output case. Note that the entire table must be considered in determining dominance, not just that part for a single function. For example, in the table of Fig. 7.28, row **d** would dominate row **i** if only function f_β were considered. But if the whole table is considered, it is seen that **i** has a check in a column in which **d** does not. Therefore, **d** does not satisfy the conditions for dominance of **i**.

In this example, we cannot eliminate any rows through dominance, so there are no secondary essential prime implicants, and we proceed to apply Petrick's method. As before, we develop a product-of-sums form, each sum listing the prime implicants available to cover one of the minterms not covered by the essential implicants *for all functions*. In this case, we have

$$(\mathbf{b} + \mathbf{i})(\mathbf{c} + \mathbf{k})(\mathbf{d} + \mathbf{i})(\mathbf{d} + \mathbf{e})(\mathbf{e} + \mathbf{k}) = 1 \qquad (7.19)$$

By successive multiplications, we convert this to the sum-of-products form:

$$(\mathbf{i} + \mathbf{bd})(\mathbf{c} + \mathbf{k})(\mathbf{e} + \mathbf{dk}) = 1$$

$$(\mathbf{ci} + \mathbf{bcd} + \mathbf{ik} + \mathbf{bdk})(\mathbf{e} + \mathbf{dk}) = 1$$

$$\mathbf{cei} + \mathbf{cdik} + \mathbf{bcde} + \mathbf{bdek} + \mathbf{eik} + \mathbf{dik} + \mathbf{bcdk} + \mathbf{bdk} = 1$$

and then eliminate redundant products by application of Lemma 4.5 $(\mathbf{a} + \mathbf{ab} = \mathbf{a})$:

$$\mathbf{cei} + \mathbf{bcde} + \mathbf{eik} + \mathbf{dik} + \mathbf{bdk} = 1 \qquad (7.20)$$

Each product represents a sufficient set of prime implicants to cover the remaining minterms, and we wish to select the set having the lowest cost. Since each literal represents a prime implicant, we consider only the products with the fewest literals, **cei**, **bdk**, **dik**, and **eik**. The first two represent two 1-cubes and one 0-cube; the latter two, one 1-cube and two 0-cubes. Thus, **cei** and **bdk** represent the lowest-cost sets, and we arbitrarily select **cei**.

From the total set of selected prime implicants, **c**, **e**, **f**, **g**, **h**, **i**, and **j**, we now determine those required for each function. If a prime implicant appears in only one function, it obviously must be used in the realization of that function. For prime implicants of function products, we must be more careful. Consider prime implicant $j = m_{12}$, which appears in f_α and f_γ. It is essential to f_γ and must be used in its realization. However, m_{12} is covered by prime implicant $c = 12,13(1)$ if f_α, so **j** would be redundant in the realization of f_α. On the basis of such considerations, we make the following selection of prime implicants for the three functions:

$$f_\alpha = \mathbf{c} + \mathbf{g} + \mathbf{h} + \mathbf{i}$$

$$f_\beta = \mathbf{e} + \mathbf{h} + \mathbf{i}$$

$$\mathbf{f}_\gamma = \mathbf{f} + \mathbf{g} + \mathbf{h} + \mathbf{j}$$

Either by locating the cubes on the maps or by the method given earlier, we convert these to algebraic notation:

$$f_\alpha = AB\bar{C} + \bar{B}C\bar{D} + A\bar{B}C + \bar{A}B\bar{C}\bar{D} \tag{7.21}$$

$$f_\beta = A\bar{B}C + \bar{B}C\bar{D} + \bar{A}B\bar{C}\bar{D} \tag{7.22}$$

$$f_\gamma = \bar{A}\bar{B}D + \bar{B}C\bar{D} + A\bar{B}C + AB\bar{C}\bar{D} \tag{7.23}$$

The realization of these equations is shown in Fig. 7.29.

7.7   Tabular Determination of Multiple-Output Prime Implicants

We have already adapted the prime implicant table to multiple output in the last section,* so all that remains to complete the tabular method is to adapt the chart method to the determination of multiple-output prime implicants. Let us consider the same set of functions used in the last section:

$$f_\alpha(A, B, C, D) = \sum m(2, 4, 10, 11, 12, 13)$$

$$f_\beta(A, B, C, D) = \sum m(4, 5, 10, 11, 13)$$

$$f_\gamma(A, B, C, D) = \sum m(1, 2, 3, 10, 11, 12)$$

---

* Such combinations of map and tabular methods are, obviously, equally valid for single functions and are sometimes the most efficient method.

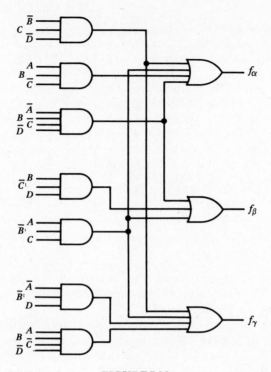

FIGURE 7.29

In the first column of Fig. 7.30 are listed all the minterms of the functions, mixed together without regard for which function they appear in, and grouped by number of 1s in their binary representations as before. To each minterm we append a *tag* to indicate in which functions it appears. We then proceed to form 1-cubes in essentially the same fashion as for the single function but with some important differences.

First, we cannot combine two cubes unless they have at least one letter in common in their tags. That is, we cannot combine a cube in one function with a cube in another. For example, in Fig. 7.30, m_1 and m_5 meet the conditions for combination, except that m_1 is in f_γ and m_5 is in f_β.

Second, when two cubes combine, the tag of the resultant larger cube consists only of those letters common to the tags of both cubes. Again referring to Fig. 7.30, when we form 5,13(8), the tag is β, since this cube is found only in f_β, as shown by the fact that β is the only letter common to the tags of 5 and 13.

Third, we check off a cube as having combined only if its entire tag appears in the tag of the larger cube. Again referring to the combination of 5 and 13 into

Minterms	1-cubes	2-cubes
1 γ ✓	*1,3(2) γ f	*2,3,10,11(1,8) γ a
2 αγ ✓	2,3(1) γ ✓	
* 4 αβ i	*2,10(8) αγ g	
	*4,5(1) β d	
3 γ ✓	*4,12(8) α b	
5 β ✓		
10 αβγ ✓	3,11(8) γ ✓	
*12 αγ j	*5,13(8) β e	
	*10,11(1) αβγ h	
11 αβγ ✓	*12,13(1) α c	
*13 αβ k		

FIGURE 7.30   Tabular determination of multiple-output prime implicants.

5,13(8), we check off 5 as having combined; but we do *not* check 13. Although 13 has combined to form a 1-cube of f_β, it has not combined in f_α and is therefore a prime implicant of the product $f_\alpha \cdot f_\beta$.

Subject to these special rules, we complete the determinations of prime implicants as described in Section 7.2. The reader will note that the prime implicants thus formed are precisely those found on the maps in the previous section. We now proceed to the prime implicant table, Fig. 7.27, and complete the design as before.

Problems

7.1. Minimize the following functions by Quine–McCluskey (sum of products).

(a) $f(W, V, X, Y, Z) = \sum m(0, 1, 3, 8, 9, 11, 15, 16, 17, 19, 24, 25, 29, 30, 31)$

(b) $f(A, B, C, D) = \sum m(0, 1, 4, 5, 6, 7, 8, 9, 10, 12, 14)$

(c) $f(a, b, c, d) = \sum m(0, 2, 3, 7, 8, 10, 11, 12, 14)$

(d) $f(A, B, C, D, E) = \sum m(0, 1, 2, 3, 4, 6, 9, 10, 15, 16, 17, 18, 19, 20, 23, 25, 26, 31)$

(e) $f(a, b, c, d, e, f) = \sum m(0, 2, 4, 5, 7, 8, 16, 18, 24, 32, 36, 40, 48, 56)$

7.2. Determine minimal POS realizations of the functions of Problem 7.1 by Quine–McCluskey.

7.3. Repeat Problem 6.3 by Quine–McCluskey.

7.4. Repeat Problem 6.4 by Quine–McCluskey.

7.5. Determine minimal SOP realizations of the following incompletely specified functions by Quine–McCluskey.

(a) $f(A, B, C, D, E) = \sum m(4, 5, 10, 11, 15, 18, 20, 24, 26, 30, 31)$
$\qquad\qquad\qquad\quad + d(9, 12, 14, 16, 19, 21, 25)$

(b) $f(A, B, C, D, E) = \sum m(0, 2, 3, 6, 9, 15, 16, 18, 20, 23, 26)$
$\qquad\qquad\qquad\quad + d(1, 4, 10, 17, 19, 25, 31)$

(c) $f(a, b, c, d, e, f) = \sum m(0, 2, 4, 7, 8, 16, 24, 32, 36, 40, 48)$
$\qquad\qquad\qquad\quad + d(5, 18, 22, 23, 54, 56)$

(d) $f(A, B, C, D) = \sum m(0, 1, 4, 6, 8, 9, 10, 12) + d(5, 7, 14)$

(e) $f(a, b, c, d) = \sum m(2, 4, 8, 11, 15) + d(1, 10, 12, 13)$

(f) $f(W, X, Y, Z) = \sum m(1, 4, 8, 9, 13, 14, 15) + d(2, 3, 11, 12)$

7.6. Repeat Problem 6.8 by Quine–McCluskey.

7.7. Determine minimal POS realizations of the functions of Problem 7.5 by Quine–McCluskey.

7.8. A circuit receives two 2-bit binary numbers $Y = y_1 y_0$ and $X = x_1 x_0$. The 2-bit output $Z = z_1 z_0$ should equal 11 if $Y = X$, 10 if $Y > X$ and 01 if $Y < X$. Design a minimal SOP realization.

7.9. Five people judge a certain competition. The vote of each is indicated by 1 (pass) or 0 (fail) on a signal line. The five signal lines form the input to a logic circuit. The rules of the competition allow only 1 dissenting vote. If the vote is 2–3 or 3–2, the competition must continue. The logic circuit is to have two outputs, xy. If the vote is 4–1 or 5–0 to pass, $xy = 11$. If the vote is 4–1 or 5–0 to fail, $xy = 00$. If the vote is 3–2 or 2–3, $xy = 10$. Design a minimal SOP circuit.

7.10. A 5-bit binary number $N = x_4 x_3 x_2 x_1 x_0$ appears at the inputs to a combinational logic circuit. The circuit has two outputs:

$\qquad\qquad$ z_1 indicates the number is evenly divisible by 6
$\qquad\qquad$ z_2 indicates the number is evenly divisible by 9

Design a minimal SOP realization.

7.11. A logic circuit has five inputs, x_4, x_3, x_2, x_1, and x_0. Output z_0 is to be 1 when a majority of the inputs are 1. Output z_1 is to be 1 when fewer than four of the inputs are 1, provided that at least one input is 1. Output z_2 is to be 1 when two, three, or four of the inputs are 1. Design a minimal SOP circuit.

7.12. Determine minimal SOP forms for the following multiple-output systems.

(a) $f_1(a, b, c) = \sum m(0, 1, 3, 5)$
$\qquad f_2(a, b, c) = \sum m(2, 3, 5, 6)$
$\qquad f_3(a, b, c) = \sum m(0, 1, 6)$

(b) $f_1(A, B, C, D) = \sum m(0, 1, 2, 3, 6, 7)$
    $f_2(A, B, C, D) = \sum m(0, 1, 6, 7, 14, 15)$
    $f_3(A, B, C, D) = \sum m(0, 1, 2, 3, 8, 9)$

(c) $f_1(A, B, C, D) = \sum m(4, 5, 10, 11, 12)$
    $f_2(A, B, C, D) = \sum m(0, 1, 3, 4, 8, 11)$
    $f_3(A, B, C, D) = \sum m(0, 4, 10, 12, 14)$

(d) $f_1(A, B, C, D, E) = \sum m(0, 1, 2, 3, 6, 7, 20, 21, 26, 27, 28)$
    $f_2(A, B, C, D, E) = \sum m(0, 1, 6, 7, 14, 15, 16, 17, 19, 20, 24, 27)$
    $f_3(A, B, C, D, E) = \sum m(0, 1, 2, 3, 8, 9, 16, 20, 26, 28, 30)$

(e) $f_1(A, B, C, D, E) = \sum m(0, 1, 2, 8, 9, 10, 13, 16, 17, 18, 19, 24, 25)$
    $f_2(A, B, C, D, E) = \sum m(0, 1, 3, 5, 7, 9, 13, 16, 17, 22, 23, 30, 31)$
    $f_3(A, B, C, D, E) = \sum m(2, 3, 8, 9, 10, 11, 13, 15, 16, 17, 18, 19, 22, 23)$

7.13. In the yard tower of a railroad yard, a controller must select the route of freight cars entering a section of the yard from point A as shown on the control panel, Fig. P7.13. Depending on the positions of the switches, a car

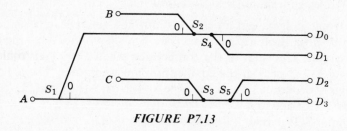

FIGURE P7.13

can arrive at any one of four destinations. Other cars may enter from points B or C. Design a circuit that will receive as inputs signals S_1 to S_5, indicating the positions of the corresponding switches, and will light a lamp, D_0 to D_4, showing which destination the car from A will reach. For the cases when cars can enter from B or C (S_2 or S_3 in the 0 position), all output lamps should light, indicating that a car from A cannot reach its destination safely.

7.14. Determine minimal SOP forms for the following multiple output systems.

(a) $f_1(a, b, c, d) = \sum m(0, 2, 9, 10) + d(1, 8, 13)$
    $f_2(a, b, c, d) = \sum m(1, 3, 5, 13) + d(0, 7, 9)$
    $f_3(a, b, c, d) = \sum m(2, 8, 10, 11, 13) + d(3, 9, 15)$

(b) $f_1(A, B, C, D) = \sum m(2, 3, 6, 10) + d(8)$
    $f_2(A, B, C, D) = \sum m(2, 10, 12, 14) + d(6, 8)$
    $f_3(A, B, C, D) = \sum m(2, 8, 10, 12) + d(0, 14)$

(e) $f_a(W, X, Y, Z) = \sum m(0, 5, 7, 14, 15) + d(1, 6, 9)$
$f_b(W, X, Y, Z) = \sum m(13, 14, 15) + d(1, 6, 9)$
$f_c(W, X, Y, Z) = \sum m(0, 1, 5, 7) + d(9, 13, 14)$

(d) $f_1(A, B, C, D, E) = \sum m(2, 8, 10, 12, 18, 26, 28, 30)$
$\qquad\qquad\qquad + d(0, 14, 22, 24)$
$f_2(A, B, C, D, E) = \sum m(2, 3, 6, 10, 18, 24, 26, 27, 29)$
$\qquad\qquad\qquad + d(8, 19, 25, 31)$
$f_3(A, B, C, D, E) = \sum m(1, 3, 5, 13, 16, 18, 25, 26)$
$\qquad\qquad\qquad + d(0, 7, 9, 17, 24, 29)$

7.15. Design a minimal second-order circuit to convert binary coded decimal input to bi-quinary (or 2-out-of-7) output. This will have four inputs and seven outputs, as shown in Fig. P7.15a. The codes for input and output for the ten decimal digits are listed in Fig. P7.15b. It may be assumed that the six possible input combinations not listed (corresponding to 10–15) will never occur.

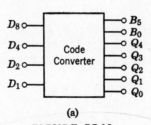

(a)

FIGURE P7.15a

Digit	BCD				Bi-Quinary						
	D_8	D_4	D_2	D_1	B_5	B_0	Q_4	Q_3	Q_2	Q_1	Q_0
0	0	0	0	0	0	1	0	0	0	0	1
1	0	0	0	1	0	1	0	0	0	1	0
2	0	0	1	0	0	1	0	0	1	0	0
3	0	0	1	1	0	1	0	1	0	0	0
4	0	1	0	0	0	1	1	0	0	0	0
5	0	1	0	1	1	0	0	0	0	0	1
6	0	1	1	0	1	0	0	0	0	1	0
7	0	1	1	1	1	0	0	0	1	0	0
8	1	0	0	0	1	0	0	1	0	0	0
9	1	0	0	1	1	0	1	0	0	0	0

(b)

FIGURE P7.15b

7.16. Repeat Problem 7.15 to convert from the excess-3 code to the 2-out-of-5 code, as listed in Fig. 8.12.

7.17. A certain digital computer uses 6-bit "OP codes," that is, a unique combination of 6 bits is used to specify each of the possible operations of the machine. The OP codes are received by a decoder, which must decide what is to be done and issue control signals accordingly. This decoding is generally done in several steps, with the codes first being classified into basic types. Thus, some operations will require a word to be read from memory, in which case the decoder must issue the signal RO (read operand). Other operations require that a word be stored in memory, in which case the decoder will issue the signal WO (write operand). Other operation categories are jump (J), zero address (ZA), iterative (I), and input/output (I/O). These categories are not mutually exclusive, i.e., an operation may fall into more than one category.

The operations falling into each category are listed in Fig. P7.17. The operations are designated by the octal equivalents of the binary codes. Thus, for example, Operation 17 has the code $\underbrace{001}_{1}\ \underbrace{111}_{7}$. Design a minimal SOP circuit to produce the control signals, RO, WO, J, ZA, I, I/O in response to the 6-bit OP code. Designate the 6 bits of the OP code as $C_5 C_4 C_3 C_2 C_1 C_0$. You may assume that 00 and 77 are not used.

Category	Signal	Operations
Read Operand	RO	12–17, 24–33, 36–46, 52, 53, 70–73
Write Operand	WO	20–23, 27, 47, 56–61, 70–73, 75, 76
Jump	J	22, 23, 75, 76
Zero Address	ZA	01–11, 34, 35, 50, 51, 54, 55
Iterative	I	24–33
Input/Output	I/O	62–67, 74

FIGURE P7.17

7.18. In *floating-point* operations, numbers are represented in a computer in the binary equivalent of the familiar "scientific notation" of a *mantissa* (or coefficient) multiplied by the radix raised to a power denoted by the *exponent*, e.g., 0.314×10^6, where 0.314 is the mantissa and 6 the exponent. If two floating-point numbers are to be added or subtracted, their exponents must be equal. Thus, if we wish to add 0.6124×10^8 to 0.4042×10^7, we first shift the mantissa of the number with the smaller exponent to the right to equalize the exponent. Here we form $0.6124 \times 10^8 + 0.0404 \times$

$10^8 = 0.6528 \times 10^8$. Thus, the first step in adding two floating-point numbers is to inspect the exponents to determine which number should be shifted. The number of digits in the mantissa is fixed by the size of the computer registers, so that the least significant digits of the mantissa shifted will be lost. In the example just given, the 2 in the smaller number

C	A_s	D_s	E_s	G	$A > D$	$A \gg D$	$D > A$	$D \gg A$
0	0	0	0	0	1	0	0	0
0	0	0	0	1	0	1	0	0
0	0	0	1	0	0	0	1	0
0	0	0	1	1	0	0	0	1
0	0	1	0	0	1	0	0	0
0	0	1	0	1	0	1	0	0
0	0	1	1	0	1	0	0	0
0	0	1	1	1	0	1	0	0
0	1	0	0	0	0	0	0	1
0	1	0	0	1	0	0	0	1
0	1	0	1	0	0	0	0	1
0	1	0	1	1	0	0	0	1
0	1	1	0	0	1	0	0	0
0	1	1	0	1	0	1	0	0
0	1	1	1	0	0	0	1	0
0	1	1	1	1	0	0	0	1
1	0	0	0	0	1	0	0	0
1	0	0	0	1	0	1	0	0
1	0	0	1	0	0	0	1	0
1	0	0	1	1	0	0	0	1
1	0	1	0	0	0	1	0	0
1	0	1	0	1	0	1	0	0
1	0	1	1	0	0	1	0	0
1	0	1	1	1	0	1	0	0
1	1	0	0	0	0	0	1	0
1	1	0	0	1	0	0	0	1
1	1	0	1	0	0	0	1	0
1	1	0	1	1	0	0	0	1
1	1	1	0	0	1	0	0	0
1	1	1	0	1	0	1	0	0
1	1	1	1	0	0	0	1	0
1	1	1	1	1	0	0	0	1

FIGURE P7.18

was lost. If the difference between the exponents is greater than the number of digits in the mantissa, all the digits of the smaller number will be lost and the sum will be equal to the larger number. For example, to four places of accuracy, $0.6124 \times 10^8 + 0.1048 \times 10^3 = 0.6124 \times 10^8$. Thus, if we can detect this condition, we can skip the actual addition of the mantissa and simply transfer the larger operand to the result register, with a resultant saving in time.

The comparison is carried out by subtracting one exponent from the other and inspecting the difference for magnitude and sign. The magnitude is determined by a decoder that puts out a signal, G, if the magnitude of the difference is greater than N, the number of digits in the mantissa. Let the exponents of the operands be A and D, let $A > D$ denote that A is larger than D by an amount no greater than N, and $A \gg D$ denote that A is larger than D by an amount greater than N. $D > A$ and $D \gg A$ will have similar meaning for D greater than A. Let A_s be the sign bit of A, D_s the sign bit of D, and E_s the sign bit of their difference, and let C be the carry from the most significant bit when the subtraction is carried out. Now it can be shown [8] that the truth table in Fig. P7.18 applies for two's complement arithmetic. Develop a minimal SOP circuit to produce two outputs, AG and MG, corresponding to the following conditions:

AG: A is equal to or greater than D, by any amount.

MG: A is larger than D by an amount greater than N, or D is larger than A by an amount greater than N.

Note that $A > D$ also includes $A = D$, so AG includes $A = D$.

Bibliography

1. Miller, R. E. *Switching Theory*, Vol. I, Wiley, New York, 1965.
2. Quine, W. V. "The Problem of Simplifying Truth Functions," *Am. Math. Monthly*, **59**: 8, 521–531 (Oct. 1952).
3. McCluskey, E. J. "Minimization of Boolean Functions," *Bell System Tech. J.*, **35**: 5, 1417–1444 (Nov. 1956).
4. ———. *Introduction to the Theory of Switching Circuits*, McGraw-Hill, New York, 1965.
5. Petrick, S. R. "On the Minimization of Boolean Functions," *Proc. Symp. on Switching Theory*, ICIP, Paris, France, June 1959.
6. Bartee, T. C. "Automatic Design of Logical Networks," *Proc. Western Joint Computer Conf.*, 1959, pp. 103–107.
7. Rhyne, V. T. *Fundamentals of Digital Systems Design*, Prentice-Hall, Englewood Cliffs, N.J., 1973.
8. Flores, I. *The Logic of Computer Arithmetic*, Prentice-Hall, Englewood Cliffs, N.J., 1963.

8

SPECIAL REALIZATIONS AND CODES

8.1 Introduction

The material in this chapter will serve as a bridge between the combinational and sequential sections of the book. So far we have approached combinational logic design on the assumption that two-level AND-OR-NOT forms will always provide the best realization. In Sections 8.2 and 8.7, we consider some important practical types of circuits for which this is often not true. All of the theory and methods so far considered were developed at a time when the only logic elements available were individual gates constructed of discrete components. The development of integrated circuits has made entirely new types of realizations practical and has radically altered the economic constraints on design. Some of the problems and implications of these developments will be treated briefly in Sections 8.4 through 8.6 and extensively in Chapters 11 and 12.

The material in the remaining sections is an introduction to certain techniques which are best realized in sequential circuits. It will thus, we hope, provide motivation for the study of sequential circuits and will also be a source of examples in succeeding chapters.

8.2 The Binary Adder

Chapters 6 and 7 dealt primarily with two-level realizations of Boolean functions. Such realizations will always be the fastest possible circuits and, for typical functions of a few variables, will have the minimal cost or a cost very close to minimal. There are, however, some very important functions that do not lend themselves to two-level realizations.

Let us consider the design of a circuit to accomplish the addition of two binary numbers. The addition of two n-bit binary numbers is depicted in Fig. 8.1. A two-level realization of this process would require a circuit with $n + 1$ outputs and $2n$ inputs. In a typical computer, n might be 32. Thus, an attempt at two-level design would seem a staggering problem, even for a computer implementation of the tabular method of minimization.

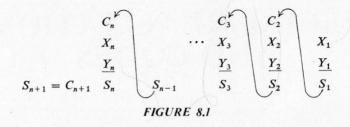

FIGURE 8.1

Fortunately, a more natural approach to the problem is possible. Each pair of input digits X_i and Y_i may be treated alike. In each case, the carry, C_i, from the previous position is added to X_i and Y_i to form the sum bit, S_i, and the carry, C_{i+1}, to the next higher-order position. If exactly one or three of the bits X_i, Y_i, and C_i is 1, then $S_i = 1$. If not, then $S_i = 0$. The carry, C_{i+1}, will be 1 if two or more of X_i, Y_i, and C_i are 1. The carry, C_i, into the first digit position, generally 0,* is available as an input at the same time X and Y become available. The addition process may thus be implemented one digit at a time, starting with the least significant digit. As soon as X, Y, and C_1 are available, C_2 may be generated, in turn making possible the generating of C_3, and so on.

A circuit that accepts one bit of each operand and an input carry and produces a sum bit and an output carry is known as a *full adder*. From the above discussion, we can set down the equations and K-maps for the full adder as shown in Fig. 8.2. Two-level realizations of these two functions require a total of nine gates with 25 inputs plus one inverter to generate $\bar{C}_i$. A NAND circuit for this form is shown in Fig. 8.3.

Since the full adder is of basic importance in digital computers, a great deal of effort has gone into the problem of producing the most economical realization. The form leading to best economy is a function of the technology used. Today

---

* In some cases involving complement arithmetic, the input carry is a 1. See, for example, Flores [1], p. 40.

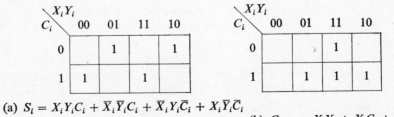

(a) $S_i = X_i Y_i C_i + \bar{X}_i \bar{Y}_i C_i + \bar{X}_i Y_i \bar{C}_i + X_i \bar{Y}_i \bar{C}_i$

(b) $C_{i+1} = X_i Y_i + X_i C_i + Y_i C_i$

FIGURE 8.2   *Sum and carry functions for the binary full adder.*

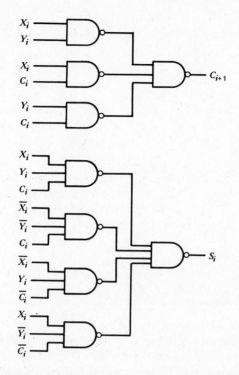

FIGURE 8.3 *Basic two-level realization of full adder.*

complete full adders are normally realized as integrated circuits, so the forms of interest are those most suitable for realization in MSI form. One interesting form can be seen on the K-maps as shown in Fig. 8.4. We generate C_{i+1} as before and use it in the generation of S_i. Note that the map of $X_i + Y_i + C_i$ in Fig. 8.4a agrees with S_i in five of the eight squares. In Fig. 8.4b, we see the intersection of this map and a map of $\bar{C}_{i+1}$ (complement of Fig. 8.2b). It is only necessary to include one more term, $X_i Y_i C_i$, to form the sum

$$S_i = \bar{C}_{i+1}(X_i + Y_i + C_i) + X_i Y_i C_i \qquad (8.1a)$$

$$= X_i \bar{C}_{i+1} + Y_i \bar{C}_{i+1} + C_i \bar{C}_{i+1} + X_i Y_i C_i \qquad (8.1b)$$

as indicated in Fig. 8.4c. A realization of Equation 8.1a is shown in Fig. 8.5a, requiring 8 gates with 19 inputs. An alternate form, realizing Equation 8.1b, requires 10 gates and 22 inputs (Fig. 8.5b). Although it appears to be more expensive, this form is better suited to integrated circuit realization and is used in the SN7480 and SN7482 full adder chips.

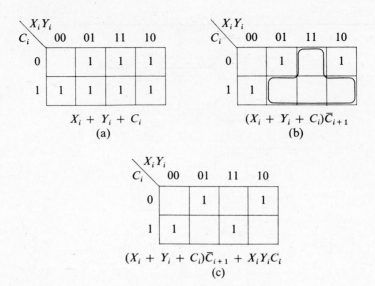

FIGURE 8.4   *K-maps for alternate realization of sum bit in full adder.*

Another popular form of full adder utilizes two *half adders*. A half adder is a circuit for adding two binary bits. The K-maps and circuit for a half-adder are shown in Figs. 8.6a and 8.6b. To construct a full adder, we note that the equations for a full adder (Fig. 8.2) can be written in the alternate forms

$$S_i = X_i \oplus Y_i \oplus C_i \tag{8.2a}$$

$$C_{i+1} = (X_i \oplus Y_i)C_i + X_i Y_i \tag{8.2b}$$

These equations can be implemented with two half adders plus an OR gate, as shown in Fig. 8.6c. Note one advantage of this circuit, that it does not require any inverted inputs.

In the preceding designs, we see examples of the usual compromise between speed and cost. The more economical circuits have more levels of gating and therefore more delay from input to output. However, before concluding that the two-level realization should be chosen if speed is the dominant consideration, the reader should note that carry delay is the limiting factor in a cascade of n full adders forming an n-bit parallel adder (Fig. 8.7). For some combinations of input bits, the presence or absence of the first carry, C_1, will determine the final two bits, S_n and C_{n+1}. In such cases, C_1 must propagate through at least two levels of logic in each adder before C_{n+1} and S_n can assume their final values. The adders of Figs. 8.5a and 8.5b have four and five levels, respectively, of logic

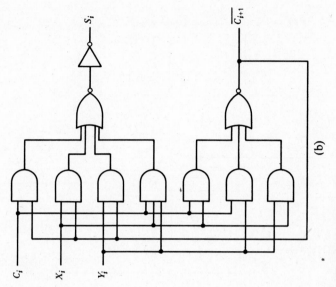

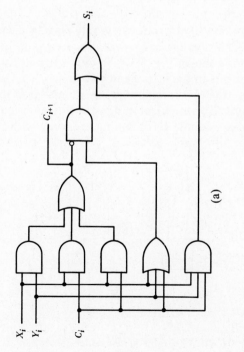

(a)

(b)

FIGURE 8.5 Circuits for alternate realizations of Fig. 8.4.

179

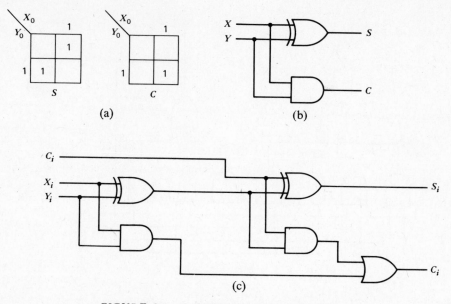

(a) (b)

(c)

FIGURE 8.6 *Full adder based on half adders.*

from input to sum but only two levels from carry-out. Thus, the carry propagation delay in a parallel adder using either of these designs would be the same as for one using the two-level design of Fig. 8.3. The only extra delay would be in the development of the sum bits in the last one or two stages. This would seldom be important, and if it were, the two-level design could be used in the last one or two stages and one of the more economical designs in the other stages.

This case provides a good example of obtaining economy without sacrificing speed by taking advantage of characteristics of the entire system. There are,

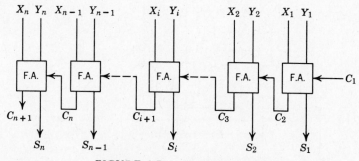

FIGURE 8.7 *n-bit binary adder.*

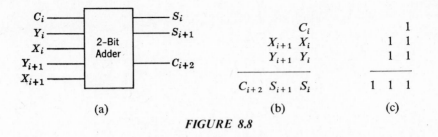

FIGURE 8.8

unfortunately, no simple rules or completely specified procedures for finding such designs. Even with computer mechanization, completely determinate design procedures, such as the Quine-McCluskey method, will always be limited to relatively simple devices or subsystems. The efficient integration of such subsystems into complete large-scale systems is dependent on the ingenuity and resourcefulness of the designer.

As a means of increasing speed, we note that it is not necessary to consider only one digit at a time. Consider the possibility of generating two sum-bits in a step, as illustrated in Fig. 8.8. The intermediate carry, C_{i+1}, is not actually generated but is implicit in the logic of the adder. The process is similar to the addition of three 2-bit binary numbers except that one of those numbers may not exceed 1. Thus, the sum will not exceed 7_{10}, as indicated in Fig. 8.8c. The function S_i may be generated as before, and C_{i+2} will be 1 whenever the sum is 4_{10} or more. Therefore,

$$C_{i+2} = X_{i+1} Y_{i+1} + (X_{i+1} + Y_{i+1})(C_i X_i + C_i Y_i + X_i Y_i) \qquad (8.3)$$

The bit S_{i+1} will be 1 whenever the sum is 2, 3, 6, or 7_{10}, leading to the map in Fig. 8.9. From this map, we determine the minimal second-order form for S_{i+1}, Equation 8.4. Thus, thirteen gates are required for a two-level realization of S_{i+1} compared to five gates for the two-level realization of S_i.

$$
\begin{aligned}
S_{i+1} =\ & \bar{X}_{i+1} Y_{i+1} \bar{X}_i \bar{C}_i + X_{i+1} \bar{Y}_{i+1} \bar{X}_i \bar{C}_i + \bar{X}_{i+1} \bar{Y}_{i+1} X_i Y_i \\
& + X_{i+1} Y_{i+1} X_i Y_i + \bar{X}_{i+1} Y_{i+1} \bar{Y}_i \bar{C}_i + X_{i+1} \bar{Y}_{i+1} \bar{Y}_i \bar{C}_i \\
& + \bar{X}_{i+1} Y_{i+1} \bar{X}_i \bar{Y}_i + X_{i+1} \bar{Y}_{i+1} \bar{X}_i \bar{Y}_i + \bar{X}_{i+1} \bar{Y}_{i+1} X_i C_i \\
& + X_{i+1} Y_{i+1} X_i C_i + \bar{X}_{i+1} \bar{Y}_{i+1} Y_i C_i + X_{i+1} Y_{i+1} Y_i C_i \qquad (8.4)
\end{aligned}
$$

For groups of three bits, a similar increase in functional complexity could be expected, and so on. Thus, as conjectured earlier, the cost of a two-level realization of the entire n-bit addition would be prohibitive. However, it should be noted that the expression for the carry (Equation 8.3) is much simpler than that for the sum (Equation 8.4), and the carry propagation is the primary source of

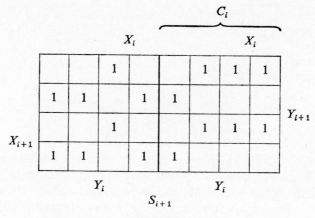

FIGURE 8.9

delay in the cascade of full adders of Fig. 8.7. It is quite practical to realize the carry over four or five stages in a two-level form similar to Equation 8.3 and thus speed the carry propagation significantly. For example, this form of carry propagation is used in the SN74283 4-bit adder to provide a carry time of 10 ηsec over 4-bits, compared the same delay for a single bit in the SN 7480 full adder. Adders using this form of carry circuit are known as *carry lookahead adders* [1, 6].

8.3 Coding of Numbers

As discussed briefly in Section 2.5, in some computers, decimal numbers are not converted directly into the corresponding binary numbers but are instead converted on a digit-by-digit basis, with each decimal digit being represented by some binary code. The most commonly-used code is the BCD code, in which each digit is represented by the corresponding 4-bit binary number. This code is repeated in Fig. 8.10.

Addition in a binary-coded decimal machine is usually accomplished in a digit-by-digit fashion. The basic configuration of a binary-coded decimal adder is shown in Fig. 8.11. The two decimal digits to be added are represented by $X_{i3} X_{i2} X_{i1} X_{i0}$ and $Y_{i3} Y_{i2} Y_{i1} Y_{i0}$. The carry from the previous stage, C_i, may take on only the values 0 or 1 as only two decimal digits are being added. The adder must be a 5-output function of nine variables in order to generate the binary representation of the sum $S_{i3} S_{i2} S_{i1} S_{i0}$, and the carry to the next stage C_{i+1}.

The discussion of the previous section would indicate that a two-level realiza-

Decimal Digit	Binary Representation $X_3\ X_2\ X_1\ X_0$
0	0000
1	0001
2	0010
3	0011
4	0100
5	0101
6	0110
7	0111
8	1000
9	1001

FIGURE 8.10 *Binary-coded decimal digits.*

tion of the circuit in Fig. 8.11 would not be simple. A computer implementation of tabular minimization would seem to be a reasonable approach.

It turns out that the sum bits differ only slightly from the output bits of a four-stage binary adder. In the usual realization, the 4-bit binary addition is accomplished first, and the sum and carry bits are then adjusted to their proper values in an additional two levels of logic.

The code of Fig. 8.10 is not the only scheme used for representing decimal digits as sets of binary bits. In Fig. 8.12 is a listing of three other possible codes. The excess-3 code is particularly useful where it is desired to perform arithmetic by the method of complements. The 9s complement of a decimal digit a is defined as 9-a. See Ware [4] for a discussion of 9s complement arithmetic. The 9s complement of a decimal digit expressed in excess-3 code may be obtained by complementing each bit individually. This fact is easily verified by the reader.

Not all codes of decimal digits use only four bits. The 2-out-of-5 code represents each decimal digit by one of the ten possible combinations of two 1s and

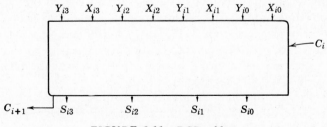

FIGURE 8.11 *BCD adder.*

Decimal Digit	Excess-3 $X_3\ X_2\ X_1\ X_0$				2-out-of-5 $X_4\ X_3\ X_2\ X_1\ X_0$					Gray Code $X_3\ X_2\ X_1\ X_0$			
0	0	0	1	1	0	0	0	1	1	0	0	1	0
1	0	1	0	0	0	0	1	0	1	0	1	1	0
2	0	1	0	1	0	0	1	1	0	0	1	1	1
3	0	1	1	0	0	1	0	0	1	0	1	0	1
4	0	1	1	1	0	1	0	1	0	0	1	0	0
5	1	0	0	0	0	1	1	0	0	1	1	0	0
6	1	0	0	1	1	0	0	0	1	1	1	0	1
7	1	0	1	0	1	0	0	1	0	1	1	1	1
8	1	0	1	1	1	0	1	0	0	1	1	1	0
9	1	1	0	0	1	1	0	0	0	1	0	1	0

FIGURE 8.12

three 0s. The third code in Fig. 8.12 utilizes the center ten characters of a 4-bit Gray code. The distinguishing feature of the Gray code is that successive coded characters never differ in more than one bit. The basic Gray code configuration is shown in Fig. 8.13a. A 3-bit Gray code may be obtained by merely reflecting the 2-bit code about an axis at the end of the code and assigning a third bit as 0 above the axis and as 1 below the axis. This is illustrated in Fig. 8.13b. By reflecting the 3-bit code, a 4-bit code may be obtained as in Fig. 8.13c, etc.

We have already employed the Gray code in the designation of rows and columns on the Karnaugh map. This code is particularly useful in minimizing errors in analog-digital conversion systems. Consider a shaft-position encoder, such as discussed in Chapter 1, that puts out a digital signal to indicate which of ten segments the shaft is in. When the shaft moves from segment seven to segment eight, for example, the code must change from that for seven to that for eight. As the shaft moves across the segment boundary, if more than one bit has to change, it is possible, due to slight mechanical inaccuracies, that not all will change at exactly the same time. If the BCD code were used, the code would have to change from 0111 to 1000, a change of all 4 bits. If the most significant digit were to change from 0 to 1 before any of the other bits changed, we would then momentarily have the code 1111, (15_{10}), and thus a very large error. With the Gray code, since only one bit is to change at a time, this sort of error cannot occur. This code may also bring about a reduction in the amount of logic required in certain types of registers and counters.

Codes are also encountered in connection with the control of output devices, or indicators. In recent years, light-emitting diodes (LEDs) and liquid crystals have become increasingly popular for use with logic circuits. Their main advantage

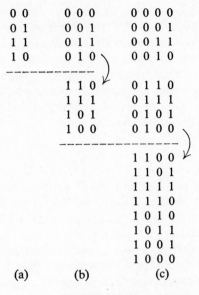

```
0 0        0 0 0        0 0 0 0
0 1        0 0 1        0 0 0 1
1 1        0 1 1        0 0 1 1
1 0        0 1 0        0 0 1 0
         -----------
           1 1 0        0 1 1 0
           1 1 1        0 1 1 1
           1 0 1        0 1 0 1
           1 0 0        0 1 0 0
                       ----------------
                        1 1 0 0
                        1 1 0 1
                        1 1 1 1
                        1 1 1 0
                        1 0 1 0
                        1 0 1 1
                        1 0 0 1
                        1 0 0 0
   (a)        (b)          (c)
```

FIGURE 8.13

is that they are directly compatible with logic circuits in contrast to incandescent or neon indicators, which usually require special drivers.

For display of decimal digits, the standard form is the seven-segment display, shown in Fig. 8.14a. Each of the seven segments is a separate LED or crystal that can be turned on or off individually. It should be apparent to the reader that each decimal digit can be formed by lighting some subset of the seven segments. To control this display, we must generate a 7-bit code to indicate

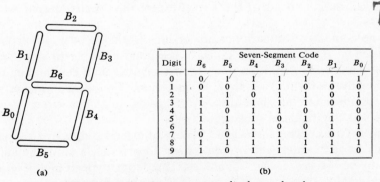

| Digit | Seven-Segment Code | | | | | | |
	B_6	B_5	B_4	B_3	B_2	B_1	B_0
0	0	1	1	1	1	1	1
1	0	0	1	1	0	0	0
2	1	1	0	1	1	0	1
3	1	1	1	1	1	0	0
4	1	0	1	1	0	1	0
5	1	1	1	0	1	1	0
6	1	1	1	0	0	1	1
7	0	0	1	1	1	0	0
8	1	1	1	1	1	1	1
9	1	0	1	1	1	1	0

(a)   (b)

FIGURE 8.14   Seven-segment display and code.

whether each segment should be on or off. If we let 0 correspond to OFF and 1 to ON, the *seven-segment code* for the decimal digits will be as shown in Fig. 8.14b. Since the seven-segment code is highly redundant, it would be unsuitable for internal use in a computer. Thus, the use of this form of display will require logic to convert from some internal code, such as BCD, to the seven-segment code. Before considering the design of a circuit for this particular application, let us consider the design of another, more general, type of circuit associated with coding and decoding problems.

8.4   The Decoder

One of the simplest and most useful types of multiple-output circuit encountered in digital systems is the *decoder*. Assume we have a small memory array of eight locations. The address of the location to be accessed is a 3-bit binary number. To access a location we need to generate a signal of one of eight lines, corresponding to the addressed location. A 3-to-8 line decoder consists simply of 8 AND gates realizing the 8 minterms of three variables (Fig. 8.15a). For each possible combination of the inputs, there will be a 1 on just one of the output lines, and we assume that the 1 on the output line will make it possible to access the desired word. Circuits of this type are known as *n-to-2^n line decoders*, or, more commonly, simply as *decoders*.

Although decoders can be constructed of discrete gates, they are generally realized in the form of MSI circuits. There are a number of different types of decoders available in MSI form, including 3-to-8 line and 4-to-16 line. There are also decoders in which the number of output lines is not 2^n, such as the SN7442A *BCD*-to-decimal decoder, shown in Fig. 8.15b. Again, the circuit simply realizes minterms, but now only ten of them, corresponding to the *BCD* codes for the decimal digits. As is usually the case in MSI decoders, NAND gates are used instead of AND gates, so that the polarity of the outputs is reversed, all outputs being at the 1-level except the selected output, which is at the 0-level.

Also note another peculiarity characteristic of MSI decoders—that the inputs are labeled A, B, C, D, but are taken in the opposite order to that used in this book, A being least significant and D most significant in determining the *BCD* code. Finally, note the two levels of inversion. The extra inversion provides power amplification to drive the NAND gates, so that a circuit driving the decoder sees the inputs as single-gate loads.

In theory, the networks of Fig. 8.15 could be extended directly to any number, n, inputs using 2^n gates with n inputs each. In practice, a fan-in limitation will eventually be reached in any technology. Thus, some type of multilevel network must be employed in large decoders. It is impossible to make an unqualified

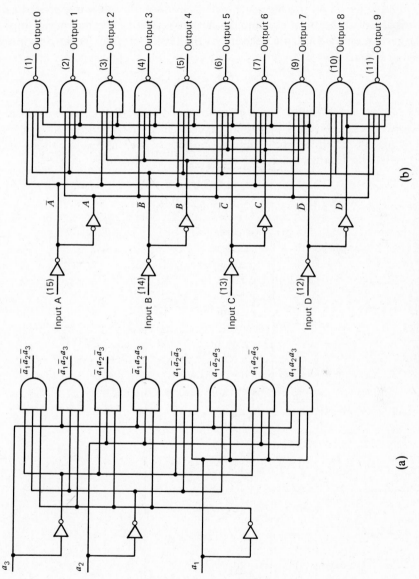

FIGURE 8.15 *Typical decoders*

187

statement as to the most economical decoder design, because cost factors vary with different technologies. If we assume that decoders of the form of Fig. 8.15a are available as single-package integrated circuits, the following procedure will generally lead to the most economical design. Given m, the number of inputs of the largest integrated circuit decoder available, determine the largest integer r satisfying Equation 8.5.

$$r \leq \frac{n}{k} \leq m \qquad k = \text{a power of two} \tag{8.5}$$

If $r = n/k$, we use k r-input decoders to form the minterms of disjoint subsets of r variables. These minterms will then be combined in as many levels of 2-input AND gates as required. For example, suppose $n = 12$ and $m = 5$; then the largest r satisfying Equation 8.5 is given by

$$r = 3 = \tfrac{12}{4} \leq 5 \tag{8.6}$$

So we use four 3-input decoders, as shown in Fig. 8.16.

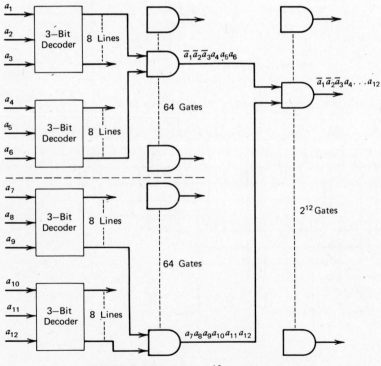

FIGURE 8.16   *12-to-2^{12} line decoders.*

To keep the diagram readable, only a few connections are actually shown. The 3-bit decoders may be considered to be copies of the circuit of Fig. 8.15a in integrated circuit form. There are 64 pairs of output lines, one from each of the upper two 3-bit decoders. These pairs form the inputs to the upper 64 second-level AND gates. The outputs of these gates are the 64 possible minterms of the variables a_1, a_2, a_3, a_4, a_5, a_6. The lower 64 second-level gate outputs are the minterms of a_7, a_8, a_9, a_{10}, a_{11}, and a_{12}. The 2^{12} 12-bit minterms are formed by using all possible pairs of outputs of the second-level gates (one from the upper 64 and one from the lower 64) to form inputs to the final 2^{12} AND gates.

Assuming the fan-in level permits, it might seem that it would be better to combine the outputs of all four decoders in 4-input AND gates, thus eliminating the second level of gates. Let us assume the cost is proportional to the number of gate inputs. The number of gate inputs, exclusive of the 3-bit decoders, is given by Equation 8.7. Clearly the dominant cost is the output gates. No matter what the form of the decoder, the cost will always be dominated by the 2^n output gates, so these should have the minimum number of inputs—two.

$$\text{2nd level} \qquad \text{output gates}$$

$$\text{number of inputs} = \overbrace{2 \cdot 2 \cdot 64}^{\vdots} + \overbrace{2 \cdot 2^{12}}^{\vdots}$$
$$= 2^8 + 2^{13} \approx 2^{13} \tag{8.7}$$

If the 2^n output gates each have two inputs, there will have to be $2 \times 2^{n/2}$ gates at the next level of gating; and by the same reasoning, these should have only two inputs. If we continue this same general reasoning, it would appear that the r-bit decoders should be multilevel, using only 2-input gates. This would be true if the decoder were constructed of discrete gates. However, input count is a significant measure of cost primarily because of the cost of wiring to the inputs. Thus, if the decoder of Fig. 8.15 is implemented as an integrated circuit, with all gate connections internal, the effective input count is only three. This is a first example of the manner in which the availability of integrated circuits may influence cost considerations.

If r in Equation 8.5 is not an integer, i.e., if n does not divide evenly by a power of two, there are several possible circuits, the choice depending on the relative cost of the gates themselves compared to the costs of interconnections. In most cases, the best design results from using as many r-bit decoders as necessary, followed by as many levels of 2-input AND gates as required (see Problem 8.7).

While fan-in problems limit the size of decoders of the form of Fig. 8.15, fan-out limits can also be a problem in very large decoders. In Fig. 8.16, for example,

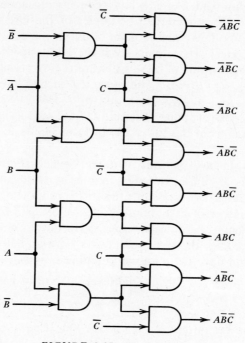

FIGURE 8.17   Tree decoder.

each of the 128 second-level gates provides an input to 64 of the output gates. Fan-out problems can be relieved by the use of a tree decoder, in which new input variables are introduced at each level of the tree. Figure 8.17 shows a tree decoder for three variables. There are still fan-out problems in that the variables introduced at the later levels must drive many gates. However, it is easier to provide heavy drive capability for a few input variables than for $2 \times 2^{n/2}$ gates at the second level in a circuit of the form of Fig. 8.16. Tree decoders, having more levels than the form of Fig. 8.16, are slower and contain almost twice as many gates, but for very large decoders they may be the only choice. Obviously, various combinations of the two basic forms could be devised.

8.5   Code Conversion and Read-Only Memories

At the close of Section 8.3, we pointed out the need for circuits to convert from one code to another. The decoder may be regarded as a special form of code converter, converting from some n-bit code to a one-out-of-2^n code. Further, the decoder can be used as the basis for more general code converters.

Example 8.1

Consider the problem of converting from a *BCD* code to a seven-segment code, as discussed in Section 8.3. The corresponding codes are tabulated in Fig. 8.18. We require a circuit simultaneously producing seven functions of four inputs. We could apply the formal multiple-output techniques of Chapter 7; but with seven functions to be realized, it is obvious that this would be rather tedious. (There turn out to be 127 maps of functions and function products required.)

A more practical design approach is found by noting that a full decoder produces all minterms of the input variables and that any function can be obtained by an ORing of minterms. In this case, since we are using *BCD* code, the decimal digits in Fig. 8.18 are the minterm numbers. So we can construct our code converter from a partial decoder, producing the ten required minterms as shown in Fig. 8.19. To complete the design, we note that the table of Fig. 8.18 effectively lists the minterms making up each output function. For example,

$$B_6 = \sum m(2, 3, 4, 5, 6, 8, 9) \tag{8.8}$$

and

$$B_0 = \sum m(0, 2, 6, 8) \tag{8.9}$$

so we OR these minterms together as shown. The OR gates developing the other output functions would be similarly connected, although we have not shown the connections in the interest of clarity. ∎

It can be seen that we have here the basis of a completely general design technique, not just for code converters but for logic circuits of any type. The resultant designs will seldom be even close to minimal in the classic sense of Chapters 6 and 7, but the simplicity of the design technique makes it quite

Dec.	BCD Input				Seven-Segment Output						
Digit	A_3	A_2	A_1	A_0	B_6	B_5	B_4	B_3	B_2	B_1	B_0
0	0	0	0	0	0	1	1	1	1	1	1
1	0	0	0	1	0	0	1	1	0	0	0
2	0	0	1	0	1	1	0	1	1	0	1
3	0	0	1	1	1	1	1	1	1	0	0
4	0	1	0	0	1	0	1	1	0	1	0
5	0	1	0	1	1	1	1	0	1	1	0
6	0	1	1	0	1	1	1	0	0	1	1
7	0	1	1	1	0	0	1	1	1	0	0
8	1	0	0	0	1	1	1	1	1	1	1
9	1	0	0	1	1	0	1	1	1	1	0

FIGURE 8.18   BCD to seven-segment code.

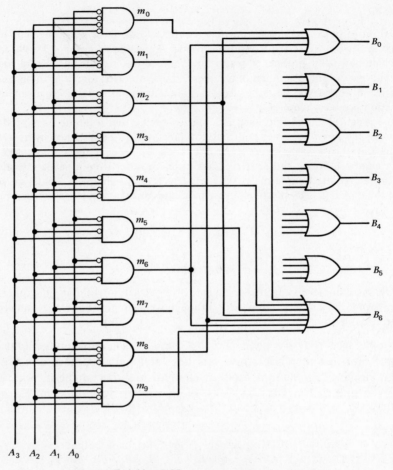

FIGURE 8.19   *BCD to seven-segment converter.*

attractive. The larger the number of output functions, the greater the savings in design effort will be and the closer to minimal the circuits will be. And as the integrated circuit decoders become ever less expensive, this approach becomes economically practical for an ever wider class of circuits.

We can carry this approach further by noting that the circuit form of Fig. 8.19 could be generalized to a collection of 4-input AND gates connected to a set of output OR gates through an interconnection matrix. If the interconnection matrix can be easily altered, we have the basis for a sort of "general-purpose" integrated circuit that can be custom-tailored to individual applications. Such a circuit is commonly known as a *read-only memory* (ROM).

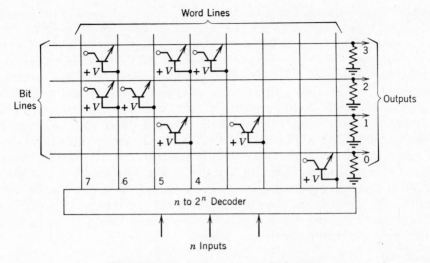

FIGURE 8.20 *Transistor-coupled ROM.*

There are many types of ROMs, but the type best suited to applications requiring logic level outputs as a function of level inputs is the transistor-coupled, or semiconductor, ROM. The basic structure of a transistor-coupled ROM is shown in Fig. 8.20. The lines making up the interconnection matrix are known as *word lines* and *bit lines*. The word lines are driven by an n-to-2^n decoder, and the outputs are taken from the bit lines. Wherever a connection is to be made between a word line and bit line, a transistor is connected, emitter to the bit line and base to the word line. All collectors are tied to a common supply voltage.

In the absence of any input, all word lines are held at a sufficiently negative level to cut off the transistors so that the bit lines are at 0 volts. When an input appears, the corresponding word line is raised to a sufficiently positive level to turn on the transistors, thus raising the connected bit lines to a positive level. For example, in Fig. 8.20, if the input were 100 (m_4), word line 4 would be turned on and the output would be $b_3 b_2 b_1 b_0 = 1010$. Comparing this to Fig. 8.19, we see that the decoder is the array of AND gates, and the bit lines with their associated transistors make up the array of OR gates. (Please note that we have described here only the basic principles involved; the precise details of the technology vary widely.)

ROMs come in a variety of sizes and in two basic types, *mask-programmable* and *field-programmable*. For the mask-programmable type, the customer indicates, on a standard form provided by the manufacturer, the desired connection matrix. The manufacturer uses this data to create a mask used in the metallization phase of fabrication to connect transistors only where required. In the field-programmable type the ROM is provided to the customer with connections at every intersection and the customer programs the ROM, using current pulses literally to "blow out" the undesired connections. Because of this process, the connections are generally referred to as *links* or *fuses*. A special type of field-programmable ROM is the erasable ROM, or *EPROM*. The choice between the types is mostly a matter of cost. In terms of direct chip cost, the field-programmable ROM will be cheaper than the mask-programmable ROM since the manufacturer has to provide only one standard type, but the user must add the cost of programming each unit. For the mask-programmable type, the initial cost of the mask is significant, but this is a one-time cost that will be cheaper than the cost of individual programming if spread over enough units. EPROMS are expensive and generally used only in development labs where they are convenient for experimenting with new designs.

Example 8.2

The *BCD*-to-seven-segment code converter is to be realized using a ROM. The closest standard size is 16 × 8. Specify the connection matrix.

Solution   Figure 8.21 shows a typical form of the type provided by ROM manufacturers for the customer to specify the desired connection matrix. All that is necessary is to fill in the truth tables of the desired functions in the output section. In this case, we simply enter functions B_0 to B_6 from Fig. 8.18 as outputs B_0 to B_6, as shown in Fig. 8.21. Since output B_7 and words 10–15 are not used in this case, ×'s (for don't-care) are entered in these squares. (The reader should note that *BCD*-to-seven-segment conversion is so common that standard circuits are available, but this basic approach can be used for any arbitrary codes.) ■

In view of the application of the ROM discussed above as a realization of custom logic in a standard package, the very name, *read-only memory*, may seem inappropriate. However, if we regard the signals appearing on the m output lines not as distinct functions of the n input variables but rather as the bits making up an m-bit binary word, then the *memory* interpretation becomes more obvious. The patterns of 0s and 1s stored on each word line by connections to the bit lines now represent individual units, or *words*, of binary information; and the input bits represent the *address* of the word to be read out on the bit lines. The term *read-only* represents the fact that, once the pattern of connections has been established, either it cannot be altered at all, or else it can be altered only by procedures that are complex, expensive, and time-consuming. This is in contrast to a *read-write* memory, such as a core memory, in which the information stored can be altered by electronic processes basically the same as those used to read out information.

CUSTOMER: _____   THIS PORTION TO BE COMPLETED BY XY MFG.

P.O. NO.: _____   PART NO.: _____

YOUR PART NO.: _____   S.D. NO.: _____

DATE: _____   DATE RECEIVED: _____

WORD	INPUTS				OUTPUTS							
	A_3	A_2	A_1	A_0	B_7	B_6	B_5	B_4	B_3	B_2	B_1	B_0
0	0	0	0	0	X	O	I	I	I	I	I	I
1	0	0	0	1	X	O	O	I	I	O	O	O
2	0	0	1	0	X	I	I	O	I	I	O	I
3	0	0	1	1	X	I	I	I	I	I	O	O
4	0	1	0	0	X	I	O	I	I	O	I	O
5	0	1	0	1	X	I	I	I	O	I	I	O
6	0	1	1	0	X	I	I	I	O	O	I	I
7	0	1	1	1	X	O	O	I	I	I	O	O
8	1	0	0	0	X	I	I	I	I	I	I	I
9	1	0	0	1	X	I	O	I	I	I	I	O
10	1	0	1	0	X	X	X	X	X	X	X	X
11	1	0	1	1	X	X	X	X	X	X	X	X
12	1	1	0	0	X	X	X	X	X	X	X	X
13	1	1	0	1	X	X	X	X	X	X	X	X
14	1	1	1	0	X	X	X	X	X	X	X	X
15	1	1	1	1	X	X	X	X	X	X	X	X

FIGURE 8.21 *ROM truth table for BCD-to-seven-segment converter.*

Used as a means of information storage, ROMs find wide application in the control units of computers, in which application they store coded instructions representing the sequences of internal operations which the computer must carry out in executing program instructions. The execution of a single instruction, such as ADD, will involve a whole sequence of register transfers, memory accesses, and logical operations. When the appropriate sequences are stored in a ROM, the computer is often said to be *microprogrammed*. ROMs used for this application are generally much larger than those used for implementing logic, typically containing several thousand words of 16–64 bits in length. ROMs also are widely applied for storing system software and other types of semipermanent programs. For example, many microcomputer systems come equipped with monitors and assemblers in ROM. This approach provides great advantages in reliability over the more traditional approach of loading software into regular read-write memory, since ROMs are immune to accidental loss of programs through power failure or user errors.

8.6   Multiplexers and PLAs

Another very important MSI circuit is the multiplexer, or selector. The basic function of this circuit is to select one of several inputs to connect to a single output line. The opposite function, demultiplexing, connects one input line to one of several output lines. In their most basic form, multiplexers and demulti-plexers simply function as multiposition switches, as shown in Fig. 8.22. Multi-plexing is used most commonly in communications systems in which you have a group of data lines in one location and another group of lines at a remote location, and you wish to connect any line in one group to any line in the other group via a single communication link between the two locations.

In Fig. 8.22, the data could be either analog or digital and the control of switches is not shown explicitly. If the signals are restricted to digital data, the multiplexing and demultiplexing functions can be realized by logic networks in which gates are used to realize both the switching of the data and the control of the switching. Digital multiplexers are available as MSI circuits in a variety of forms. Figure 8.23 shows the logic diagram of one type, the SN74153 dual 4-input multiplexer. The upper and lower portions function in the same way, so let us consider just the upper portion. Note that there are four data inputs, 1C0, 1C1, 1C2, 1C3, and two address, or control inputs, A and B. The control signals are inverted internally and each of the four possible combinations of values A and B is used to select one of the four AND gates. For example, if A, $B = 0, 0$, the top AND gate is selected and 1C0 is gated through to the output. Any other combination of values on A and B will select one of the other data inputs to be gated through to the output. The STROBE (enable) input is used to turn off the output. If STROBE $= 0$, the circuit functions as described above; if STROBE $= 1$, the output is held low regardless of the values of the address and data inputs. Standard MSI multiplexers are available with 8 data lines and 3 control lines, and with 16 data lines and 4 control lines. The logic is the same as for the 4-input circuit, except that there are 8 or 16 AND gates.

From Fig. 8.22, we can see that the demultiplexing function has some resemb-lance to decoding, in the sense of having few inputs and many outputs, and the demultiplexing function can indeed be performed by slightly modified versions of

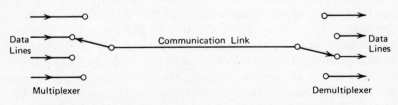

FIGURE 8.22   Multiplexing and demultiplexing with switches.

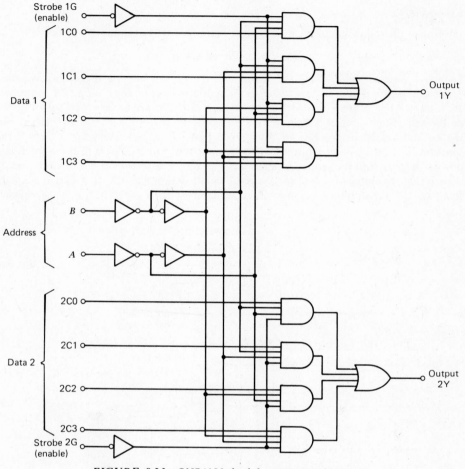

FIGURE 8.23 *SN74153 dual four-input multiplexer.*

the decoder circuits discussed earlier. Consider the basic 3-line to 8-line decoder of Fig. 8.15a. To convert this to a demultiplexer, we add a data line that drives all 8 output gates. The three inputs, a_1, a_2, a_3, select one of the gates, and the output of that gate then follows the data line, providing 1-line to 8-line demultiplexing.

Although multiplexers were originally developed to perform the function of multiplexing, they have also found widespread application as general-purpose logic devices. For this type of application, we regard the multiplexer as a general-purpose AND-OR circuit with considerable flexibility as to what is connected to the AND gates.

Example 8.3

Implement the following function, using one-half of a 74153 multiplexer.

$$f(X, Y, Z) = \sum m(3, 4, 6, 7)$$

Solution We start by drawing the Karnaugh map of the function (Fig. 8.24). We then connect X to the B address line and Y to the A address line on the multiplexer. With that connection each column of the Karnaugh map corresponds to one AND gate in the multiplexer. For example, if $B, A = X, Y = 0, 0$, the top AND gate will be selected and the output will be determined by input 1C0. For this set of values for X and Y the function is to be 0, so we connect logic-0 to input 1C0. For $X, Y = 0, 1$, the second gate down is selected and the output is determined by 1C1. For this set of values of X and Y, we see that the output is equal to Z so we connect Z to input 1C1. In a similar manner, for $XY = 10$, the output is equal to $\bar{Z}$, so we connect $\bar{Z}$ to input 1C2, and we connect

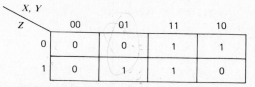
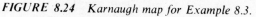

Z \\ X, Y	00	01	11	10
0	0	0	1	1
1	0	1	1	0

FIGURE 8.24 *Karnaugh map for Example 8.3.*

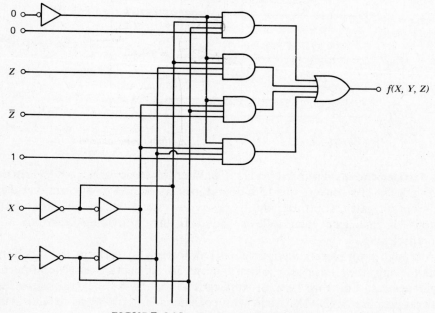

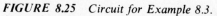

FIGURE 8.25 *Circuit for Example 8.3.*

logic-1 to input IC2 to provide the outputs required for $XY = 11$. The complete circuit is shown in Fig. 8.25, with the strobe tied low so that the output will follow the address and data inputs. ∎

It can be seen from the above example that any function of three-variables can be realized with a 4-input multiplexer. In the example we have all four possible combinations of 1 and 0 in the K-map columns, so this form can be used no matter where 1 and 0 appear on the map. In the same manner, any 4-variable function can be realized with an 8-input multiplexer, and any 5-variable function can be realized with a 16-input multiplexer. For an n-variable function realized with a multiplexer, $n - 1$ variables are connected to the address lines, and the data lines are driven by the remaining variable or its complement, or 0 or 1.

The above rules provide for the realization of *any* 3-variable function with a 4-input multiplexer, *any* 4-variable function with an 8-input multiplexer, etc., but sometimes larger functions can be realized with smaller multiplexers and a small amount of external logic.

Example 8.4

Realize the function below, using a 4-input multiplexer.

$$F(x_1, x_2, x_3, x_4) = \sum m(4, 5, 6, 7, 8, 13, 14, 15)$$

Solution The procedure described above would require an 8-input multiplexer, but let us see if we can do it with half of a SN74153. We connect x_3 to B and x_4 to A, and draw the map of the function with these two variables across the top, as shown in Fig. 8.26. Assume we are going to use the top half of the SN74153, as shown in Fig. 8.23. The first column of the map corresponds to B, $A = 0, 0$, for which the top AND gate is selected and the output is controlled by 1C0. From the map, we see that the output should be 1 in this column if x_1, $x_2 = 0$, 1, or if x_1, $x_2 = 1$, 0, so that these 1s will be realized by connecting 1C0 to $x_1 \oplus x_2$. The second column represents B, $A = 0, 1$, for which 1C1 controls the output. In this column the output is to be 1 if $x_2 = 1$, so we connect 1C1 to x_2. In a similar manner, we see that 1C2 and 1C3 should be connected to x_2. The complete circuit, utilizing one 4-input multiplexer and one exclusive-OR gate, is shown in Fig. 8.27. ∎

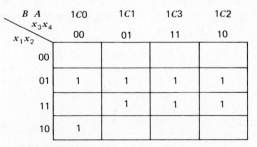

FIGURE 8.26 *Map of example function.*

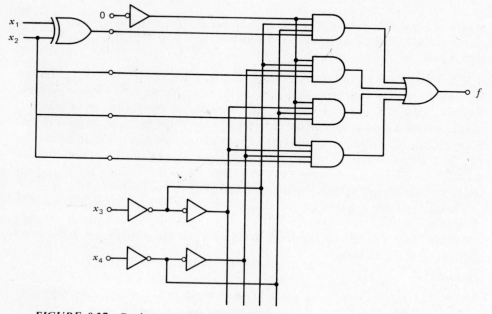

FIGURE 8.27   *Realization of four-variable function with four-input multiplexer.*

An obvious question in connection with the above design is why we chose to use x_3 and x_4 to drive the address lines. Why not x_1 and x_2, or some other combination? If we chose some other combination, would the resultant realization be more or less complex? As it turns out, the only way to find out which variables should be connected to the address lines to achieve the simplest circuit is to try all possibilities. This can be done quite conveniently through the use of *decomposition charts*, the theory of which is treated in Chapter 16. Figure 8.28 shows the decomposition chart forms for 4-variable functions. These charts are nothing more than K-maps showing the positions of the minterms for all possible combinations of variables along the tops and sides of the maps. Although the variables are rearranged on the three maps of Fig. 8.28, in every case the minterm numbers correspond to the ordering of the variables as x_1, x_2, x_3, x_4.

$x_3 x_4$	00	01	11	10
00	0	1	3	2
01	4	5	7	6
$x_1 x_2$ 11	12	13	15	14
10	8	9	11	10

$x_2 x_4$	00	01	11	10
00	0	1	5	4
01	2	3	7	6
$x_1 x_3$ 11	10	11	15	14
10	8	9	13	12

$x_2 x_3$	00	01	11	10
00	0	2	6	4
01	1	3	7	5
$x_1 x_4$ 11	9	11	15	13
10	8	10	14	12

FIGURE 8.28   *Decomposition chart forms for four-variable functions.*

$\bar{x_2}\,x_4 + \bar{x_3}\,x_4$

$x_3\,x_4$

$10 \quad \bar{x_2} + x_3$

$x_3 x_4$	00	01	11	10
00	0	1	3	2
$x_1 x_2$ 01	④	⑤	⑦	⑥
11	12	⑬	⑮	⑭
10	⑧	9	11	10

$x_2 x_4$	00	01	11	10
00	0	1	⑤	④
$x_1 x_3$ 01	2	3	⑦	⑥
11	10	11	⑮	⑭
10	⑧	9	⑬	12

$x_2 x_3$	00	01	11	10
00	0	2	⑥	④
$x_1 x_4$ 01	1	3	⑦	⑤
11	9	11	⑮	⑬
10	⑧	10	⑭	12

(a) $x_3 + x_4$ (b) $x_2 \oplus x_4$ (c)

FIGURE 8.29 *Decomposition charts for example function.*

To use the charts we circle the minterms of the function on all three charts, as shown in Fig. 8.29. We see that the first chart (Fig. 8.29a) corresponds to the K-map of Fig. 8.26. With B, A connected to x_3, x_4, the columns of this chart correspond to 1C0, 1C1, 1C3, and 1C2, in that order from left to right. Each circled minterm corresponds to a 1 of the function and we read off the functions to be connected to the data lines just as we did from the map. If we connect B, A to $x_1 x_2$, the *rows* of this same chart correspond to 1C0, 1C1, 1C3, 1C2, in that order from top to bottom. From this we read off the following functions to be connected to the data lines if $x_1 x_2$ are connected to the address lines:

$$1C0 = 0 \qquad 1C1 = 1 \qquad 1C3 = x_3 + x_4 \qquad 1C2 = \bar{x}_3 \cdot \bar{x}_4$$

Moving to the second chart (Fig. 8.29b), if we connect B, A to x_2, x_4, the columns specify the data functions; if we connect B, A to $x_1 x_3$, the rows specify the data functions. For B, $A = x_2 x_4$ we read off the following data functions:

$$1C0 = x_1 \bar{x}_3 \qquad 1C1 = 0 \qquad 1C3 = 1 \qquad 1C2 = \bar{x}_1 + x_3$$

For B, $A = x_1 x_3$ we read off the following data functions:

$$1C0 = x_2 \qquad 1C1 = x_2 \qquad 1C3 = x_2 \qquad 1C2 = \overline{x_2 \oplus x_4}$$

On the third map (Fig. 8.29c) the pattern of circled minterms is the same as on the second, so the form of the data functions will be the same. We can see that our original choice of B, $A = x_3 x_4$ does lead to the simplest circuit.

We have now seen two types of circuits, ROMs and multiplexers, that can be used as generalized logic circuits to provide single-chip realizations of logic functions. A third approach is the use of a decoder package together with separate OR gates to form a "user-made" multipackage ROM. It is appropriate at this point to consider the relative merits of these approaches. Consider the 3-variable function of Example 8.3 which we realized with a single 4-input multiplexer. Ignoring the Strobe line, which is not necessary to the multiplexing function and is not provided in all multiplexers, we have achieved our realization with four 3-input AND gates plus a 4-input OR gate. If we were to realize the same function with a decoder (multichip ROM), we would need a 3-to-8-line decoder, such as shown in Fig. 8.15a. In this event, we would be using eight 3-input AND gates plus a separate (off the chip) OR gate. One might argue that not all eight gates are needed since there are only 4 minterms, but when you buy a decoder in IC form, you get all the minterms whether you need them or not. Thus, even without considering the separate OR gate, we see that the multiplexer form makes much more efficient use of the on-chip logic than does the decoder form. The reason for the inefficiency of the decoder (ROM) form is simple; the decoder (ROM) realization implements the canonic form, the ORing of the minterms of the function. As we saw in Chapter 6 when minimization was first discussed, the canonic form is usually just about the worst possible realization in terms of gate count. A decoder (ROM) form may still be more economical than a discrete gate realization because of the reduction in package count, but a multiplexer form will usually be even more economical for realizing single functions.

The decoder (ROM) is most appropriate when realizing multiple functions. Since the decoder realizes all the minterms, it can be used to realize any number of functions, just by adding another OR gate for each function. By contrast, a multiplexer can realize only one function at a time. Consider the 7-segment decoder of Fig. 8.19. Here we make very efficient use of the decoder since each minterm is used several times. Each individual function could be realized more efficiently by minimization, but the total number of gates for all functions would be considerably greater since there would be few shared terms. The choice between a single-chip ROM and a multichip ROM (decoder plus OR gates) will usually depend strictly on production volume. To justify an ROM, the production volume must be enough so that the initial cost of programming will be offset by the savings in wiring costs.

Summarizing the discussion to this point, if we are interested in single-chip or "few-chip" realizations, the multiplexer form will usually be preferable for single functions of up to 6 variables. For multiple functions of up to 6 variables the decoder/ROM form is likely to be more economical. Above 10 variables, the situation becomes more complex. Suppose we have a system with 10 input variables in which 10 different functions are to be realized. None are functions of more than 6 variables, but every variable appears in several functions, in such a way that there is no way to partition the functions into groups dependent on a smaller number of variables. We could use 10 multiplexers to realize the functions independently, but a single-chip realization might be more economical. For that purpose, one could use a 10-input, 10-output ROM. Such a ROM would require a decoder realizing 1024 minterms, in turn requiring at least 1024 AND gates, but many of these minterms, probably the majority, would not be needed. In order to obtain more efficient realization in such cases, the programmed logic

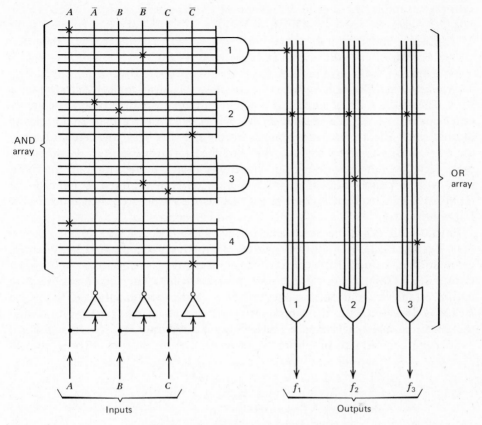

FIGURE 8.30   Basic logic model of PLA.

array (PLA) has been developed. In a ROM, we have a fixed array of AND gates (the decoder) and an array of OR gates, with programmable connections between the AND gate outputs and the OR gate inputs. In a PLA, we have an array of inputs and an array of AND gates with programmable connections between the inputs and the AND gates, plus the array of OR gates with programmable connections from the AND gates. Figure 8.30 shows the basic organization of a PLA with 3 input variables and 3 output lines. The 3 input variables and their complements provide the inputs to the AND array. There are four AND gates, each with 6 inputs, for the 3 variables and their complements, so that any AND function of the 3 variables can be realized. The outputs of the AND gates are connected to the three OR gates through another array. In each of the arrays, an X marks a point where a connection is made.

In this example, AND-gate 1 realizes $A\bar{B}$, AND-gate 2 realizes $\bar{A}B\bar{C}$, AND-gate 3 realizes $\bar{B}C$, and AND-gate 4 realizes $A\bar{C}$. OR-gate 1 is connected to AND-gates 1 and 2 and so realizes $A\bar{B} + \bar{A}B\bar{C} = f_1$. Similarly, the other two OR-gates realize $f_2 = \bar{A}B\bar{C} + \bar{B}C$ and $f_3 = \bar{A}B\bar{C} + A\bar{C}$.

Note that in both arrays, lines that are not connected have no effect. The diagram of Fig. 8.30 is intended only to show the basic logical character of a PLA. The AND and OR functions are generally realized directly in the connection matrices, as the case in the ROM of Fig. 8.20, rather than in physically distinct gates. Figure 8.30 does represent accurately the basic logical character of the PLA, but it is unwieldy, and the simplified form of Fig. 8.31 is commonly used. Here we have just one input line to each gate, representing all the input lines, with X connections at as many intersections as required to indicate the inputs of that gate. The diagram of Fig. 8.31 is thus exactly equivalent to that of Fig. 8.30.

Practical PLAs will be much larger than the simple example we have shown here. Typical PLAs have 10 to 20 inputs, 30 to 60 AND gates, and 10 to 20 OR gates, so that a single PLA can realize a considerable number of complex functions. Like ROMs, PLAs can be mask-programmable, with the interconnection matrix established during manufacturing, or field-programmable (FPLAs). An FPLA is delivered with all the connections made (X at every intersection) and the undesired *links*, or *fuses*, are "blown out" to program the desired functions.

Still another form of programmed logic is available, known as programmed

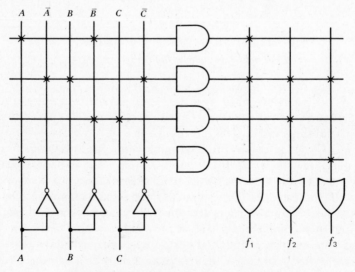

FIGURE 8.31   *Simplified diagram of PLA.*

array logic (PAL). In a ROM we have a fixed AND array and a programmable OR array; in a PLA, we have a programmable AND array and a programmable OR array. In a PAL circuit, we have the other possible combination of these features, a programmable AND array and a fixed OR array. Figure 8.32 shows the simplified diagram of a typical PAL circuit. Here we have 12 inputs and 16 AND gates with a connection array permitting any gate to realize any AND function of any of the 12 variables. The AND gates are connected to 6 OR gates in a fixed pattern with no overlap, i.e., no AND gate can be used in more than one function. The rationale behind the PAL is that there are few situations where it is necessary to realize a large number of functions of exactly the same large set of variables. The more common situation is one in which a large number of functions of a large number of variables must be realized, but there is little, if any, overlap between the functions. The PAL of Fig. 8.32 could be used in situations where there are six functions to be realized involving a total of 12 variables, whether or not any of the variables occur in more than one function. In situations with little overlap, it is unlikely that any product terms can be used in more than one function, so that the flexibility of the OR array in allowing us to connect any AND gate to any OR gate is wasted.

Example 8.5

Realize the following functions using the PAL of Fig. 8.32.

$$f_1(A, B, C, D, E, F) = AB\bar{D}F + C\bar{D}E\bar{F} + AC\bar{E}F + A\bar{B}CD\bar{F}$$

$$f_2(D, E, F, G, H, K) = D\bar{E}K + E\bar{F}G\bar{H}K + D\bar{G}\bar{H}K + \bar{D}FGH\bar{K}$$

$$f_3(A, B, C, F, G, H, K, M, N) = A\bar{B}C\bar{G}H\bar{K}N + \bar{B}CG\bar{H}KM\bar{N}$$

$$f_4(G, H, K, M, N) = G\bar{H}KM\bar{N} + GH\bar{K}\bar{M}N$$

$$f_5(A, B, G, H, J, K, M, N) = A\bar{B}G\bar{H}JK\bar{M}N + \bar{A}BG\bar{H}JKM\bar{N}$$

$$f_6(A, C, D, E, F, G, K, M, N) = \bar{A}C\bar{D}E\bar{F}GK\bar{M}N + A\bar{C}DEFGKMN$$

Although there is overlap, none of the functions involve the same set of variables, and two functions, f_1 and f_4, are completely disjoint. Nevertheless, all six functions can be realized in a single PAL, as shown in Fig. 8.33. By comparison, a realization in discrete NANDs would require 20 chips. ■

PAL circuits generally have from 10 to 16 inputs and come in a variety of configurations in terms of the number of OR gates and the number of AND gates driving each. The relative economy of this approach will depend on how closely an available PAL matches the requirements of a particular application.

ROMs, PLAs and PALs are all suitable in situations requiring realization of several functions of the same set of variables, so it is appropriate to consider the factors that must be evaluated in choosing among them. Many of the factors are

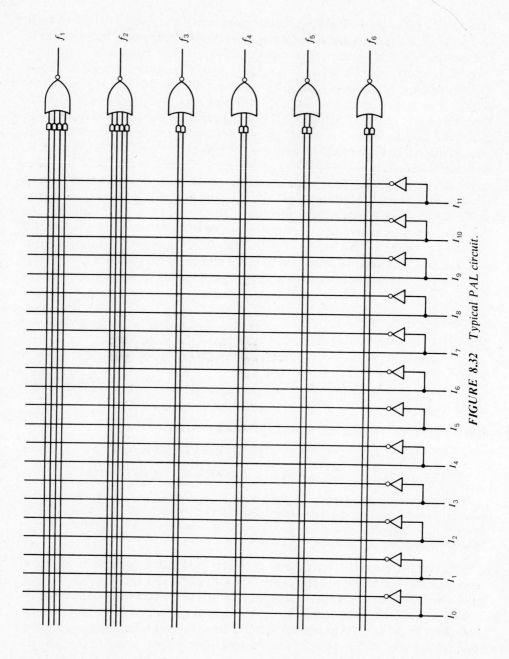

FIGURE 8.32   Typical PAL circuit.

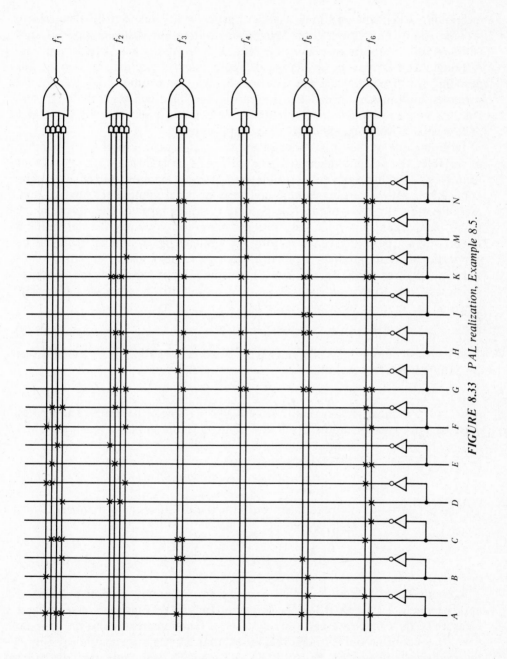

FIGURE 8.33   PAL realization, Example 8.5.

technology-dependent and may change drastically in time, so all we can do is provide some rough guidelines. The first consideration is the number of variables. ROMs are currently available with 4 to 10 inputs and 4 to 16 outputs. Although ROMs may be available for more than 10 inputs, the cost of full decoding will probably rule out their use for logic realization. For fewer than 6 variables, multiplexers, decoders, or even discrete logic will generally be preferable. PLAs are avaiable with 6 to 16 inputs and a like number of outputs. PALs are available with 10 to 16 inputs and 2 to 8 outputs.

Assuming any of the three forms might be suitable with respect to the number of variables, the primary determining factor will be cost. Costs depend greatly on technology, production volume, and type of programming. Given a specific application, with a certain volume and type of programming (i.e., mask, field, erasable, etc.), the basic production costs of the three types of circuits are likely to be comparable, in which case programming cost becomes the determining factor. Again, assuming other factors are similar, programming cost will be proportional to the number of links (fuses), so let us consider the number of links in the three types. In all cases we assume n input variables, and m output functions. In a ROM, we have 2^n decoder outputs to be connected to m OR gates, so the number of links is given by

$$\text{ROM links} = 2^n \times m \tag{8.10}$$

In a PLA the number of AND gates will always be less than in an ROM, so let the number of AND gates be

$$2^n/k$$

where k is a number greater than 1, usually in the range of 10 to 100. These gates must be linked to $2n$ inputs, (true and complemented values of each variable) and m OR gates, so the total number of links is given by

$$\text{PLA links} = \frac{2^n}{k}(2n + m) \tag{8.11}$$

To make it easier to compare them, we show the ratios of these numbers.

$$\frac{\text{ROM}}{\text{PLA}} = \frac{2^n \times m}{\frac{2^n}{k}(2n + m)} = \frac{k}{\frac{2n}{m} + 1} \tag{8.12}$$

In many cases, $n \cong m$, in which case the programming cost of a ROM is approximately $k/3$ times that of a corresponding PLA. Since k is essentially a measure of how much we can simplify a set of functions relative to the canonic form, this confirms our earlier observation that ROMs are economical only if the minterms are needed. In the BCD-to-7-segment code converter, discussed earlier, there are only 16 minterms and all are needed so the ROM is clearly preferable. By contrast, consider the conversion of Hollerith card code to 6-bit ASCII code. The standard Hollerith code is a 12-bit code with 46 characters, so the conversion to ASCII requires the realization of 6 functions of 12 bits. Since there are only 46 characters, a maximum of 46 AND-gates will suffice, even if no simplification is possible. A 12-input ROM, if one were available, would have 4096 AND gates, so that a PLA or PAL would be clearly preferable. A standard PLA suitable for this application is the 82S100, with 16 inputs, 48 AND-gates, and 8 outputs, a total of 1920 links, compared with 24,576 links in a 12-input, 6-output ROM.

Comparing the PAL to the other two forms is difficult because the PAL usually finds application in very different kinds of problems. Because of the possible sharing of AND gates, the ROM and PLA are best suited to true multiple-output problems, involving the realization of a number of functions of the same variables. In a PAL there is no sharing of terms, and a PAL is best regarded as a set of independent AND-OR networks on a single chip with programmable connections to a large number of input variables. In most cases few, if any, of the functions will involve all n variables, so it is not meaningful to relate the number of AND gates to 2^n. If we assume an average of g gates for each of the m functions, each to be connected to $2n$ inputs, the total number of links for a PAL is given by

$$\text{PAL links} = 2 \times n \times m \times g \qquad (8.13)$$

If we have an application for which a PAL is suitable, it will usually be one of which few, if any, terms can be shared. If no AND gates are shared between functions, the number of AND gates in the PLA, $2^n/k$, will be the same as the number of AND gates in the PAL, $g \times m$. In such a case, and again assuming that $n \cong m$, the programming cost of a PLA will be $1\frac{1}{2}$ times that of a PAL. This is not a large advantage and may be offset by the greater flexibility of the PLA unless there is a standard PAL that matches the requirements of a particular design more closely than any standard PLA. Consider the design of Example 8.5, which exactly matches a standard PAL with 12 inputs and 16 AND gates, requiring only 384 links. If the 82S100 with 1920 links were the closest match in a standard PLA, the PAL would clearly be preferable.

8.7 Multilevel Circuits

While ROMs, multiplexers, and PLAs offer attractive means of realizing complex logic functions, there will still be many situations where technical or economic constraints dictate the use of individual gates. As we have seen, logic circuits are usually laid out initially in terms of AND, OR, and NOT gates, most commonly in two-level form. However, as we saw in the case of the adder, some functions can be more economically realized in multilevel form. Sometimes a multilevel network will arise out of an attempt to minimize the number of SSI or MSI packages used to realize a function. Whatever the number of levels in a circuit, economic factors may restrict us to realization in terms of NAND or NOR. We saw in Chapter 3 that any switching function can be realized in terms of NAND or NOR; we further saw, in Chapter 5, that, for two-level forms, the translation from AND-OR to NAND-NAND or OR-AND to NOR-NOR simply requires a direct one-for-one replacement of gates. For multilevel circuits the situation is more complex; in this section we wish briefly to consider methods for designing multilevel circuits, with emphasis on NAND realizations. Analogous techniques for NOR design are fairly obvious.

The most common technique for multilevel design is to work initially in AND-OR-NOT form, and then convert to all-NAND. Consider the function

$$f(A, B, C, D) = A\bar{C}D + AC\bar{D} + B\bar{C}D + BC\bar{D} \qquad (8.14)$$

$$= \bar{C}D(A + B) + C\bar{D}(A + B) \qquad (8.15)$$

This function is, in fact, the function realized in Fig. 6.21, with Equation 8.14 corresponding to Fig. 6.21b and Equation 8.15 to Fig. 6.21a. The circuit of Fig. 6.21a, with three levels of gating, is clearly preferable if gate count is our only concern. The basic technique used to find such forms is, as suggested by the above equations, algebraic factoring.

The circuit of Fig. 6.21a requires both two-input and three-input gates, which will not be available on the same chip in SSI form. When dealing with large production volumes, significant economies may be realized by reducing the number of different types of SSI chips. For example, a manufacturer might specify that all functions are to be realized using only two-input gates. When AND and OR are involved such a restriction is no problem, since AND or OR of three or more inputs can always be broken up into a cascade of two-input operations. For example, this function can be factored in alternate ways. Breaking the three-input ANDs into two-input ANDs leads to the form

$$f(A, B, C, D) = (A + B)(\bar{C}D) + (A + B)(C\bar{D}) \qquad (8.16)$$

leading to the realization of Fig. 8.34a. A different factoring leads to the form

$$f(A, B, C, D) = (A + B)(\bar{C}D + C\bar{D}) \qquad (8.17)$$

leading to the realization of Fig. 8.34b. Finally, in Equation 8.17 we recognize the exclusive-OR of C and D, giving the form

$$f(A, B, C, D) = (A + B)(C \oplus D)$$

and the circuit of Fig. 8.34c.

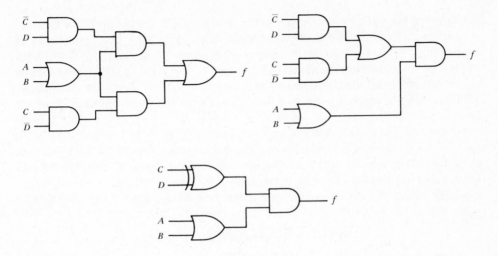

FIGURE 8.34   *Additional realizations of example function.*

The circuits of Fig. 8.34a and 8.34b are both more complex than Fig. 6.21a in terms of numbers of gates, but have the potential advantage of using only two-input gates. This sort of uniformity can be particularly important in all-NAND or all-NOR designs.

The circuit of Fig. 8.34c introduces a number of important factors in evaluating circuit cost. The most obvious factor is the direct cost of the gates. The circuit of Fig. 8.34c has the fewest gates, but exclusive-OR gates are more expensive than other types, typically costing about three times as much as two-input AND, OR, NAND, or NOR. As in this example, an exclusive-OR will usually replace three simpler gates, so that direct gate cost is about the same. A second factor is the number of chips (IC packages) required to realize a circuit. Each chip will typically contain several gates, e.g., four gates per chip for two-input gates. For each chip there is a basic mounting and wiring cost that is essentially independent of the complexity of the chip. A four-gate Ex-OR chip may cost as much as three four-gate NAND chips, but if that one chip can replace three, the total cost will be much lower because only one chip must be mounted and wired. A third factor is the availability of single-rail or double-rail inputs. Up to this point we have assumed *double-rail* inputs, i.e., we have assumed all signals to be equally available in either true or complemented form. In some circuits, however, we may have *single-rail* inputs, i.e., each signal is available in only one form, true or complemented. If the other form is needed, as in the AND-OR realization of exclusive-OR, inverters must be added, increasing circuit cost. With the exclusive-OR gate, no external inverters are needed.

For all of the above reasons, the use of exclusive-OR gates can often result in significant reductions in circuit cost. Sometimes, as in this case, exclusive-OR factors may be recognized in the course of algebraic manipulation. Sometimes the K-map of a function may help the designer to find such factors. These factors are often associated with patterns of alternating rows or columns on the map, or with a strongly diagonal pattern, as is the case with the parity check function to be discussed in Section 8.8.

Having found a multilevel design by factoring, it may then be desired to convert the design to all-NAND form. The basic technique is to break the circuit up into sections that have the AND-OR form, convert these to NAND-NAND form, and then reconnect the sections, adding inversions where necessary. Where the circuit does not divide conveniently into AND-OR sections, we use DeMorgan's law,

$$\overline{A \cdot B} = \bar{A} + \bar{B}$$

which, in terms of gate realization, tells us that a NAND gate is equivalent to an OR gate with inversion at the inputs (Fig. 8.35).

As a first example of AND-OR to NAND conversion, consider the circuit of Fig. 6.21a. In Fig. 8.36a we show this circuit divided into two sections, an AND-OR section and an isolated OR gate. Wherever we cut lines in dividing the circuit we assign symbols to those lines to keep track of them for future interconnection. In Fig. 8.36b we convert the AND-OR section to NAND-NAND and replace the OR gate with one inverted inputs, complementing A and B to compensate. Finally, in Fig. 8.36c, we replace the OR gate with NAND and reconnect to complete the conversion.

As a second example, let us convert the alternate form of the same function shown in Fig. 8.34b. In Fig. 8.37a we first divide the circuit into three sections, an AND-OR section, an isolated OR gate, and an isolated AND gate. In Fig. 8.37b we convert the AND-OR to NAND-NAND, convert the OR as above, and convert the AND to NAND, giving $\bar{f}$, rather than f at the output. In Fig. 8.37c we convert the OR gate, add inversion at the output and reconnect. When we have an isolated AND, realization in NAND requires use of a NAND gate followed by inversion. We use a NAND with both inputs tied together rather than an inverter, since our objective is all-NAND design.

In comparing these two designs, the design of Fig. 8.36 would appear to be more economical because it uses only 4 gates compared to 6 in Fig. 8.37, but we

FIGURE 8.35   *DeMorgan's law in terms of gates.*

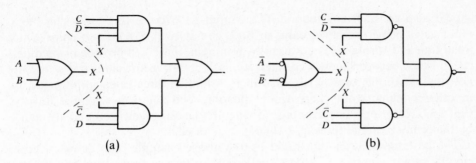

(a)   (b)

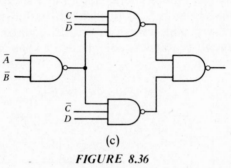

(c)

FIGURE 8.36

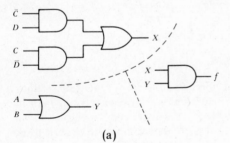

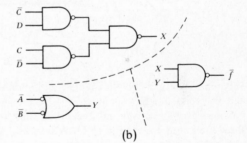

(a)   (b)

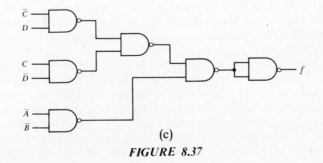

(c)

FIGURE 8.37

should also consider chip count. Three-input gates come three to a chip, two-input gates come four to a chip, so both circuits will require two chips. Since mounting and wiring costs are usually dominant, there will be little or no difference in cost between these two realizations. Another factor is the gates that are not used. With the form of Fig. 8.36 there would be three gates left over for use in other circuits, while there would be only two left over with the form of Fig. 8.37. However, this advantage for the first circuit might well be outweighed by the economies due to using only one type of chip.

The basic procedures illustrated by the above examples can be extended to circuits of any complexity. Break the circuit up into sections small enough that the conversion to NAND form is obvious, keep careful track of the identity of signals between sections, add inversions where required, and reconnect. The following example circuit is about as complicated as one is likely to see.

Example 8.6

Convert the circuit shown in Fig. 8.38 to all-NAND form.

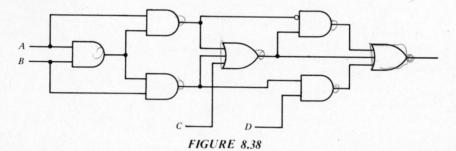

FIGURE 8.38

Solution   In Fig. 8.39a we divide the circuit into three sections, two AND-OR sections and an isolated AND. Note that there are more connections between sections than in previous examples and that every line that has been cut is identified. In Fig. 8.39b we show the initial conversion to NAND form. In this case there are inputs to the OR gates in the AND-OR sections other than directly from the AND gates. We therefore initially convert these gates to the OR-with-inversion form, complementing the direct inputs to compensate. When we convert the upper-left AND to NAND we delete the input inverter and complement the signal. The outputs of the ANDs in the leftmost AND-OR group go to other parts of the circuit, so we identify these outputs as complements of the signals developed by the corresponding AND gates. At this point we see the importance of keeping track of the identity of the intermediate signals. We know exactly what signals are required at the inputs of the various gates and we know what signals are available at the outputs of gates. For example, we need $\bar{X}$ at the input of the upper-right NAND gate and $\bar{X}$ is available from the upper-left NAND gate, so we can make a direct connection here. By contrast, we need Y at the input of the lower-right NAND gate but have $\bar{Y}$ from the lower-left NAND, so an inverter will be needed here. In a similar manner we see that two other inverters are needed in connecting the final circuit, Fig. 8.39c.   ∎

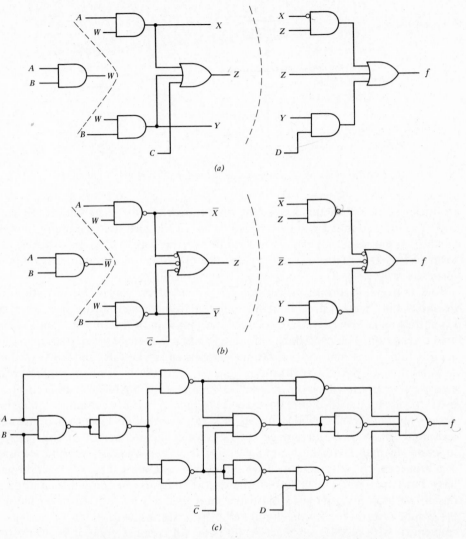

FIGURE 8.39 *AND-OR to NAND conversion, Example 8.6.*

Gate conversion can obviously be done in the opposite direction, from all-NAND form to AND-OR form. This can be useful in analyzing an existing circuit that is realized in all-NAND form. In this connection it should be noted that in many industrial drawings NAND gates that replace OR gates in the equivalent AND-OR form will be drawn in the equivalent form of an OR gate with inverted inputs. For example, the all-NAND circuit of Fig. 8.36c might appear as shown in Fig. 8.40. This form is used because many feel that it is easier

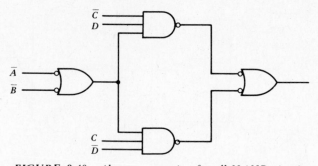

FIGURE 8.40   *Alternate notation for all-NAND circuits.*

to interpret. It will be understood by any knowledgeable user that an OR gate with all inputs inverted will be realized by a NAND gate, but the use of the OR symbol emphasizes that the basic function of the gate is OR. In this case the form of Fig. 8.40 makes it very obvious that the function of the input gate is to form the term $(A + B)$.

The conversion process discussed above involves, except for insertion or deletion of inverters, a direct gate-for-gate replacement, so that the basic form, or organization, of the circuit is preserved. In mathematical terms, it is an *isomorphic* translation. The procedure will always work, but there is no guarantee that a form that is optimal for realization in terms of AND, OR, and NOT will be optimal for all-NAND realization. This is particularly true in circuits with single-rail inputs, in which the inherent inversion of the NAND can sometimes be used to advantage to eliminate extra inverters. A useful technique for direct design for multilevel NAND circuits is *map factoring*, an extension of K-map techniques that takes advantage of the special characteristics of the NAND function. In map factoring we make use of the Venn diagram interpretation of AND in terms of intersection; if we AND two functions, their intersection will have 1s where both have 1s and 0s where either have 0s. To apply this concept to design we look for groups of 1s on the map that can be realized in terms of a large cube of 1s with a few 1s eliminated by intersection with a cube of 0s. This is convenient with NAND since a NAND gate will create a cube of 0s where the corresponding AND gate would create a cube of 1s.

Example 8.7

Find an all-NAND design for the function represented by the map of Fig. 8.41, subject to a fan-in limit of 3, assuming single-rail inputs.

Solution   The minimal SOP form for this function is

$$f(A, B, C, D) = \bar{C}\bar{D} + \bar{B}\bar{C} + \bar{A}BD + AB\bar{D}$$

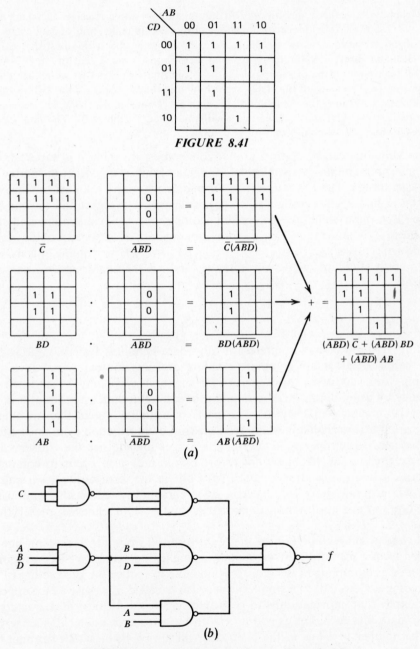

FIGURE 8.41

(a)

$(\overline{ABD})\,\overline{C} + (\overline{ABD})\,BD + (\overline{ABD})\,AB$

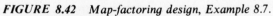

(b)

FIGURE 8.42   Map-factoring design, Example 8.7.

which cannot be realized directly with a fan-in limit of three because there are four products. One rather obvious factoring will lead to a 3-chip realization, and an exclusive-OR realization can be done in two chips. Before concluding that is the best that can be done, let us try direct NAND design by map factoring. Assuming that the final NAND gate will realize an ORing of groups of 1s, the fan-in limit requires that we realize the 1s in three groups. The maps of Fig. 8.42a show how this can be done. All but two of the 1s are realized by ANDing the 3-cube of 1s, $\bar{C}$, with the 1-cube of 0s, $\overline{ABD}$. The remaining 1s are realized by ANDing 2-cubes of 1s with the same 1-cube of 0s. The final circuit (Fig. 8.42b) requires two chips of 3-input gates.   ∎

This technique can be applied in many ingenious ways, but it is important for the reader to maintain perspective on the value of this or any other technique for multilevel design. There is a tendency for people who have a knack for this sort of thing to find a stimulating intellectual challenge in trying to find *the one best* design. But these techniques are often tedious and time-consuming and it is important to balance the value of your time against possible economies in the final design. Unless a circuit is to be produced in very large volume, it is usually best to apply standard two-level techniques and use the first design you can find that works.

8.8 Parity

Avoiding error is of major interest in any data-handling system. Among the major causes of error are component failure and intermittent signal deviation resulting from additive noise. Errors of the first type may be reduced in frequency by using duplicate or redundant circuits. Random errors due to noise occur with greater frequency in some parts of a digital system than in others. Errors are particularly likely where transmission of information from one system or subsystem to another is involved. The longer and noisier the transmission path, the greater the likelihood of error. Errors may also occur in computer memories where information is often read out in the form of low-level pulses. Noise will generally have less effect in an integral subsystem, such as an arithmetic unit, where information is primarily in the form of levels rather than pulses.

If it were possible to determine that data received from a transmission line or sampled from a memory were erroneous, it might be possible to effect a retransmission or a resampling of the data. It is possible to reduce the probability of an error going undetected. The simplest approach is called a *parity check*. Suppose, for example, that information is to be stored on a magnetic tape in characters of seven binary bits each. Let us add an eighth bit to each character in such a way that the number of 1-bits in the character will always be even. We say that the character is coded with even parity. After a character has been read from the magnetic tape, a parity check is made to see that the number of 1-bits is still even.

Example 8.8

Establish even parity by adding a bit to each of the 7-bit characters 1101001 and 0101111. These even-parity characters are to be transmitted through a medium in which an error occurs once in every 10^4 bits (one out of 10^4 bits is received in complemented form). Determine the probability that an erroneous character will be developed that will go undetected by a parity check at the receiver.

Solution Determination of the parity bits is immediate.

Data Character								Parity Bit Coded Character							
X_6	X_5	X_4	X_3	X_2	X_1	X_0		X_7	X_6	X_5	X_4	X_3	X_2	X_1	X_0
1	1	0	1	0	0	1		0	1	1	0	1	0	0	1
0	1	0	1	1	1	1		1	0	1	0	1	1	1	1

For the above, or any even-parity character, complementing two bits will result in a character that still has even parity, as will the complementation of any even number of bits. The parity checker at the receiver would be unable to distinguish such characters from valid coded characters. There are

$$_8C_2 = \frac{8!}{6! \cdot 2!}$$

combinations of 2 of the 8 bits. For each of these combinations, the probability that those 2 bits are in error and the remaining 6 are correct is $10^{-8}(1 - 10^{-4})^6$. Proceeding similarly for 4, 6, and 8 bits, we arrive at

$$P(\text{even number of bits in error}) = \frac{8!}{6! \cdot 2!} 10^{-8}(1 - 10^{-4})^6$$

$$+ \frac{8!}{4! \cdot 4!} 10^{-16}(1 - 10^{-4})^4 + \frac{8!}{6! \cdot 2!} 10^{-24}(1 - 10^{-4})^2 + 10^{-32} \quad (8.18)$$

as the expression for the probability of an undetectable erroneous character. Neglecting the obviously insignificant terms yields the close approximation.

$$P(\text{even number of bits in error}) = 28(10^{-8}) \quad (8.19)$$

This is very significantly less than $7 \cdot 10^{-4}$, which is the approximate probability of a single error in a 7-bit character transmitted without a parity bit. ■

The Boolean function for checking odd parity is the classic example of a function that is not efficiently realized in two levels. The function $f(X_1, X_2, X_3, X_4)$, which is 1 when an odd number of the variables X_1, X_2, X_3, or X_4 are 1, is

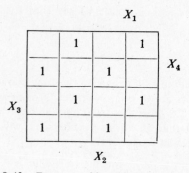

FIGURE 8.43   *Four-variable odd-parity check function.*

depicted in Fig. 8.43. An odd-parity check over any number of variables will appear as a "checkerboard" pattern on a Karnaugh map. A two-level parity check over eight variables would require 129 NAND gates.

Let us consider an alternative arrangement. The circuit for detecting odd parity over two bits is simply exclusive-OR. In Fig. 8.44a we show two exclusive-OR gates with two distinct pairs of inputs. If exactly one of the four variables X_1, X_2, X_3, and X_4 is 1, then exactly one of the outputs, Z_1 or Z_2, must be 1. This must also be true if three of the four X's are 1. If all of the X's are 0 or if all are 1, clearly $Z_1 = Z_2 = 0$. If the two inputs to one gate are 1 while the two inputs to the other gate are 0, both Z_1 and Z_2 must be 0. If one input to each gate is 1, then $Z_1 = Z_2 = 1$. This discussion is summarized in the table of Fig. 8.44b.

From the table, the number of outputs equal to 1 is odd if and only if an odd number of inputs are 1. Thus, a check of odd parity may be accomplished by adding a third exclusive-OR gate, as shown in Fig. 8.44c. Extending the argument to eight variables leads to the circuit of Fig. 8.45, which may be realized by

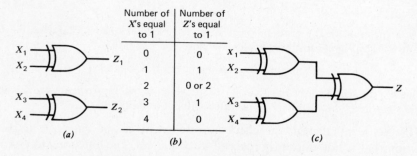

FIGURE 8.44   *Four-variable odd parity checker.*

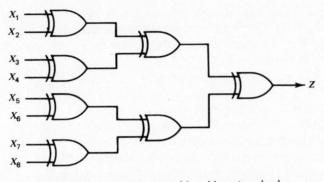

FIGURE 8.45   *Eight-variable odd parity checker.*

using 21 NAND gates and 14 inverters. We could have arrived at this same result by verifying that $X_1 \oplus X_2$ is an associative operation. This task is left to the reader.

8.9 Error-Detecting and Correcting Codes

Noting the decrease in probability of error resulting from the addition of a single parity bit, we are led naturally to consider the possibility of adding more than one redundant bit. This approach will lead to a further reduction in the probability of error and, as we shall see, can even facilitate error correction without retransmission.

Let us suppose that the four coded characters in Fig. 8.46 are the only ones that will be transmitted over some communications system. Note that each of the four coded characters differs from the other coded characters in at least 3 of the 5 bits. We say that the *minimum distance* of the code is 3. In general, the minimum distance, M, of a code is defined as the minimum number of bits in which any two characters of a code differ.

Because there are four distinct 2-bit coded characters, we see that $M = 3$ has been achieved at the price of adding 3 redundant bits. Suppose that character D is transmitted as shown but is received as 11000. Although the last 2 bits are in error, 11000 will not be confused with any of the bits A, B, and C. Indeed, 2

A	0	0	0	0	0
B	1	1	1	0	0
C	0	0	1	1	1
FIGURE 8.46 D	1	1	0	1	1

erroneous bits in any character of the code will not distort it into any other character of the code. It should be apparent to the reader that we may generalize our observation to the statement: errors in 2 or fewer bits may be detected in any minimum-distance-3 code.

Errors in 3 or more bits cannot always be detected in a minimum-distance 3-code. Note, for example, that errors in the first 3 bits of *B* in Fig. 8.46 will distort this character into the code for *A*.

The probability of an error in 2 bits may be sufficiently small in some applications to be disregarded. In this case, it is possible to employ a minimum-distance-3 code as an error-correcting code. Suppose that the character 11000 is received in a communications system employing the code of Fig. 8.46. If it is assumed that no more than 1 bit will be in error, we may deduce that the transmitted character was 11100. Thus, circuitry at the receiving end may be provided to convert 11000 directly to 11100. In general, we may say that a received character with a single-bit error will differ in only 1 bit from the transmitted character but in at least 2 bits from an other character in the code. Thus, circuitry may be provided to correct all single-bit errors without retransmission. The above is true for any minimum-distance-3 code.

Suppose a 2-bit error actually occurs in a character of a minimum-distance 3-code set up to correct single-bit errors. The correction process is still carried out on the erroneous character, but it will be corrected to the wrong coded character. Thus, the 2-bit error will in effect go undetected.

We have discussed above the special case of the general relation

$$M - 1 = C + D, \qquad C \leq D \tag{8.20}$$

where *C* is the number of erroneous bits that will be corrected and *D* is the number of errors that will be detected. For a specified minimum distance, *M*, various combinations of values of *C* and *D* will satisfy Equation 8.20. A partial list is given in Fig. 8.47. The single-parity-bit code and the 2-out-of-5 code are examples of minimum-distance-2 codes for detecting single-bit errors. As discussed above, a minimum-distance-3 code may be used either for detecting 2-bit errors or for detecting and correcting single-bit errors. It is necessary to choose between rejecting all erroneous characters and requiring retransmission or cor-

M	1	2	3	4	5	6
D	0	1	2 1	3 2	4 3 2	5 4 3
C	0	0	0 1	0 1	0 1 2	0 1 2

FIGURE 8.47

recting single-bit errors on the spot and passing 2-bit errors. We cannot have the benefit of both modes of operation.

A minimum-distance-4 code may be used for detecting triple errors. Alternatively, it may be assumed that triple errors are sufficiently improbable to be ignored. A 2-bit error will result in a character differing in 2 bits from the correct and in only 2 bits from at least one other character in the code. Thus, 2-bit errors cannot be corrected. Single-bit errors, however, may be corrected. This correction process will alter characters with 3-bit errors to the wrong coded character. Two-bit errors, however, will not be mistaken for single-bit errors. Thus, it is again necessary to choose between the two modes of operation listed in Fig. 8.47.

The solutions of Equation 8.20 for still larger M may be similarly justified.

8.10 Hamming Codes

One of the most convenient minimum-distance-3 codes was devised by R. W. Hamming [5]. We number the bit positions in sequence from left to right. Those positions numbered as a power of 2 are reserved for parity check bits. The remaining bits are information bits.

The 7-bit code is shown in Fig. 8.48, where P_1, P_2, and P_4 indicate parity bits. The bits X_3, X_5, X_6, and X_7 make up the information character to be transmitted. From these bits, the parity bits are determined as follows:

P_1 is selected so as to establish even parity over bits 1, 3, 5, and 7.
P_2 is selected so as to establish even parity over bits 2, 3, 6, and 7.
P_4 is selected so as to establish even parity over bits 4, 5, 6, and 7.

1	2	3	4	5	6	7
P_1	P_2	X_3	P_4	X_5	X_6	X_7

FIGURE 8.48

Example 8.9

Determine the Hamming-coded character corresponding to the information character $X_3 X_5 X_6 X_7 = 1010$.

Solution We may write

$$P_1 = X_3 \oplus X_5 \oplus X_7 \tag{8.21}$$

The above extension of the exclusive-OR notation of Chapter 3 is defined to mean that P_1 will be 1 if there are an odd number of 1s among X_3, X_5, and X_7 and that P_1 will be zero otherwise. Thus

$$P_1 \oplus X_3 \oplus X_5 \oplus X_7 = 0 \tag{8.22}$$

establishing even parity over bits 1, 3, 5, and 7. In this case,

$$P_1 = 1 \oplus (0 \oplus 0) = 1 \oplus 0 = 1 \tag{8.23}$$

Similarly,

$$P_2 = X_3 \oplus X_6 \oplus X_7 = 1 \oplus (1 \oplus 0) = 0$$

and

$$P_4 = X_5 \oplus X_6 \oplus X_7 = 0 \oplus (1 \oplus 0) = 1 \tag{8.24}$$

The final coded character is 1011010.

∎

The correction process at the receiving end is very convenient in the case of a Hamming code. Since we must assume that only 1 bit is in error, it is only necessary to locate that bit. This may be accomplished by checking for odd parity over the same three combinations of bits for which even parity was established at the transmitting end as follows:

$$C_1 = P_1 \oplus X_3 \oplus X_5 \oplus X_7$$
$$C_2 = P_2 \oplus X_3 \oplus X_6 \oplus X_7 \tag{8.25}$$
$$C_4 = P_4 \oplus X_5 \oplus X_6 \oplus X_7$$

If $C_1 = 1$, there must have been an error in one of the 4 bits, 1, 3, 5, and 7, and so on.

The erroneous bit may be determined directly from Fig. 8.49. If all 3 check bits indicate 0 or even parity, then no single bit is in error. This was the situation upon transmission, and we again remind ourselves that we have assumed that there are no 2-bit errors.

If $C_1 = 1$ but $C_2 = C_4 = 0$, we may conclude that one of the bits 1, 3, 5, or 7 is in error but that 4, 5, 6, and 7 as well as 2, 3, (6, 7) are all correct. Thus, bit 1 must be in error as indicated in the second column of Fig. 8.49. Should $C_4 = C_2 = C_1 = 1$, we may conclude that bit 7 is in error because this is the only bit influencing all three parity checks. The entries at the bottom of remaining columns of Fig. 8.49 may be similarly determined.

C_4	(odd parity over 4 5 6 7)	0	0	0	0	1	1	1	1
C_2	(odd parity over 2 3 6 7)	0	0	1	1	0	0	1	1
C_1	(odd parity over 1 3 5 7)	0	1	0	1	0	1	0	1

Erroneous bit   none 1 2 3 4 5 6 7

FIGURE 8.49

Example 8.10

Suppose the character $P_1 P_2 X_3 P_4 X_5 X_6 X_7 = 1101101$ is received in a communications system employing the Hamming code. In this case,

$$C_1 = 1 \oplus 0 \oplus 1 \oplus 1 = 1$$

$$C_2 = 1 \oplus 0 \oplus 0 \oplus 1 = 0$$

$$C_4 = 1 \oplus 1 \oplus 0 \oplus 1 = 1$$

Thus, bit 5 is in error and the correct character is 1101001. ■

If the 7-bit Hamming code is employed for single-bit detection and correction, the probability of an undetected error is the same as the original probability of error in 2 or more of the 7 bits. This is approximately*

$$P_2 = \frac{7!}{5! \, 2!} (P)^2 (1 - P)^5 \tag{8.26}$$

where P is the probability of an error in any one bit.

By rejecting and calling for the retransmission of any impossible characters, the 7-bit Hamming code may be used for double-error detection. In this case, the probability of an undetected error would be the probability of a 3-or-more-bit error, which is approximately

$$P_3 = \frac{7!}{4! \, 3!} P^3 (1 - P)^4 \tag{8.27}$$

If $P = 10^{-4}$, then $P_2 = 20.9(10^{-8})$ and $P_3 = 34.9(10^{-12})$.

Minimum-distance-3 Hamming-codes are possible for any $2^n - 1$ bits, where $2^n - n - 1$ of these are information bits. In these cases, all bits numbered as a power of 2 will be parity-check bits. The groups of bits checked for parity continue in the same binary pattern. For example, in the 15-bit code, even parity is achieved over the last 8 bits through P_8. The bit P_4 falls in the group consisting of every other series of 4 bits, beginning at the left, and so on. Each of these codes may be extended to minimum-distance 4 by adding one more bit to provide even parity over the entire character.

Example 8.11

Suppose information is to be transmitted in blocks of several coded characters. Each row of Fig. 8.50, for example, is one character of a five-character block. Establishing even parity over the corresponding bits of each character, as well as over the individual characters, results in a minimum distance of two between valid blocks.

---

* This expression neglects the probability that more than 3 bits may be in error, which is considerably smaller still.

$$\begin{array}{ccccc} X_{11} & X_{12} & X_{13} & X_{14} & P_{15} \\ X_{21} & X_{22} & X_{23} & X_{24} & P_{25} \\ X_{31} & X_{32} & X_{33} & X_{34} & P_{35} \\ X_{41} & X_{42} & X_{43} & X_{44} & P_{45} \\ P_{51} & P_{52} & P_{53} & P_{54} & P_{55} \end{array}$$

FIGURE 8.50

Even parity is established over the corresponding bits of the several characters of Fig. 8.50 by $P_{51} \ldots P_{55}$, respectively. The single parity bits of each character are $P_{15} \ldots P_{55}$. Neglecting 2-bit errors, if odd parity is determined for row i and column j of a received block, the bit ij is in error. ■

Problems

8.1 Design a combinational circuit that will accomplish the multiplication of the 2-bit binary number $X_1 X_0$ by the 2-bit binary number $Y_1 Y_0$. Is a two-level circuit the most economical?

8.2. Verify that the circuit of Fig. 8.6c is a full adder.

8.3. The bits of two binary-coded decimal digits $X_3 X_2 X_1 X_0$ and $Y_3 Y_2 Y_1 Y_0$ serve as the input to a 4-bit binary adder. The output bits of the adder are $Z_4 Z_3 Z_2 Z_1 Z_0$. Let the bits $S_3 S_2 S_1 S_0$ be the binary-coded decimal representation of the least significant bit of the decimal sum. Let $S_4 = 1$ if the most significant bit of the sum is 1, and let $S_4 = 0$ otherwise. Determine a minimal multiple-output realization of S_4, S_3, S_2, S_1, and S_0 as functions of Z_4, Z_3, Z_2, Z_1, and Z_0.

8.4. Design a circuit to convert from the 2-out-of-5 code of Fig. 8.12 to the seven-segment code, using discrete gates.
(a) Assume only the codes shown in Fig. 8.12 will occur.
(b) Include provision for detection of illegal codes.

8.5. Develop an expression for the number of gate inputs required in a tree-type decoder. Assume two input gates and assume one variable is introduced at each succeeding level of the tree.

8.6. Assume a 4-input decoder is required. There is a choice between realizing it from discrete gates in the general form of Fig. 8.15a or using a complete decoder in a single integrated circuit package. The cost of the decoder is $1.00. Dual 4-input AND gates (two 4-input gates/package) are available at $0.20 per package. The mounting and wiring costs are $0.25 per package plus $0.05 per input or output connected. Determine the costs of the two methods of realizing the decoder.

8.7. A 12-input decoder is to be realized with a structure similar to that of Fig. 8.16, except that the only decoders available are 4-input decoders at \$1.00 each. These decoders could be used in the circuit of Fig. 8.16, leaving one input unused; or three decoders could be used, requiring a different, and possibly more complex, arrangement of AND gates. Available gates that might be useful are dual 4-input AND, triple 3-input AND, and quad 2-input AND, all at \$0.25/package. Mounting and wiring costs are as given in Problem 8.6. Determine the most economical design.

8.8. Repeat Problem 8.4 using a 32 × 8 ROM. Use a form similar to Fig. 8.21.

8.9. Given the function $f(A, B, C, D) = \sum m(1, 2, 4, 5, 6, 7, 8, 9, 10, 11, 14, 15)$
   (a) Realize this function using an 8-input multiplexer.
   (b) Realize this function using a 4-input multiplexer plus external gates as required.

8.10. Realize the following two functions, using both halves of a SN74153 multiplexer and as little external gating as possible.

$$f_1(A, B, C, D) = \sum m(2, 4, 6, 10, 12, 15)$$
$$f_2(A, B, C, D) = \sum m(3, 5, 9, 11, 13, 14, 15)$$

8.11. Draw a diagram similar to Fig. 8.31 showing a PLA realization of the following four functions, assuming a PLA with 6 inputs, 4 outputs, and 10 AND gates.

$$f_1(A, B, C, D, E, F) = \sum m(0, 1, 2, 3, 7, 8, 9, 10, 11, 15, 19, 23, 26, 27, 31,$$
$$32, 33, 34, 35, 39, 40, 41, 42, 43, 44, 45,$$
$$46, 47, 51, 55, 58, 59, 63)$$

$$f_2(A, B, C, D, E, F) = \sum m(0, 1, 2, 3, 7, 8, 9, 10, 11, 15, 32, 33, 34, 35, 39,$$
$$40, 41, 42, 43, 45, 47)$$

$$f_3(A, B, C, D, E, F) = \sum m(8, 9, 10, 11, 12, 14, 21, 25, 27, 40, 41, 42,$$
$$43, 44, 46, 57, 59)$$

$$f_4(A, B, C, D, E, F) = \sum m(0, 1, 2, 3, 7, 8, 9, 10, 11, 15, 32, 33, 34, 35, 39,$$
$$40, 41, 42, 43, 44, 45, 46, 47, 56, 57, 58,$$
$$59, 60, 61, 62, 63)$$

8.12. Draw a diagram similar to Fig. 8.31 showing a PLA realization of the code converter of Problem 8.4a. Assume a PLA is available with 5 inputs and 7 outputs and as many AND gates as needed. Compare the number of links with the number of links in the ROM realization found in Problem 8.8.

8.13. The network of Fig. P8.13 has been designed for comparing the magnitudes of two 2-bit binary numbers, a_2, a_1 and b_2, b_1. If $z = 1$ and $y = 0$,

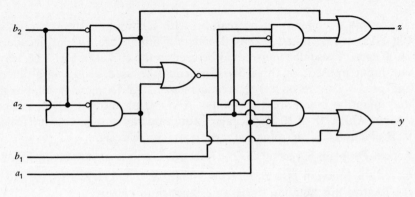

FIGURE P8.13

a_2, a_1 is larger. If $z = 0$ and $y = 1$, b_2, b_1 is larger. If $z_2 = z_1 = 0$, the two numbers are equal. Translate the network into a form that uses only NAND gates. Do not attempt to minimize the network.

8.14. Determine NAND equivalents of the circuits of Fig. P8.14.

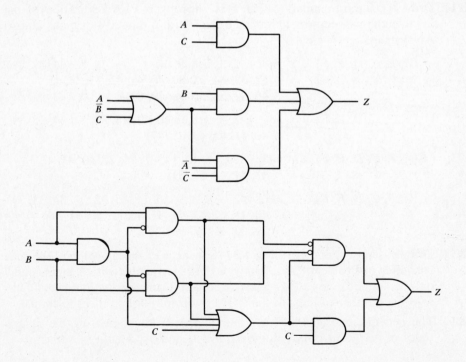

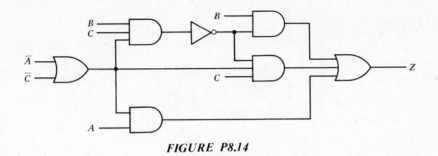

FIGURE P8.14

8.15. Convert the circuit of Fig. P8.15 to NOR form.

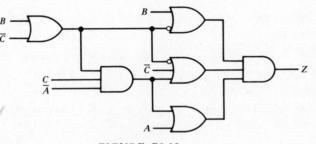

FIGURE P8.15

8.16. Realize the following function in all-NAND form with a fan-in limit of two. Assume double-rail inputs.

$$f(A, B, C) = \sum m(0, 3, 4, 6, 7)$$

8.17. Realize the following function on a single-chip, single-rail input, all-NAND.*

$$f(A, B, C) = \sum m(1, 2, 3, 5)$$

8.18. Find a two-chip realization of the function of Example 8.7, using NAND gates and exclusive-OR.*

8.19. Find an all-NAND realization of the function below, fan-in limit of 3, single-rail inputs. What is the chip count?*

$$f(A, B, C, D) = \sum m(0, 1, 2, 3, 7, 8, 9, 10, 13)$$

---

* For Problems 8.17 to 8.19, assume NAND gates are available in the following packages: 2-input—4 per chip; 3-input—3 per chip; 4-input—2 per chip; 8-input—1 per chip.

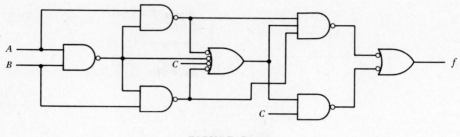

FIGURE P8.20

8.20. Determine a minimal SOP expression for the all-NAND circuit shown in Fig. P8.20.

8.21. Determine a circuit that checks for odd parity over 7 bits using exclusive-Or gates.

8.22. Show that $X_1 \equiv X_2$ and $X_1 \oplus X_2$ are associative operations.

8.23. Encode the information character 0 1 1 0 1 1 1 0 1 0 1 according to the 15-bit Hamming code.

8.24. Determine which bit, if any, is in error in the Hamming-coded character 1 1 0 0 1 1 0

8.25. Determine the exact versions of the approximate expressions given by Equations 8.26 and 8.27.

8.26. Suppose the 5×5 cross-parity configuration of Fig. 8.50 is used for single-bit error detection and correction. If the probability that any 1 bit will be in error is 10^{-8}, determine the approximate probability of an undetected error.

Bibliography

1. Flores, I. *The Logic of Computer Arithmetic*, Prentice-Hall, Englewood Cliffs, N.J., 1963.
2. Blakeslee, T. R. *Digital Design with Standard MSI and LSI*, 2nd Ed., Wiley-Interscience, New York, 1979.
3. Maley, G. R. and J. Earle, *The Logic Design of Transistor Digital Computers*, Prentice-Hall, Englewood Cliffs, N.J., 1963.
4. Ware, W. H. *Digital Computer Technology and Design*, Wiley, New York, 1963.
5. Hamming, R. W. "Error Detecting and Error Correcting Codes," *BSTJ*, **29**: 2, 147–160 (April, 1950).
6. Hill, F. J. and G. R. Peterson, *Digital Systems: Hardware Organization and Design*, 2nd Ed., Wiley, New York, 1978.

7. Barna, A. and D. Porat. *Integrated Circuits in Digital Electronics*, Wiley-Interscience, New York, 1973.
8. Sandige, R. S. *Digital Concepts Using Standard Integrated Circuits*, McGraw-Hill, New York, 1978.
9. Kraft, G. D. and W. N. Toy. *Mini/Microcomputer Hardware Design*, Prentice-Hall, Englewood Cliffs, N.J., 1979.

9

INTRODUCTION TO SEQUENTIAL CIRCUITS

9.1 General Characteristics

Our attention up to this point has been directed to the design of *combinational* logic circuits. Consider the general model of a switching system in Fig. 9.1. Here we have n input, or excitation, variables, $x_k(t)$, $k = 1, 2, \ldots, n$ and p output, or response, variables, $Z_i(t)$, $i = 1, 2, \ldots, p$, all assumed to be functions of time. If at any particular time the present value of the outputs is determined solely by the present value of the inputs, we say that the system is *combinational*. Such a system can be totally described by a set of equations of the form

$$Z_i = F_i(x_1, x_2, \ldots, x_n) \tag{9.1}$$

where the dependence on time need not be explicitly indicated, since it is understood that the values of all the variables will be those at some single time. If, on the other hand, the present value of the outputs is dependent not only on the present value of the inputs but also on the past history of the system, then we say that it is a *sequential system*.

As an example to clarify the distinction, consider two types of combination* locks (Fig. 9.2). One type, the conventional combination padlock (Fig. 9.2a), has

---

* The word *combination* as used to identify the type of lock has no connection with the technical word *combinational* under discussion here.

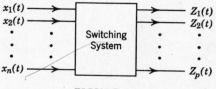

FIGURE 9.1

232

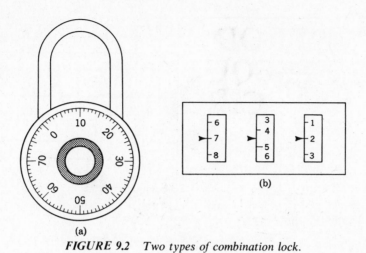

FIGURE 9.2 *Two types of combination lock.*

only one dial. The second type, sometimes used in luggage, has several dials (Fig. 9.2b). The inputs are the dial settings, the outputs the condition of the locks, open or closed. For the first type, the present condition of the lock (open or closed) depends not only on the *present* setting of the dial but also on the manner in which the dial has previously been manipulated. The second type will open whenever the three dials are set to the correct numbers; it does not matter what the previous settings were or in what order the dials were set. Obviously, then, the single-dial lock is a sequential device, the multiple-dial lock a combinational device.

The telephone system is another example of a sequential system. Suppose you have dialed the first six digits of a telephone number and are now dialing the last digit. This seventh digit is the present input to the system, and the output is a signal causing you to be connected to the desired party. Obviously, the present input is not the only factor determining which number you reach, since the other six digits are equally important.

Digital computers are very important examples of sequential systems. There are many levels of sequential operation in a digital computer involving sequential use of subsystems that are themselves sequential. Consider the addition of a number into the accumulator of a digital computer; the resultant sum depends both on the number and on what has been put into the accumulator previously.

9.2 Flip-flops

Clearly evident in the sequential devices and systems discussed above is the requirement to store information about previous events. In the combination

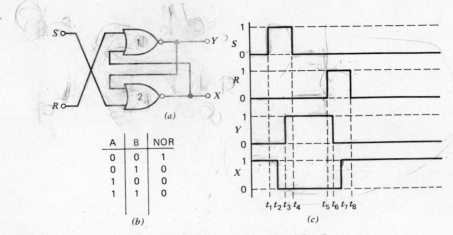

A	B	NOR
0	0	1
0	1	0
1	0	0
1	1	0

(b)

(c)

FIGURE 9.3   *Operation of flip-flop.*

lock, information about the previous dial settings is stored in the mechanical position of certain parts. In the dial telephone system, each number is stored as it is dialed in relays or electronic devices. In a computer, many different operations involve the storage of information.

There are many types of storage, or memory, devices. In electronic sequential circuits or systems, the most common memory device is the flip-flop.* Figure 9.3 shows the circuit for a flip-flop constructed from two NOR gates and the timing diagram for a typical operation sequence. We have also repeated the truth table for NOR for convenience in explaining the operation.

At the start, both inputs are at 0, the Y output is at 0, and the X output at 1. Since the outputs are fed back to the inputs of the gates, we must check to see that the assumed conditions are consistent. Gate 1 has inputs of $R = 0$ and $X = 1$, giving an output $Y = 0$, which checks. Similarly, at gate 2, we have $S = 0$ and $Y = 0$, giving $X = 1$. At time t_1, input S goes to 1. The inputs of gate 2 are thus changed from 00 to 01. After a delay (as discussed in Chapter 5), X changes from 1 to 0 at time t_2. This changes the inputs of gate 1 from 01 to 00, so Y changes from 0 to 1 at t_3. This changes the inputs of gate 2 from 01 to 11, but this has no effect on the outputs. Similarly, the change of S to 0 at t_4 has no effect. When R goes to 1, Y goes to 0, driving X to 1, thus "locking-in" Y so that the return of R to 0 has no further effect.

A flip-flop operating in the above fashion is known as a *reset-set* or *R-S* flip-flop and has the block diagram symbol in Fig. 9.4a.

The S and R inputs correspond to those in Fig. 9.3, and the Q and $\bar{Q}$ outputs

---

* The more formal name is *bistable multivibrator*. The origin of the name *flip-flop* is uncertain, but the name has become standard.

$S-0$
$R=0$

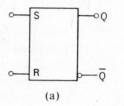

(a)

S^v	R^v	Q^v	Q^{v+1}
0	0	0	0
0	0	1	1
0	1	0	0
0	1	1	0
1	0	0	1
1	0	1	1
1	1	0	x
1	1	1	x

(b)

FIGURE 9.4   *R-S flip-flop.*

correspond to the Y and X outputs, respectively. A pulse on the S input will "set" the flip-flop—that is, drive the Q output to the 1 level and the $\bar{Q}$ output to the 0 level. A pulse on the R line will "reset" the flip-flop—that is, drive the Q output to the 0 level and the $\bar{Q}$ output to the 1 level. This behavior is summarized in tabular form in the *transition table*, shown in Fig. 9.4b. S^v, R^v, *and* Q^v indicate the values of the inputs and output at some arbitrary time t_v, and Q^{v+1} indicates the value to which the output will go as a result of an input at t_v. The reader should be satisfied that this table can be represented by the equation

$$Q^{v+1} = S^v + Q^v \bar{R}^v \qquad \text{(where } S^v \cdot R^v = 0) \qquad (9.1)$$

The don't-care entries for Q^{v+1} in the last two rows reflect the fact that in normal operation both inputs should not be permitted to be 1 at the same time. There are two reasons for this restriction. First, if both inputs are 1, both outputs will be driven to 0, which violates the basic definition of flip-flop operation, which requires that the outputs should always be the complements of each other. Second, if both inputs are 1 and then go to 0 at the same time, both NOR gates will see 00 at the inputs and both outputs will try to go to 1. However, because of the feedback, it is impossible for both outputs to be 1 at the same time (the reader should convince himself of this), with the result that the flip-flop will switch unpredictably and may even go into oscillation. With this restriction on the inputs, the R-S flip-flop functions reliably as a memory device in which the state of the outputs indicates which of the inputs was last at the 1 level.

In the discussion above we referred to *pulses* on the R and S lines, and it is important that the reader understand clearly what is meant by this term. Up to this point, except for a brief consideration of delay in Chapter 5, time has not been an explicit factor in our study of digital circuits. All signals were assumed to be *level signals*, i.e., signals that can remain at either value (0 or 1) for any length of time and can change from one value to the other at any arbitrary time. We have considered signals only when stable, and any previous values or transitions have been of no interest. With sequential circuits the situation is different. Consider the R-S flip-flop and assume that $R = 0$ and $S = 0$ at a particular time. If this were a combinational circuit we could determine the outputs from that information. But in this circuit we also need to know which input was most recently at the 1-level. The behavior in time has thus become a significant factor.

Again assuming $R = S = 0$ and also assuming $Q = 0$ (the flip-flop is reset), assume we wish to set the flip-flop. From Fig. 9.3 we see that this requires that S be raised to 1 while R is held at 0. If the circuit were combinational, we could return S to 0 at an arbitrary time, but this case S must be held at 1 for some minimum time if a change in the flip-flop state is to be accomplished reliably. If S returned to 0 before time t_3 (Fig. 9.3c), both inputs to gate-2 would again be 0 at least momentarily, tending to cause X to return to 1. In this situation the operation of the circuit would be unpredictable. Once the transition to the Set state has taken place, S can go back to 0 at any time since that transition will cause no further changes. We note, however, that S must go back to 0 before the flip-flop can be reset by a 1 on the R line, and it will be simplest to take S to the 1-level only long enough to assure reliable setting and then return it to the 0-level.

A signal of this sort that normally remains at one level and goes to the other level only for short periods of time is called a *pulse signal*. Pulses may be periodic, i.e., occurring at regular intervals (Fig. 9.5a), or aperiodic (Figs. 9.5b and 9.5c); they may be positive (Figs. 9.5a and 9.5b) or they may be negative (Fig. 9.5c). One problem with the above definition is the difficulty of specifying just what is meant by a "short" period of time. "Short" compared to what? As suggested in the discussion of the R-S flip-flop, pulses may be of the order of several gate delays, but they may be much longer and still be properly considered pulses. The important distinction between pulses and levels is in the way circuits respond to them. Although a pulse involves two transitions, regardless of its duration, it is considered to be a *single event* leading to a single response. The response may occur after the first transition, the *leading edge*, as is the case in the R-S flip-flop, or it may occur after the second transition, the *trailing edge*.

R-S flip-flops can also be constructed from NAND gates as shown in

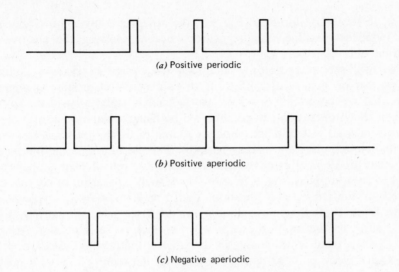

(a) Positive periodic

(b) Positive aperiodic

(c) Negative aperiodic

FIGURE 9.5   *Different types of pulse signals.*

Fig. 9.6a. In the timing diagram of Fig. 9.6c we see that the operation is similar to that of the NOR flip-flop of Fig. 9.3 except that the inputs are normally at the 1-level and negative pulses set or reset the flip-flop. The inversion circles at the inputs on the block diagram symbol (Fig. 9.6b) indicate that the state transitions occur following the negative-going transitions of the inputs, i.e., the transitions from the more positive value of the input to the more negative value. By

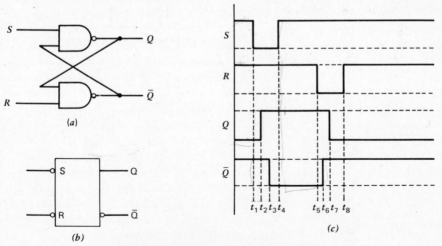

FIGURE 9.6   *Negative-input R-S flip-flop.*

contrast, in the NOR flip-flop (Fig. 9.3) the absence of inversion circles at the inputs indicates that the state transitions occur following the positive-going transitions of the inputs.

As we will see in succeeding chapters, the design of systems containing flip-flops can be greatly simplified if we can synchronize state changes, i.e., arrange that whenever any flip-flops are to change state, all will do so at the same time. Such operation can be achieved by using a common source of pulses to synchronize all state changes. Such a source of timing pulses is known as a *clock*, and clock pulses are usually periodic. The use of a separate timing signal implies that flip-flops must have separate inputs to control *what* happens and to control *when* it happens. Fig. 9.7a shows how such separation of control can be achieved in the NOR *R-S* flip-flop, and Fig. 9.7b shows a typical timing sequence. Note that all we have done is to add AND gates so that S and R will have no effect on the flip-flop except when a clock pulse is present. At t_1 both inputs are 0 so there is no change in the state of the flip-flop. At t_2, $S = 1$ and $R = 0$, so the clock pulse at t_2 sets the flip-flop. Between t_2 and t_3, R goes to 1

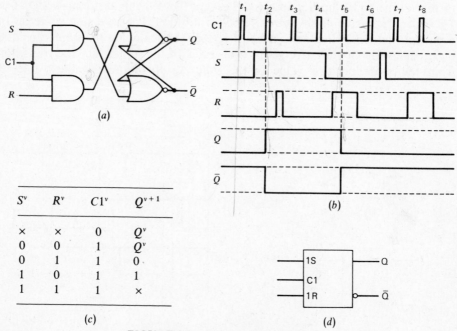

S^v	R^v	$C1^v$	Q^{v+1}
×	×	0	Q^v
0	0	1	Q^v
0	1	1	0
1	0	1	1
1	1	1	×

(c)

(d)

FIGURE 9.7   *Clocked R-S flip-flop.*

momentarily but this has no effect since the clock is 0 at this time. Between t_4 and t_5, S goes back to 0 and R goes to 1 so the pulse at t_5 resets the flip-flop.

Figures 9.7c and 9.7d show the transition table and block diagram symbol for this flip-flop, respectively. The first row of the transition table shows that the flip-flop will not change state as long as the clock is 0, regardless of S and R. When the clock is 1, the output will be determined by S and R in the same manner as in the circuit of Fig. 9.3. In normal clocked operation it is assumed, as in Fig. 9.7b, that S and R do not change during the clock pulse. The block diagram symbol (Fig. 9.7d) is an example of dependency notation. The input $C1$ is a *control input*, controlling the action of any other input with a label starting with the numeral 1. In this case, the effect of inputs $1S$ and $1R$ is *dependent* on the state of $C1$. When $C1$ is low, $1S$ and $1R$ have no effect on the flip-flop; when $C1$ is high, $1S$ and $1R$ control the operation in the normal manner associated with Set and Reset inputs to a flip-flop.

The block diagram symbol of Fig. 9.7d and those for other types of flip-flops to be discussed in this chapter are not a part of the 1973 IEEE standard mentioned in Chapter 3. That standard is not a complete standard, particularly in the area of sequential circuits, such as flip-flops. Since that time, efforts to develop a comprehensive set of standards for logic diagrams have continued; the resulting standards have been published, and there is little doubt that they will be formally adopted shortly. The gate symbols in the new standard are the same as those in the 1973 standard. The symbols for flip-flops introduced in this chapter and used in the remainder of the book follow the new standard.

The clocked R-S flip-flop is rarely encountered in practice, having been almost totally supplanted by the clocked J-K flip-flop. In the J-K flip-flop the restriction that both inputs may not be 1 at the same time, which may be inconvenient for the designer, is removed. The transition table and block diagram symbol for a J-K flip-flop are shown in Fig. 9.8. The behavior of the J-K is the same as the R-S (Fig. 9.7c), with J corresponding to S and K to R, except that, when

J^v	K^v	$C1^v$	Q^{v+1}
$\times$	$\times$	0	Q^v
0	0	1	Q^v
0	1	1	0
1	0	1	1
1	1	1	$\bar{Q}^v$

FIGURE 9.8 J-K flip-flop.

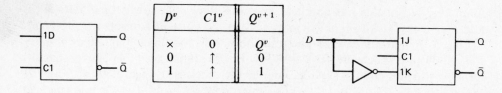

FIGURE 9.9   D flip-flop.

$J = K = 1$ at the time of a clock pulse, the flip-flop changes state. The equation for this flip-flop is

$$Q^{v+1} = \bar{K}^v Q^v + J^v \bar{Q}^v \tag{9.2}$$

Note that J-K flip-flops are always clocked.

Another type of flip-flop is the D flip-flop, the block diagram symbol and transition table for which are shown in Figs. 9.9a and 9.9b. This flip-flop has only a single data input, D, and is the most direct realization of a memory device, in that it simply stores the current value of D whenever it is clocked. The equation for the D flip-flop is simply

$$Q^{v+1} = D^v \tag{9.3}$$

D flip-flops are available in IC form and can also be constructed by connecting a J-K as shown in Fig. 9.9c.

A fourth type of flip-flop is the T or trigger flip-flop (Fig.9.10), which has only a single input, T. Whenever T is pulsed, the flip-flop changes state, so that the equation is simply

$$Q^{v+1} = \bar{Q}^v \tag{9.4}$$

T flip-flops are not available in IC form and are realized by connecting J-K flip-flops as shown in Fig. 9.10b.

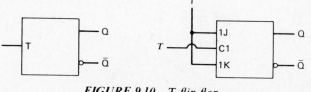

FIGURE 9.10   T flip-flop.

9.3 Why Sequential Circuits?

We have seen some examples of systems that are sequential, and it is obvious that many existing systems behave in a similar manner. As designers, however, our job is not merely to analyze but also to choose the best possible approach to a synthesis. Given a certain logical task to perform, we are faced with a specific decision with regard to implementation: should we use a combinational or sequential circuit? Now that we know how information can be stored in sequential circuit, let us consider some of the factors that should be considered in making this decision.

Consider the design of a tic-tac-toe machine. The overall "game system" may be represented as shown in Fig. 9.11. The game board might consist of switches by which the human player could indicate moves and lights to indicate the present state of the game. The machine player will consist of some sort of logic system to decide what moves should be made in response to the moves of the human player.

Clearly, the overall game progresses sequentially, but this does not mean that the machine player has to be a sequential device. It could be a combinational circuit, receiving from the game board information as to the state of the board at any given time and deciding on a move solely on the basis of that current information. There are nine squares on the board and three possible conditions (blank, O, X) for each square, so there are $3^9 = 19,683$ possible game situations. Thus, the combinational approach involves a circuit to implement the following logical requirement: given any one of 3^9 possible input combinations, choose the proper one of nine possible responses. It is fairly evident that satisfaction of this requirement is going to involve a massive combinational logic circuit.

As an alternative approach, let us provide the machine player with some means of altering its structure as the game progresses to take advantage of the fact that not all of the 3^9 situations can occur at any particular point in the game. Let us assume that the human player moves first. There are only nine possible moves, so the machine will start in a configuration designed to select the proper response to whichever move is made. Furthermore, when the machine selects its response, it then knows which seven responses are available to the human on the next move, and it can alter its circuitry to take this into account. A similar sequence of events is repeated at each successive move.

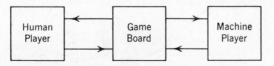

FIGURE 9.11   Model of tic-tac-toe game.

In this sequential approach, we are simply breaking a large problem into a number of smaller problems. Rather than building one large circuit capable of dealing with all the situations that might occur at any point in the game, we build a number of smaller circuits,* each capable of dealing with the limited number of situations possible at some particular point in the game. The machine player then sequences from one smaller circuit to another as the game progresses.

In this case, the problem is inherently sequential, since it involves a continuous interchange of information between the two players. Interchanges closely resembling a question-and-answer format are common between individual units of a digital system. Such inherently sequential processes lead naturally to a sequential design approach, which is generally more economical than a purely combinational approach. The methods to be developed in the following chapters will enable the reader, should he be so inclined, to design his own tic-tac-toe machine and check the statement that a sequential design is more economical.

As another example, consider an 8-bit odd-parity checker. If all 8 bits are available simultaneously, the desired function can be realized by a six-level combinational circuit described by the equation

$$Z = [(x_1 \oplus x_2) \oplus (x_3 \oplus x_4)] \oplus [(x_5 \oplus x_6) \oplus (x_7 \oplus x_8)] \tag{9.5}$$

requiring 21 gates in a NAND realization. It will be recalled that in designing the similar even-parity checker of Fig. 8.44 we broke it down into parts, considering each pair of digits, then each pair of pairs, etc.

Next, let us assume that the 8-bit word is a unit of data that must be transmitted over a long distance—in an airline reservation system, for example. We could use eight lines, one for each bit, but this would be expensive. Instead, let us use a single line and send the 8 bits in sequence, x_1 at time t_1, x_2 at t_2, etc. On a second line, we send *clock*, or synchronizing, pulses to mark these times t_1, t_2, etc.† These two lines will provide the inputs to a circuit that is to put out a signal if the number of 1s in each group of 8 bits is not even (Fig. 9.12).

Each bit could be stored as it arrived until all 8 were present. They could then be applied to the six-level combinational circuit; but this would be very expensive, requiring eight flip-flops for storage plus the 21-gate network. Instead, we note that the exclusive-OR is associative, and rewrite Equation 9.5 in the form

$$z = \left(\left\{ \left[(\{[(x_1 \oplus x_2) \oplus x_3] \oplus x_4\} \oplus x_5) \oplus x_6 \right] \oplus x_7 \right\} \oplus x_8 \right) \tag{9.6}$$

---

* As will become evident in succeeding chapters, these circuits will usually not be completely independent.

† The synchronizing line would also be needed with eight-line approach to indicate when a data word was present.

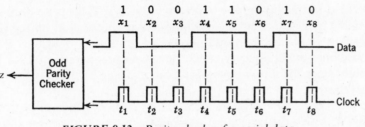

FIGURE 9.12 *Parity checker for serial data.*

Functionally, this means that first 2 bits are checked for parity, then the third bit is checked against the results of the first check, etc. (The reader should try this method on a specific example to make certain that it works.)

We do not suggest a combinational realization of Equation 9.6, which would be very inefficient. Instead, the first 2 bits are checked in an exclusive-OR network and the results stored until the third bit arrives. Then the third bit is compared to the stored results *in the same exclusive-OR* network. This result is stored in turn until the fourth bit arrives, etc. Figure 9.13 shows a simple circuit for realizing this method.

Initially, the flip-flop must be cleared to $z = 0$ before x_1 arrives so that the first operation is

$$x_1 \oplus z = x_1 \oplus 0 = x_1$$

This is stored in the flip-flop when a clock pulse arrives at time t_1. That is, if $x_1 \oplus z = 1$ when the clock pulse arrives, a pulse will be gated to the S input setting the flip-flop to 1. Similarly, the flip-flop will be set to 0 if $x_1 \oplus z = 0$ when the clock pulse arrives. Then x_2 arrives, and the exclusive-OR forms

$$x_2 \oplus Z = x_2 \oplus x_1$$

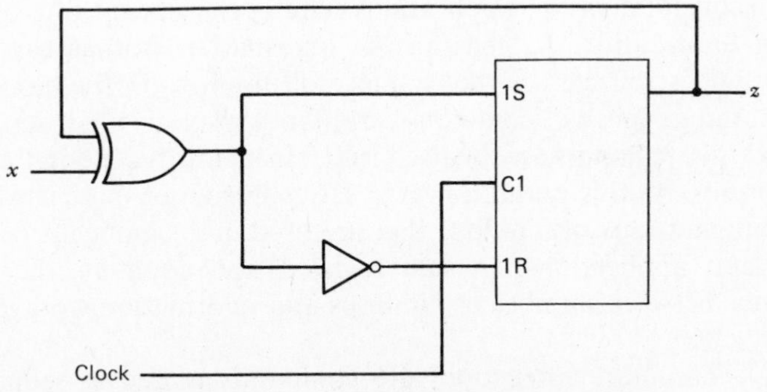

FIGURE 9.13 *Sequential parity checker.*

which is stored in the flip-flop by the clock pulse at t_2. This procedure continues until, after the clock pulse at t_8, the output z is the desired parity check, i.e., $z = 1$ only if there has been an odd number of 1s in the 8 data bits. In fact, the output z provides an odd-parity check on whatever number of data bits have been received since the flip-flop was last cleared. Thus, Fig. 9.13 is a general n-bit odd-parity checker.

The tic-tac-toe machine and the serial parity checker are examples of the type of situation in which the sequential approach is most often preferred. In neither case is there any speed advantage to using a combinational approach. In the tic-tac-toe machine, the speed of the game is dictated by the speed with which the human player makes his moves. The sequential parity checker is slower than the combinational circuit if we measure the speed of the combinational circuit *from the time that all 8 bits are available*; but if the bits are going to arrive in sequence anyway, there is no difference in speed. The sequential circuit, however, is simpler, and therefore cheaper, than the combinational circuit.

Data always arrive sequentially at some level or another, usually for reasons of cost or hardware limitations rather than logical necessity. This is particularly true for data modulated on a carrier and transmitted over a long distance. Disc and drum memories generally put out data 1 bit at a time, while data from magnetic or paper tape generally arrive 1 *character* at a time (i.e., 6 or 8 bits at a time). Data from punched cards may arrive 1 character at a time or 80 characters at a time. The output of a core memory will arrive in groups (words) of anywhere from 12 to 100 bits at a time. Whatever the size of the data groupings, there will certainly be no speed disadvantage to processing the data as it arrives, and there may be a definite cost advantage.

9.4 Shift Registers and Counters

In Fig. 9.13, we saw a sequential circuit employing a single flip-flop. As we might expect, most sequential circuit applications require the storage of more than one binary bit of information. In general, the information storage capability of a sequential circuit is reflected by the number of flip-flops in the circuit. Thus, a typical sequential circuit will contain several flip-flops.

In the next five chapters, we will consider in detail the design of arbitrary sequential circuits. At this point, however, let us introduce in general terms two particular configurations of flip-flops that are used in a significant percentage of sequential-circuit applications. In shift registers and counters, the one-to-one correspondence between number of flip-flops and information storage capability is particularly clear.

In a digital computer, instructions are commonly stored in sequential locations in memory. Thus, if the first instruction were in location 501, for example,

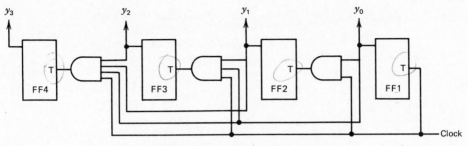

FIGURE 9.14 *Modulo-16 counter.*

the second instruction would be in 502, the third in 503, etc. To control the sequencing of the program, the computer uses an *instruction counter*. At the start of the program, the contents of the counter are set to the address of the first instruction. As each instruction is executed, the counter is incremented by one, thus changing the address to that of the next instruction. This is just one example of the use of counters in digital systems.

Although counters may be designed by using any kind of flip-flop, the functioning of a trigger flip-flop counter is most quickly understood. A modulo-16 counter is shown in Fig. 9.14. By modulo-16 we mean that the counter counts from 0 through 15 and then back to 0. In Fig. 9.14, y_0 represents the least significant bit.

Suppose that initially the count is 0, that is, $y_3 y_2 y_1 y_0 = 0000$. With the arrival of the first input pulse, flip-flop 1 will change state. No pulses will appear at the output of the AND gates, so the other three flip-flops will remain in the 0-state. Thus, the count will have advanced to $y_3 y_2 y_1 y_0 = 0001$. With y_0 now 1, the second input pulse will propagate through the first AND gate, triggering y_1 to 1 while at the same time triggering y_0 to 0. Thus, the count will advance to $y_3 y_2 y_1 y_0 = 0010$. In general, a flip-flop will change state at the time of an input pulse if all less significant flip-flops are in the 1 state, so that the count progresses as shown in Fig. 9.15.

Most counters will sometimes have to be reset to zero or set to some specified starting value. Circuitry for this purpose will generally be superimposed on that shown in Fig. 9.14.

In addition to counting, a counter also *stores* the current count in the intervals between input pulses. In general, any set of flip-flops used to store related bits of information is known as a *register*. A counter may thus be referred to as a counting register. Often a register is used to assemble and store information arriving from a serial source. The circuit shown in Fig. 9.16, called a *shift register*, is set up to accomplish this task. It is assumed that a new bit of information may be sampled from line *X* with the arrival of each clock pulse. If

| Count Before Clock Pulse | Count After Clock Pulse |
$y_3 y_2 y_1 y_0$	$y_3 y_2 y_1 y_0$
0 0 0 0	0 0 0 1
0 0 0 1	0 0 1 0
0 0 1 0	0 0 1 1
0 0 1 1	0 1 0 0
0 1 0 0	0 1 0 1
0 1 0 1	0 1 1 0
0 1 1 0	0 1 1 1
0 1 1 1	1 0 0 0
1 0 0 0	1 0 0 1
1 0 0 1	1 0 1 0
1 0 1 0	1 0 1 1
1 0 1 1	1 1 0 0
1 1 0 0	1 1 0 1
1 1 0 1	1 1 1 0
1 1 1 0	1 1 1 1
1 1 1 1	0 0 0 0

FIGURE 9.15

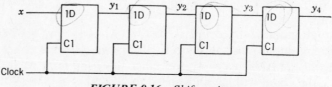

FIGURE 9.16   Shift register.

the first bit is 1, flip-flop 1 will be set to 1 by the clock pulse. If the first bit is 0, flip-flop 1 will be reset to 0. The second clock pulse will similarly shift the second bit into flip-flop 1 while simultaneously shifting the first bit into flip-flop 2 and so on as each successive bit is sampled.

9.5   Speed-Versus-Cost Trade-off

In situations such as those described earlier, in which the information to be processed arrives in time sequence, sequential circuits are generally the natural choice. Since sequential circuits often use the same equipment repetitively on successive pieces of information, they generally provide the most economical means of accomplishing a specified task. If all the information required for a certain task is available simultaneously, it is obviously possible to process it all simultaneously. However, if the nature of the task is such that it may be broken up into identical (or at least similar) segments, then it may be advantageous to serialize the process deliberately in order to gain the economy of repetitive application of a single unit of equipment. In such situations, a trade-off between speed and cost may exist.

A number of complex factors are involved in deciding whether serialization is desirable, some of which are illustrated by the following example. Let us assume that a 32-bit word is to be read out of memory into a 32-bit flip-flop register and checked for even parity. Since the bits are all present at once, a combinational circuit might seem logical. If we follow the same approach as used in the 8-bit checker of Fig. 8.36, we find that a rather expensive 10-level, 93-gate circuit is required. We recall that the sequential checker, Fig. 9.13, can be used for any number of bits, so perhaps it would be cheaper to provide circuitry to sequence the 32-bits into this checker one at a time. The system of Fig. 9.17 provides a straightforward realization of this approach.

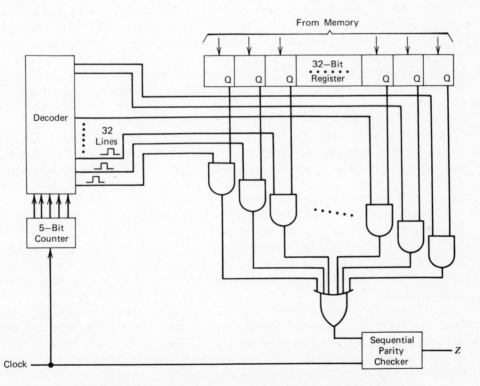

FIGURE 9.17 Parity check by serialization of parallel data.

The clock-pulses drive a 5-bit (modulo-32) counter, which in turn drives a decoder. The decoder produces gating pulses one at a time, in sequence, on 32 output lines. These gating pulses are applied to AND gates to sequence the 32 bits into the parity checker. Using the simplest possible design for the counter and decoder, this circuit requires 6 flip-flops and 72 gates so that the cost of the two methods would be roughly comparable. However, in a computer, the counter and decoder might be used for other purposes, in which case only a part of their cost should be charged to the parity checker.

We have a general and perhaps obvious principle that the evaluation of any circuit must consider its relation to the overall system. A cost evaluation certainly must take into account any possible sharing of circuitry with other parts of the system, as above. Similarly, an evaluation of speed advantage must consider the value of speed to the overall system. The combinational circuit in the above example would be about 32 times as fast as the sequential, but perhaps the parity check can be carried out simultaneously with some other operation in such a manner that the speed advantage is of little value.

In the general case, we have a device, or system, that is presented with n units of information all at the same time. The units may be bits, words, characters, or any other convenient grouping. In a fully *parallel* system, a separate circuit, or subsystem, is provided for each of the n units of information, and the n units of information are supplied to all circuits simultaneously. Because of differences between the various circuits or logical dependence of one on another, not all circuits may complete their operations at the same time. But they all start together, working on all the available information at once.

In a fully *serial* system, all n units of information are applied to a single circuit, one at a time, in sequence. The parity checker of Fig. 9.13 is thus fully serial. Assuming circuit elements of the same speed and cost, the fully parallel system will be the fastest and most expensive approach, and the fully serial will be the slowest and least expensive. Generally, there are alternatives available between these two extremes. We might divide the n units of information into two equal groups and process the first half in one time period, followed by the second half in a second time period. This approach would generally be less expensive than the fully parallel and about half as fast. Similarly, we could divide the information into three groups, four groups, etc., until we reached the fully serial approach.

If we were to plot cost versus speed for all these possible alternatives, we would get a plot looking something like that shown in Fig. 9.18. Starting with the fully parallel system, the cost goes down as the speed goes down. The cost may continue to decrease until the fully serial system is reached. More often, however, the curve levels off before reaching the limit of fully serial operation. This occurs because the cost of sequencing the data tends to cancel out the savings resulting

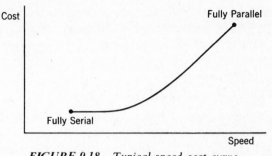

FIGURE 9.18   *Typical speed-cost curve.*

from the reduction in the number of circuits required for the actual processing. We saw this to be the case in the parity checker of Fig. 9.17. Just where this leveling off occurs depends on many factors, as indicated in the parity checker example. Once a cost-speed relationship has been determined, the designer must then decide how much speed he can afford.

9.6   A General Model for Sequential Circuits

The general model for sequential circuits, shown in Fig. 9.19, consists of two parts, a combinational logic section and a memory section. The memory stores information about past events required for proper functioning of the circuit. This information is represented in the form of r binary outputs, $y_1, y_2, \ldots, y_r$, known as the *state variables*. Each of the 2^r combinations of state variables defines a *state* of the memory; in general, each state corresponds to a particular combination of past events.

The combinational logic section receives as inputs the state variables and the n circuit inputs, $x_1, x_2, \ldots, x_n$. It generates m outputs, $z_1, z_2, \ldots, z_m$ and p *excitation variables*, $Y_1, Y_2, \ldots, Y_P$, which specify the next state to be assumed by the

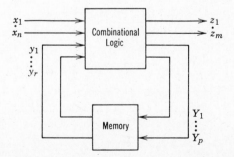

FIGURE 9.19   *General model of sequential circuits.*

memory. The number of excitations is often equal to the number of state variables, but this depends on the type of memory. The combinational logic network may be defined by Boolean equations of the standard forms, i.e.,

$$z_i = f_i(x_1, x_2, \ldots, x_n, y_1, y_2, \ldots, y_r) \quad (i = 1, 2, \ldots, m) \tag{9.7}$$

$$Y_j = f_j(x_1, x_2, \ldots, x_n, y_1, y_2, \ldots, y_r) \quad (j = 1, 2, \ldots, p) \tag{9.8}$$

These are time-dependent equations, i.e., they are valid at any time all the inputs and outputs are stable, as for any combinational circuit.

We have not yet specified the relationship between the inputs and outputs of the memory portion of Fig. 9.19. This relationship will depend on whether the memory is realized in terms of an array of memory elements or flip-flops or is merely a set of feedback loops with time delay. These two points of view will be considered in Sections 9.7 and 9.9.

9.7   Clock-Mode and Pulsed Sequential Circuits

A model for the *clock-mode*, or simply *clocked*, sequential circuit is shown in Fig. 9.20. The memory consists of r clocked flip-flops, which may be any type, although we show *J-K*. The excitations thus consist of r J-signals and r K-signals. We show the state variables coming from the center of the flip-flops, because either the true or complemented outputs, or both, may be used.

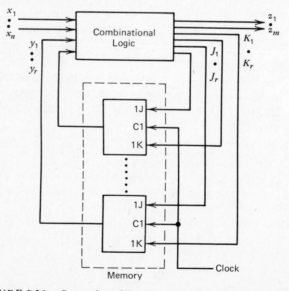

FIGURE 9.20   *General model of clock mode sequential circuit.*

All inputs and outputs of the combinational logic section are *level* signals, i.e., signals that may take on either the 1 or 0 values for arbitrary periods of time. As discussed above, the combinational logic is time-independent, so that the outputs and excitations may change at any time in response to changes in the circuit inputs. However, the state variables will change only when the flip-flops are triggered by a clock pulse. The J and K excitation variables may change at any time in response to changes in the x input variables, but no change in state will occur until a clock pulse arrives. The clock pulse contains no information in the sense of determining *what* state change will take place, it simply times, or synchronizes,* the state change. The values of the J and K signals at the time of the clock pulse determine *what* transition takes place. The times of arrival of the clock pulses are denoted by $t_v (v = 1, 2, 3, \ldots)$, and the values of the variables at these times are denoted by the superscript v. At time t_v, the excitation variables, $J_1, K_1, J_2, K_2, \ldots, J_r, K_r$, must be at stable levels determined by the current values of the inputs and state variables. The outputs are similarly stable at t_v and may be used as the inputs to other circuits synchronized by the same clock. The clock pulse arriving at t_v will cause a transition of the state variables from y_i^v to y_i^{v+1}, thus placing the circuit in a new state. The excitation and output variables will in turn adjust to this state change, as well as any input changes, reaching new stable levels before the next clock pulse, at t_{v+1}.

This circuit behavior may be described by the equations

$$z_i^v = f_{zi}(x_1^v \cdots x_n^v, y_1^v \cdots y_r^v) \qquad (i = 1, 2, \ldots, r) \qquad (9.9)$$

$$J_i^v = f_{Ji}(x_1^v \cdots x_n^v, y_1^v \cdots y_r^v) \qquad (i = 1, 2, \ldots, r) \qquad (9.10a)$$

$$K_j^v = f_{Kj}(x_1^v \cdots x_n^v, y_1^v \cdots y_r^v) \qquad (j = 1, 2, \ldots, r) \qquad (9.10b)$$

$$y_j^{v+1} = \bar{K}_j^v y_j^v + J_j^v \bar{y}_j^v \qquad (j = 1, 2, \ldots, r) \qquad (9.11)$$

Equation 9.11 characterizes the behavior of a J-K flip-flop, as discussed in Section 9.2. In that discussion, we used the notation Q^v and Q^{v+1} to refer to the "present" and "next" values of the flip-flop output. Here y^v refers to the value of y at the time of the clock time t_v and y^{v+1} to the value at clock time t_{v+1}. This does not imply that y does not change to y^{v+1} until t_{v+1}; it generally changes immediately in the conclusion of the clock pulse, t_v. However, this value (y^{v+1}) is the value it will have at t_{v+1}, and that is the value that is significant in terms of the next transition.

To illustrate the fact that the inputs, outputs, and state variables in a clock-mode circuit are only of interest at the time of a clock pulse, let us consider the circuit of Fig. 9.21a. This 2-bit shift register circuit will have an output of 1 if and

---

* Systems of this type are sometimes referred to as *synchronous* systems, but the word has been used so many different ways as to be virtually meaningless.

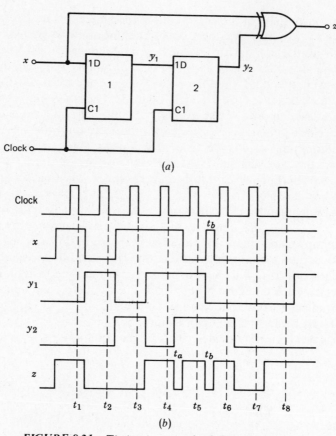

FIGURE 9.21   *Timing in a simple shift register circuit.*

only if the present input value is *not* the same as the input two clock periods earlier, which is stored in flip-flop 2. Note first in Fig. 9.21b that y_1 is a near replica of the input x delayed by one clock period. At time t_1, for example, $x = 1$, so that the pulse at time t_1 causes y_1 to take on this value and remain 1 until after the clock pulse at time t_2. Succeeding values of x are similarly shifted into flip-flop 1 by succeeding clock pulses. Note next the brief pulse appearing on input line x at time t_b. This pulse does not overlap a clock pulse so it has no effect on y_1.

At time t_2, $y_1 = 1$, so that the corresponding clock pulse triggers this value into flip-flop 2. The value of y_2 remains 1 until after the pulse at time t_3. With each succeeding clock pulse, a new value is shifted from y_1 to y_2. Thus, y_2 is a replica of y_1 delayed one more clock period.

The output z may be found by continuously computing the exclusive-OR function of the values of y_2 and x. The resulting values of z are plotted as the last line of Fig. 9.21b. Note the brief zero-going pulse at time t_a. This pulse occurs because both x and y_2 change values after time t_4, but y_2 changes first. Also note that the narrow input pulse at time t_b causes another zero-going output pulse at the same time. Neither of these output pulses overlaps a clock pulse. Suppose that this output serves as the input to another clock-mode sequential circuit. Just as the input pulse at time t_b did not affect the state of the circuit of Fig. 9.21a, the two output pulses will not affect the state of succeeding clock mode circuits because they end before the occurrence of the next clock pulse.

Closely related to clock-mode circuits are *pulsed* sequential circuits. Pulsed circuits are distinguished by the absence of a separate clock line. They are similar to clock-mode circuits in that every state change must coincide with an input pulse. The pulses that trigger state changes must appear on the circuit input lines and must satisfy timing restrictions similar to any set forth for a clock-pulse line. They must be of sufficient duration, and the circuit must have time to stabilize from an input pulse on one line before a second pulse can occur on that or any other line. Pulsed circuits will be discussed in more detail in Chapter 11.

9.8 Timing Problems in Flip-flops

We noted earlier a first timing restriction on flip-flops—that the clock or trigger pulses must have a certain minimum duration. In this section, we wish to investigate some additional restrictions, some applying to any type of circuit, some peculiar to circuits employing flip-flops. One fairly obvious restriction, which applies to all types of logic circuits, combinational as well as sequential, is that the interval between successive changes of a single variable must not be less than the basic delay time of the logic circuits. If the basic delay time of a gate is 10 ηsec, for example, it certainly cannot respond reliably to signal changes only 5 ηsec apart. In simpler terms, the designer must select a family of logic circuits fast enough to keep up with expected signal rates.

In discussing flip-flops, we introduced clocking as a means of minimizing timing problems in flip-flops, but this should not be taken to mean that clocked flip-flops are free of timing problems. They are not, but it is often simpler to enforce timing restrictions on a single source of control pulses than on a variety of level signals. In discussing the R-S flip-flop with reference to the timing of Fig. 9.3, we noted that, if S changed back to 0 before the resultant change of Y at t_3, the flip-flop would not trigger reliably. This is simply a specific example of the restriction noted above, that the interval between two successive changes must not be too short. In terms of flip-flops, this means there is a minimum duration for trigger pulses, primarily determined by the basic gate delay but also affected by other complex electronic factors. This minimum trigger duration will be a basic specification provided by the flip-flop manufacturer.

Next, if one pulse is followed immediately by another, the leading edge comprises a third successive change in the same variable and, by the same argument as above, this cannot occur too soon. Thus, there is a minimum interval between trigger pulses, most frequently stated in terms of a maximum clock rate. For example, the minimum interval between trigger pulses for a 10 megahertz flip-flop is 100 ηsec. Because of electronic factors too complex to discuss here, the minimum trigger interval is generally much longer than the minimum trigger duration. For example, a typical 10-MHz flip-flop would have a minimum pulse duration of 20 ηsec, one fifth the minimum interval.

The above restrictions on trigger duration and interval apply to all flip-flops regardless of how they are used. When flip-flops are so connected that their outputs affect the inputs of other flip-flops or even their own inputs, still other problems arise. The nature of these problems can be seen very clearly in considering the design of a *J-K* flip-flop. In the clocked *R-S* flip-flop of Fig. 9.7a, if both *R* and *S* are 1 at clock time, the clock pulse will be gated to both NOR gates, tending to drive both to the same level and resulting in improper operation. The basic technique in building a *J-K* flip-flop is to start with an *R-S* and feed the outputs back to the inputs in such a way as to "steer" the clock pulse to the correct NOR gate, as shown in Fig. 9.22. Note that the circuit is the same as for the *R-S* of Fig. 9.7a except that the outputs are fed back to the input AND gates. Assume that the flip-flop is reset and $J = K = 1$ when a clock pulse arrives. The combination of $Q = 0$ and $\bar{Q} = 1$ will steer the clock pulse, via the upper AND gate to the lower NOR gate, to set the flip-flop. If the flip-flop is set, the clock pulse will be gated through the lower AND gate to the upper NOR gate, to reset the flip-flop.

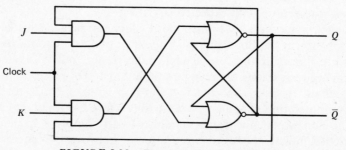

FIGURE 9.22   Basic circuit of J-K flip-flop.

This all seems reasonable at first glance, but let us consider the operation more carefully. If the circuit is reset, the 0 at Q will block the clock signal at the lower AND gate, preventing it from reaching the upper NOR gate, and the flip-flop will set. When it sets, Q will change to 1. If the clock is still high when this new value of Q feeds back to the lower AND, this signal to the upper NOR will go high, driving the flip-flop back to reset. Indeed, this circuit will oscillate between set and reset as long as the clock is held high with $J = K = 1$. The same sort of problem can occur when one flip-flop drives another that is controlled by the same clock. Consider the shift register of Fig. 9.21, which is repeated as Fig. 9.23 with a slightly different clock source. Note that clock pulse 5 is significantly wider than the other pulses, which were tuned to the delays in the two D flip-flops shown. We see that $x = 1$ at the onset of clock pulse 5. This value is shifted into flip-flop 1 and appears as output, y_1, while clock pulse 5 is still 1. This means that the input to flip-flop 2 is now 1, while clock pulse 5 is still on. Unfortunately, this causes y_2 to go to 1, also. This was *not* the desired behavior. Note that the value of x *was shifted two places in the shift register by 1*

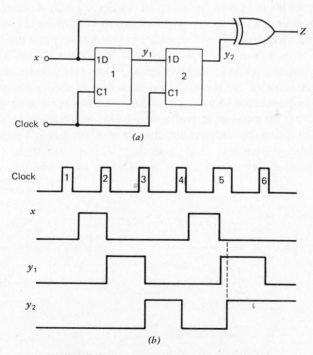

(a)

(b)

FIGURE 9.23   *Double transition problem.*

clock pulse. Proper behavior calls for the shift of each bit in the register 1 place by each clock pulse.

Let us consider a flip-flop to be *enabled* any time the clock signal is such that the flip-flop can respond to the data inputs, i.e., *J-K* or *R-S* or *D* signals, and *disabled* when it cannot. Problems of the sort described above can occur when the outputs change while the flip-flop is still enabled. One way to eliminate these double transitions is to find some way to ensure that output changes do not occur until after the flip-flop has returned to the disabled state. All of the flip-flops we have considered so far are enabled as long as the clock pulse is present, so that the only remedy for the problems discussed above is to make the clock pulse short enough to ensure that the flip-flop is disabled before the outputs change. However, as we have seen, if the pulse is too short the flip-flop will not trigger reliably, so we have a difficult problem: to ensure that the pulse is long enough, but not too long. It may be possible to do this for individual flip-flops, but it is virtually impossible when using a number of flip-flops, each with different delay characteristics. Reliable correction of these problems requires modifying the design of the flip-flops to ensure that the outputs do not change while the flip-flop is enabled. There are many possible ways to do this, but we will consider only two of the most widely used approaches, one or the other of which is used in the majority of standard IC flip-flops available today.

The first type of circuit we will consider is the *master-slave* flip-flop, which is nothing more than a cascade of two flip-flops, as shown in Fig. 9.24a. The two *R-S* flip-flops are elementary clocked flip-flops, such as shown in Fig. 9.7. The clock is applied directly to the first flip-flop, the *master*, and inverted before being applied to the second flip-flop, the *slave*. When the clock is low the master is disabled so that changes in *R* and *S* will have no affect on the flip-flop. The slave is enabled at this time but cannot change because it is driven by the master, which is disabled. When the clock goes high the master is enabled and will behave as determined by the values of *R* and *S* at that time. At the same time,

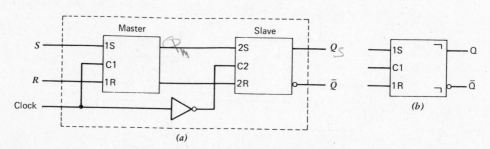

FIGURE 9.24 *Master-slave R-S flip-flop.*

the inverted clock disables the slave, so it will not be affected by any changes at the output of the master. This requires that the inverted clock disable the slave before the outputs of the master can change, but this is easy to arrange by controlling the internal delays in the flip-flop. When the clock goes back down the master is again disabled, isolating the flip-flop from changes in R and S. The inverted clock enables the slave, and the setting of the master, determined by R-S when the clock went high, is copied into the slave. Because the outputs of the complete flip-flop are taken from the slave, we are thus assured that the outputs cannot change until *after* the clock pulse has ended.

Although the master-slave flip-flop does consist of two pairs of cross coupled gates, it is considered to be a single flip-flop for logical purposes, with the block diagram symbol (Fig. 9.24b) indicating only its signal characteristics, not its internal structure. The symbol ($\sqsupset$) at the outputs indicates that a pulse on the $C1$ line is required to activate a transition in this flip-flop and that the outputs will change on the falling edge of the pulse.

The master-slave concept can be applied to any type of flip-flop and can be physically realized in a variety of circuits. Fig. 9.25 shows a typical TTL realization of a master-slave J-K flip-flop in all-NAND form. This circuit functions in the same manner as the elementary J-K shown in Fig. 9.22 except that it cannot change more than once in response to a single clock pulse, no matter how long. The outputs that are fed back to steer the clock to the proper side of the master are taken from the slave and thus cannot change until after the clock pulse has ended. In this form there is no separate inverter for the clock, the necessary, inversion being provided by the input NAND gates of the master, the outputs of which go directly to the slave as well as to the master.

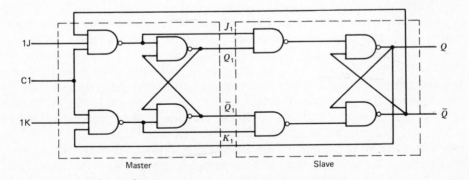

FIGURE 9.25 *Master-slave J-K flip-flop.*

9.9   The Standard Edge-Triggered D Flip-flop

A second type of flip-flop that can be used to assure that all state changes are synchronized at a single point in time is the edge-triggered D flip-flop, as shown in Fig. 9.26. The circuit is complex and is most effectively described using the level-mode sequential circuit techniques to be developed in Chapter 14. For now consider Fig. 9.26b, which defines the input-to-output behavior of the circuit.

Note that Q goes to 1 immediately following the first positive transition on the clock line, as marked by the arrow. The circuit is designed so that this clock transition triggers the state change, and state changes cannot take place except at the time of a positive clock transition. We see in fact that D returns to 0 while the clock remains high, but no further change in Q occurs. Similarly D is ignored while the clock is 0. A second change in Q does not take place until the second positive clock transition, also marked by an arrow. In order for the circuit to work reliably, the data input, D, must always be stable throughout a short interval around each positive clock transition. The "triangle" at the $C1$ input of the block diagram symbol (Fig. 9.26c) indicates that triggering takes place on the rising edge of clock signal. A triangle preceded by an inversion bubble indicates triggering on the falling edge of the clock signal.

The transition period during which the input must be stable is a factor of both the circuit parameters and the rise time of the clock pulse. The value of D must be "set up," i.e., established at a stable value a specified time before the clock transition starts, and must "hold" at that value until a specified time after the clock transition is complete. The manufacturer will provide specifications for t_{setup} and t_{hold}. The timing diagram of Fig. 9.27 will clarify the meaning of these terms. The total time during which the D input must be held stable is the sum of the setup time, the rise time, and the hold time, and is typically about 35 ηsec for standard TTL.

Although the edge-triggered and master-slave flip-flops are similar in that both delay the switching of the outputs until a time when further changes at the inputs can have no effect, there are important differences. In a master-slave flip-flop, depending on the current state of the outputs, changes at the input *during* the clock pulse may or may not cause changes at the output. To ensure reliable operation, the data inputs must be held stable throughout the clock pulse. In the edge-triggered flip-flop, by contrast, the inputs need be held stable only at the time of the triggering-edge. As a general rule, an increase in the time inputs must be held stable will require a slower clock, so that edge-triggered flip-flops are generally preferable. D flip-flops are almost invariably edge-triggered, but J-K flip-flops may be of either type, and the designer should check the data sheets carefully to be sure which type he is getting. For example, the 74107 J-K flip-flop is master-slave, but the 74LS107 is falling-edge-triggered.

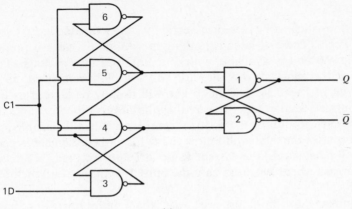

(a)

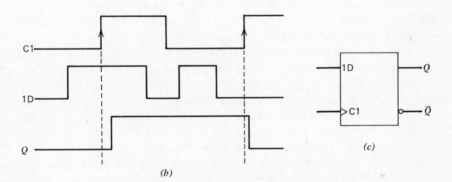

(b)

(c)

FIGURE 9.26 *Edge-triggered D flip-flop.*

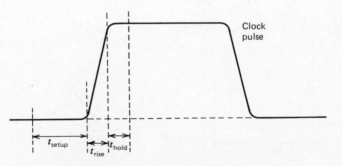

Clock
pulse

t_{setup} t_{rise} t_{hold}

FIGURE 9.27 *Transition time specifications.*

Elementary flip-flops such as discussed in Section 9.2 are available in IC form, but are listed as *latches*. The $\bar{S}$-$\bar{R}$ latch is nothing more than the negative-input R-S flip-flop of Fig. 9.6 and is available with four flip-flops in a package. The D-latch consists of the clocked R-S flip-flop of Fig. 9.7 connected as a D-flip-flop, as shown in Fig. 9.28. The clock line will usually be labeled as the *enable* line. In contrast to the edge-triggered D flip-flop, output changes in latches are not restricted to occurring on clock transitions. When the enable line is held high the output of the latch will follow the D line; when the enable goes low the flip-flop will *latch* (store) the current value of D. For this reason latches are used in applications where synchronizing the output changes of all memory elements is not critical.

As a final comment on flip-flops, we note that many flip-flops have direct-reset and direct-set lines by means of which the flip-flop can be set or reset independently of the values on the data or clock lines. These lines are generally used to establish initial values in flip-flops when systems are turned on or reset. For example, if we have a counter in a system we might want to initialize the counter to zero when the system is started up. For this purpose we could tie all the clear inputs together and apply the appropriate voltage to set all the flip-flops to 0. Figure 9.29 shows the D flip-flop of Fig. 9.26 with set and reset added. As shown in the schematic (Fig. 9.29a), the set and reset lines are applied directly to the output flip-flop. These lines are normally held high, in which case they have no effect on the operation of the flip-flop. If the set line goes low it drives gate 1 high, setting the flip-flop; if the reset line goes low it drives gate 2 high, resetting the flip-flop. The set and reset lines are also applied to gates 3, 5, and 6 to prevent the flip-flop from switching to the opposite state if the clock is high when the set and reset signal goes back to 1. Figure 9.29b shows the transition table including the set and reset lines, and Fig. 9.29c shows the block diagram symbol. The inversion bubbles on the R

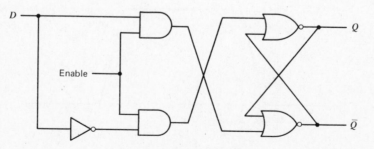

FIGURE 9.28   D-latch.

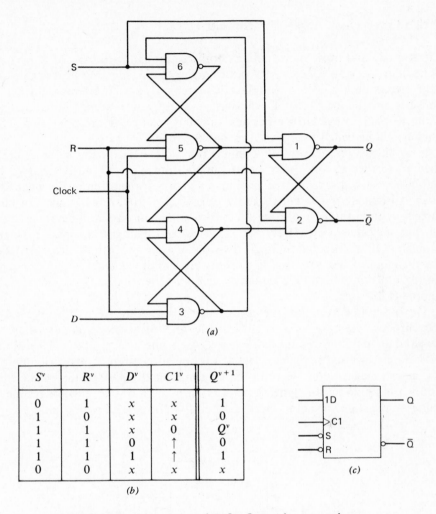

(a)

S^v	R^v	D^v	$C1^v$	Q^{v+1}
0	1	x	x	1
1	0	x	x	0
1	1	x	0	Q^v
1	1	0	$\uparrow$	0
1	1	1	$\uparrow$	1
0	0	x	x	x

(b)

(c)

FIGURE 9.29   *Edge-triggered D flip-flop with reset and set.*

and *S* inputs indicate that these lines are *active low*; i.e., the low level causes the action associated with that input line. The first two lines of the transition table show that the reset and set lines override the other inputs. The next three lines indicate that the flip-flop acts as a normal *D* flip-flop when the reset and set are high. The last line indicates that the action is unpredictable if both go low at the same time. Most standard IC flip-flops include reset or set or both, and most, as in this case, are active low.

9.10   Level-Mode Sequential Circuits

Clock-mode and pulse-mode circuits both require pulses to control state transitions, and the very concept of a pulse implies a considerable degree of control over signal timing. We now relax these timing restrictions even further and consider circuits for which all inputs are levels: signals that may remain at either value (0 or 1) for arbitrary periods and may change at arbitrary times. Such circuits will be called *level-mode* sequential circuits.

If we place a restriction on the inputs that only one may change at a time and that no further changes may occur until all outputs and state variables have stabilized, we will say that the level mode circuit is operating in the *fundamental mode*. Circuit design is considerably simplified if these restrictions can be enforced on circuit inputs. The reason for this will be discussed in detail in Chapter 14, but we can see intuitively that simultaneous changes of inputs may cause problems. If we use fine enough division of time, no events are ever exactly simultaneous. Thus, response to presumably simultaneous events might depend on the order in which they actually occur, an order that may be determined by unpredictable circuit delays.

The basic mode for the level-mode circuit is shown in Fig. 9.30. We see that the flip-flop memory of the clock-mode and pulse-mode circuits has been replaced by delay in the feedback loops. As a rule, this is not a physical delay deliberately inserted in the feedback loops but instead represents a "lumping" of the inherent distributed delay in the combinational logic developing the excitation variables. With this type of memory, the excitations are themselves the next values of the state variables. The excitations will be labeled with capital Ys, while

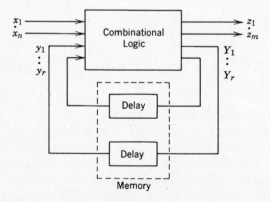

FIGURE 9.30   Basic model of level-mode sequential circuits.

the theoretically delayed values of the excitations that serve as inputs to the combinational logic network will be called secondaries and be labeled with lower case *y*s. Where the distinction is not important, the excitations and secondaries will be called state variables.

The equations for the *z* and *Y* variables are the same as for the general model (Equations 9.7 and 9.8). The equation describing the state change is simply

$$y_j(t + \Delta t) = Y_t(t) \qquad (j = 1, 2, \ldots, r) \tag{9.11}$$

The meaning of these expressions will become clear as we consider the design of level-mode circuits in Chapter 14.

It should be noted that the level-mode circuit is the most general type of sequential circuit in the sense that it can respond to any type of signal, pulse or level. As pointed out earlier, the distinction between a pulse and a level is a matter of definition. Physically, a pulse may be considered as two successive changes in a level signal occurring in a short period of time. The most important distinction lies in the manner in which the signal is interpreted. In a clock-mode or pulse-mode circuit, a pulse is considered as a single event that can trigger only a single change of state. In a level-mode circuit, a pulse is considered as two events, two level changes, each of which may cause a change in state.

Recognition of this dual nature of pulses enables us to apply level-mode circuits to pulse situations in which pulse-mode circuits cannot be used. Any pulse may cause a change in state, and a second pulse arriving while a circuit is still in transition as a result of a first pulse clearly may cause unreliable operation. As noted earlier, no two events are ever exactly simultaneous, so we must therefore limit pulse mode circuits to situations where only one pulse can occur at a time. But there are many practical situations where we must deal with pulses from unrelated sources, so that such a restriction cannot be enforced. In such cases, we must apply level-mode techniques, with each pulse edge considered a separate event. Simultaneous changes can cause problems in level-mode circuits, as noted, but they can be resolved by level-mode techniques.

9.11   Conclusion

In this chapter, we have attempted to give the reader an overview of the general characteristics of sequential circuits, of some of the factors governing their use, and of some of their special problems. We have treated the timing of clock-mode circuits in more detail to prepare the reader for Chapter 10. The reader should be able to consider the logical analysis and synthesis of such sequential circuits with confidence that physical relizations will work as postulated.

We should note that nomenclature regarding sequential circuits is not standardized. As we noted, clocked circuits are often referred to as *synchronous* circuits, leaving all other types lumped in the category of *asynchronous* circuits. But this division groups pulse-mode and level-mode circuits, which are very dissimilar, and separates pulse-mode and clock-mode circuits, which are quite similar. McCluskey [1] first proposed the division into *pulse-mode* and *fundamental-mode* circuits, with clock-mode as subdivision of pulse-mode circuits. Unfortunately, this leaves the important group of circuits that are neither pulse mode nor fundamental mode in limbo, with no specific name. We have therefore adopted the name *level mode* for all nonpulsed circuits, with fundamental mode as a subdivision of this category. But names are not that important; the important thing is that you fully understand the characteristics of and the differences between the various types of circuits.

Problems

9.1. By means of a timing diagram similar to Fig. 9.3c, investigate the behavior of the R-S flip-flop when R and S pulses occur at the same time. Assume simultaneous 100-ηsec pulses on both R and S. Assume the delays through the gates are 20 ηsec. Continue the timing diagram for a long enough period after the end of the pulses to determine what the flip-flop will do.

9.2. Construct a timing diagram similar to Fig. 9.7b showing the behavior of the flip-flop if the R and S inputs are both 1 during the clock pulses. Assume a 100-ηsec clock pulse and 20-ηsec delays through the gates. Continue the diagram at least 100 ηsec after the end of the clock pulse.

9.3. Figure 9.13 shows a sequential parity checker using an R-S flip-flop. Determine the design for a sequential parity checker using a J-K flip-flop. Verify its operation by means of a timing diagram showing a sequence of possible inputs and resulting outputs.

9.4. Assume the sequence of signals shown in Fig. P9.4 is applied to the shift register of Fig. 9.16. Complete the timing diagram by showing the resulting values at y_1, y_2, y_3, y_4, assuming that all four outputs are 0 prior to the arrival of the clock pulse at t_1.

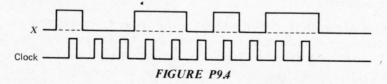

FIGURE P9.4

9.5. By means of a timing diagram demonstrate how improper operation can occur in the J-K flip-flop of Fig. 9.22 if the clock pulse is still high when the change at the outputs is seen at the input AND gates.

9.6. In the master-slave *R-S* flip-flop of Fig. 9.24 assume that both the master and the slave are clocked *R-S* flip-flops as shown in Fig. 9.7a. By means of a timing diagram show a typical sequence of state changes in the flip-flop. This diagram should be similar to Fig. 9.7b, except that you should show S, R, Q_1, $\bar{Q}_1$, Q_2, $\bar{Q}_2$, CK_1, CK_2. Assume clock pulses are 100 ηsec in duration and the delays in all gates are 10 ηsec.

9.7. Repeat Problem 9.6 for the master-slave *J-K* flip-flop of Fig. 9.25.

9.8. Figure P9.8 shows the circuit diagram for the SN74109 *J-K* flip-flop, which is listed by the manufacturer as positive edge-triggered. By means of a timing diagram determine when this flip-flop triggers and when it is enabled. Assume the delay through each gate is 10 ηsec.

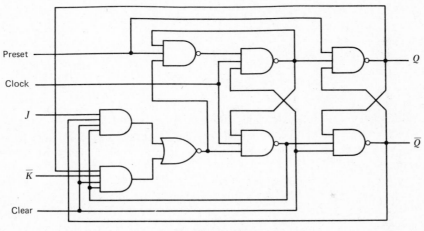

FIGURE P9.8

9.9. By means of timing diagrams, illustrate the behavior of the *D*-latch of Fig. 9.27. Assume all gate delays are 10 ηsec.

9.10. Consider the circuit realization of an edge-triggered *D* flip-flop given in Fig. P9.10. Suppose that, prior to the occurrence of a clock pulse, the circuit output Q and the circuit input D are stable at the values 0 and 1, respectively.

(a) Under the above conditions, what must be the logical values at points *a*, *b*, *c*, and *d*?

(b) Start with the initial conditions determined above and let the clock go to 1 and remain at that value for 200 ηsec. Construct as accurately as possible a timing diagram of *D*, *a*, *b*, *c*, *d*, and *Q* beginning 50 ηsec prior to

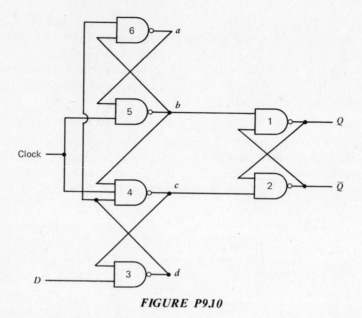

FIGURE P9.10

the clock pulse and extending 100 ηsec after the trailing edge of the pulse. Assume that the delay associated with each gate is 20 ηsec. Be sure to verify that changes at the D input while the clock is at the 1 level do not affect the output.

Bibliography

1. McCluskey, E. J. "Fundamental Mode and Pulse Mode Sequential Circuits," *Proc, IFIP Congress 1962*, North-Holland Publ. Co., Amsterdam, 1962.
2. Mealy, G. H. "A Method for Synthesizing Sequential Circuits," *Bell System Tech J.*, **34**: 5, 1045–1080 (Sept. 1955).
3. Moore, E. F. *Sequential Machines: Selected Papers*, Addison-Wesley, Reading, Mass., 1964.
4. Peatman, J. B. *Digital Hardware Design*, McGraw-Hill, New York, 1980.

10

SYNTHESIS OF CLOCK-
MODE SEQUENTIAL
CIRCUITS

10.1 Analysis of a Sequential Circuit

In this chapter, we will develop a complete design procedure for clock-mode sequential circuits. In Chapters 11 and 13, we will develop similar procedures for pulse-mode and fundamental-mode (pulseless) sequential circuits. The clock mode will be considered first because the clock-mode design process is the most straightforward throughout. A disadvantage of the clock mode as the initial topic might be the difficulty in constructing practical examples. Most real-life systems that include a clock contain a large number of states and are often better described by a system design language than a state diagram. Sequential circuits that are both interesting and limited in size tend to be pulse or fundamental mode. We have chosen several clock-mode examples that we hope the reader will not regard as overly artificial.

As is the case with electrical networks, the analysis of a sequential circuit is easier than the synthesis. To provide the reader with insight into the goal of the synthesis process, we will first analyze an already designed sequential circuit. Consider the circuit given in Fig. 10.1. Note that this is an example of the general clock-mode model presented in Chapter 9.

The *memory element input equations*, 10.1 and 10.2, can be determined by tracing connections in the simple combinational logic block.

$$J_{y_1} = x \qquad K_{y_1} = \bar{x} \tag{10.1}$$

$$J_{y_0} = z = (x + y_1)\bar{y}_0 \qquad K_{y_0} = \overline{(x + y_1)\bar{y}_0} \tag{10.2}$$

From these expressions we can use the general next state equation for the J-K flip-flop, $y^{v+1} = J \cdot \bar{y}^v + \bar{K}^v \cdot y^v$, to determine the *next state equations* for the memory elements y_0 and y_1 as given by Equations 10.3a and 10.3b.

$$y_1^{v+1} = x^v \cdot \bar{y}_1^v + x^v \cdot y_1^v = x^v \tag{10.3a}$$

$$y_0^{v+1} = [(x^v + y_1^v)\bar{y}_0^v] \cdot \bar{y}_0^v + [\overline{(x^v + y_1^v)\bar{y}_0^v}]y_0^v$$

$$= (x^v + y_1^v)\bar{y}_0^v \tag{10.3b}$$

267

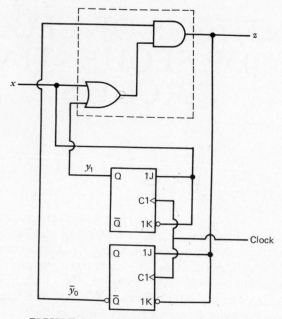

FIGURE 10.1   *Sequential circuit example.*

In both of these expressions $K = \bar{J}$, leading to a considerable simplification. In effect, the circuit could have been implemented equally well using D flip-flops instead of J-K flip-flops. Although it will usually not be possible to simplify the expressions so readily, the use of $J \cdot \bar{y}^v + \bar{K} \cdot y^v = y^{v+1}$ to determine next state equations is a completely general procedure where J-K flip-flops are used.

From Equations 10.3a and 10.3b, we can tabulate what we will define as the *transition table* for the sequential circuit. This is merely an evaluation of y_1^{v+1}, y_0^{v+1}, and z^v for all possible combinations of values of x^v, y_0^v, and y_1^v. The transition table for the circuit under discussion is given in Fig. 10.2. For example, if $x^v = y_0^v = 0$ and $y_1^v = 1$, then y_1^{v+1} and y_0^{v+1}, as given by Equations 10.3a and b, are 0 and 1, respectively. These values are entered in the square enclosed by the dashed line in Fig. 10.2. The reader can easily determine the remaining values in the table. The values for z^v are similarly determined. In this case, $z^v = y_0^{v+1}$.

Transition tables of the form of Fig. 10.2 will tend to become rather bulky as the number of variables increases. To simplify the tables, we will assign a shorthand notation, as follows. First, we represent the states by the letter q, with arbitrarily assigned decimal subscripts. For example, we will represent $y_1 y_0 = 00$ by q_1, $y_1 y_0 = 01$ by q_2, $y_1 y_0 = 11$ by q_3, and $y_1 y_0 = 10$ by q_4. In this example,

y_1^ν	y_0^ν		$y_1^{\nu+1}, y_0^{\nu+1}$		z^ν	
		x^ν	0	1	0	1
0	0		0 0	1 1	0	1
0	1		0 0	1 0	0	0
1	1		0 0	1 0	0	0
1	0		0 1	1 1	1	1

FIGURE 10.2   *Transition table for circuit of Fig. 10.1.*

$Z = \overline{(y_1 + x)\, y_0}$

there is only one input and one output. If there are multiple inputs and outputs, these values may be coded also, in which case we represent the set of input variables $(x_1, x_2, \ldots, x_n)$ by X and the set of output variables $(z_1, z_2, \ldots, z_q)$ by Z. Particular combinations of values will be represented by the decimal equivalents to the binary interpretations of the combinations. Thus, $x_2 x_1 = 00$ will be represented by $X = 0$, $x_2 x_1 = 01$ by $X = 1$, $z_2 z_1 = 11$ by $Z = 3$, etc. We also combine the output and next-state portions of the transition table. This notation results in the *state table*, as shown in Fig. 10.3a. Now the entry in the dashed square, for example, may be interpreted as follows: if the circuit is in state q_4 and receives input 0, the output at clock time will be 1 and the next state will be q_2.

The same information contained in the state table may be displayed more graphically by a *state diagram*. The state diagram corresponding to the state table of Fig. 10.3a is shown in Fig. 10.3b. There is a circle for each state and an arrow leaving each state for each input. The arrows terminate at the appropriate next states. Each is labeled with the number of the input that causes that particular transition. The inputs are indicated on the left of each slash and the

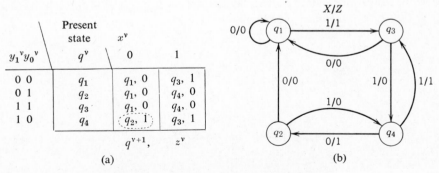

$y_1^\nu y_0^\nu$	Present state q^ν	x^ν 0	1
0 0	q_1	q_1, 0	q_3, 1
0 1	q_2	q_1, 0	q_4, 0
1 1	q_3	q_1, 0	q_4, 0
1 0	q_4	q_2, 1	q_3, 1
		$q^{\nu+1}$,	z^ν

(a)   (b)

FIGURE 10.3   *State table and state diagram.*

output corresponding to that particular input present state combination on the right, i.e., X/Z. For example, the square enclosed by the dashed circle in Fig. 10.3a corresponds to the arrow from q_4 to q_2. The square is in the column corresponding to $X = 0$, and the output entry is 1. Hence, the arrow is denoted $0/1$. While state tables and state diagrams contain the same information, both are useful at different steps in the synthesis process.

10.2 Design Procedure

The analysis process of the previous section followed a sequence of steps that successively transformed the circuit description as follows:

$$\text{Circuit} \rightarrow \begin{array}{c} \text{Memory} \\ \text{element} \\ \text{input} \\ \text{equations} \end{array} \rightarrow \begin{array}{c} \text{Next} \\ \text{state} \\ \text{equations} \end{array} \rightarrow \begin{array}{c} \text{Transition} \\ \text{table} \end{array} \rightarrow \begin{array}{c} \text{State} \\ \text{table} \end{array} \rightarrow \begin{array}{c} \text{State} \\ \text{diagram} \end{array}$$

The synthesis process resembles an analysis in reverse. The principal distinction is that synthesis begins prior to the existence of a state diagram. This diagram must be obtained from some other description of the problem. Usually this is merely an unambiguous (hopefully) natural language description of the intended circuit function. The synthesis procedure is depicted in Fig. 10.4.

We will consider the complete process in this chapter. For the reader impatient to get back to the circuit, we first remark that considerable attention must be paid to steps 1 and 2. Sections 10.3 and 10.4 are concerned with various approaches to obtaining a state table. The next two sections are devoted to finding an equivalent state table with a minimum number of states. A circuit can be obtained from the minimal state table without great difficulty. Obtaining the most economical circuit is not so easy. The complexity of the memory element input equations will vary with the assignment of combinations of flip-flop output values to states in the state table. The attempt to optimize this correspondence is called the *state assignment problem*.

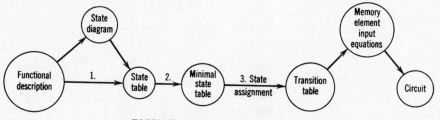

FIGURE 10.4 Synthesis process.

The state diagram and state table can become cumbersome when a large number of input lines are present. If only a small subset of these inputs are sensed while the circuit is in any given state, a more convenient notation, called an *algorithmic state machine* (ASM) chart, can be used in place of the state diagram. The ASM chart [13] resembles a program flow chart, but can represent the same information as a state diagram. We shall feel free to make use of an ASM chart where convenient in a later example.

10.3 Synthesis of State Diagrams

As depicted in Fig. 10.4, one approach to obtaining the state table involves the use of a state diagram. This method is particularly convenient for certain types of circuits that possess a readily identifiable state called a *reset state*. Where such a state exists, there is a mechanism by which the circuit may be returned to the reset state from any other state in a single operation. This operation may be specified as a special input column in the state table, or it may function independently of the state table. For example, a special line may be connected to all flip-flops that will clear them to zero. This line may be connected to separate input points rather than those used in the implementation of the state table. Example 10.1 involves a simple circuit with a reset state.

Example 10.1

The beginning of a message in a particular communications system is denoted by the occurrence of three consecutive 1s on a line x. Data on this line have been synchronized with a source of clock pulses. A clock-mode sequential circuit is to be designed that will have an output of 1 only at the clock time coinciding with the third of a sequence of three 1's on line x. The circuit will serve to warn the receiving system of the beginning of a message. It will be provided with a separate reset mechanism to place it in state q_0 following the end of a message. Typical behavior of the circuit is depicted in Fig. 10.5.

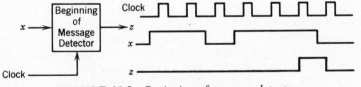

FIGURE 10.5 Beginning-of-message detector.

Solution The synthesis procedure begins by designating the reset state as q_0, as shown in Fig. 10.6a. Essentially the circuit must keep track of the number of consecutive 1s it has received. It does this by moving to a different state with each input of 1. As shown in Fig. 10.6b, the circuit goes to state q_1 with the first 1 input and to state q_2 with the

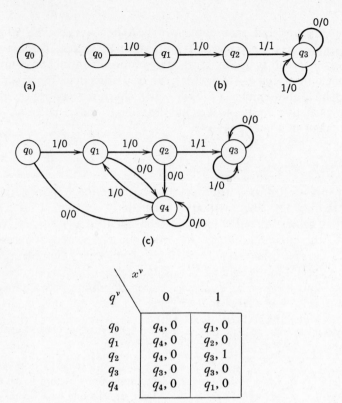

(d) State Table

FIGURE 10.6 *State diagram for beginning-of-message detector.*

second. The outputs associated with both of these inputs are 0, as shown. The third consecutive input of 1 generates a 1 output and causes the circuit to go to q_3. Once in q_3, the circuit will remain in this state emitting 0 outputs until it is externally reset to q_0. This reset mechanism is not shown in the state table.

It is possible for an input of 0 to appear during any clock period, possibly interrupting a sequence of 1s. A state q_4 is provided for this eventuality. Every input of 0 (unless the circuit is already in q_3) generates a 0 output and causes the circuit to go to q_4, as shown in Fig. 10.6c. A second 0 merely causes the circuit to remain in state q_4. An input of 1 while the circuit is in q_4 may be the first of a sequence of three 1s. Thus, a 1 will cause the circuit to change state from q_4 to q_1.

All possible input sequences that can occur have been considered. The reader may verify this by noting that arrows for both 0 and 1 inputs have been defined leaving each of the five states. Thus, Fig. 10.6c is the completed state diagram. This translates directly to the state table of Fig. 10.6d.   ■

The perceptive reader may have noticed that the circuit could be returned to q_0 for inputs of 0. This would eliminate the need for state q_4. As a general rule, already existing states should be used as next states wherever possible. At some point, the designer must stop defining new states. However, since state diagram construction is an intuitive procedure, there may be doubt as to whether an existing state will serve. *When in doubt, define a new state,* thus avoiding serious error.

Example 10.2

Numbers between 0 and 7 expressed in binary form are transmitted serially on line x from a digital measuring device. The bits appear in descending order of significance, synchronized with a clock. Design a circuit that will have an output $z = 1$ at the clock time of the third bit of a binary number representing an extreme value (0 or 7) on line x. Otherwise, $z = 0$. The circuit will have two inputs, x and d, in addition to the clock. If $d = 1$ at a given clock time, then the most significant bit of a measurement will appear on line x at the next clock time.

Assume that $d = 1$ will cause the circuit to assume a reset state a. Construct a state table regarding line d as an external reset mechanism. Then revise the state table to include input d as part of the state table.

Solution Beginning with state a, we construct a tree of input arrows as shown in Fig. 10.7a. Each arrow leads to a new state from which two new arrows (in general, one for each possible input) emerge. In this problem, it is only necessary to generate two levels of the tree. The numbers 0 and 7 are detected at the time of the third input. The binary number is zero if an input of 0 occurs when the circuit is in state d. An input of 1 for state g indicates 7. Outputs of 1 are shown for these cases.

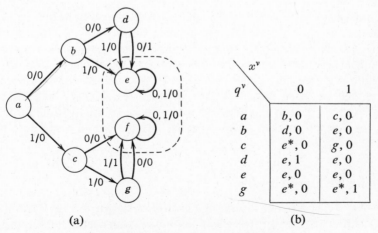

(a)   (b)

FIGURE   10.7   Detector of 0 and 7.

dx	X

00	0
01	1
10	2
11	3

(a)

q^v \ X	0	1	2	3
a	$b, 0$	$c, 0$	$a, 0$	$a, 0$
b	$d, 0$	$e, 0$	$a, 0$	$a, 0$
c	$e, 0$	$g, 0$	$a, 0$	$a, 0$
d	$e, 1$	$e, 0$	$a, 0$	$a, 0$
e	$e, 0$	$e, 0$	$a, 0$	$a, 0$
g	$e, 0$	$e, 1$	$a, 0$	$a, 0$

(b) q^{v+1}, z

FIGURE 10.8 *Including input* d *in the state table.*

The third input must take the circuit to a state for which subsequent outputs of 1 are impossible. Notice that states e and f satisfy this requirement. Once the circuit is in either of these states, it remains there permanently; the outputs are always 0. In fact, the circuit will behave in the same manner whether it is in state e or state f. Therefore, only one of these states is necessary, as emphasized by the dashed closed curve in Fig. 10.7a. In translating the state diagram to the state table in Fig. 10.7b, state f is omitted. Where f would be a next state, it is replaced by e, as indicated by asterisks.

The state table with the effect of the reset input d included is given in Fig. 10.8b. The four combinations of the inputs x and d are coded as shown in Fig. 10.8a. The two columns for $X = 0$ and $X = 1$ correspond to $d = 0$. These two columns are a direct copy of Fig. 10.7b. Anytime $d = 1$, the circuit is reset to state a, as shown. If the sequential circuit is implemented as given by Fig. 10.8b, no external reset mechanism is required.

■

The response tree of Fig. 10.7 illustrates a general procedure that can be used in the absence of other insight. Where no specific reset state exists, an arbitrary state may be used as a starting point. Preferably this state will have some past history to distinguish it clearly from other states. In the absence of a well-defined starting state, the tree may grow very large before the branches are terminated. In general, the designer should prune the tree steadily by routing arrows to existing states that lead to the same future behavior as might be defined for new states. If the designer is slow in applying insight, the size of a response tree can get out of hand rapidly.

10.4 Finite Memory Circuits

An equivalent version of the circuit of Fig. 10.1 is shown in Fig. 10.9a. The J-K flip-flops with inverters on the clear input are replaced with equivalent D flip-flops, with the clock input omitted for clarity.

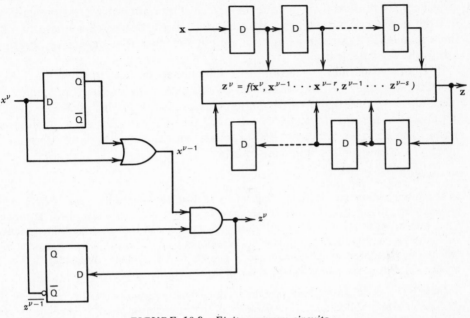

FIGURE 10.9 Finite memory circuits.

Recalling that the output of a D flip-flop is equal to the value of the input at the immediately previous clock time, t^{v-1}, we see that in this case the present output is a function only of the present input and the immediately previous input and output, i.e.,

$$z^v = f(x^v, x^{v-1}, z^{v-1}) \tag{10.4}$$

This circuit is an example of a *finite memory* circuit, *finite* in the sense that it need "remember" only a finite number of past inputs and outputs in order to determine the appropriate present output and next state. Note that *finite* here has nothing to do with the memory capacity in bits, which is, of course, always finite.

A generalized version of a finite memory circuit is shown in Fig. 10.9b. Note that r previous inputs and s previous outputs are stored. The current output is a function of this stored information and the current input. The symbols X and Z are vectors that may represent several input and output lines. The blocks labeled D are not individual flip-flops but represent one-dimensional arrays of flip-flops, one for each input or output line.

Not all sequential circuits are *finite memory*. In Examples 10.1 and 10.2, the circuits were to issue an output on the occurrence of some particular input sequences. They were then to issue no further output after being reset, regardless of what additional inputs (other than RESET) might occur. Thus, these circuits must remember a particular input or output sequence that may have occurred arbitrarily far back in time.

It is often possible to recognize a finite memory circuit from the initial specifications, and this knowledge can often simplify the synthesis procedure. If the designer knows the circuit is finite memory, she or he can generally specify an upper bound on the memory size, i.e., the maximum number of internal states. The designer can then set up either the transition table or state table directly, as shown in the following example.

Example 10.3

Information bits are encoded on a single line x so as to be synchronized with a clock. Bits are encoded so that two or more consecutive 1s or four or more consecutive 0s should never appear on line x. An error-indicating sequential circuit is to be designed to indicate an error by generating a 1 on output line z coinciding with the fourth of every sequence of four 0s or the second of every sequence of two 1s. If, for example, three consecutive 1s appear, the output remains 1 for the last two clock periods.

Solution The circuit to be designed may be immediately recognized as finite memory. In this case, we may call it *finite input memory* in that only inputs need be remembered. In

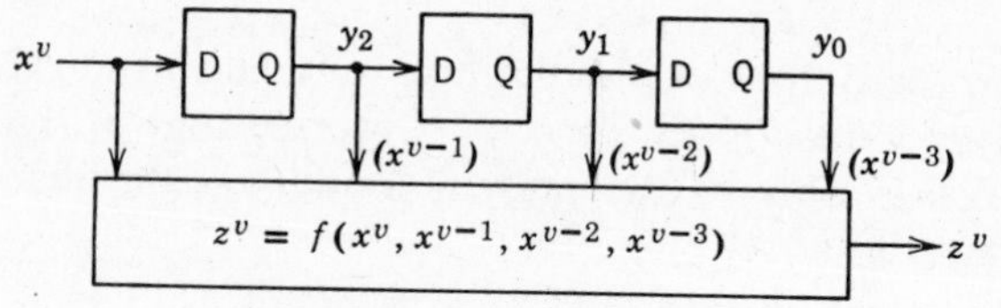

(a)

$(y_2 y_1 y_0)^v$	x^v 0	x^v 1	0	1
000	000	100	1	0
001	000	100	0	0
011	001	101	0	0
010	001	101	0	0
100	010	110	0	1
101	010	110	0	1
111	011	111	0	1
110	011	111	0	1

$(y_2 y_1 y_0)^{v+1}$ z^v

(b)

q^v	x^v 0	x^v 1
a	$a, 1$	$e, 0$
b	$a, 0$	$e, 0$
c	$b, 0$	$f, 0$
d	$b, 0$	$f, 0$
e	$d, 0$	$h, 1$
f	$d, 0$	$h, 1$
g	$c, 0$	$g, 1$
h	$c, 0$	$g, 1$

q^{v+1}, z^v

(c)

FIGURE 10.10 *Synthesis of circuit with finite input memory of length three.*

the worst case, the circuit will have to remember the value of three previous inputs. If these stored inputs are all 0 and the present input is 0, then the circuit output is 1. A circuit of the form shown in Fig. 10.10a can be made to do the job. As the input bits arrive, they are shifted through the string of flip-flops so that all times $y_2 = x^{v-1}$, $y_1 = x^{v-2}$, $y_0 = x^{v-3}$. The combinational logic then determines the output on the basis of these stored inputs and the present input.

The transition table for this circuit can be set up directly, as shown in Fig. 10.10b. First list all eight combinations of state variables as the present state entries. The next state entries, $(y_2 y_1 y_0)^{v+1}$, follow directly from noting that the state variables shift to the right each clock time. For example, if the present values are 000 and the input is 1, the next values are 100. To complete the transition table, the appropriate outputs should be entered. The output is to be 1 if the three previous inputs and the present input are all 0 or if the immediately previous and present inputs are both 1. If the designer wishes to use the circuit form shown in Fig. 10.10a, the only step left is design of the combinational logic to generate the output, z. For this case,

$$z = xy_2 + \bar{x}\bar{y}_2\bar{y}_1\bar{y}_0 \tag{10.5}$$

■

Although the circuit form of Fig. 10.10a will work, it may not be minimal. In only one case is it necessary for the circuit to "remember" three inputs back. If $x^{v-1} = 1$, it does not matter what x^{v-2} and x^{v-3} were. Thus, it appears that less than eight internal states may be sufficient. To check on this, the designer may convert the transition table to a state table simply by assigning an arbitrary state designation to each combination of state variable in the table of Fig. 10.10b. The resultant state table, Fig. 10.10c, may then be minimized by the formal techniques to be developed in later sections.

In a finite input memory circuit of the general form of Fig. 10.10a, the memory essentially acts as a serial-to-parallel converter. The input information required to make an output decision is arriving in serial, i.e., in time sequence, so the circuit simply stores the inputs until enough have been received to make the output decision and then applies them all at once, i.e., in parallel, to the combinational logic implementing the output function. Any time it can be determined that the output of a sequential circuit depends on no more than some maximum number of previous inputs, then the approach discussed in the previous example can be used.

If the longest sequence has length n, we say that the finite input memory circuit has *memory of length* n, and n memory elements will be required for each input line. Assume there is only one input line. Then the memory will have 2^n possible states, i.e., 2^n distinct input messages can be stored. If the output is to be different for each of these 2^n messages, then the general form of Fig. 10.10a is the most economical, and the designer's task reduces to nothing more than realizing the output function.

If the outputs are the same for many of the input messages, then fewer memory elements may be sufficient. In the above example, we have a memory of length 3 so that eight distinct messages can be stored. As we have already seen, however, all messages with $x^{v-1} = 1$ produce the same output, so the memory need not distinguish among the four messages with this characteristic. Thus, by processing, or classifying, the messages as they arrive, we may be able to reduce the memory requirements.

If we reduce the memory in this manner, the circuit form will not be the same as Fig. 10.10a, in that the outputs of the memory flip-flops will not be equal to previous inputs but will be functions of them. Nevertheless, we may still use the state table derived from this basic circuit form as a starting point in the design process. The situation is somewhat analogous to first specifying a combinational function as a standard or expanded sum of products. This may not be minimal,

| | | x^v | |
$x^{v-1}x^{v-2}$	q^v	0	1
00	a	a	d
01	b	a	d
11	c	b	c
10	d	b	c

q^{v+1}

(a) $n = 2$

| | | x^v | |
$x^{v-1}x^{v-2}x^{v-3}$	q^v	0	1
000	a	a	e
001	b	a	e
011	c	b	f
010	d	b	f
100	e	d	h
101	f	d	h
111	g	c	g
110	h	c	g

q^{v+1}

(b) $n = 3$

| | | x^v | |
$x^{v-1}x^{v-2}x^{v-3}x^{v-4}$	q^v	0	1
0000	a	a	i
0001	b	a	i
0011	c	b	j
0010	d	b	j
0100	e	d	l
0101	f	d	l
0111	g	c	k
0110	h	c	k
1000	i	e	m
1001	j	e	m
1011	k	f	n
1010	l	f	n
1100	m	h	q
1101	n	h	q
1111	p	g	p
1110	q	g	p

q^{v+1}

(c) $n = 4$

FIGURE 10.11   *Standard-state tables for single-input finite input memory circuits of length n.*

but it allows us to express the function in Karnaugh map form, from which a minimal expression can be determined.

Once we have any state table describing the appropriate behavior, we can derive from it any number of equivalent tables, including a unique minimal table. For an assumed circuit form such as Fig. 10.10a, the transition table, with regard to next-state entries, is unique for a given memory length and number of inputs since the next state entries are simply derived by shifting the inputs through the memory. We may thus set up standard transition tables for finite input memory circuits. These tables are shown in Fig. 10.11 for single-input circuits of length 2, 3, and 4. To form the complete transition tables for any specific circuit, the designer need only fill in the appropriate outputs. If the number of distinct outputs is more than half the number of states, the number of memory devices cannot be reduced, so the designer should use the standard circuit form and proceed directly to the design of the output logic. If the number of distinct outputs is less than half the number of states, the designer should generally minimize the corresponding state table before deciding on a circuit form.

Although *finite input memory* circuits are possibly the easiest to recognize and design, general finite memory circuits (of the form of Fig. 10.9b) are also important, and recognizing their finite memory character can often simplify the design process. This is illustrated in the next example.

Example 10.4

Design a 3-bit (modulo-8) up-down counter. The count is to appear as a binary number on three output lines, z_2, z_1, and z_0. The most significant bit is z_2. The count will change with each clock pulse. If the input $x = 1$, the count will increase. If $x = 0$, the count will decrease. Modulo-8 implies that increasing the count from 7 results in a count of 0, and decreasing it from 0 gives a count of 7.

Solution   One can immediately recognize the up-down counter as a *finite output memory* machine of memory length 1. The value of the output z is a function of only the immediately previous output, z^{v-1}, and the present input, x. z, however, is a 3-bit vector with components z_2, z_1, and z_0, so that three flip-flops are required to represent one past output.

A circuit configuration and the corresponding transition table are shown in Fig. 10.12. The transition table is constructed directly by noting that the desired present values of the outputs are the next values of the state variables. For example, for the circled entry in Fig. 10.13b, the previous count (present state) was 011, so if $x = 1$, the output (next state) should be 100.

Clearly, eight states are needed for a modulo-8 counter, so the table of Fig. 10.12b is minimal. If the circuit form of Fig. 10.12a is satisfactory, the designer can proceed directly to the design of the combinational logic circuits to realize the required outputs. However, even though the table is minimal, the form of Fig. 10.12 is not the only possible circuit

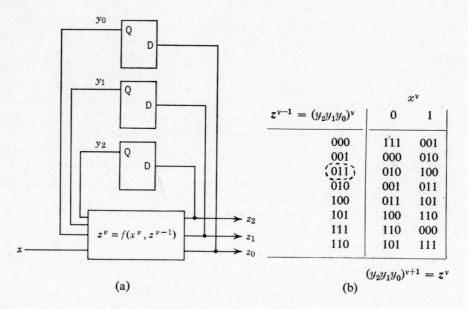

FIGURE 10.12   *Up-down counter.*

form. Different types of flip-flops might be used, in which case the state variables need not be equal to the most recent outputs. For completeness, we assign state numbers to the transition table to obtain the state table of Fig. 10.13. The corresponding state diagram is given in Fig. 10.14.

q^ν	x^ν	
	0	1
q_0	$q_7, 7$	$q_1, 1$
q_1	$q_0, 0$	$q_2, 2$
q_2	$q_1, 1$	$q_3, 3$
q_3	$q_2, 2$	$q_4, 4$
q_4	$q_3, 3$	$q_5, 5$
q_5	$q_4, 4$	$q_6, 6$
q_6	$q_5, 5$	$q_7, 7$
q_7	$q_6, 6$	$q_0, 0$

$$q^{\nu+1}, z$$

FIGURE 10.13   *State table for up-down counter.*

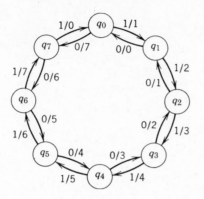

FIGURE 10.14 *State diagram for up-down counter.* ■

Two distinct approaches to state table synthesis have been described, both of which should be mastered by the student. Sometimes the most difficult task for the novice designer is deciding which method to use in a specific case. If the problem can be identified as finite memory, the direct synthesis of the state table or transition table is generally preferable. The state diagram approach works best when a reset or starting state can be identified; as it happens, machines with reset states are often nonfinite memory machines. A nonfinite memory machine will usually be driven to some holding state or caused to cycle within a subset of its states after it has made the desired response to some input sequence. Examples 10.1 and 10.2 illustrate machines with holding states. If such a machine is to be reusable, there must be an input sequence that will return it to its initial state. A fixed sequence that always will return a sequential machine to its initial state is called a *synchronizing sequence*. An initial state that can be reentered in this manner will be called a *reset state*. Often the synchronizing sequence will be a single input on a special line called the reset line.

Example 10.5

Repeat the design of the 0, 7 detector of Example 10.2, except that now the reset line *d* is to be omitted. Instead, a second 7 (three 1s in a row), coming any time after a 3-bit group has been tested, will signal that a new character will start arriving at the next clock time.

Solution The state diagram (Fig. 10.15a) through state *g* is the same as Fig. 10.7a with *e* and *f* combined, as discussed in Example 10.2. We now recognize *e* as the holding state characteristic of nonfinite memory circuits. The reset is now provided by a sequence of

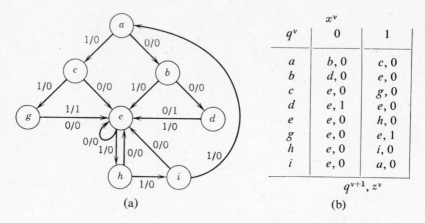

FIGURE 10.15   State diagram and state table, Example 10.6.

three 1s taking the circuit from e back to a via h and i. The resultant state table is shown in Fig. 10.15b. Since the sequence of three 1s must be applied after the circuit has reached state e, it is not strictly speaking a synchronizing sequence.   ∎

In some problems, it might not be possible to identify either a reset state or a bound on the memory. In such cases, one must arbitrarily choose some state whose past history can be specified as closely as possible to use as an initial state. From here, special care should be applied in the generation of a state table. Occasionally the reader may encounter the dilemma of a finite memory machine with a reset state. In that case, either approach may be used.

10.5   Equivalence Relations

The procedures discussed in Section 10.4 provide means of obtaining a state table for a given specification, but the resultant table is not unique and may not be optimum in terms of the ultimate realization. Since the number of flip-flops in a circuit increases as the number of states increases, we may say that a minimal state table is one with a minimum number of states. In some cases, different states in the original state table will be found to perform the same function. Or more specifically, it may not be possible, by means of output measurements, to distinguish between two or more states. In such cases, the states are said to be *equivalent* and may be replaced by a single state. In order to develop a more concrete definition of the equivalence of states, let us first pursue the concept of an *equivalence relation*.

When an ordered pair of elements, (x, y), possesses some relating property, we may speak of x as being *R-related* to y. This notion is symbolized by xRy. The relation, R, is defined as the set of all ordered pairs that possess this same particular property, whatever it may be. It is perhaps worth remarking that the elements x and y might be statement variables, real numbers, integers, or most anything. For our purposes, it can be assumed that R is a relation on a set of elements such that either x or y may represent any element in the set.

If xRx for every x in the set of interest, then the relation R is said to be *reflexive*. The relation "is greater than or equal to" is reflexive, while the relation "is greater than" is not reflexive.

If yRx whenever xRy, then the relation R is *symmetric*. On sets of usual interest, a symmetric relation will be reflexive as well. An exception is a relation such as "is a cousin of," which is symmetric. In the usual sense, one is not his or her own cousin, so this relation is not reflexive. Neither of the relations in the previous paragraph ($x \geq y$ or $x > y$) is symmetric.

Finally, if xRy and yRz imply xRz, then R is a *transitive* relation. A relation satisfying all three of the above criteria is called an *equivalence* relation. An example of an equivalence relation might be "is a member of the same team as."

An equivalence relation on a set will always partition the set into disjoint subsets, known as *equivalence classes*. For example, the set

$$\{A, a, B, b, C, c, D, d, E, e\}$$

is partitioned into the subsets

$$\{A, B, C, D, E\} \quad \text{and} \quad \{a, b, c, d, e\}$$

if xRy is interpreted to mean that x and y are the same case. A different group of disjoint subsets is obtained if xRy is taken to mean that x and y are the same letter.

Our intuitive concept of equality is an example of an equivalence relation since

$$x = x \qquad \qquad \text{(Reflexive)}$$

and

$$y = x \quad \text{if} \quad x = y \qquad \text{(Symmetric)}$$

$$\left. \begin{array}{l} x = y \quad \text{and} \quad y = z \\ \\ \\ x = z \end{array} \right\} \qquad \text{(Transitive)}$$

then

10.6 Equivalent States and Circuits

In order to define the notion of equivalent states and equivalent sequential circuits, we must introduce some new notation. State tables comprise tabular representations of two functions, the output function and the next-state function. If the outputs and next states are specified for every combination of inputs and present states, the circuits are called *completely specified* circuits. Let us denote the next-state function as δ and the output function as λ. For example, in the circuit of Fig. 10.16,

$$\lambda(q_2, 3) = 0 \qquad\qquad\qquad (10.6)$$

and

$$\delta(q_2, 3) = q_4 \qquad\qquad\qquad (10.7)$$

If a circuit is in some initial state, it will respond to a sequence of inputs with a specific sequence of outputs. Consider the case where the circuit of Fig. 10.16 is initially in state q_1 and is subjected to the series of inputs 0 2 3 0 0 1. In order to determine the output sequence in response to this input sequence, it is also necessary to determine the next-state sequence.

$$\lambda(q_1, 0) = 0 \qquad \text{and} \qquad \delta(q_1, 0) = q_3$$

$$\lambda(q_3, 2) = 2 \qquad \text{and} \qquad \delta(q_3, 2) = q_1$$

$$\lambda(q_1, 3) = 0 \qquad \text{and} \qquad \delta(q_1, 3) = q_2$$

$$\lambda(q_2, 0) = 0 \qquad \text{and} \qquad \delta(q_2, 0) = q_3$$

$$\lambda(q_3, 0) = 0 \qquad \text{and} \qquad \delta(q_3, 0) = q_3$$

and, finally,

$$\lambda(q_3, 1) = 1 \qquad \text{and} \qquad \delta(q_3, 1) = q_1$$

The output sequence and final state may now be summarized as a function of the initial state and the input sequence, as follows:

$$\lambda(q_1, 023001) = 020001$$

$$\delta(q_1, 023001) = q_1 \qquad\qquad\qquad (10.8)$$

x^v Present State	0	1	3	2
q_1	$q_3,0$	$q_1,0$	$q_2,0$	$q_2,0$
q_2	$q_3,0$	$q_3,0$	$q_4,0$	$q_4,0$
q_3	$q_3,0$	$q_1,1$	$q_1,3$	$q_1,2$
q_4	$q_4,0$	$q_4,0$	$q_2,0$	$q_2,0$

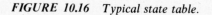

$$q^{v+1}, z$$

FIGURE 10.16   Typical state table.

Now let us take a different point of view. Suppose that the output sequence for a given initial state and input sequence is known. Is the sequence of internal states then of any real interest? The answer is "No." To the "outside world," only the output response of a circuit to a sequence of inputs is significant. In a sense, the function of a sequential circuit may be regarded as the translation of one sequence of signals to a second sequence.

We are thus led toward a meaningful definition of the equivalence of two sequential circuits. First, let us define the equivalence of two states within these circuits.

Definition 10.1   Let S and T be completely specified circuits, subject to the same possible input sequences. Let $(X_1, X_2, ..., X_n)$ represent a sequence of possible values of the input set X, of arbitrary length. States $p \in T$ and $q \in S$ are *indistinguishable* (equivalent), written $p \equiv q$, iff (if and only if) $\lambda_T(p, X_1, X_2, ..., X_n) = \lambda_S(q, X_1, X_2, ..., X_n)$ for every possible input sequence.

This definition may be applied equally well to different states of a single circuit. Less formally, p and q are equivalent if there is no way of distinguishing between them on the basis of any output sequences starting from these states. We use the terms *indistinguishable* and *equivalent* almost synonymously. *Indistinguishable* will be used where necessary to emphasize that indistinguishability is only one of many equivalence relations that might be defined on the states of a state table.

The equivalence of states in a single circuit is an equivalence relation. This can be easily verified as follows. Since the machine is completely specified, the output sequence resulting from the application of a given input sequence to a particular initial state q will always be the same. Therefore, $q \equiv q$, satisfying the reflexive law. Clearly, no distinction can be made between $p \equiv q$ and $q \equiv p$, so the symmetric law is satisfied. Furthermore, if p and q lead to identical output sequences for every input sequence and q and r lead to identical output sequences for every input sequence, certainly p and r must lead to identical output sequences for every input. That is, if $p \equiv q$ and $q \equiv r$, then $p \equiv r$; and the transitive law is satisfied.

Definition 10.2   The sequential circuits S and T are said to be equivalent, written $S \equiv T$ iff for each state p in T there is a state q in S such that $p \equiv q$, and, conversely, for each state q in S there is a state p in T such that $q \equiv p$.

Less formally, two circuits are equivalent if there is no way to distinguish between them simply by observing the response to input sequences.

As an example, consider the simple circuits described by the state tables of Fig. 10.17. We first examine circuit S. Suppose that this circuit is initially in state q_1. An input of 0 will take the circuit to state q_3 with an output of 0. As long as 0

S	$X = 0$	$X = 1$
q_1	$q_3,0$	$q_2,1$
q_2	$q_1,1$	$q_2,0$
q_3	$q_1,0$	$q_2,1$

T	$X = 0$	$X = 1$
p_1	$p_1,0$	$p_2,1$
p_2	$p_1,1$	$p_2,0$

FIGURE 10.17

inputs continue, the circuit will cycle back and forth between q_3 and q_1 with 0 outputs. Should a 1 input occur, the circuit will go to q_2, and the output will be 1. A 0 input would then send the circuit back to q_1.

Note that if the circuit were initially in state q_3, exactly the same output sequences as outlined above would occur. This is illustrated in Fig. 10.18 for two input sequences. These two input sequences are, in fact, sufficient to demonstrate that states q_1 and q_3 are equivalent. Readers who remain in doubt should test the circuit with input sequences of their own design until convinced.

Note now that these or any other input sequences will also result in the same outputs for initial state p_1 in circuit T. Thus, p_1 is equivalent to both q_1 and q_3. Similarly, p_2 is equivalent to q_2. Thus, corresponding to every state in T, there is an equivalent state in S, and vice versa. Therefore, by Definition 10.2, S and T are equivalent.

Our basic problem is to find an optimum state table. In view of the above definitions, the problem can now be stated in more specific terms, as follows: given a state table, find an equivalent state table with as few states as possible. A table with fewer states may require less memory, thus generally leading to a

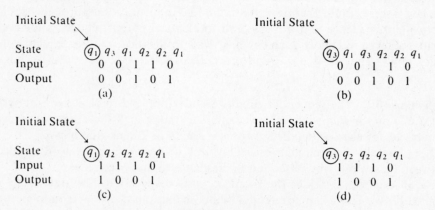

Initial State

State	q_1	q_3	q_1	q_2	q_2	q_1
Input		0	0	1	1	0
Output		0	0	1	0	1

(a)

Initial State

State	q_3	q_1	q_3	q_2	q_2	q_1
Input		0	0	1	1	0
Output		0	0	1	0	1

(b)

Initial State

State	q_1	q_2	q_2	q_2	q_1
Input		1	1	1	0
Output		1	0	0	1

(c)

Initial State

State	q_3	q_2	q_2	q_2	q_1
Input		1	1	1	0
Output		1	0	0	1

(d)

FIGURE 10.18

more economic realization. For example, circuit S of Fig. 10.17 has three states and would therefore require two flip-flops for realization. Circuit T, having only two states, would require only a single flip-flop.

10.7   Determination of Classes of Indistinguishable States

Our basic approach to finding an optimum state table will be to partition the state table into the smallest possible number of *equivalence classes* of indistinguishable states. We will then show that an equivalent sequential circuit can be formed by defining one state corresponding to each class of indistinguishable states. There are many ways to partition the states of a sequential circuit into disjoint classes. Not all such partitions result in equivalence classes of indistinguishable states. The following theorem provides a test for this condition.

THEOREM 10.1   Let the states of a sequential circuit be partitioned into disjoint classes. $p \triangleq q$ denotes that states p and q fall into the same class in the partition. The partition is composed of equivalence classes of indistinguishable states (two indistinguishable states must be in the same class) if and only if the following two conditions are satisfied for every pair of states p and q in the same class ($p \triangleq q$) and every single input X.

(1) $\lambda(p, X) = \lambda(q, X)$
(2) $\delta(p, X) \triangleq \delta(q, X)$

Proof   A. (Necessity.) Assume that the partition consists of equivalence classes of indistinguishable states. Let p and q be any pair of states such that $p \triangleq q$. Therefore, $p \equiv q$. By Definition 10.1, we can write

$$\lambda(p, XX_1 X_2 \cdots X_k) = \lambda(q, XX_1 X_2 \cdots X_k) \tag{10.9}$$

for any sequence $X_1 X_2 \cdots X_k$ and any single input X. Both sides of Equation 10.9 may equally well be written as the concatenation of a single output and the remaining sequence of outputs.

$$\lambda(p, X)\lambda[\delta(p, X), X_1 X_2 \cdots X_k] = \lambda(q, X)\lambda[\delta(q, X), X_1 X_2 \cdots X_k] \tag{10.10}$$

Therefore,

$$\lambda(p, X) = \lambda(q, X)$$

and

$$\lambda[\delta(p, X), X_1 X_2 \cdots X_k] = \lambda[\delta(q, X), X_1 X_2 \cdots X_k] \tag{10.11}$$

so that

$$\delta(p, X) \equiv \delta(q, X) \quad \text{and} \quad \delta(p, X) \triangleq \delta(q, X) \qquad \text{Q.E.D.}$$

B. (Sufficiency.) Assume that the states of a sequential circuit are partitioned into a set of equivalence classes such that two states fall into the same class if and only if conditions

(1) and (2) are satisfied. Let p and q be any two states in the same equivalence class, i.e., $p \triangleq q$. Suppose that $\delta(p, X_1 X_2 \cdots X_i) \triangleq \delta(q, X_1 X_2 \cdots X_i)$. Therefore, by condition (2),

$$\delta(p, X_1 X_2 \cdots X_{i+1}) \triangleq \delta(q, X_1 X_2 \cdots X_{i+1})$$

so that, by induction,

$$\delta(p, X_1 X_2 \cdots X_R) \triangleq \delta(q, X_1 X_2 \cdots X_R) \tag{10.12}$$

for all R. Since

$$\lambda(p, X_1 \cdots X_R) = \lambda(p, X_1)\lambda(\delta(p, X_1), X_2) \cdots \lambda(\delta(p, X_1 \cdots X_{R-1}), X_R)$$

and

$$\lambda(q, X_1 \cdots X_R) = \lambda(q, X_1)\lambda(\delta(q, X_1), X_2) \cdots \lambda(\delta(q, X_1 \cdots X_{R-1}), X_R \tag{10.13}$$

and the sequences of symbols on each side of these expressions are identical, we have

$$\lambda(p, X_1 \cdots X_R) = \lambda(q, X_1 \cdots X_R)$$

for any k. Therefore, by Definition 10.1, $p \equiv q$, and the partition is composed of equivalence classes of indistinguishable states.   Q.E.D.

We are now ready to use Theorem 10.1 to partition example sequential circuits into equivalence classes of indistinguishable states. As a first example, we will use the partitioning technique to reduce the number of states in the error-detecting sequential circuit of Example 10.3. A technique usually referred to as the Huffman-Mealy method will be illustrated in Example 10.6.

Example 10.6

Partition the states of the sequential circuit in Fig. 10.19 into equivalence classes of indistinguishable states.

The method consists of two steps. The first step is to partition the states into the smallest possible number of equivalence classes so that states in the same class have the same outputs, i.e., satisfy condition (1) of Theorem 10.1. This is done by inspection of the state table. For the state table in Fig. 10.19, we see that three classes are required. Class **a** will include states whose output is 1 for input 0 and 0 for input 1. Class **b** will include states whose output is 0 for either input. Class **c** will include states whose output is 0 for input 0 and whose output is 1 for input 1.

$$\begin{array}{ccc} \textbf{a} & \textbf{b} & \textbf{c} \\ (q_0) & (q_1, q_2, q_3) & (q_4, q_5, q_6, q_7) \end{array}$$

The classes of indistinguishable states will always be composed of subclasses of classes **a**, **b**, and **c**. These classes are the largest classes that will satisfy condition (1). Insisting on the satisfaction of condition (2) will only further subdivide the classes.

The partition satisfies the first condition of Theorem 10.1. If the next states for each state in class **b** (and for the states in class **c**) are in the same class for every possible single input, the second condition of Theorem 10.1 would be satisfied. In such a case, classes **a**, **b**, and **c** would be the desired equivalence classes. Let us determine for each state whether

q^v \ x^v	0	1
q_0	$q_0, 1$	$q_4, 0$
q_1	$q_0, 0$	$q_4, 0$
q_2	$q_1, 0$	$q_5, 0$
q_3	$q_1, 0$	$q_5, 0$
q_4	$q_2, 0$	$q_6, 1$
q_5	$q_2, 0$	$q_6, 1$
q_6	$q_3, 0$	$q_7, 1$
q_7	$q_3, 0$	$q_7, 1$

q^{v+1}, z^v

FIGURE 10.19   State table for error detector.

the next states are in classes **a**, **b**, or **c**. Below, the next-state class for a 0 input is written to the left of each state; the next-state class for a 1 input is written to the right.

Class	a		b				c		
States	0	1	2	3	4	5	6	7
Next Class	a c	a c	b c	b c	b c	b c	b c	b c

Clearly, class **b** contains pairs of states that do not satisfy condition (2) and are therefore not indistinguishable. We must separate class **b** into two classes **b** and **d**. The new partition may or may not consist of equivalence classes of indistinguishable states. To find out, the process of listing the classes containing the next states is repeated.

Class	a	b			c			d
States	0	2	3	4	5	6	7	1
Next Class	a c	d c	d c	b c	b c	b c	b c	a c

In this case, the four classes **a**, **b**, **c**, and **d** are equivalence classes of indistinguishable states. Note that all states in a class have next states in the same class for inputs or 0 and 1. Thus, condition (2) is finally satisfied. ∎

The partitioning of states into classes whose outputs agree and whose next states are in the same class is graphically illustrated in Fig. 10.20 for the above example. It should be clear to the reader that the same inputs applied to two states in the same class will result in the same output and send the circuit to next

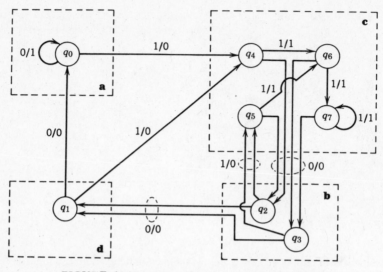

FIGURE 10.20   *Illustration of equivalence classes.*

states in the same class. Any input applied to both of these states will yield the same output and again send the circuit to next states in the same class, and so on indefinitely. Therefore, the class containing the two initial states is an equivalence class. Similar arguments apply to the other classes.

Having arrived at a set of equivalence classes of states within a circuit, S, as in the previous example, it remains to determine from them a minimal state circuit equivalent to S.

Equivalence Class Circuit   Corresponding to a completely specified sequential circuit, S, form a circuit, T, with one state, p_j, corresponding to each equivalence class, C_j, of the states of S. As each state in C_j has the same output, $\lambda(C_j, X)$, for any input X, let $\lambda(p_j, X) = \lambda(C_j, X)$. Similarly, each state of C_j has, for any input X, a next state in the same class, i.e., $\delta(C_j, X) \equiv q_k$ is any member of the equivalence class C_k. Therefore, we let $\delta(p_j, X) \equiv p_k$. Given this construction it can be shown by an argument similar to the proof of Theorem 10.1 that for every class C_k, $p_k \equiv q_{ki}$ for every $q_{ki} \in C_k$. A circuit T, formed in this manner, is known as an equivalence class circuit.

THEOREM 10.2   The circuit, T, as defined above is equivalent to S. Furthermore, no other circuit equivalent to S has fewer states than T, and any circuit equivalent to S having the same number of states as T must be T.

Proof   Let q_j be any state in S. Since it falls in some equivalence class C_j, there is a corresponding state, p_j, in T such that, according to the above definition, $q_j = p_j$. Conversely, for any state p_j in T, there is a corresponding equivalence class C_j such that, by the above definition, p_j is equivalent to any state in the class. Thus, by Definition 10.2, the circuits are equivalent.

No single state of any circuit can be equivalent to two states, q_1 and q_2 of circuit S, which are in different equivalence classes. By definition, these states must have different output sequences for some input sequence. Thus, the minimal state equivalent of S can have no fewer states than T, and one state must be assigned to each class.   Q.E.D.

Example 10.6 (continued)

Determine the minimal state equivalent of the sequential circuit in Fig. 10.19.

Solution   Corresponding to each of the classes **a**, **b**, **c**, and **d**, we define states p_1, p_2, p_3, p_4 in a circuit T. The outputs and next states, defined according to Theorem 10.2, are listed in the state table of Fig. 10.21a. The corresponding state diagram is given in Fig. 10.21b. It is interesting to observe that an eight-state circuit, apparently requiring three flip-flops in the direct realization of Fig. 10.10, has been reduced to a four-state machine. Conceivably one might have arrived at this four-state circuit directly by syn-

p^v	x^v 0	1	Class	Equivalent States of S
p_1	$p_1, 1$	$p_3, 0$	a	q_0
p_2	$p_4, 0$	$p_3, 0$	b	q_2, q_3
p_3	$p_2, 0$	$p_3, 1$	c	q_4, q_5, q_6, q_7
p_4	$p_1, 0$	$p_3, 0$	d	q_1

$$p^{v+1}, z^v$$

FIGURE 10.21a

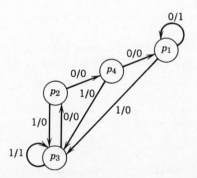

FIGURE 10.21b   *Minimal state equivalent of Fig. 10.19.*

thesizing the state diagram. Fig. 10.21b has been deliberately arranged to illustrate this possibility. The pair of states p_2 and p_3 may have been set down as a starting point. These will be the states of the machine following a single 0 or a single 1, respectively. It is not always possible to be clever, and arriving at Fig. 10.21 after a minimization process would probably be the normal circumstance.   ∎

The state minimization process resulted in the saving of a flip-flop in the previous example and will do so again in Example 10.7. This will not always happen. It may be possible to reduce the number of states without reducing the number of flip-flops. Assume that the original number of states, N_0, that satisfies Equation 10.15 is reduced to N_1. The number of states, N_1, in the minimal circuit must be less than or equal to 2^{r-1} if a saving in memory elements is to be achieved. A designer who is certain that sufficient states cannot be eliminated may decide to skip the state minimization process.

$$2^r \geq N_0 > 2^{r-1} \tag{10.15}$$

Example 10.7

The state table for the "beginning of message" detector is repeated as Fig. 10.22a. Obtain a minimal state equivalent of this table.

q^v \ x^v	0	1
q_0	$q_4, 0$	$q_1, 0$
q_1	$q_4, 0$	$q_2, 0$
q_2	$q_4, 0$	$q_3, 1$
q_3	$q_3, 0$	$q_3, 0$
q_4	$q_4, 0$	$q_1, 0$

q^{v+1}, z^v

(a)

q^v \ x^v	0	1
a	$a, 0$	$c, 0$
b	$a, 0$	$d, 1$
c	$a, 0$	$b, 0$
d	$d, 0$	$d, 0$

q^{v+1}, z^v

(b)

FIGURE 10.22 *State tables for beginning-of-message detector.*

Solution   We first partition the states into classes satisfying condition (1) and then find the classes containing the next states. We note that state q_1 must be removed from class **a**.

Class	a				b
States Next Class	q_0 a a	q_1 a b	q_3 a	q_4 a a	q_2 a a

We thus define class **c** and repeat the process of classifying next states. This time we find that state q_3 must be removed from class **a**, so we define class **d**. After another pass, we

Class	a			b	c
States Next Class	q_0 a c	q_3 a a	q_4 a c	q_2 a a	q_1 a b

find that states q_0 and q_4 are equivalent, and the process terminates. States a, b, c, and d are defined corresponding to the respective classes. Using the already tabulated informa-

Class	a		b	c	d
States Next Class	q_0 a c	q_4 a c	q_2 a d	q_1 a b	q_3 d d

tion, the minimal state table of Fig. 10.22b is easily generated according to Theorem 10.2. Once again, one might have recognized at the outset that q_0 and q_4 are equivalent.   ∎

10.8   Simplification by Implication Tables

An alternative method for determining the equivalence classes involves the use of the *implication table*, based on the work of M. C. Paull and S. H. Unger [2]. This method will probably seem more time-consuming than the partitioning method. However, unlike the partitioning method, it can be extended to the incompletely specified case, as discussed in Chapter 12.

Definition 10.3   A set of states **P** is *implied* by a set of states **R** if, for some specific input X_j, **P** is the set of all next states $\delta(r, X_j)$ for all present states r in **R**.

Consider the next-state table shown in Fig. 10.23. Assume that $\mathbf{R}_1 = (q_1, q_2)$. Then the sets *implied* by $\mathbf{R}_1$ and $\mathbf{R}_2 = (q_3, q_4)$ and $\mathbf{R}_3 = (q_2, q_3)$, since $\delta(q_1, 0) = q_3$ and $\delta(q_2, 0) = q_4$; and $\delta(q_1, 2) = q_2$ and $\delta(q_2, 2) = q_3$. Similarly,

Present State q^v	Input X^v			
	0	1	2	3
1	3	4	2	4
2	4	4	3	4
3	1	1	3	4
4	1	2	1	4
	q^{v+1}			

FIGURE 10.23   Next-state table.

set $R_2 = (q_3, q_4)$ *implies* $R_1 = (q_1, q_2)$ and $R_4 = (q_1, q_3)$. Note that a given set, e.g., R_1, can both *imply* and *be implied*. By Theorem 10.1, the states in a set R are equivalent only if all the states in any set P implied by R are equivalent. This follows directly from the definition of implication and condition (2) of Theorem 10.1. Again, referring to Fig. 10.23, it is apparent that the states in $R_1 = (q_1, q_2)$ cannot be equivalent unless the states in $R_2 = (q_3, q_4)$ and $R_3 = (q_2, q_3)$ are also equivalent. The implication table provides a systematic procedure for determining the equivalence classes of the states of the circuit.

Example 10.8

Use an implication table to find a minimal state equivalent to the state table given in Fig. 10.24.

q^v	$x^v = 0$	$x^v = 1$	q^v	$x^v = 0$	$x^v = 1$
1	2, 0	3, 0	7	10, 0	12, 0
2	4, 0	5, 0	8	8, 0	1, 0
3	6, 0	7, 0	9	10, 1	1, 0
4	8, 0	9, 0	10	4, 0	1, 0
5	10, 0	11, 0	11	2, 0	1, 0
6	4, 0	12, 0	12	2, 0	1, 0

$$q^{v+1}, z \qquad\qquad q^{v+1}, z$$

FIGURE 10.24   State table, Example 10.8.

Solution   For completely specified state tables, it is sometimes possible to reduce the number of states informally before using a formal minimization technique. This is worthwhile since considerable labor is involved in both formal methods. Note in Fig. 10.24 that outputs and next states of states 11 and 12 are identical. Thus, these two states are indistinguishable, and state 12 may be eliminated from the table. Since state 12 is a next state of states 6 and 7, we replace these entries with 11. The state table thus modified is given in Fig. 10.25a. Note that only next-state information is provided in this table. State 9, which has different outputs than the other states, is listed separately at the bottom of the table.

In Fig. 10.25a states 5 and 7 are now seen to be indistinguishable. Eliminating state 7 and replacing all next state entries of 7 with 5 yield the state table of Fig. 10.25b. No further reductions are apparent. This version may be further simplified using an implication table.*

The implication table is shown in Fig. 10.25. Vertically down the left side are listed all the states of the reduced state table except the first, and horizontally across the bottom are listed all but the last. The table thus contains a square for each pair of states. We start

---

* This reduced table could equally well form the starting point for the partitioning method.

by placing an X in any square corresponding to a pair of states that have different outputs and therefore cannot be equivalent. In this case, q_9 has a different output than any other state, so we place an X in every square in row 9. The other states all have the same output, so no more squares are marked X.

We now enter in each square the pairs of states implied by the pair of states corresponding to that square. We start in the top square of the first column, corresponding to the pair of sets (q_1, q_2). From the state table, we see that this pair implies the sets (q_2, q_4) and (q_3, q_5), so we enter 2–4 and 3–5 in this square. Proceeding to the next lower square, we see that (q_1, q_3) implies (q_2, q_6) and (q_3, q_5). We proceed in this manner to complete the first column. Note that there is only one entry in the square for (q_1, q_{11}). Since one of the implied sets is the single set, (q_2), only one equivalence pair, (q_3, q_1), is implied. We next proceed to the second column, the third, etc., until the table is complete (Fig. 10.25c).

The next step is to make a second "pass" through the table to see if any of the implied sets are ruled out as possible equivalence pairs by the squares marked X. In general, if there is an X in a square with "coordinates" j-k, then any square containing j-k as an implied set may be X'd out. In this case, q_9 cannot be equivalent to any other state, so any implied sets containing 9 are also ruled out. We place an X in every square containing a 9 (Fig. 10.25d). When this has been done, every square with coordinate 4 has been crossed out, indicating that state q_4 cannot be equivalent to any other state. We therefore proceed to cross out every square containing a 4 entry (Fig. 10.25d).

No other state is ruled out completely, so we now make a systematic check of all squares still not crossed out, moving from right to left. The first such square encountered is 8–11, which contains the entry 8–2. Square 8–2 has already been crossed out, so we must cross out square 8–11 (Fig. 10.25e). Next, we find entry 1–11 in square 6–10. Square 1–11 is not crossed out, so we leave square 6–10 as is. In a similar fashion, we find that squares 5–8, 3–8, and 1–8 must be crossed out (Fig. 10.25e). When we have completed this "pass," we must make still another check of all remaining squares, since, for example, the elimination of square 1–8 might require the elimination of one of the remaining squares to the right. In general, this process must be repeated until a "pass" is completed without the elimination of any more squares. In this case, the final form is that shown in Fig. 10.25e. We have recopied the implication table twice for clarity in explaining the method, but in practice the complete procedure can be carried out on a single table.

In the completed implication table, each square not crossed out represents a pair of equivalent states. The equivalence classes may now be determined as follows. We first list the states corresponding to the columns of the implication table in reverse order (Fig. 10.26). We then check each column of the final implication table for any squares not crossed out, working from right to left. In Fig. 10.25e, there are no such squares in columns 11, 10, and 8, so dashes (—) are placed opposite 11, 10, and 8 in Fig. 10.26.

In column 6, we find square 6–10 not crossed out, so we enter the equivalence pair (6, 10) opposite 6 in Fig. 10.26. In column 5, we find the square 5–11 not crossed out, so we enter (5, 11) opposite 5 in Fig. 10.25 and recopy the previously determined pair (6, 10). In column 4, there are no squares not crossed out, so we simply recopy the already selected pairs opposite 4. In column 3, we find that q_3 is equivalent to both q_5 and q_{11}, and since q_5 and q_{11} are already equivalent, we add 3 to this set opposite 3 in Fig. 10.26. Similarly, column 2 adds q_2 to the set (q_6, q_{10}) and column 1 adds q_1 to the set $(q_3, q_5,$

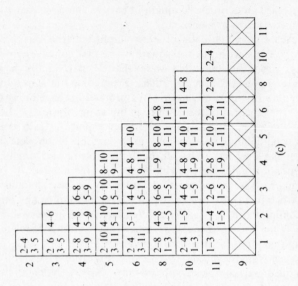

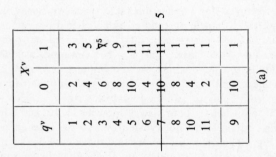

FIGURE 10.25 Simplification by inspection and implication.

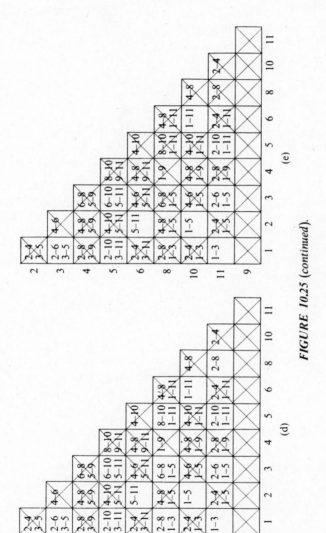

FIGURE 10.25 (continued).

297

11	—
10	—
8	—
6	(6,10)
5	(5,11) (6,10)
4	(5,11) (6,10)
3	(3,5,11) (6,10)
2	(3,5,11) (2,6,10)
1	(1,3,5,11) (2,6,10)

Equivalence Classes	(1,3,5,11) (2,6,10) (4) (8) (9)

FIGURE 10.26   *Determination of equivalence classes from the implication table.*

q_{11}). The resultant list opposite 1 in Fig. 10.26 comprises all equivalence classes with more than one member. Any states in the state table but not already on this list are not equivalent to any other state and must be included as single-state equivalence classes. These are added in the last line of Fig. 10.26.

Since there are five equivalence classes, the original state table of Fig. 10.24 reduces to the equivalent five-state table of Fig. 10.27.

q^v \ x^v	0	1
(1, 3, 5, 11) a	b, 0	a, 0
(2, 6, 10) b	c, 0	a, 0
(4) c	d, 0	e, 0
(8) d	d, 0	a, 0
(9) e	b, 1	a, 0

$$q^{v+1}, z$$

FIGURE 10.27   *Minimal-state table for Example 10.8.*

∎

Example 10.9

Find the minimal table equivalent to the state table of Fig. 10.28.

Solution   First we recopy the next-state portion of the state table, partitioning the states into groups with like output, as shown in Fig. 10.29a.

We next check for states identical by inspection. States 3 and 11, which are in the same output group, have identical next-state entries, so 11 may be replaced by 3. State 10 has 11 as a next-state entry, so we replace it with 3, after which 10 is seen to be identical to

Present State q^v	Input x^v							
	0	1	2	3	0	1	2	3
1	6	2	1	1	0	0	0	0
2	6	3	1	1	0	0	0	0
3	6	9	4	1	0	0	1	0
4	5	6	7	8	1	0	1	0
5	5	9	7	1	1	0	1	0
6	6	6	1	1	0	0	0	0
7	5	10	7	1	1	0	1	0
8	6	2	1	8	0	0	0	0
9	9	9	1	1	0	0	0	0
10	6	11	1	1	0	0	0	0
11	6	9	4	1	0	0	1	0
	q^{v+1}				z^v			

FIGURE 10.28   *State table, Example 10.9.*

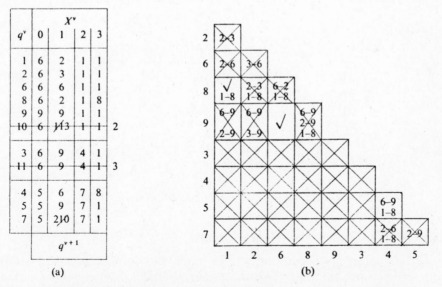

(a)   (b)

FIGURE 10.29   *Simplification of Example 10.9 by implication table.*

```
5    ----
4    (4,5)
3    (4,5)
9    (4,5)
8    (4,5)
6    (4,5)   (6,9)
2    (4,5)   (6,9)
1    (4,5)   (6,9)   (1,8)
```

| Equivalence Classes | (4,5) | (6,9) | (1,8) | (2) | (3) | (7) |

(c)

FIGURE 10.29 *(Continued)*

state 2. We then replace the next-state entry of 10 with 2 at state 7, but no further identities are found, so we proceed to the implication table, Fig. 10.29b.

Note the $(\sqrt{})$ in squares 1–8 and 6–9 of Fig. 10.29b. These indicate sets that imply only themselves and are therefore equivalent by inspection. Consider set (q_1, q_8), which has identical next-state entries except for $X = 3$, for which the implied set is also (q_1, q_8). Thus, (q_1, q_8) does not imply any other sets; and since they have the same output, they are equivalent. Squares marked with a check will never be crossed out in the processing of the implication table. If any pairs with identical next-state entries and outputs should be overlooked in the preliminary reduction by inspection, they should also be marked with a $(\sqrt{})$ in the implication table. The minimal equivalent table is shown in Fig. 10.30.

q^v	x^v 0	1	2	3
(4, 5) a	a, 1	b, 0	f, 1	c, 0
(6, 9) b	b, 0	b, 0	c, 0	c, 0
(1, 8) c	b, 0	d, 0	c, 0	c, 0
(2) d	b, 0	e, 0	c, 0	c, 0
(3) e	b, 0	b, 0	a, 1	c, 0
(7) f	a, 1	d, 0	f, 1	c, 0

q^{v+1}, z^v

FIGURE 10.30 *Minimal-state table, Example 10.9.*   ■

10.9 Mealy Circuits and Moore Circuits

In the state tables developed thus far, the output has been a function of both the input and the present state. This most general model of a sequential circuit is called a *Mealy model*. As an example, the state table of Fig. 10.13 is a Mealy

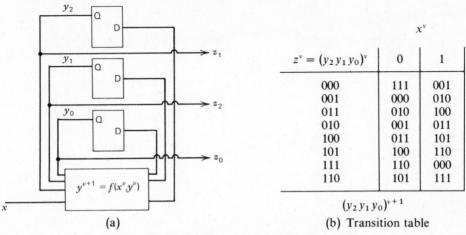

$z^v = (y_2 y_1 y_0)^v$	x^v	
	0	1
000	111	001
001	000	010
011	010	100
010	001	011
100	011	101
101	100	110
111	110	000
110	101	111

$(y_2 y_1 y_0)^{v+1}$

(a)   (b) Transition table

FIGURE 10.31   *Moore model up-down counter.*

model. This state table was a representation of an up-down counter deliberately configured to illustrate a finite-output memory circuit. Suppose the up-down counter circuit of Fig. 10.12 is adjusted slightly so that the outputs z_2, z_1, and z_0 are connected to the flip-flop outputs as shown in Fig. 10.31. This circuit remains an up-down counter, with outputs delayed one clock period compared to those of Fig. 10.12. This form of counter closely resembles the T flip-flop realization of Fig. 9.12.

The up-down counter transition table remains as shown in Fig. 10.12b except that $(z_2, z_1, z_0) = (y_2, y_1, y_0)$. Using this relationship, the state table for Fig. 10.31 may be represented as shown in Fig. 10.32a. Note that the output values for each state are the same for $x = 1$ as for $x = 0$. That is, the output is a function of the present state only and not of the inputs. Thus, it is possible to express the state table in the more compact form of Fig. 10.32b on which the outputs are listed in a separate column, one entry for each state. A sequential circuit whose outputs are a function of present states only and can be represented by a state table of the form of Fig. 10.32b is a *Moore model*. The state diagram for a Moore model also differs from the Mealy model. As shown in Fig. 10.32c, output values are included within the circles containing the state identifiers. Only inputs are associated with the arrows interconnecting the states.

In the case of the two counters discussed above, we have Mealy and Moore circuits that perform the same function (except for the timing difference inherent in the Mealy and Moore models) and have the same number of states. However, Mealy and Moore circuits performing the same functions will not always have the same number of states. Let us translate the Mealy circuit for which the state

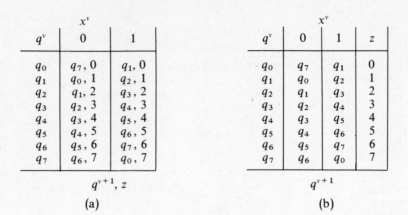

Table (a):

q^v	x^v = 0	x^v = 1
q_0	$q_7, 0$	$q_1, 0$
q_1	$q_0, 1$	$q_2, 1$
q_2	$q_1, 2$	$q_3, 2$
q_3	$q_2, 3$	$q_4, 3$
q_4	$q_3, 4$	$q_5, 4$
q_5	$q_4, 5$	$q_6, 5$
q_6	$q_5, 6$	$q_7, 6$
q_7	$q_6, 7$	$q_0, 7$

q^{v+1}, z

(a)

Table (b):

q^v	x^v = 0	x^v = 1	z
q_0	q_7	q_1	0
q_1	q_0	q_2	1
q_2	q_1	q_3	2
q_3	q_2	q_4	3
q_4	q_3	q_5	4
q_5	q_4	q_6	5
q_6	q_5	q_7	6
q_7	q_6	q_0	7

q^{v+1}

(b)

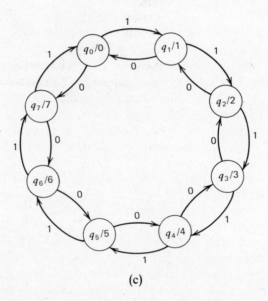

(c)

FIGURE 10.32   *Moore model state tables and state diagrams.*

diagram is shown in Fig. 10.33 to a Moore circuit with the same input-output function. Every output that is associated with a transition between states in the machine of Fig. 10.33 must be associated with a state in the proposed Moore circuit. Since the inputs cannot be anticipated, the outputs corresponding to inputs in Fig. 10.33 must be associated with the next states following the same inputs in the Moore circuit.

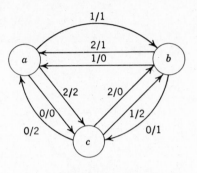

FIGURE 10.33

Consider some particular state, p, of the Mealy circuit. Let $\{\lambda_i\}_p$ be the set of all distinct outputs, λ_i, of the Mealy circuit given by

$$\lambda_i = \lambda(q, I_i) \tag{10.16}$$

for any input I_i and any state q satisfying

$$\delta(q, I_i) = p \tag{10.17}$$

For example, in Fig. 10.33, we have three different transitions to state c. That is,

$$\delta(a, 0) = c \quad \text{where} \quad \lambda(a, 0) = 0$$
$$\delta(a, 2) = c \qquad\qquad \lambda(a, 2) = 2 \tag{10.18}$$
$$\delta(b, 0) = c \qquad\qquad \lambda(b, 0) = 1$$

Therefore, $\{\lambda_i\}_c = \{0, 1, 2\}_c$.

We may now define a set of states $\{p_{\lambda_i}\}$ in the Moore circuit corresponding to the single state, p, of the Mealy circuit. That is, for every $\lambda_i \in \{\lambda_i\}_p$, let p_{λ_i} be a distinct state in the Moore circuit such that

$$\lambda(p_{\lambda_i}) = \lambda_i \tag{10.19}$$

By definition, $\lambda(p_{\lambda_i})$ is the output associated with p_{λ_i} in the Moore circuit.

In Fig. 10.33, for example, the set of states $\{C_0, C_1, C_2\}$ corresponds to the output set $\{0, 1, 2\}_c$. Similarly, for states a and b, the output sets are $\{0, 1, 2\}_a$ and $\{0, 1, 2\}_b$; the sets of states in the Moore circuit are $\{A_0, A_1, A_2\}$ and $\{B_0, B_1, B_2\}$. These states are listed in the state table of Fig. 10.34.

The next-state function for any input, I, in the Moore circuit may be defined as follows. Suppose

$$\delta(q, I) = p \tag{10.20}$$

	$X = 0$	$X = 1$	$X = 2$	λ
A_0	C_0	B_1	C_2	0
A_1	C_0	B_1	C_2	1
A_2	C_0	B_1	C_2	2
B_0	C_1	A_0	A_1	0
B_1	C_1	A_0	A_1	1
B_2	C_1	A_0	A_1	2
C_0	A_2	B_2	B_0	0
C_1	A_2	B_2	B_0	1
C_2	A_2	B_2	B_0	2

FIGURE 10.34   *Moore circuit corresponding to the Mealy circuit of Fig. 10.33.*

for two states p and q in the Mealy circuit. Let $\{p_{\lambda_j}\}$ and $\{p_{\lambda_i}\}$ be the sets of states in the Moore circuit corresponding to p and q, respectively. Let p_{λ_j} be the member of $\{p_{\lambda_i}\}$ such that

$$\lambda_j = \lambda(q, I) \tag{10.21}$$

Then we may define the next-state function in the Moore circuit for input, I, and any state $q_{\lambda_i} \in \{q_{\lambda_i}\}$ as

$$\delta(q_{\lambda_i}, I) = p_{\lambda_j} \tag{10.22}$$

Suppose the input to the Moore circuit of Fig. 10.34 is $X = 0$ when the circuit is in one of the set of states $\{A_0, A_1, A_2\}$. Recall from Fig. 10.33 that

$$\delta(A, 0) = C \qquad \text{and} \qquad \lambda(A, 0) = 0 \tag{10.23}$$

Therefore, we conclude that for present state, a, $\lambda_j = 0$, as defined by Equation 10.21, and that from Equation 10.22

$$\delta(A_0, 0) = \delta(A_1, 0) = \delta(A_2, 0) = C \tag{10.24}$$

Initial State

Input	↓ 1 1 2 1 0 0 2	A_0 ⎫	1	1	2	1	0	0	2
New State	@ b a c b c a c	A_1 ⎬	B_1	A_0	C_2	B_2	C_1	A_2	C_2
Output	1 0 2 2 1 2 2	A_2 ⎭	1	0	2	2	1	2	2
	(a)				(b)				

FIGURE 10.35   *Input-output sequences.*

$$
\begin{array}{c|c|c|c}
A_0 & C_0 \ \ 0 & B_1 \ \ 1 & C_2 \ \ 2 \\
A_1 & C_0 \ \ 0 & B_1 \ \ 1 & C_2 \ \ 2 \\
A_2 & C_0 \ \ 0 & B_1 \ \ 1 & C_2 \ \ 2 \\
B_0 & C_1 \ \ 1 & A_0 \ \ 0 & A_1 \ \ 1 \\
B_1 & C_1 \ \ 1 & A_0 \ \ 0 & A_1 \ \ 1 \\
B_2 & C_1 \ \ 1 & A_0 \ \ 0 & A_1 \ \ 1 \\
C_0 & A_2 \ \ 2 & B_2 \ \ 2 & B_0 \ \ 0 \\
C_1 & A_2 \ \ 2 & B_2 \ \ 2 & B_0 \ \ 0 \\
C_2 & A_2 \ \ 2 & B_2 \ \ 2 & B_0 \ \ 0 \\
\end{array}
$$

FIGURE 10.36

The remaining next state entries in Fig. 10.34 may be determined in a similar manner.

So far we have only indicated that the circuits of Figs. 10.33 and 10.34 should have the same output sequence for any input sequence. The further refinement in notation required to complete a proof of this fact would take us too far afield. The reader who remains unconvinced can easily check the outputs of the circuit for a typical input sequence. In Fig. 10.35a, we see the sequence of outputs and sequence of states assumed by the Mealy circuit when subjected to the input sequence

$$1021002 \tag{10.25}$$

when initially in state a. Similarly, we see in Fig. 10.35b the output and next-state sequences of the Moore circuit when subjected to the same input sequences. In the latter case, the initial state may be any member of A_0, A_1, A_2.

Note in this case that a 9-state Moore circuit was required to imitate a 3-state Mealy circuit. This Mealy circuit was deliberately chosen to have distinct outputs for each transition to each of its three states. As we have seen, the situation is not this bad for many examples of practical interest.

The Moore circuit of Fig. 10.34 may be translated to an equivalent Mealy circuit by merely assigning an output to each input-present-state combination equal to the output associated with the next state for that combination in the Moore circuit. The result is Fig. 10.36.

Using the minimization procedure of the previous section, we find that the states of the circuit in Fig. 10.36 fall into the following equivalence classes:

$$(A_0 A_1 A_2)(B_0 B_1 B_2)(C_0 C_1 C_2) \tag{10.26}$$

The circuit formed from these equivalence classes is the original circuit of Fig. 10.33.

10.10   State Assignment and Memory Element Input Equations

Once a minimal-state table has been obtained, it remains to design a circuit realizing that table. It must be recognized that the question of economics has not been exhausted with the minimization of the number of states. It is true that the number of memory elements required by such a state table is minimum, but we also require combinational logic to develop the output and memory element input equations. It may be that some state table other than a minimal table would require significantly less combinational logic and, therefore, less overall hardware than the latter. However, we will ignore this problem for now and concentrate our attention on the problem of realizing a given state table as economically as possible.

As shown in Fig. 10.4, the next step in the design process after obtaining a minimal-state table is state assignment, the assignment of a specific combination of values of the state variables to each state, leading to the transition table. In some cases this step is trivial, the transition table following directly from the state table. If the original state table was obtained by the finite memory approach and the number of states has not reduced sufficiently to eliminate any flip-flops, then the basic finite memory form of Fig. 10.9b should be used. Since the state variables are simply stored values of inputs and outputs, the transition table follows directly from the specification of the desired sequence of inputs and outputs, as seen in the examples of finite memory circuits. This approach not only eliminates the state assignment problem, but also leads to an economical circuit. Since the inputs and outputs are simply stored, no input logic is required for the flip-flops. Another form might result in simpler output logic, but it is unlikely that this saving would outweigh the cost of memory element input logic.

Another type of circuit for which the state assignment problem is trivial is a Moore circuit in which each state has a distinct output. In such cases it is best to set the state variables equal to the outputs, in which case the transition table follows directly from the desired sequence of outputs. We have already seen an example of this in the counter of Fig. 10.31, and most circuits of this type are counters. Such a design will usually be economical since no output logic is required.

If neither of the above methods is appropriate, the problem of choosing the best circuit realization is unfortunately a difficult problem in its own right. We recall that each of the states will correspond to one of the 2^r combinations of the r state variables. The first problem in designing a circuit to realize a given state table is to decide which of the 2^r combinations shall be assigned to each state. If the number of states m satisfies

$$2^{r-1} < m \le 2^r$$

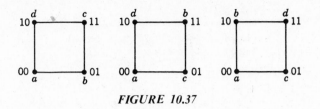

FIGURE 10.37

then r state variables will be required. There will be

$$\frac{2^r!}{(2^r - m)!} \tag{10.27}$$

ways to assign the 2^r combinations of state variables to the m states.

The problem of assigning state-variable combinations to four states, say a, b, c, d, can be visualized as the assignment of the four states to the four vertices of the 2-cube. Three distinct assignments are shown in this fashion in Fig. 10.37. Any other possible assignments would amount to rotations or reversals of these three assignments and would thus correspond either to reversing the order of the variables or complementing one or both variables. Such changes will not affect the form of any Boolean function and are thus irrelevant with respect to cost. The three distinct assignments for three rows can be obtained in a similar manner. The distinct assignments for three and four rows are listed in Table 10.1. The number of distinct assignments for various numbers of states is tabulated in Table 10.2 [2].

For circuits with only two states, there is no choice, and therefore no assignment problem. For three and four states, the simplest thing is to try all three possible assignments and see which produces the most economical circuit. For more than four states, complete enumeration is obviously impossible, so we will need to develop methods of picking a good assignment, preferably on the basis of the final state table. Let us start by taking a four-state example and trying all

Table 10.1

States	3-state assignments			4-state assignments		
	1	2	3	1	2	3
a	00	00	01	00	00	00
b	01	11	11	01	11	10
c	11	01	00	11	01	01
d	—	—	—	10	10	11

Table 10.2

No. of states m	No. of state variables r	No. of distinct assignments
2	1	1
3	2	3
4	2	3
5	3	140
6	3	420
7	3	840
8	3	840
9	4	10,810,800

three possible assignments to see if we can gain some insight into the characteristics of a good assignment.

Figure 10.38a shows the minimal-state table for the beginning-of-message detector, Examples 10.1 and 10.7. Figures 10.38b,c,d show the resultant next-state portion of the transition tables for assignments 1, 2, 3 of Table 10.2. Since there is only one output, the assignment will not affect the cost of realizing z, so

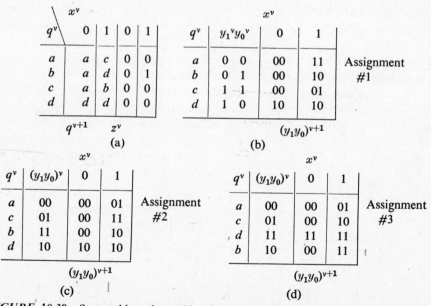

FIGURE 10.38 *State table and possible transition tables, beginning-of-message detector.*

the output portions of the transition tables have been omitted for clarity. Note that only the order of the states is changed; the state variables are always listed in K-map order, as required for transition tables. This is important, because the next step is to translate the transition tables into K-maps, known as the excitation maps, representing the memory element input equations.

In Chapter 9, we developed K-maps representing the next states as a function of the present state and inputs for the common types of flip-flops. As we synthesize, we will use this information in a slightly different form. In a J-K flip-flop, for example, we must have at our fingertips the values of J and K that will cause any particular state transition to take place. The values of J and K required to effect each of the four input transitions are given as a *transition list* in Fig. 10.39b. This table can be derived immediately from the K-map of Fig. 10.39a. Transition lists for RS, T, and D flip-flops are given in Fig. 10.39c,d, and e, respectively.

The left column in each diagram lists the four possible combinations of present value and desired next value for the output, Q. The entries give the necessary values at the inputs to achieve these combinations. Consider for example, the diagram for the J-K flip-flop. If the present value is 0, i.e., the flip-flop is

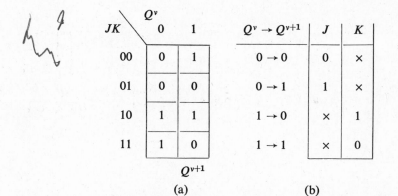

FIGURE 10.39   *Transition lists.*

cleared, and it is to remain that way, then J must equal 0 and K can be either 0 or 1 (don't care). If Q is now 0 and is to be 1, i.e., the flip-flop is to be set, then $J = 1$ and $K = 0$ or 1. The other entries follow by similar reasoning.

To develop the excitation maps for assignment #1, we start with partial transition tables, each showing the transition in only a single-state variable (Figs. 10.40a, 10.40d). From these, we generate the excitation maps for the flip-flops controlling each state variable by reference to the transition diagram for the type of flip-flop to be used. (JK, Fig. 10.39b.)

Consider the entries on the second row. For $(y_1 y_0 x)^v = 010$, we see from Fig. 10.40a that $y_1^v = 0$, $y_1^{v+1} = 0$, so $J_{y_1} = 0$ and $K_{y_1} = \times$. For $(y_1 y_0 x)^v = 011$, $y_1^v y_1^{v+1} = 01$, so $J_{y_1} = 1$ and $K_{y_1} = \times$. Similarly, from Fig. 10.40d, for both $(y_1 y_0 x)^v = 000$ and $(y_1 y_0 x)^v = 001$, $y_0^v = 1 \rightarrow y_0^{v+1} = 0$, so $J_{y_0} = \times$ and $K_{y_0} = 1$. This process can be speeded up by filling in all the don't cares first. Note that if the present value of the state variable is 0, $K = \times$ regardless of the next value, while $J = \times$ if the present value is 1.

From these maps (Figs. 10.40b,c,e,f), we read off the minimal realizations, noting that the problem is multiple output, so products should be shared where

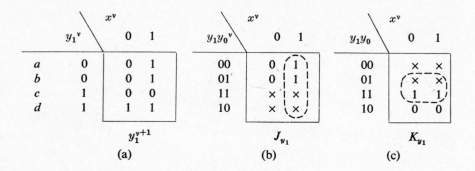

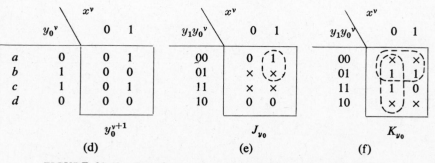

FIGURE 10.40   *Translation of transition table to excitation maps.*

possible. The minimal groupings are as shown. The equations for the assignment are

$$J_{y_1} = x \qquad J_{y_0} = x\bar{y}_1 \qquad K_{y_1} = y_0$$

$$K_{y_0} = \bar{x} + \bar{y}_1 \qquad z = x\bar{y}_1 y_0 \tag{10.28}$$

Figure 10.41 shows the maps for Assignments 2 and 3. The resultant equations are

Assignment #2

$$J_{y_1} = xy_0 \qquad K_{y_1} = \bar{x}y_0 \qquad z = xy_1 y_0$$

$$J_{y_0} = x\bar{y}_1 \qquad K_{y_0} = \bar{x} + y_1 \tag{10.29}$$

Assignment #3

$$J_{y_1} = xy_0 \qquad K_{y_1} = \bar{x}\bar{y}_0 \qquad z = xy_1 \bar{y}_0$$

$$J_{y_0} = x \qquad K_{y_0} = \bar{y}_1 \tag{10.30}$$

Assignments #1 and #3 are seen to have the same basic gate cost, but #3 is preferable because it uses only AND gates while #1 requires one OR gate. The circuit for #3 is shown in Fig. 10.42.

Before analyzing the final assignment in detail, let us consider another example to be sure the basic process is clear.

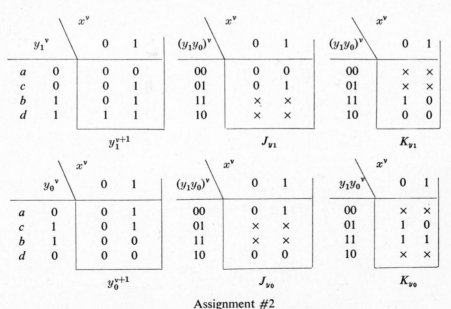

Assignment #2

FIGURE 10.41a   *Excitation maps for alternate assignments.*

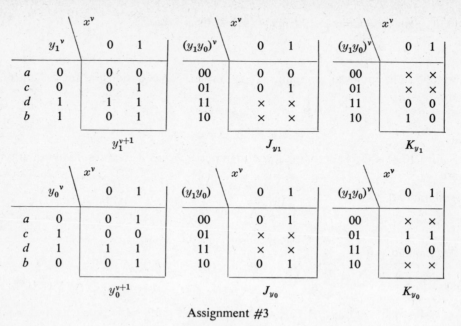

y_1^v		x^v 0	1
a	0	0	0
c	0	0	1
d	1	1	1
b	1	0	1

y_1^{v+1}

$(y_1y_0)^v$	x^v 0	1
00	0	0
01	0	1
11	×	×
10	×	×

J_{y_1}

$(y_1y_0)^v$	x^v 0	1
00	×	×
01	×	×
11	0	0
10	1	0

K_{y_1}

y_0^v		x^v 0	1
a	0	0	1
c	1	0	0
d	1	1	1
b	0	0	1

y_0^{v+1}

(y_1y_0)	x^v 0	1
00	0	1
01	×	×
11	×	×
10	0	1

J_{y_0}

$(y_1y_0)^v$	x^v 0	1
00	×	×
01	1	1
11	0	0
10	×	×

K_{y_0}

Assignment #3

FIGURE 10.41b

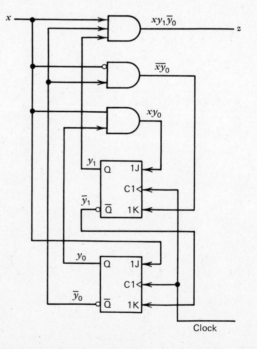

FIGURE 10.42 *Final circuit, beginning-of-message detector.*

312

Example 10.10

Determine a circuit realization for the error detector circuit of Examples 10.3 and 10.6 using *S-R* flip-flops.

Solution The minimal state table for the circuit is shown in Fig. 10.43a, and the resultant transition table for Assignment #2 is shown in Fig. 10.43b.

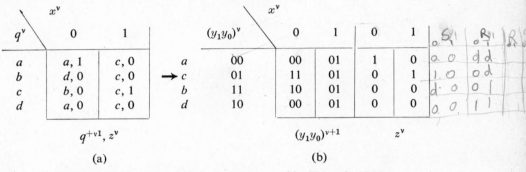

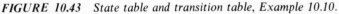

q^v	x^v = 0	1
a	$a, 1$	$c, 0$
b	$d, 0$	$c, 0$
c	$b, 0$	$c, 1$
d	$a, 0$	$c, 0$

q^{+v1}, z^v

(a)

$(y_1 y_0)^v$	x^v = 0	1	0	1
a	00	00	01	1	0
c	01	11	01	0	1
b	11	10	01	0	0
d	10	00	01	0	0

$(y_1 y_0)^{v+1}$ z^v

(b)

FIGURE 10.43 *State table and transition table, Example 10.10.*

The translation of the partial transition tables into the excitation maps for *S-C* flip-flops is shown in Fig. 10.44. Breaking up the transition tables into partial tables for each variable is not absolutely necessary. One could work directly from the complete

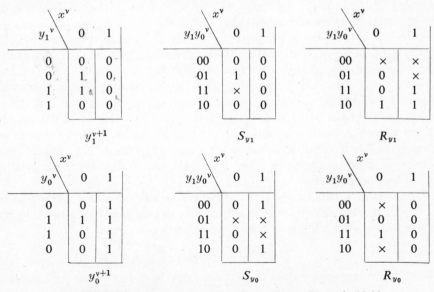

$y_1{}^v$	x^v = 0	1
0	0	0
0	1	0
1	1	0
1	0	0

y_1^{v+1}

$y_1 y_0{}^v$	x^v = 0	1
00	0	0
01	1	0
11	×	0
10	0	0

S_{y_1}

$y_1 y_0{}^v$	x^v = 0	1
00	×	×
01	0	×
11	0	1
10	1	1

R_{y_1}

$y_0{}^v$	x^v = 0	1
0	0	1
1	1	1
1	0	1
0	0	1

y_0^{v+1}

$y_1 y_0{}^v$	x^v = 0	1
00	0	1
01	×	×
11	0	×
10	0	1

S_{y_0}

$y_1 y_0{}^v$	x^v = 0	1
00	×	0
01	0	0
11	1	0
10	×	0

R_{y_0}

FIGURE 10.44 *Derivation of excitation maps, Example 10.10.*

transition table, but separating the variables reduces the likelihood of error. The equations for this assignment are

Assignment #2

$$S_{y_1} = \bar{x}y_0 \qquad R_{y_1} = x + \bar{y}_0 \qquad S_{y_0} = x \qquad R_{y_0} = \bar{x}y_1$$

$$z = \bar{x}\bar{y}_1\bar{y}_0 + x\bar{y}_1 y_0 \tag{10.31}$$

We will leave it to the reader to verify that the equations for assignment #1 and #3 are as follows:

Assignment #1

$$S_{y_1} = x + \bar{y}_1 y_0 \qquad R_{y_1} = \bar{x}y_1 \qquad S_{y_0} = x \qquad R_{y_0} = \bar{x}\bar{y}_1$$

$$z = \bar{x}\bar{y}_1\bar{y}_0 + xy_1 y_0 \tag{10.32}$$

Assignment #3

$$S_{y_1} = \bar{x}\bar{y}_1 y_0 \qquad R_{y_1} = x + y_1 y_0 \qquad S_{y_0} = x + y_1\bar{y}_0 \qquad R_{y_0} = \bar{x}y_0$$

$$z = \bar{x}\bar{y}_1\bar{y}_0 + xy_1 y_0 \tag{10.33}$$

Assignment #2 is seen to lead to the simplest equations. We will leave it to the reader to draw the final circuit. ∎

When there are only three or four states, all three possible assignments can be tried, as in the previous examples. The process may seem very complex and time consuming to the reader at this point, but it becomes surprisingly efficient as one gains experience. As we see from Table 10.2, however, if there are five or more states, trying all possible assignments is clearly out of the question. We need some rules or guidelines that will enable us to find assignments that are at least close to minimal with a reasonable amount of effort.

In an effort to find such rules, let us look more closely at the assignments in the previous examples to see if there are any characteristics that might have enabled us to predict in advance which assignment would lead to the minimal realization. In Fig. 10.45 are shown the state tables for the three assignments in Examples 10.7 and 10.10, with the states arranged in the K-map orders corresponding to each assignments. The minimal assignments for each case are enclosed by dashed lines.

A comparison of these maps leads to the observation that the selected assignments tend to group like next-state entries. For example, in Fig. 10.45a, we note that d is the next state for $x^v = 1$ for both states b and d, i.e.,

$$d = \delta(b, 1) = \delta(d, 1)$$

As a result, the next-state values of the state variables will, of course, be identical. If we give b and d adjacent assignments (i.e., place them on adjacent rows on the

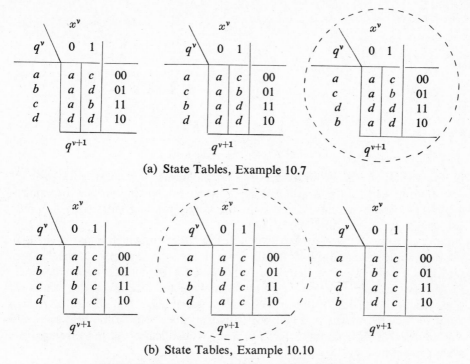

(a) State Tables, Example 10.7

(b) State Tables, Example 10.10

FIGURE 10.45 *State tables for possible assignments.*

K-map), the present values of one of the state variables will be identical. Thus, the transitions in that state variable will be the same for both entries, and the corresponding entries on the excitation maps will be the same.

These identical transitions are important because the excitation maps will lead to the simplest equations if 1s, 0s, and don't-cares are grouped as much as possible. In this case, the selected assignments assign like values of y_1 to b and d, so the corresponding transitions in y_1 are identical. This effect is seen on the excitation maps (Fig. 10.41b), where the corresponding entries are also identical—don't cares for J_{y_1}, 0s for K_{y_1}.

Our basic rule, then, will be to group the next-state entries as much as possible. But there may be several ways of grouping like entries, and this basic rule can be expanded and made more specific. Note that in the state table for Example 10.10 (Fig. 10.45b), states a and d have identical next-state entries in both columns, which is likely to produce a simpler map than identical entries in only one column, as was the case in Example 10.7. Further, if these two next-states, a and c, are given adjacent assignments, then the next-state entries for one

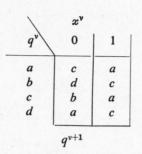

FIGURE 10.46   *Example state table.*

of the variables will be the same in both columns of a given row, probably resulting in even further simplification. This can be seen in the excitation map of Fig. 10.44, where we have four 0s in the corresponding entries for S_{y_1} and a group of four 1s and don't cares for C_{y_1}.

Next consider the state table shown in Fig. 10.46, in which a and c are again the next-states of a and d but this time not in the same columns. Nevertheless, if we make a and c adjacent, one state variable will have identical next-state entries in both columns of a given row, and if we make a and d adjacent, these rows will then pair off. A check of the three possible assignments will show that Assignment #2, which makes a, c and a, d adjacent, does lead to the best circuit for this table. Our discussion so far has been in justification of the three parts of Rule I given below.

In some tables, it will be possible to simplify one particular row of each excitation map by giving the various next states on a given row of the state table adjacent-state variable assignments. This observation is formalized as Rule II. If further flexibility remains in the assignment after applying Rules I and II, the assignment may be completed in a way that will simplify the output map as stated in Rule III.

Rule I a   Check for rows of the state table that have identical next-state entries in every column. Such rows should be given adjacent assignments. If possible, the next-state entries in these rows should be given adjacent assignments according to Rule II.

b   Check for rows of the state table that have the same next-state entries but in different column order. Such rows should be given adjacent assignments if the next-state entries can be given adjacent assignments.

c   Rows with identical next-state entries in some, but not all, columns should be given adjacent assignments, with rows having more identical columns having the higher priority.

Rule II   Next-state entries of a given row should be given adjacent assignments.

Rule III   Assignments should be made so as to simplify the output maps.

It is important to note that these rules are listed in decreasing order of their importance in producing good assignments. For tables of any complexity, they will generally lead to conflicting requirements, but an assignment made to satisfy a higher-priority rule should not be changed to satisfy a lower-priority rule.

The reader should note that Rule II has lower priority than all three parts of Rule I. It might seem surprising, because column-to-column adjacency would seem to be just as useful as row-to-row adjacency. The problem is that we have no control over column order, this being a function of the input code. Thus, when we group the entries in one row, we are just as likely to "ungroup" the entries in another row. In general, Rule II produces useful results only if the adjacency requirements of several rows can be satisfied at the same time.

Let us now apply these rules to another example to clarify their use.

Example 10.11

Design a serial parity checker for 4-bit binary characters. Starting from reset, the circuit receives 4-bit characters serially on a single input line x. At the time of the fourth bit, the output is to be 1 only if the total number of 1s in the character is even. At all other times the output is to be 0. Following receipt of the fourth bit, the circuit should return to reset to receive the next character.

Solution   At any time, the only important factors are how many bits have been received and whether the number of 1s has been even or odd. This analysis leads directly to the state diagram of Fig. 10.47, which is clearly minimal.

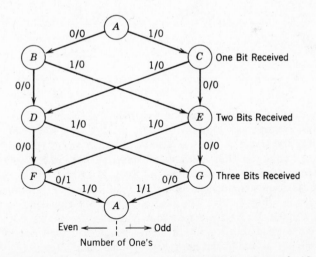

FIGURE 10.47   *State diagram for parity checker, Example 10.11.*

q^v	x^v 0	1
A	$B, 0$	$C, 0$
B	$D, 0$	$E, 0$
C	$E, 0$	$D, 0$
D	$F, 0$	$G, 0$
E	$G, 0$	$F, 0$
F	$A, 1$	$A, 0$
G	$A, 0$	$A, 1$

q^{v+1}, z^v

(a)

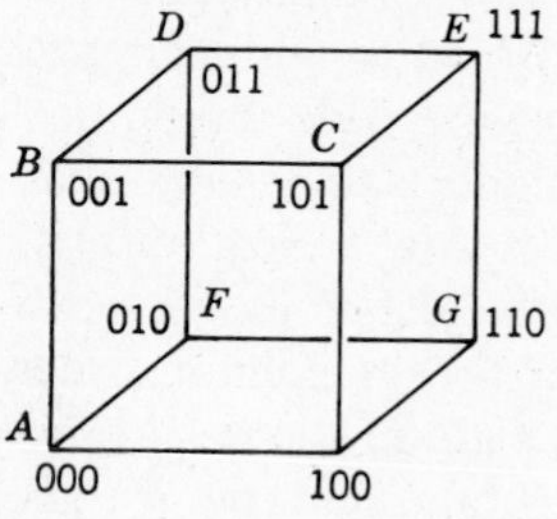

(b)

$(y_2 y_1 y_0)^v$	x^v 0	1	0	1
A 000	001	101	0	0
B 001	011	111	0	0
D 011	010	110	0	0
F 010	000	000	1	0
100	× × ×	× × ×	×	×
C 101	111	011	0	0
E 111	110	010	0	0
G 110	000	000	0	1

$(y_2 y_1 y_0)^{v+1}$ z^v

(c)

FIGURE 10.48   State table, assignment cube, and transition table, Example 10.11.

The corresponding state table is shown in Fig. 10.48a. Inspecting this table, we see first that Rule Ia requires that states F and G be adjacent. Rule Ib in turn requires that D and E be adjacent if F and G can be made adjacent and that B and C be made adjacent if D and E can be made adjacent. Thus adjacent assignments for B and C, D and E, and F and G will simultaneously satisfy all requirements of Rules Ia and b.

In setting up adjacent assignments for up to eight states, the Boolean hypercube is useful for visualization (Fig. 10.48b). The location of state A is apparently arbitrary, so we start by placing it at 000. Unless there are strong reasons to the contrary, it is best to have the reset state correspond to all flip-flops being cleared. This will usually simplify any separate reset circuit, as many flip-flops have special reset inputs available. With A located, we then place B and C, D and E, and F and G on succeeding edges of the cube, as

x^v

$y_2{}^v$	0	1
0	0	1
0	0	1
0	0	1
0	0	0
1	×	×
1	1	0
1	1	0
1	0	0

y_2^{v+1}

x^v

$y_2 y_1 y_0$	0	1
000	0	1
001	0	1
011	0	1
010	0	0
100	×	×
101	×	×
111	×	×
110	×	×

J_{y_2}

x^v

	0	1
	×	×
	×	×
	×	×
	×	×
	×	×
	0	1
	0	1
	1	1

K_{y_2}

x^v

$y_1{}^v$	0	1
0	0	0
0	1	1
1	1	1
1	0	0
0	×	×
0	1	1
1	1	1
1	0	0

y_1^{v+1}

x^v

$y_2 y_1 y_0$	0	1
000	0	0
001	1	1
011	×	×
010	×	×
100	×	×
101	1	1
111	×	×
110	×	×

J_{y_1}

x^v

	0	1
	×	×
	×	×
	0	0
	1	1
	×	×
	×	×
	0	0
	1	1

K_{y_1}

x^v

$y_0{}^v$	0	1
0	1	1
1	1	1
1	0	0
0	0	0
0	×	×
1	1	1
1	0	0
0	0	0

y_0^{v+1}

x^v

$y_2 y_1 y_0$	0	1
000	1	1
001	×	×
011	×	×
010	0	0
100	×	×
101	×	×
111	×	×
110	0	0

J_{y_0}

x^v

	0	1
	×	×
	0	0
	1	1
	×	×
	×	×
	0	0
	1	1
	×	×

K_{y_0}

$Q^v \to Q^{v+1}$	J	K
$0 \to 0$	0	×
$0 \to 1$	1	×
$1 \to 0$	×	1
$1 \to 1$	×	0

FIGURE 10.49   *Excitation maps, Example 10.11.*

shown. The corresponding transition table is shown in Fig. 10.48c. Note that one combination of state variables is not used since there are only seven states, so all entries for this row are don't cares.

The translation of transition table to excitation maps for J-K flip-flops is shown in Fig. 10.49. The final equations are found to be

$$J_{y_2} = x\bar{y}_1 + xy_0 \qquad K_{y_2} = x + \bar{y}_0 \qquad J_{y_1} = y_0 \qquad K_{y_1} = \bar{y}_0$$
$$J_{y_0} = \bar{y}_1 \qquad K_{y_0} = y_1 \qquad z = xy_2\bar{y}_0 + \bar{x}\bar{y}_2 y_1 \bar{y}_0 \tag{10.34}$$

We note that our rules have led us to an assignment that could hardly be simpler. We will leave it as an exercise for the reader to draw the final circuit. ∎

It is important that the reader recognize that these rules for state assignment do not describe a complete algorithmic procedure leading directly to the best assignment. They reduce the number of alternative assignments we need consider and enable us to find a reasonably good assignment in a practical length of time. Consider again the possible assignments for Example 10.7, (Fig. 10.39c). Rule Ia tells us that b and d should be adjacent, but Assignments #2 and #3 both meet this requirement. So how do we know that #3 is best? The rules give us no guidance here, but at least Rule Ic enables us to eliminate Assignment #1 from consideration.

For Example 10.10 (Fig. 10.39b), Rule Ia is fully satisfied only by Assignment #2, and this does produce the best circuit of S-C flip-flops. But if we use J-K flip-flops. Assignment #1 turns out to be better. The reason is that on the excitation maps for J-K flip-flop, half the entries are always don't-cares, no matter what the assignment. If the present value of a state variable is 0, the K input will have no bearing on the next value. Similarly, if the present value is 1, the J input doesn't matter. The large blocks of don't cares combine with the remaining 1s and 0s in unpredictable ways that the rules cannot completely take into account. However, in this example, Assignment #1 is better than #2 for J-Ks by only one gate, so the rules still get us close to the best assignment.

10.11 Partitioning and State Assignment*

In Section 10.7, we encountered the concept of partitioning the states of a sequential circuit into equivalence classes such that two states, p and q, were in the same class ($p \triangleq q$) if and only if the following two conditions were satisfied for all inputs X.

(1) $\lambda(p, X) = \lambda(q, X)$
(2) $\delta(p, X) \triangleq \delta(q, X)$

---

* The material of this section is not essential to subsequent developments and may be omitted from a first reading.

q	$y_0^v y_1^v y_2^v$	$x^v = 0$		$x^v = 1$	
q_0	0 0 0	0	01	1	01
q_1	0 0 1	0	10	1	10
q_2	0 1 0	0	11	1	11
q_3	0 1 1	0	00	0	00
q_4	1 0 0	—		—	
q_5	1 0 1	1	10	0	10
q_6	1 1 0	1	11	0	11
q_7	1 1 1	0	00	0	00

$$(y_0 y_1 y_2)^{v+1}$$

(a)

q^v	$x^v = 0$	$x^v = 1$
q_0	$q_1,$ 0	$q_5,$ 0
q_1	$q_2,$ 0	$q_6,$ 0
q_2	$q_3,$ 0	$q_7,$ 0
q_3	$q_0,$ 1	$q_0,$ 0
q_4	—	—
q_5	$q_6,$ 0	$q_2,$ 0
q_6	$q_7,$ 0	$q_3,$ 0
q_7	$q_0,$ 0	$q_0,$ 1

$$q^{v+1}, z^v$$

(b)

Present states	Next-states	
	$x^v = 0$	$x^v = 1$
(04)	(15)	(15)
(15)	(26)	(26)
(26)	(37)	(37)
(37)	(04)	(04)

(c)

FIGURE 10.55 *State table for sequential parity checker.*

variables to perform distinct functions, the circuit breaks down into distinct subcircuits, as shown in Fig. 10.54.

Now that we have obtained a realization, let us examine the state table of the sequential parity checker. Figures 10.53a and b may be combined to form the transition table of Fig. 10.55a and the corresponding state table of Fig. 10.55b. Since y_0 is not an input to the 2-bit counter, we would expect to find a closed partition of the states into four classes, one for each combination of values of $y_1^v y_2^v$. This partition is (04) (15) (26) (37). That this partition is closed may be verified from Fig. 10.55c, in which the classes of next states for each class are tabulated.   ∎

The reader will note that Expressions 10.24 are somewhat simpler than those generated by Rules I, II, and III. If a student is perceptive (and lucky), partitioning can sometimes be a short route to an efficient realization.

Obtaining a successful partition depends on the natural structure of the machine and the designer's ability to recognize this structure. As with many aspects of engineering judgment, this ability grows with experience. We do not suggest that the reader fall back on intuitive design techniques, hoping blindly for a partitionable structure. Indeed, in many types of problems, none will exist. Instead we suggest that the reader be alert for a partitionable structure while synthesizing each state diagram. If no such structure is evident, the reader will have to do the best possible with Rules I–III.

In some cases, the reader will be able to use natural structure to specify the state assignment only partially. The reader must then rely on Rules I–III to complete the realization. We leave it to the reader to integrate the two approaches as experience is gained.

We conclude this section with one more example.

Example 10.12

Let us consider again the modulo-8 up-down counter of Example 10.4, for which the state table is repeated in Fig. 10.56a. We noted at the beginning of this section that there are two nontrivial closed portions in this table, but let us assume that we are unaware of these and consider how our perception of the natural structure of this circuit might lead us to these partitions.

A natural partition for any counter is into blocks of states corresponding to odd counts and even counts. Such a partition will be closed if the number of states is even, since an odd count will always be followed by an even count, and vice versa. For this circuit, this corresponds to the partition $(0, 2, 4, 6)$, $(1, 3, 5, 7)$; as already suggested, we will assign y_1 to this partition. We can then set up a transition table for the subcircuit developing y_1, as shown in Fig. 10.56b. The function of this subcircuit is to identify which class the overall circuit is in, $(0, 2, 4, 6)$ or $(1, 3, 5, 7)$. As shown, if the present class is $(0, 2, 4, 6)$, the next will be $(1, 3, 5, 7)$, and vice versa. From this, we read off the equation

$$y^{v+1} = \overline{y_1^v} \qquad (10.36)$$

as noted earlier. The corresponding circuit using a D flip-flop is shown in Fig. 10.56c.

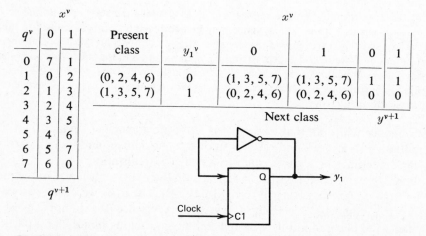

FIGURE 10.56 *State table and design of y_1 subcircuit for up-down counter.*

Each state must be identified by a unique combination of state variables. If we are to identify eight states with three state variables, the second variable must divide the two four-state blocks in half, giving four two-state blocks, and the third variable must divide these in half, giving eight blocks of one state each. Following our odd-even approach, we will let $y_2 = 0$ specify the first and third states in each block of the y_1 partition and $y_1 = 1$ specify the second and fourth, i.e., y_2 induces the partitions $(0, 1, 4, 5)$ $(2, 3, 6, 7)$. The variables y_1 and y_2 together then induce the partition $(0, 4)$ $(1, 5)$ $(2, 6)$ $(3, 7)$, as shown in Fig. 10.57a.

Since this is a closed partition, as noted earlier, we know that the next values of the state variables inducing it will not be functions of any other state variables. We can thus develop a transition table for the y_2 subcircuit, as shown in Fig. 10.57b. We do this by first using Fig. 10.56a to find the next-state pairs corresponding to each pair of present

y_2y_1	
00	$(0, 1, 4, 5) \wedge (0, 2, 4, 6) = (0, 4)$
01	$(0, 1, 4, 5) \wedge (1, 3, 5, 7) = (1, 5)$
11	$(2, 3, 6, 7) \wedge (1, 3, 5, 7) = (3, 7)$
10	$(2, 3, 6, 7) \wedge (0, 2, 4, 6) = (2, 6)$

		x^v			
Present class	y_2y_1	0	1	0	1
$(0, 4)$	00	$(3, 7)$	$(1, 5)$	1	0
$(1, 5)$	01	$(0, 4)$	$(2, 6)$	0	1
$(3, 7)$	11	$(2, 6)$	$(0, 4)$	1	0
$(2, 6)$	10	$(1, 5)$	$(3, 7)$	0	1
		Next class		y_2^{v+1}	

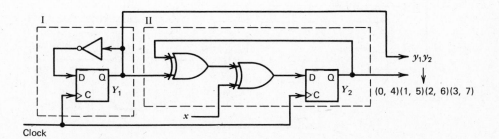

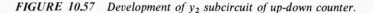

FIGURE 10.57 *Development of y_2 subcircuit of up-down counter.*

states. Note that we do not show the next values of y_1 in this table because we have already designed a separate circuit for y_1. From this table, we can read off the equation

$$y_2^{v+1} = x^v \cdot (y_1^v \oplus y_2^v) + \bar{x}^v \cdot \overline{(y_1^v \oplus y_2^v)} = \bar{x}^v \oplus (y_1^v \oplus y_2^v) \qquad (10.37)$$

The resultant interconnection of the two subcircuits inducing the partition $(0, 4)$ $(1, 5)$ $(2, 6)$ $(3, 7)$ is shown in Fig. 10.57c.

Finally, y_3 must divide those four blocks in half requiring that y_3 induce the partition

		x^v			
q^v	$(y_3 y_2 y_1)^v$	0	1	0	1
0	000	7	1	1	0
1	001	0	2	0	0
3	011	2	4	0	1
2	010	1	3	0	0
4	100	3	5	0	1
5	101	4	6	1	1
7	111	6	0	1	0
6	110	5	7	1	1
		q^{v+1}		y_3^{v+1}	

FIGURE 10.58a Development of y_3 subcircuit of up-down counter.

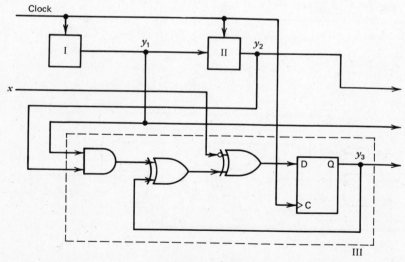

FIGURE 10.58b

(0, 1, 2, 3) (4, 5, 6, 7). Since we have already designed subcircuits to develop y_1 and y_2, it is obvious that the final subcircuit will develop only y_3. The transition table for y_3 is given in Fig. 10.58a. The complete counter circuit is shown in Fig. 10.58b with the already developed realizations of y_1 and y_2 represented by blocks I and II, respectively. It will be left to the reader to verify that the transition table of Fig. 10.58a leads to the equation

$$y_3^{v+1} = ((y_1^v \cdot y_2^v) \oplus y_3^v) \oplus \bar{x}^v \tag{10.38}$$

The reader may not feel that this procedure has produced a particularly economical design in view of the number of exclusive-OR gates, but this is the best assignment using D flip-flops. Any other assignment will make y_1 and y_2 functions of more variables, leading to a more complex design. We will see in Chapter 11 that the use of T flip-flops will lead to a more economical design, but this design will be based in this same partition.

∎

We hope that the above example, illustrating as it does some of the more interesting and challenging aspects of partition theory, may motivate some of our readers to explore this fascinating subject in more depth (see References 8, 12).

The relationship between state assignment partitions and the natural structure of the circuit function raises questions as to the desirability of state table minimization. In Example 10.5, we arrived at an eight-state table, which would require three state variables. The reader who considers the basic structure will note two distinct three-state counting sequences. Perhaps a simpler and more natural structure would result with four state variables, two for each sequence. The extra state variables would require an extra flip-flop, but the simpler structure might simplify the excitation and output logic sufficiently to result in a net decrease in cost.

A special case where the use of extra flip-flops is very likely to lead to lower cost is the finite-memory circuit. Recall the error detector of Example 10.3, which was originally formulated as a finite-memory circuit, as shown in Fig. 10.10. If we realize the circuit in this form, the only cost, in addition to the three D flip-flops, is for logic to realize the output equation, Equation 10.5. We minimized this same table for four states in Example 10.6 and made a state assignment in Example 10.10. The equations for the selected assignment are shown in Equation 10.31. The output equations for the two cases are comparable in cost, so the saving of the flip-flop is offset by the cost of three gates to realize the excitation equations of Equation 10.31. As a rough rule of thumb, one flip-flop is equal in cost to two to four gates, so the basic logic cost for the two forms would be comparable. On the other hand, the four-state version will have many more interconnections between logic packages and will thus be more expensive to manufacture. In addition, use of the finite memory form initially would have saved the time spent in state reduction and assignment.

10.12 Conclusion

In this chapter, we have developed a design procedure that is considerably more dependent on the intuition and ingenuity of the designer than was the case for combinational logic circuits. This seeming lack of formality and rigor may mislead the novice designer into thinking the procedure is basically trial-and-error. But it is a carefully organized procedure, designed to lead the designer from an initial, imprecise specification to a final circuit with minimum likelihood of error along the way.

The first step, setting up the state diagram or state table, is the most important step. The initial specification of a circuit function is almost invariably vague and incomplete. The procedures developed in Sections 10.3 and 10.4 help clarify your understanding of the circuit function and force you to specify precisely the desired behavior in every possible circumstance. Possibly because it requires considerable time and effort to think the problem through carefully and thoroughly, there is a tendency for designers to skip this step. Instead, they try to apply the sort of intuition about structure that we discussed in connection with state assignment at the very beginning. They try to interconnect a set of hazily perceived subcircuits, such as counters, parity checkers, etc., in a trial-and-error fashion, usually producing a design that does many unexpected things. The time to apply intuition about circuit structure is after you thoroughly understand the circuit function.

If the circuit is finite memory and the transition table has been formulated accordingly, the only step left is to read the output equations directly from the transition table. As discussed above, where the finite memory form is applicable, it almost invariably produces the most economical designs and should be used.

If the circuit is not finite memory, the next step is to minimize the state table. As discussed, a minimal table does not necessarily result in the most economical design. However, it is best to start with a minimal table and consider adding more state variables only if there seems to be an obvious natural structure requiring more variables.

The final step is state assignment, which is a difficult and sometimes frustrating process. Our basic approach is to use a set of semiempirical rules combined with a healthy dose of intuition to try to find desirable partitions. It is quite natural to wish for rigorous procedures that will lead to guaranteed results. A great deal of research has gone into the effort to find such procedures, and papers on state assignment appear regularly in the journals reporting computer research. Perhaps some day a really good state assignment procedure will be found. For the present, we believe that the sort of "guided intuition" we have tried to illustrate in the last two sections remains the most satisfactory method.

Finally, we hope that the reader will acquire some perspective about where best to apply efforts in this rather elaborate design process. The first step warrants the greatest attention because no amount of later effort can produce a good design from an initially incomplete or erroneous specification. The second step, state reduction, is rather tedious but has the virtue of producing guaranteed results (assuming no clerical errors) and is usually time well spent. It is in the state assignment step that judgment must be applied. It is natural to seek simple, elegant designs and easy to spend more time on state assignment than on other parts of the design process. But engineering time is expensive, and many hours spent eliminating a gate or two can be a very bad investment unless production of the circuit is to be enormous.

The above paragraph and indeed most of this chapter is most pertinent to machines with relatively few states (≤ 32). As the total amount of information to be stored increases, so will the likelihood of uncovering a partitionable structure. For large circuits, some formal system description other than the state table would be very desirable. Such formal descriptions exist in the form of hardware register transfer languages. Such a language will be the subject of Chapter 15. As we will see, a register transfer language description provides a starting point from which a digital system can be partitioned into subsystems. These subsystems may then be optimized using the techniques of this chapter or realized in terms of standard MSI components.

Problems

10.1. Analyze the sequential circuit given in Fig. P10.1 in the manner discussed in Section 10.1. Obtain first the flip-flop input equations and then the state diagram representation of the circuit.

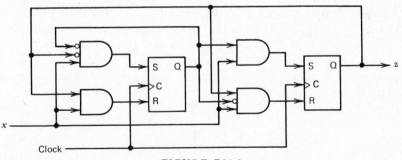

FIGURE P10.1

10.2. A clock-mode sequential circuit is to be designed featuring an external reset mechanism that will on occasion reset the circuit to state q_0. Determine a state diagram of the circuit so that it will generate an output of 1 for one clock period only coinciding with the second 0 input of a sequence consisting of exactly two 1s (no more than two) followed by two 0s. Once the output has been 1 for one clock period, the output will remain 0 until the circuit is externally reset to q_0.

10.3. Derive the state diagram and state table of a circuit featuring one input line, X_1, in addition to the clock input. The only output line is to be 0 unless the input has been 1 for four consecutive clock pulses or 0 for four consecutive pulses. The output must be 1 at the time of the fourth consecutive identical input.

10.4. A sequential circuit is to be designed with two input lines x_1 and x_2 and a single output z. If a clock pulse arrives when $x_2 x_1 = 00$ the circuit is to assume a reset state that may be labeled q_0. Suppose the next 6 clock pulses following a resetting pulse coincide with the following sequence of input combinations: 01—10—11—01—10—11. The output, z, is to be 1 coinciding with the sixth of such a string of six clock pulses but is to be 0 at all other times. The circuit cannot be reset to q_0 except by input 00. Define a special state q_1 to which the circuit will go once it becomes impossible for an output-producing sequence to occur. The circuit will thus wait in q_1 until it is reset. Determine a state diagram for this circuit.

10.5. A sequential circuit is to be designed in which the circuit output, z^v, is a function of only the current input x^v and the three previous inputs x^{v-1}, x^{v-2}, and x^{v-3}. A Boolean expression for z^v is given by

$$z^v = x^v \cdot x^{v-1} \cdot x^{v-2} \cdot x^{v-3} + \overline{x^v} \cdot \overline{x^{v-1}} \cdot \overline{x^{v-2}} \cdot \overline{x^{v-3}}$$

Treat the circuit as a finite-input memory circuit and determine an 8-state state table representation.

10.6. A variable-delay circuit is to provide a two or three clock period delay of an input signal, x_i. That is, the circuit output x_d^v is to be equal to x_i^{v-3} if the control input $r = 1$ and is to be equal to x_i^{v-2} if $r = 0$. A simplified block diagram of the circuit is shown in Fig. P10.6. Treat it as a finite memory circuit and obtain a state diagram for the variable-delay sequential circuit.

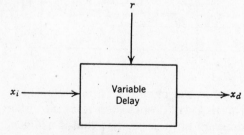

FIGURE P10.6

10.7. A sequential circuit has two inputs, x_1 and x_2. Five-bit sequences representing decimal digits coded in the 2-out-of-5 code appear from time to time on line x_1, synchronized with a clock pulse on a third line. Each five consecutive bits on the x_1 line, which occur while $x_2 = 0$ but immediately following one or more inputs of $x_2 = 1$, may be considered a code word. The single output z is to be 0 except upon the fifth bit of an invalid code word. Determine a state table of the above sequential circuit.

10.8. A clock-mode sequential circuit has two input lines a and b. During any clock period, $a^v a^{v-1} a^{v-2}$ and $b^v b^{v-1} b^{v-2}$ may be regarded as 3-bit binary numbers. The output z is to be 1 if $a^v a^{v-1} a^{v-2} \geq b^v b^{v-1} b^{v-2}$. The inputs a^v and b^v are the most significant bits as implied by the format shown. Obtain a state table for this sequential circuit.

10.9. Let aRb if and only if $a = b \pm 8X$ where X is any integer. (We say that a is congruent to b modulo-8.) Show that R is an equivalence relation on the set of all integers.

10.10. Determine a minimal-state table equivalent to the state table in Fig. P10.10. [*Answer:* 5 states are required.]

$$q^{v+1}, Z^v$$

q^v	$x = 0$	$x = 1$
1	1,0	1,0
2	1,1	6,1
3	4,0	5,0
4	1,1	7,0
5	2,0	3,0
6	4,0	5,0
7	2,0	3,0

FIGURE P10.10

10.11. Determine a minimal state table equivalent to the state table shown in Fig. P10.11.

Present State	Next State, Output			
	$x = 0$	$x = 1$	$x = 2$	$x = 3$
A	E,1	C,0	B,1	E,1
B	C,0	F,1	E,1	B,0
C	B,1	A,0	D,1	F,1
D	G,0	F,1	E,1	B,0
E	C,0	F,1	D,1	E,0
F	C,1	F,1	D,0	H,0
G	D,1	A,0	B,1	F,1
H	B,1	C,0	E,1	F,1

FIGURE P10.11

10.12. Determine a circuit whose output is equivalent to the circuit shown in Fig. P10.12 but that has a minimum number of states.

q^v	q^{v+1}, Z	
	$x = 0$	$x = 1$
q_1	$q_2, 0$	$q_3, 0$
q_2	$q_4, 0$	$q_5, 0$
q_3	$q_6, 0$	$q_7, 0$
q_4	$q_8, 0$	$q_9, 1$
q_5	$q_{10}, 0$	$q_{11}, 0$
q_6	$q_{12}, 1$	$q_{13}, 0$
q_7	$q_{14}, 0$	$q_{15}, 0$
q_8	$q_1, 0$	$q_3, 0$
q_9	$q_1, 0$	$q_3, 0$
q_{10}	$q_1, 0$	$q_3, 0$
q_{11}	$q_1, 0$	$q_2, 0$
q_{12}	$q_2, 0$	$q_1, 0$
q_{13}	$q_2, 0$	$q_1, 0$
q_{14}	$q_2, 0$	$q_1, 0$
q_{15}	$q_2, 0$	$q_1, 0$

FIGURE P10.12

10.13. The minimal-state circuits satisfying the functions of previous problems are listed as follows. In each case where the circuit synthesized by the reader was not minimal, obtain a minimal machine by the formal techniques of Sections 10.7 and 10.8.

> 10.2   6 states
> 10.5   6 states
> 10.7 14 states

10.14. Determine a minimal-state equivalent of the state table found in Problem 10.8.

10.15. Determine a realization of a clocked J-K flip-flop in terms of a clocked S-C flip-flop and appropriate combinational logic.

10.16. Obtain a circuit realization of the state table found in Problem 10.5:
(a) Using S-R flip-flops.
(b) Using D flip-flops.
(c) Using J-K flip-flops.

10.17. Obtain a circuit realization of the state table given in Fig. P10.17 using J-K flip-flops.

q^v	q^{v+1}, Z	
	$x = 0$	$x = 1$
A	B,0	B,1
B	F,0	D,1
C	E,1	G,1
D	A,0	C,0
E	D,1	G,0
F	F,0	A,0
G	C,1	B,0

FIGURE P10.17

10.18. Repeat Problem 10.17 using *S-R* flip-flops.

10.19. Repeat Problem 10.17 using *D* flip-flops.

10.20. Obtain circuit realizations for the minimal state tables for:
(a) Problem 10.2
(b) Problem 10.6
(c) Problem 10.14
using whichever type of flip-flop appears most suitable in each case.

10.21. A circuit with an external reset mechanism, a single input x, and a single output z is to behave in the following way. At the first clock time following reset, the output is to be $z = 0$. At the second clock time, $z^v = x^v \cdot x^{v-1}$. At the third clock time, $z^v = x^v + x^{v-1}$, and at the fourth clock time, $z^v = x^v \oplus x^{v-1}$. This sequence of output functions is then repeated every four clock periods until the circuit is reset. The circuit can be readily partitioned into two subcircuits, a 2-bit counter, and a 1-bit shift register. Take advantage of this fact to determine a state diagram and a realization of the circuit in terms of *J-K* flip-flops.

10.22. Obtain the state table of a serial translating circuit that performs the translations tabulated in Fig. P10.22. As indicated, there is only one input line, X, and one output line, Z, in the circuit. The inputs occur in characters of three consecutive bits. One character is followed immediately by another. A reset signal, which need not be included in the state table, will cause the translator to begin operation synchronized with the input. A translated output character must be delayed by only two clock periods with respect to an input character. [*Hint:* A 12-state minimal-state table can be obtained by partitioning the circuit into a 3-state counter and a 2-bit shift register.]

X^v	X^{v-1}	X^{v-2}	Z^{v+2}	Z^{v+1}	Z^v
0	0	0	0	0	0
0	0	1	0	1	1
0	1	0	1	0	0
0	1	1	1	0	1
1	0	0	1	1	0
1	0	1	0	0	1
1	1	0	0	1	0
1	1	1	1	1	1

FIGURE P10.22

10.23. Obtain a realization of the state table derived in Problem 10.22 using J-K flip-flops.

10.24. Repeat Problem 10.7 using an approach that partitions the circuit into a modulo-5 counter and a 2-bit counter for counting 1s appearing on line x_1. Determine a 5 flip-flop (J-K) realization of this circuit without minimizing the state table. Compare the overall complexity without minimizing the state table. Compare the overall complexity of this realization with a realization of a minimal equivalent of the state table determined in Problem 10.7.

Bibliography

1. Huffman, D. A. "The Synthesis of Sequential Switching Circuits," *J. Franklin Inst.*, **257**, No. 3, 161–190; No. 4, 275–303 (March–April, 1954).
2. Paul, M. C. and S. H. Unger, "Minimizing the Number of States in Incompletely Specified Sequential Switching Functions," *IRE Trans. on Electronic Computers*, **EC-8**: 3, 356–357 (Sept. 1959).
3. Mealy, G. H. "A Method for Synthesizing Sequential Circuits," *Bell System Tech. J.*, **34**: 5, 1045–1080 (Sept. 1955).
4. Moore, E. F. "Gedanken Experiments on Sequential Machines," in C. E. Shannon and J. McCarthy (Eds.), *Automata Studies*, Princeton Univ. Press, Princeton, N.J., 1956.
5. McCluskey, E. J. and S. H. Unger. "A Note on the Number of Internal Variable Assignments for Sequential Switching Circuits," *IRE Trans. on Electronic Computers*, **EC-8**, No. 4, 439–40 (Dec. 1959).
6. Hartmanis, J. "On the State Assignment Problem for Sequential Machines, I," *IRE Trans. on Electronic Computers*, **EC-10**: 2, 157–165 (June 1961).
7. Stearns, R. E., and J. Hartmanis. "On the State Assignment Problem for Sequential Machines, II," *IRE Trans. on Electronic Computers*, **EC-10**: 4, 593–603 (December 1961).

8. Hartmanis, J. and R. E. Stearns. *Algebraic Structure Theory of Sequential Machines*, Prentice-Hall, Englewood Cliffs, N.J., 1966.

9. Karp, R. M. "Some Techniques of State Assignment for Synchronous Sequential Machines," *IEEE Trans. of Electronic Computers*, **EC-13:** 5, 507–518 (Oct. 1964).

10. Dolotta, T. A. and E. J. McCluskey. "The Coding of Internal States of Sequential Circuits," *IEEE Trans. on Electronic Computers*, **EC-13:** 5, 549–562 (Oct. 1964).

11. Harrison, M. A. *Introduction to Switching and Automata Theory*, McGraw-Hill, New York, 1965.

12. Kohavi, Z. *Switching and Finite Automata Theory*, 2nd Edition, McGraw-Hill, New York, 1978.

13. Wiatrowski, C. A. and House, C. H., *Logic Circuits and Microcomputer Systems*, McGraw-Hill, New York, 1980.

11

CLOCK-CONTROL AND PULSE-MODE CIRCUITS

11.1 Introduction

The standard model of a clock-mode sequential circuit was given in Fig. 9.20. It included an array of clocked flip-flops and a combinational logic section whose inputs were state variables and external input levels. The only source of timing pulses was a single clock tied directly to the clock input of each flip-flop. Figure 11.1 shows the general model for a second class of circuit, the *pulsed sequential circuit*. This model differs from the clock-mode model in several respects. First, no separate source of timing signals (clock) is applied directly to the memory

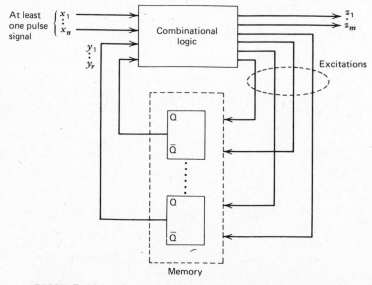

FIGURE 11.1   *General model of pulsed sequential circuits.*

flip-flops. Second, the explicit external input lines may carry either levels or pulses, although at least one must be a pulse input. Third, the memory flip-flops may be either clocked or unclocked.

In this chapter we will deal with two specific types of circuits that fall into this category. One is the *controlled clock* circuit, in which the flip-flops are clocked and there is only one pulse input line, the clock. The clock line is not applied directly to flip-flops, as in the clock-mode circuit, but is an input to the combinational logic section, where it is logically combined with level signals to provide separate sources of control pulses to the clock inputs of various subsets of flip-flops. This is a very important class of circuit, because many complex digital systems are built this way.

If there is no single source of periodic pulses among the inputs in a circuit of the form of Fig. 11.1, we call the circuit *pulse mode*. Pulse inputs in pulse-mode circuits are often aperiodic and unrelated with respect to timing. For example, we might have a digital system with several different sections operating independently and each sending pulses to a central control unit when certain events occur. That central control unit would have to operate in the pulse mode with respect to these signals. The flip-flops in a pulse-mode circuit may be either clocked or unclocked.

These changes from the clock-mode model necessitate changes in design procedures, but the basic concepts developed in the last chapter still apply and we will see that the methods developed there can be extended to accommodate this new class of circuits. In Chapter 10 it was pointed out that certain constraints on the timing of clock pulses and level-input changes were required, if a circuit was to be considered as clock mode. These constraints may be interpreted for pulsed circuits as follows.

1. All input pulses must be sufficiently wide to trigger a flip-flop. All flip-flops must be edge-triggered or master-slave, or it must be ensured in some other way that only one state change can take place in response to each input pulse.
2. The triggering edges of input pulses (irrespective of input line) will never be separated in time by less than the period corresponding to the maximum pulse repetition rate for the flip-flops used.
3. Level inputs must be stable when pulse inputs occur.

The second criterion implies that pulses may occur on only one input line at a time. Were it possible for two input pulses to occur exactly simultaneously, the inputs could be considered as an AND-coding of these pulses. As a practical matter, this is impossible. Unpredictable delays would always separate the pulses slightly, and unreliable operation would result. Where pulses arrive from completely separate and independent sources, special effort may be required to synchronize them so as to satisfy the second criterion. This problem will be considered in Chapter 14.

The concept of controlled clocking will extend our design capability to much larger digital systems than were considered in Chapter 10. Treatment of this topic will occupy most of this chapter and Chapter 12. Occasionally it is necessary to design a circuit in the more general pulse mode. The approach for this case requires the use of state tables similar to those defined in Chapter 10. For this reason we consider pulse-mode design in the next two sections and then turn our attention to controlled clocking.

11.2   General Pulse Mode

Nothing was said about the outputs in considering Fig. 11.1. In clocked circuits, the outputs, generally functions of both the inputs and state variables, are levels, since the inputs and state variables are levels. On the assumption that these outputs would in turn be inputs to other circuits subject to the same clock, it was appropriate to consider the value of these levels as meaningful only at clock time. The situation is a bit more complex for pulse-mode circuits.

If the outputs are to be functions of the inputs as well as the state variables, they must be pulses, obtained by gating the state variables with the input pulses. In that case, a distinct output may be defined for every possible combination of states and inputs.

If the outputs are functions of the state variables only, they will be levels whose value will be defined in the intervals *between* input pulses rather than at the time of the input pulses. Furthermore, the number of distinct outputs can be no greater than the number of states.

State tables and state diagrams for both types of pulse-mode circuits are shown in Fig. 11.2. The reader will notice a marked similarity between these two circuits. In fact, they represent solutions to essentially the same problem. The function of the circuits is to identify each sequence $x_1 x_2 x_1$ following an x_3 input. The x_3 pulse behaves essentially as a start, or reset, signal. In the pulse-output circuit, q_3 is reached by an input sequence $x_1 x_2$. If the next pulse to arrive is x_1, the circuits puts out a pulse and returns to state q_2 to await another start signal. If the next input is other than x_1, there is no output pulse.

The circuit in Fig. 11.2b operates in a similar fashion, except that the x_1 pulse arriving when the circuit is in state q_3 takes the circuit to a new state, q_4, which produces a 1 output *until* another pulse input occurs. Thus, one circuit produces a pulse at the time of the last pulse of the specified sequence, while the other produces a level after the sequence has been completed.

The reader will observe that the pulse-input, level-output circuit of Fig. 11.2b is a **Moore Model**. This must always be the case, since the output levels cannot

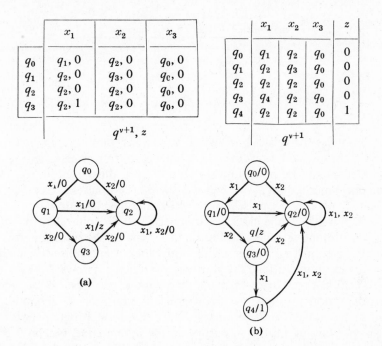

	x_1	x_2	x_3
q_0	$q_1, 0$	$q_2, 0$	$q_0, 0$
q_1	$q_2, 0$	$q_3, 0$	$q_c, 0$
q_2	$q_2, 0$	$q_2, 0$	$q_0, 0$
q_3	$q_2, 1$	$q_2, 0$	$q_0, 0$

q^{v+1}, z

	x_1	x_2	x_3	z
q_0	q_1	q_2	q_0	0
q_1	q_2	q_3	q_0	0
q_2	q_2	q_2	q_0	0
q_3	q_4	q_2	q_0	0
q_4	q_2	q_2	q_0	1

q^{v+1}

FIGURE 11.2   (a) *Pulse input-pulse output* (*Mealy*). (b) *Pulse input-level output* (*Moore*).

be functions of pulses. If two or more inputs can cause the same transition, they are separated by commas, e.g., x_1 or x_2 from q_2 will keep the circuit in q_2. Note carefully that this does not imply that both can occur at the same time, only that both cause the same transition.

The restrictions that must be observed in clock-mode design are rigid. They make it possible to design large systems without separately considering timing details at every point in the circuit. For example, all inputs to a clock-mode circuit must originate in other clock-mode circuits synchronized to the same clock. (Exceptions are possible if special synchronizing circuits are employed. See Chapter 9 of Reference 1.)

There will always be situations where parts of systems cannot be made to satisfy the rules for clock mode or controlled clock design. This may occur at the interface between systems designed by separate individuals or separate organizations. It may also arise because systems to be interconnected are implemented using different families of logic circuits with significantly different speed characteristics. These small parts of systems can often be formulated in the pulse mode.

Every pulse-mode circuit can be considered independently. The designer is free to designate which input lines are to be regarded as levels and which are to be

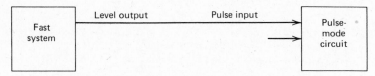

FIGURE 11.3 *Designating inputs of a pulse-mode circuit.*

regarded as pulses. The designer must only ascertain that the three rules of Section 11.1 are satisfied. For example, a level output of some faster system may be designated as a pulse input to a slower circuit as illustrated in Fig. 11.3.

In our pulse-mode designs we will use the clocked master-slave J-K flip-flop of Fig. 9.24. If we connect pulses to the clock line the element will function as a clocked J-K flip-flop. If the clock input is permanently connected to logic 1, the flip-flop will respond to pulses on the J and K inputs. The transition lists for this flip-flop are given in Fig. 11.4, showing both clocked and pulse-triggered operation. For clocked operation, the downward-pointing arrows indicate that the output transition takes place on the falling edge of the clock pulse. For pulse-triggered operation we show a symbol for a pulse on J or K, to emphasize that pulses must be applied to these inputs if the clock is held high. As with clocked operation, the output transition occurs on the falling edge, but both edges are needed, the rising edge to trigger the master, the falling edge to trigger the slave. Also note that the flip-flop is effectively an R-S flip-flop in the pulse-triggered mode, since we cannot predict what it will do if we try to pulse both J and K at the same time. In the clocked mode we have taken into account the possibility of controlling the clock. We can keep the setting the same $(0 \to 0$ or $1 \to 1)$ either by controlling J and K, or by not clocking $(CK = 0)$.

$Q^{\nu} \to Q^{\nu+1}$	J	K
$0 \to 0$	0	x
$0 \to 1$	⊓	0
$1 \to 0$	0	⊓
$1 \to 1$	x	0

(a) Pulse-triggered

$Q^{\nu} \to Q^{\nu+1}$	J	K	CK
$0 \to 0$	x	x	0
	0	x	x
$0 \to 1$	1	x	↓
$1 \to 0$	x	1	↓
$1 \to 1$	x	x	0
	x	0	x

(b) Clocked

FIGURE 11.4 *Transition lists for J-K flip-flop.*

Example 11.1

A major cycle of a certain digital system contains three subcycles that must be completed in a certain order. To check this, a sequence checker will receive a completion pulse from each subcycle and a check pulse when the major cycle is complete. When the check pulse R arrives, the sequence checker should reset and put out an error pulse if the three completion pulses A, B, C, were not received in that order. Each completion signal will occur exactly once in each major cycle. Design the sequence checker, using J-K flip-flops and AND-OR logic. Assume these pulses are synchronized to meet the criteria of Section 11.1.

Solution   There are six possible sequences, *ABC*, *ACB*, *BAC*, *BCA*, *CBA*, *CAB*, of which only the first is correct. We see that, once an incorrect sequence has been identified, the circuit can go to a state awaiting the arrival of the check pulse. For example, if the first pulse is B or C, the order in which the other two pulses arrive does not matter. This leads to the partial-state diagram of Fig. 11.5a, where state q_2 is the state of the machine after the first symbol of a correct sequence and state q_5 is the state of the machine as soon as the sequence has been identified as incorrect. From q_2, the only possible inputs are B and C. Since B represents the continuation of a correct sequence, it leads to a new state q_3 as shown in Fig. 11.5b.

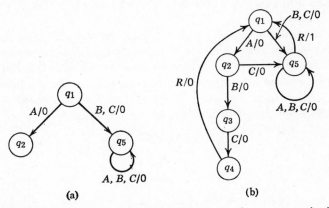

(a)   **(b)**

FIGURE 11.5   Developing the state diagram for the sequence checker.

From state q_3, the only possible input is C leading to a new state q_4. Since a pulse on line R will always follow the three-pulse sequence to be checked, this input will occur only when the circuit is in state q_4 or q_5. Adding the arrows corresponding to R with an output from state q_5 completes the state diagram of Fig. 11.5b.   ∎

The reader should notice that Fig. 11.5b differs from the state diagrams considered thus far in one important respect—not all possible input arrows are included for every present state. That is, some combinations of inputs and present states will never occur. For this reason, it does not matter what next states

q^v	Inputs			
	A	B	C	R
1	2, 0	5, 0	5, 0	–, –
2	–, –	3, 0	5, 0	–, –
3	–, –	–, –	4, 0	–, –
4	–, –	–, –	–, –	1, 0
5	5, 0	5, 0	5, 0	1, 1
	q^{v+1}, z^v			

(a)

q^v	Inputs			
	A	B	C	R
1	2, 0	5, 0	5, 0	–, –
2	–, –	3, 0	5, 0	–, –
3	–, –	–, –	3, 0	1, 0
5	5, 0	5, 0	5, 0	1, 1
	q^{v+1}, z^v			

(b)

FIGURE 11.6   State table for sequence checker.

or outputs are specified for these combinations. That is, they are don't-cares in much the same sense as don't cares in combinational logic. In translating the state diagram to a state table, we enter blanks in place of the outputs and next states for these don't-care combinations so that Fig. 11.5b translates to the state table of Fig. 11.6a.

A state table containing don't-care entries is defined as *incompletely specified*. In general, state table minimization is more difficult for incompletely specified state tables. This topic will be treated in detail in Chapter 13. For now, let us minimize Fig. 11.6a by intuitively extending the techniques of Chapter 10.

Consider states 3 and 4. Since $\delta(3, R)$, $\lambda(3, R)$ is a don't-care, we may enter what ever values we choose. Therefore, let

$$\delta(3, R), \lambda(3, R) = 1, 0$$

and similarly let

$$\delta(4, C), \lambda(4, C) = 4, 0$$

Now rows 3 and 4 of the state table are identical. Thus, we combine the two rows replacing all next-state values of 4 in the table with 3. This leads to the table of Fig. 11.6b. We conclude by inspection that this table cannot be further simplified regardless of how the remaining don't-care entries might be specified.

As we will see, the relation between columns in the transition table of a pulse-mode circuit is not the same as for a clock-mode circuit. In particular, Rule II of Section 10.9 is not very helpful for a pulse-mode state assignment. In this case, Rule Ic suggests that q_1 should be adjacent to q_2 and q_5, and the assignment shown in Fig. 11.7 satisfies these adjacency requirements.

Let us first assume the use of flip-flops in the pulse-triggered mode of Fig. 11.4a. The corresponding excitation tables are shown in Fig. 11.8.

q^v	$(y_1y_2)^v$	Inputs			
		A	B	C	R
1	0 0	01, 0	10, 0	10, 0	–, –
2	0 1	–, –	11, 0	10, 0	–, –
3	1 1	–, –	–, –	11, 0	00, 0
5	1 0	10, 0	10, 0	10, 0	00, 1
		$(y_1y_2)^{v+1}, z^v$			

FIGURE 11.7   *Transition table for sequence checker.*

Note that the tables of Fig. 11.8 are not conventional K-maps. Since there are four input variables, a Karnaugh map would have sixteen columns. The four columns in the tables of Fig. 11.8 represent the columns $ABCR = 1000$, $ABCR = 0100$, $ABCR = 0010$, and $ABCR = 0001$. Since only one pulse can occur at a time, all the other columns except $ABCR = 0000$ of the K-map would

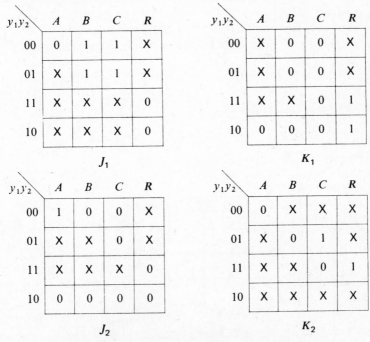

FIGURE 11.8   *Excitation tables, Example 11.1.*

have only don't-care entries. Thus, we can let the A columns of the excitation tables represent the $A = 1$ columns of the corresponding K-map, the B columns represent the $B = 1$ columns, etc. On this basis, we can write the excitation equation directly from these tables and save the labor of drawing four 4×16 K-maps. The equations for J and K for the two flip-flops are:

$$J_1 = B + C \tag{11.1}$$

$$K_1 = R \tag{11.2}$$

$$J_2 = A\bar{y}_1 \tag{11.3}$$

$$K_2 = R + C\bar{y}_1 \tag{11.4}$$

From the transition table we can read off the output equation

$$Z = R\bar{y}_2 \tag{11.5}$$

The realization of these equations is shown in Fig. 11.9.

In the above example, we saw that there are no basically new problems in designing a Mealy-model circuit. The restrictions on simultaneous inputs reduce the number of columns required in the tables, but otherwise there is little real difference in procedure from clocked design. One precaution we might note is that every product term in the excitation and output equations must contain a pulse variable in order to generate the desired pulses. Suppose that a certain design resulted in the excitation table shown in Fig. 11.10. It would appear that the best equation for J_1 would be

$$J_1 = \bar{y}_1\bar{y}_2 \tag{11.6}$$

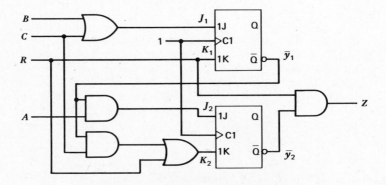

FIGURE 11.9   *Final circuit for Example 11.1.*

y_1y_2	A	B	C	D
00	1	1	X	X
01	X	X	X	0
11	X	0	X	X
10	X	X	0	0

FIGURE 11.10 *Example excitation table, J_1.*

However, this would not work since J_1 would then be a level, while a pulse is required to trigger the flip-flop. The equation that should be used is

$$J_1 = A + B\bar{y}_1 \qquad (11.7)$$

Looking at the problem another way, we recall that Fig. 11.10 represents only four columns of a 16-column K-map. The column $ABCD = 0000$ is a column of 0s, since no transitions should take place in the absence of an input pulse. Thus, all product terms in the excitation equations (assuming SOP forms) must include at least one input variable. Furthermore, no input variable should ever appear complemented in the excitation equations since, for pulse inputs, a complemented variable corresponds to the absence of the input.

Example 11.2

Obtain a realization of the state table developed in Example 11.1 using *J-K* flip flops for which the *J* and *K* inputs are level inputs only. The flip-flops can change state only in response to a pulse on the input *C1*.

Solution We begin with the four row transition table already given as Fig. 11.7. Consider first the input expression for memory element y_1. From the transition list of Fig. 11.4 we see that the values of *J* and *K* must be specified in the same way as in clock-mode circuits and pulses must be gated to the *C1* input whenever a change in flip-flop value is to take place. Specifying values to accomplish these transitions and carrying the don't-cares directly from Fig. 11.7 accounts for those squares in Fig. 11.11 containing single entries.

y_1y_2	A	B	C	R
00	x/0	1	1	x
01	x	1	1	x
11	x	x	x	x
10	x	x	x	x

J_1

y_1y_2	A	B	C	R
00	x	x	x	x
01	x	x	x	x
11	x	x	x/0	1
10	x/0	x/0	x/0	1

K_1

y_1y_2	A	B	C	R
00	0/x	1	1	x
01	x	1	1	x
11	x	x	0/x	1
10	0/x	0/x	0/x	1

$C1_1$

FIGURE 11.11 *Memory element input options.*

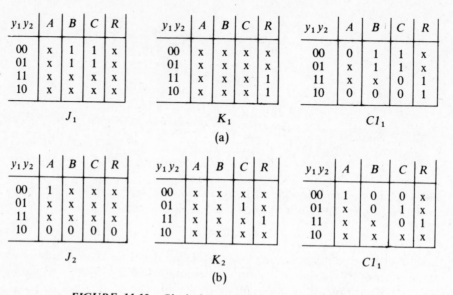

FIGURE 11.12 Clocked J-K realizations, Example 11.2.

As indicated by the four squares with dual entries in the lower half of the maps for K_1 and Cl_1, we can hold y_1 at 1 either by inhibiting the clock or by making $K_1 = 0$. The situation is similar if y_1 is to remain 0 as in the upper left squares of J_1 and Cl_1.

Since the J and K signals must be levels, they cannot be functions of the pulse inputs. With reference to the maps for J_1 and K_1 in Fig. 11.11, this means that we cannot have both 0s and 1s in the same row. Thus, in all five squares of the J_1 and K_1 maps where two entries are possible, we must choose the don't-cares. In the corresponding five squares of the Cl_1 map we must then choose the 0s. The resulting maps are given in Fig. 11.12a, leading in turn to expressions 11.6 and 11.7.

$$Cl_1 = (B + C)\bar{y}_1 + R \tag{11.8}$$

$$J_1 = K_1 = 1 \tag{11.9}$$

A similar analysis for y_2 leads to the maps of Fig. 11.12b. From these we obtain the following expressions:

$$Cl_2 = A + R + C\bar{y}_1 \tag{11.10}$$

$$J_2 = \bar{y}_1 \tag{11.11}$$

$$K_2 = 1 \tag{11.12}$$

The circuit corresponding to these equations is shown in Fig. 11.13. In this particular case the pulse triggered form (Fig. 11.9) is slightly simpler, but the clocked form might be preferable in another case.

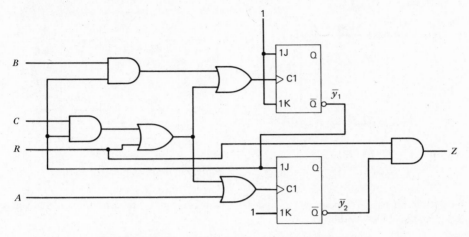

FIGURE 11.13   *Clocked circuit for Example 11.2.*   ∎

Example 11.3

A monorail system provides shuttle service between an airport and a downtown terminal. There are two cars on the system, but only one track, except for a siding at the halfway point, as shown in Fig. 11.14. The two cars will always start from the two stations at the same time, one inbound and one outbound. The outbound car will pull into siding and wait there until the inbound car has passed by on the main track. The signal and switches controlling the siding will be controlled by three track sensors, x_1, x_2, x_3, which emit pulses when a car passes over them. When the cars at the stations, both signals will be red, switch S_1 will be switched to the main line, and switch S_2 will be switched to the siding. The outbound car will go onto the siding and stop at signal light L_1. This signal will remain red until the inbound car passes over both x_1 and x_2, at which time it will turn green, and switch S_1 will switch to the siding. The inbound car will stop at signal L_2 until the outbound car moves off the siding and over x_1, at which time L_2 will turn green and S_2 will switch to the main line. When the inbound car passes over x_3 when arriving at the downtown terminal, the system will return to the original condition. Design a level

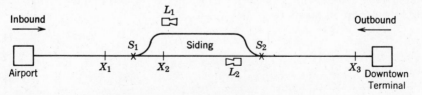

FIGURE 11.14   *Monorail shuttle system.*

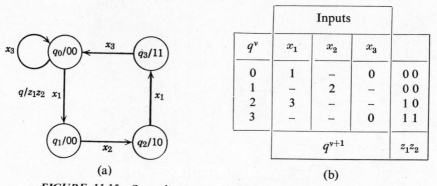

(a)   (b)

FIGURE 11.15   *State diagram and table, monorail system controller.*

output circuit to develop signals z_1 and z_2, which correspond to conditions of the signal lights and switches as follows:

$$z_1 = 0 \leftrightarrow L_1 = \text{red} \qquad S_1 \text{ to main line}$$

$$z_1 = 1 \leftrightarrow L_1 = \text{green} \qquad S_1 \text{ to siding}$$

$$z_2 = 0 \leftrightarrow L_2 = \text{red} \qquad S_2 \text{ to siding}$$

$$z_2 = 1 \leftrightarrow L_2 = \text{green} \qquad S_2 \text{ to main line}$$

Solution   The placement of the sensors and the sequence of operations are such that x pulses cannot coincide, so the Moore model is applicable. The state diagram for the system is shown in Fig. 11.15a. Starting in state q_0 (both signal lights red), the outbound car first passes over the x_3 sensor, but this causes no change of state. The inbound car then passes over x_1 and x_2 in sequence, moving the circuit to states q_1 and q_2, at which time z_1 goes to 1, causing L_1 to change to green and S_1 to switch siding. The outbound car then passes over x_1, taking the circuit to q_3 and changing z_2 to 1. The cycle is completed when the inbound car passes over x_3, returning the circuit to q_0. The corresponding state table is shown in Fig. 11.15b.

It is obvious that the only possible simplification is the combination of q_0 and q_1 into a single state. Since two state variables are required and there are two output variables, the obvious state assignment is to let the state variables correspond to the outputs, i.e., $y_1 = z_1$ $y_2 = z_2$. We will use J-K flip-flops in the pulse-triggered connection. The resultant transition table and excitation tables are shown in Fig. 11.16. From the excitation table, we read the excitation equations

$$J_1 = x_2 \tag{11.13}$$

$$K_1 = K_2 = x_3 \tag{11.14}$$

$$J_2 = x_1 y_1 \tag{11.15}$$

The corresponding circuit is shown in Fig. 11.17.

(y_1y_2)	x_1	x_2	x_3	
0 0	00	10	00	0 0
0 1	—	—	—	– –
1 1	—	—	00	1 1
1 0	11	—	—	1 0
		$(y_1y_2)^{+1}$		z_1z_2

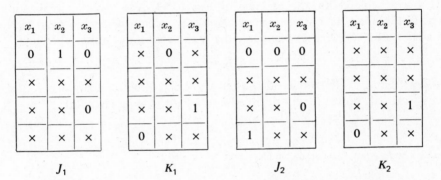

x_1	x_2	x_3
0	1	0
×	×	×
×	×	0
×	×	×

J_1

x_1	x_2	x_3
×	0	×
×	×	×
×	×	1
0	×	×

K_1

x_1	x_2	x_3
0	0	0
×	×	×
×	×	0
1	×	×

J_2

x_1	x_2	x_3
×	×	×
×	×	×
×	×	1
0	×	×

K_2

FIGURE 11.16 Transition table and excitation tables, Example 11.3.

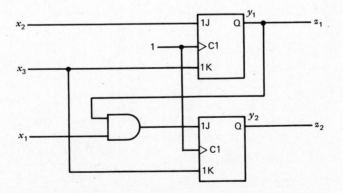

FIGURE 11.17 Final circuit, monorail system controller.

11.3   Counters Revisited*

Counters are most commonly formulated as pulse-input, level-output Moore model circuits. The state table of an elementary *decade* or modulo-10 counter is given Fig. 11.18a as an example. Note the single next-state column corresponding to pulse line P. In effect, the counter counts (modulo-10) the number of pulses appearing on this line. This pulse line might be the system clock. If so, Fig. 11.18a could equally well describe a clock-mode sequential circuit. This interpretation is somewhat less natural, however, since the clock is not counted as an input in clock-mode circuits. Thus, the circuit would have no inputs. It would, however, still have $2^0 = 1$ next-state columns, as indicated in the figure.

Assigning states in binary-coded decimal order yields the transition table of Fig. 11.18b. This state assignment effectively permits a partition very similar to the one discussed for the counter of Fig. 10.50. A T flip-flop realization of Fig. 11.18b may be obtained very simply by noting when the four flip-flops change state. Note that y_0 changes state with every input pulse. Therefore,

$$T_0 = P \tag{11.16}$$

Notice that $y_1^{\nu+1}$ (the value of y_1 after a pulse on line P) differs from y_1^{ν} if and only if $y_0^{\nu} = 1$, except for $9 \rightarrow 0$ transition, where $y_3^{\nu} = 1$. Therefore,

$$T_1 = y_0 \cdot \overline{y_3} \cdot P \tag{11.17}$$

* This section is not essential to continuity and may be omitted.

q^{ν}	P		$(y_3 y_2 y_1 y_0)^{\nu}$	P	
0	1	0	0 0 0 0	0 0 0 1	0 0 0 0
1	2	1	0 0 0 1	0 0 1 0	0 0 0 1
2	3	2	0 0 1 0	0 0 1 1	0 0 1 0
3	4	3	0 0 1 1	0 1 0 0	0 0 1 1
4	5	4	0 1 0 0	0 1 0 1	0 1 0 0
5	6	5	0 1 0 1	0 1 1 0	0 1 0 1
6	7	6	0 1 1 0	0 1 1 1	0 1 1 0
7	8	7	0 1 1 1	1 0 0 0	0 1 1 1
8	9	8	1 0 0 0	1 0 0 1	1 0 0 0
9	0	9	1 0 0 1	0 0 0 0	1 0 0 1
$q^{\nu+1}$	z^{ν}			$(y_3 y_2 y_1 y_0)^{\nu+1}$	$(z_3 z_2 z_1 z_0)^{\nu}$

(a)   (b)

FIGURE 11.18   Elementary decade counter.

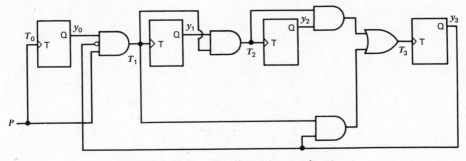

FIGURE 11.19 Decade counter realization.

Similarly, y_2 will change state $(0 \rightarrow 1$ or $1 \rightarrow 0)$ if and only if $y_1^v = y_0^v = 1$. Therefore,

$$T_2 = y_0 \cdot y_1 \cdot P \tag{11.18}$$

For a modulo-16 counter the expression for T_3 would be analogous. While y_3 changes from 0 to 1 in the case of the decade counter when $y_2 = y_1 = y_0 = 1$, we have a change from 1 to 0 when $y_3 = y_0 = 1$. Therefore,

$$T_3 = (y_2 \cdot y_1 \cdot y_0 \cdot P) + (y_3 \cdot y_0 \cdot P) \tag{11.19}$$

We leave it as an exercise for the reader to obtain these same expressions using K-maps and transition lists.

As mentioned in Chapter 9, T flip-flops are not commonly available as standard integrated circuit components, but a J-K flip-flop becomes a T if the J and K inputs are permanently connected to the 1-level. A T flip-flop realization of the decade counter is given in Fig. 11.19.

Sometimes counters which count to different values, depending on one or more level inputs, are available in integrated circuit form. We conclude this section with an example of this type of counter.

Example 11.4

Design a counter that will count modulo-5 if a level input $L = 1$ or will count modulo-8 if $L = 0$. The count is to be incremented each time a pulse appears on line P_1 and decremented each time a pulse appears on line P_2.

Solution A state table for the counter is given in Fig. 11.20. Note that there are $n \times 2^m = 4$ input columns, where the number of pulse inputs is $n = 2$ and the number of level inputs is $m = 1$. In the modulo-5 columns, we enter don't-cares on the last row, assuming that the value of L will be changed only occasionally and never when the output count is being monitored. An external reset line is also assumed.

q^v	Inputs				z
	$L = 0$		$L = 1$		
	P_1	P_2	P_1	P_2	
0	1	7	1	*4	0
1	2	0	2	0	1
2	3	1	3	1	2
3	4	2	4	2	3
4	5	3	*0	3	4
5	6	4	✕	✕	5
6	7	5	✕	✕	6
7	0	6	✕	✕	7
	q^{v+1}				z

FIGURE 11.20 *Level-controlled counter.*

Because of the modulo-5 columns, the closed partitions leading to the binary state assignment in the counters considered thus far are not valid for this state table. However, the closure property for partitions (0246) (1357) and (04) (15) (26) (37) is violated only for the two entries marked by asterisks in Fig. 11.20. Therefore, the straight binary assignment should still yield relatively simply flip-flop input equations as well as no output logic. Assigning states in this way and using K-maps along with the T flip-flop transition list will yield the following input equations:

$$T_1 = P_1 \cdot \overline{(y_2 L)} + P_2 \cdot \overline{(\bar{y}_2 \bar{y}_1 \bar{y}_0 L)} \tag{11.20}$$

$$T_2 = P_1 \cdot y_0 + P_2 \cdot \bar{y}_0 \cdot \overline{(L \bar{y}_2 \bar{y}_1)} \tag{11.21}$$

$$T_3 = P_1 \cdot (y_0 y_1 + L y_2) + P_2 \cdot (\bar{y}_0 \bar{y}_1) \tag{11.22}$$

We leave the verification of these equations as a homework problem for the reader. ■

11.4   Clock Input Control

Until now, we have inferred that a sequential circuit only operates in the clock mode if the clock or toggle inputs of all flip-flops on the circuit are tied *directly* to the system clock, as depicted in Fig. 11.21a. Under certain circumstances, particularly when D-type flip-flops are used, this approach to clocking can be unduly restrictive. A simple illustration of a sequential circuit with direct clocking is given in Fig. 11.21b. If control signal $CS_1 = 1$, x_1 is to be shifted into the flip-flop shown. If $CS_2 = 1$, x_2 is shifted in. Regardless of the control signals which might be present, it is always necessary that the value to be stored in the D flip-flop after a clock pulse be present at the flip-flop input before the clock

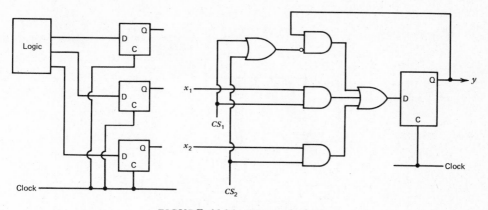

FIGURE 11.21   Direct clocking.

pulse arrives. Therefore, if both control signals are 0, the current value of the flip-flop must be routed back to the D input. This is accomplished by the upper AND gate in Fig. 11.21b.

The network in Fig. 11.21b can be simplified by a departure from the direct clocking procedure. Suppose that a clock pulse is allowed to reach the toggle input of a flip-flop only if the contents of the flip-flop are to be modified. For the example under discussion, then, a change will occur only if $CS_1 \vee CS_2 = 1$. In Fig. 11.22a, this signal is used to gate the clock pulse into the flip-flop toggle. Thus, the circuits of Figs. 11.21b and 11.22a accomplish the same function. In this case the one AND-gate and one OR-gate saved in the data network of Fig. 11.21b were added back into the clocking network of Fig. 11.22a. A significant savings can be achieved in more complicated systems, particularly where several flip-flops are combined together to form a register. Often the same control signals are used to gate data into the different flip-flops of a register. In this case, the clock control gating can be common to all flip-flops of the register, as illustrated in Fig. 11.22b. (The gating of the data is not shown.) If the approach of Fig. 11.21 were used, this saving could not be achieved.

Clearly the signal on the line labeled *control pulse* in Fig. 11.22b is a *pulse* synchronized with the clock. This is in contrast to the signals on the control lines entering the OR-gate of the same figure, which carry levels that are necessarily stable during the occurrence of a clock pulse. In Fig. 11.22 as well as in Sections 11.2 and 11.3 we observed individual AND-gates with both level and pulse inputs. Before we have further occasion to perform logic on levels and pulses, let us examine what controls must be placed on such operations. The restrictions tabulated in Fig. 11.23 partially define the distinction between levels and pulses and will be accepted throughout the remainder of the book.

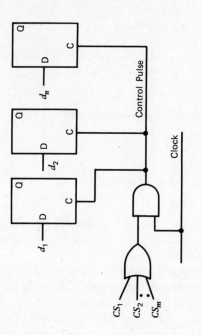

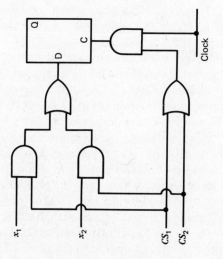

FIGURE 11.22

A	B	$A \cdot B$	$A + B$
level	level	level	level
level	pulse	pulse	undefined
pulse	pulse	undefined	pulse

FIGURE 11.23 *Interaction between levels and pulses.*

The case in which a level is ANDed with a pulse is very useful and was illustrated in Fig. 11.22. We will not, however, allow the AND-operation where more than one input line could carry a pulse. We assume that pulses are of short duration relative to the time frame of interest and that two pulses can never be guaranteed to occur simultaneously. It can be seen that the output of an OR-gate with one pulse input and one level input might be either a level or a pulse. *Every Boolean variable (or network wire) must be classified as either a pulse variable or a level variable.* We therefore disallow the ORing of a level variable with a pulse variable since the output could not be so classified.

The operation of ORing two pulse lines may be useful in a controlled clock network as well as other pulse-mode circuits. This is illustrated in Fig. 11.24.

In addition to the *J-K* flip-flop, which is the usual choice for pulse-mode circuits, the most commonly available SSI memory element is the edge-triggered *D* flip-flop. We note that the clocked response of the *D* flip-flop is independent of the present value of its output, *y*. Providing for retaining the old value of *y* in the event of a transition on line *CLK* requires an external feedback loop, as already illustrated in Fig. 11.21. In that figure the status of *y* does not differ from that of the input variables, x_1 and x_2. In a controlled clock circuit the value of *y* remains the same as long as no pulse is allowed to reach *CLK*. In most digital system applications a memory element either is left unchanged or is conditionally loaded with a data function. The function may have the memory element itself as an argument, but as an argument of no special significance. For

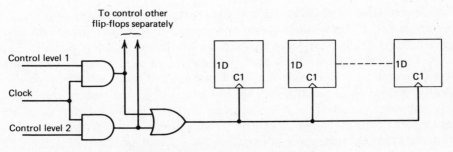

FIGURE 11.24 *ORing of control pulses.*

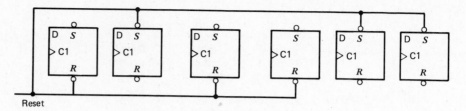

FIGURE 11.25   Circuit initialization.

this reason the D flip-flop is the memory element most commonly used in controlled-clock systems.

It will be recalled from Chapter 9 that many flip-flops have direct set (S) and direct reset (R) inputs. With the concept established of controlling some flip-flops independently of others through control of the clock, it might seem that these direct inputs would offer additional flexibility. The R and S inputs are, however, unclocked, i.e., when either goes to the 0-level, any corresponding output change will occur *immediately*, whether or not there is a clock pulse present. Suppose that we have two D flip-flops and that the output of the first is an input to the logic developing the D input to the second. If we apply a R or S signal to the first while the second is being clocked, we violate the third basic requirement for proper pulsed operation, that the level inputs be stable while the pulse inputs are present. Thus, in clocked or pulsed circuits, proper operation requires that *no memory element output change in response to a level signal at the same time that any other flip-flops whose inputs are influenced by that output are being triggered.* For this reason the application of S and R is usually restricted to initializing all or some of the flip-flops of a network prior to the start of any sequential activity. Fig. 11.25 illustrates the initialization of a set of flip-flops to 010011.

11.5   Extended State Tables

Digital computers and other large digital systems are examples of sequential circuits, but they clearly differ in size and complexity from the circuits discussed up to this point. Such systems typically include many registers of several bits each. The number of states in such systems, taken in the formal sense of all possible combinations of values of all flip-flops, can reach into the thousands. Clearly, state diagrams and tables are not practical for the description of such large systems. We need to develop more suitable techniques for describing sequential circuits composed primarily of large numbers of registers.

In the previous section we introduced the concept of controlling the clock signals to memory elements. Let us now explore the distinctions between the

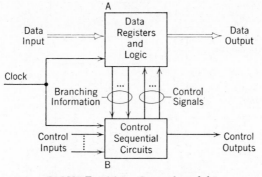

FIGURE 11.26   Control model.

control levels and pulses and the logic that develops them on the one hand, and the data networks controlled by these signals on the other. We begin by partitioning the clock mode model into two separate sequential circuits as shown in Fig. 11.26. The upper portion consists of data registers together with interconnecting combinational logic. The lower sequential circuit provides a sequence of control signals that cause the appropriate register transfers to take place with the arrival of each clock pulse. The numbers of control signal lines and control input lines are normally small in comparison to the numbers of data input and output lines and interconnecting data lines within block A. In Fig. 11.27 is shown a set of three data flip-flops, B_0, B_1, B_2, interconnected by data transfer lines that are controlled by signals C_1, C_2, C_3, C_4, developed by a control sequential circuit. The nature of the control circuit is such that only one of the four control signals can be 1 at any time. A careful examination will reveal that four sets of data transfers can take place in the lower portion of Fig. 11.27. If $C_1 = 1$, the inputs X_0 and X_1 are loaded into flip-flops B_0 and B_1, respectively. The notation we will use for this transfer is given by Equation 11.23. (The contents of B_2 are unchanged.)

$$1. \quad B_0 \leftarrow X_0; \qquad B_1 \leftarrow X_1 \qquad\qquad (11.23)$$

If $C_2 = 1$, the information in the three flip-flops is shifted right with the contents of the rightmost flip-flop shifted around into B_0. For the present, we will denote this transfer similarly as given in

$$2. \quad B_1 \leftarrow B_0; \qquad B_2 \leftarrow B_1; \qquad B_0 \leftarrow B_2 \qquad (11.24)$$

If $C_3 = 1$, the transfer is

$$3. \quad B_1 \leftarrow B_0 \wedge B_1 \qquad\qquad (11.25)$$

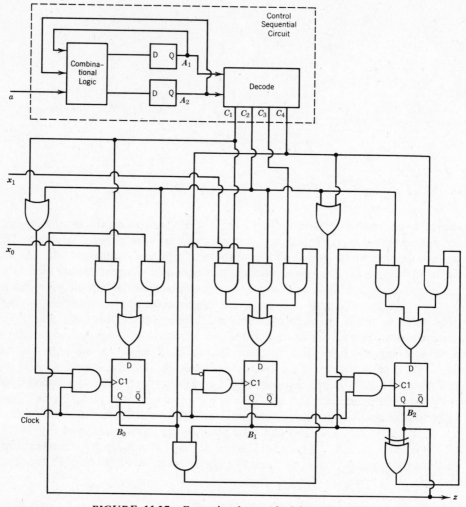

FIGURE 11.27 *Example of control of data transfers.*

and if $C_4 = 1$ the transfer is

$$4. \quad \boldsymbol{B_2 \leftarrow B_1 \oplus B_2} \qquad\qquad (11.26)$$

Note the use of the symbol $\wedge$ for the AND operation. We will find this symbol convenient as we expand our transfer notation.

The above describes the transfers that can take place, but the sequence of transfers, and therefore the sequence of outputs, will depend on the sequence of

q^v	q^{v+1}		Transfer corresponding to q^v
	$a = 0$	$a = 1$	
C_1	C_3	C_2	$B_0 \leftarrow X_0;\ B_1 \leftarrow X_1$
C_2	C_1	C_1	$B_1 \leftarrow B_0;\ B_2 \leftarrow B_1;\ B_0 \leftarrow B_2$
C_3	C_4	C_4	$B_1 \leftarrow B_0 \wedge B_1$
C_4	C_2	C_1	$B_2 \leftarrow B_1 \oplus B_2$

FIGURE 11.28 Extended state table.

control signals. Since there are four possible combinations of control signals, we may characterize the control sequential circuit as a Moore circuit with four states, which we may designate as C_1, C_2, C_3, C_4, each producing a 1 on the corresponding control output line. Let us further assume that the sequencing between these states is controlled by a sequence control signal, a, as shown in the state table of Fig. 11.28. This is consistent with Fig. 11.27, where a is shown as an input to this 4-state control sequential circuit. To this state table, we add a list of the transfers initiated by the control signals corresponding to each state.

Together with the output equation $z = B_2$, this extended state table provides a complete description of the desired sequential circuit behavior from which the complete circuit could be generated. There are three data flip-flops, and at least two control flip-flops will be required. Therefore, a conventional description in the manner of Chapter 10 would require a 32-state table, which would clearly be far less convenient as a design tool than the compact table of Fig. 11.28. We will leave it as a homework problem for the reader to obtain such a table.

11.6 A Program Description

Although Fig. 11.28 seems to be a convenient description of the corresponding circuit, a similar description for a circuit with a large number of control states and many control inputs could still be unwieldy. It is the purpose of this section to develop a still briefer sequential circuit description to take the form of a computer program.

The list of transfers in Fig. 11.28 very much resembles the transfer or replacement statements in a computer program. Individual flip-flops correspond to scalar variables. New values of these variables are the expressions at the right of each transfer arrow. Transfers listed on the same line will be assumed to take place simultaneously. A vector notation to be introduced in the next section will serve to describe simultaneous transfers more compactly.

Missing from the list of transfer statements of Fig. 11.28 are the branch statements required to sequence through the transfer statements in a data-dependent order. However, while the branching information is not found in the transfer list, it is included in Fig. 11.28 in the form of the state table. As each transfer is accomplished, the control circuit goes to its next state to enable the proper transfer at the next clock pulse. From states C_1 and C_4, a two-way branch (two possible next states) is executed as a function of input a. From state C_2, the next state is always C_1, much like a GO TO statement in Fortran.

With the addition of a branching capability we will have the beginning of a *hardware description language*. Several such languages have been widely used as a means of describing large digital systems [2], [3]. One such language, AHPL, is developed by the present authors in Reference 1. In AHPL branching capability is provided by specifying that each statement may consist of two parts, a transfer statement and a branch statement. Transfer statements, as we have seen, simply specify what transfer of data should take place; a branch statement specifies the statement to be executed next. An AHPL statement need not include both parts. If the branch statement is omitted, the next sequential statement is the next to be executed. If the transfer statement is omitted, a branch is made to another statement but nothing else is done.

As in conventional programming we have two types of branches, conditional and unconditional. The unconditional branch has the form

$$\rightarrow (S)$$

where S is the number of the statement to be executed next. The conditional branch has the form

$$\rightarrow (f_1, f_2, \ldots, f_n)/(S_1, S_2, \ldots, S_n) \tag{11.27}$$

The functions f_1, f_2, etc., are logic functions of the system variables that take on the values 0 or 1. If a given $f_i = 1$, then S_i, the statement number in the corresponding position to the right of the slash, is the number of the statement to be executed next. Normally only one of the f_i may be 1 at a time, i.e.,

$$f_i \wedge f_j = 0 \qquad \text{where } i \neq j \tag{11.28}$$

If none of the $f_i = 1$, the next statement in sequence is executed.

Following the conventions described above, Fig. 11.27 may be translated into the following *control sequence*:

1. $B_0 \leftarrow X_0; B_1 \leftarrow X_1$

   $\rightarrow (\bar{a}, a)/(3, 2)$

2. $\quad B_1 \leftarrow B_0\, ; \; B_2 \leftarrow B_1\, ; \; B_0 \leftarrow B_2$

$\qquad \rightarrow (1)$

3. $\quad B_1 \leftarrow B_0 \wedge B_1$

4. $\quad B_2 \leftarrow B_1 \oplus B_2$

$\qquad \rightarrow (\bar{a},\, a)/(2,\, 1)$

Where we have a two-way branch with one destination the next statement, we can take advantage of the default option (that the next statement is executed if all $f_i = 0$) to simplify the branch statement. For example, statement 1 in the above sequence can be simplified to the form

1. $\quad B_0 \leftarrow X_0\, ; \; B_1 \leftarrow X_1$

$\qquad \rightarrow (\bar{a})/(3)$

This form of branch statement can be interpreted to mean, "IF $\bar{a} = 1$, THEN go to 3, ELSE go to 2." This "IF-THEN-ELSE" form is standard in many programming languages. If there are more than two branch destinations the default option should not be used because it is too easy to overlook the default possibility.

Throughout the rest of this chapter, we will use the term *control sequence* to identify hardware descriptions of this form. This is a complete description of the circuit of Fig. 11.27 containing all information found in Fig. 11.28. The efficiency of this new method of circuit description is not particularly apparent for the above example. Note that only one branch statement (Step 3) was omitted because the next step (Step 4) is always taken in sequence. This situation will be much more common in larger systems. Similarly, where there are a large number of inputs to the control unit, the f_i at any given step may be a function of only a few of these inputs.

11.7 Synthesis

A principal advantage of a program description of hardware must necessarily be its use as a synthesis tool. For large systems, it is the only approach other than trial-and-error manipulation of the logic block diagram. Lanuage is used as a design tool for large systems such as computers and peripheral equipments [1]. In this chapter, we will show that it can be used effectively over a range of less complex systems that are still too bulky to describe in state-table form. We will even find it useful as a means of quickly obtaining unambiguous descriptions of certain smaller sequential circuits for which we obtained state tables in Chapter 10.

Before we move to discard all of our state-table-related tools, let us remind ourselves that our new medium is a *language and only a language*. It will have no apparent algebraic structure in terms of which state minimization can be performed. Nor does the language itself provide any constraints that will guide the designer to a minimal circuit. With practice, the designer will readily arrive at an optimal list of program steps for certain types of examples. For other examples, particularly where certain flip-flops can serve a variety of functions, it would be unrealistic to expect to obtain a minimal realization without resorting to state-table minimization. Even if the minimal number of flip-flops are used, the fact that don't-care conditions are not reflected in the language may cause a realization to be more complicated than necessary. With the recent trend toward very low-cost integrated circuits, minimization would seem to become progressively less important. This observation should not be overemphasized, however, or the result can be sloppy design requiring many more gates than necessary. The notion of state-table minimization should always be kept waiting in the wings for occasional use, at least on parts of sequential circuits.

As a first example, let us obtain a design of the 3-bit serial code translater of Problem 10.22. This problem is presented below as Example 11.5.

Example 11.5

Obtain the control sequence of a serial translating circuit that performs the translations tabulated in Fig. 11.29. As indicated, there is one data input line, x, and one output line, z, in the circuit. The inputs occur in characters of three consecutive bits. One character is followed immediately by another. A reset input, $r = 1$, will cause the translator to begin operation synchronized with the input. A translated output character will be delayed by only two clock periods with respect to an input character.

x^v	x^{v-1}	x^{v-2}	z^{v+2}	z^{v+1}	z^v
0	0	0	0	0	0
0	0	1	0	1	1
0	1	0	1	0	0
0	1	1	1	0	1
1	0	0	1	1	0
1	0	1	0	0	1
1	1	0	0	1	0
1	1	1	1	1	1

FIGURE 11.29   Code translation.

Solution The output translation cannot be performed until all three input bits have been received, at time t_v. Therefore, as the first two bits arrive, we shift them into a 2-bit shift register B consisting of bits, B_0 and B_1. At time t_v, x^{v-2} will be in B_1 and x^{v-1} will be in B_0. When x_v arrives, all three output bits will be computed simultaneously. The immediate output will be z^v, and outputs z^{v+1} and z^{v+2} will be stored in B_1 and B_0, respectively. At time t^{v+1}, z^{v+1} can be read from B_1, z^{v+2} will be shifted from B_0 to B_1, and the first input bit of the next character will be shifted into B_0. At t^{v+2}, the output z^{v+2} will be read from B_1, the previous input bit will be shifted into B_1, and the next input bit will be shifted into B_0, setting the state for the translation of the next character at the following clock time. The operation of this circuit can thus be described in terms of a repetitive three-step sequence, the first two steps being shifts, the third a translation and parallel register entry, as shown below.

$$1. \quad B_0, B_1 \leftarrow x, B_0; \qquad z = B_1$$
$$\rightarrow (r)/(1)$$

$$2. \quad B_0, B_1 \leftarrow x, B_0; \qquad z = B_1$$
$$\rightarrow (r)/(1)$$

$$3. \quad B_0, B_1 \leftarrow f_2(x, B_0, B_1), f_1(x, B_0, B_1); \qquad z = f_0(x, B_0, B_1)$$
$$\rightarrow (1)$$

The three output functions of step 3 may be obtained by converting the table of Fig. 11.29 into three K-maps, one for each output, as shown in Fig. 11.30. In correlating these maps to the table, note that x^{v-2} is stored in B_1, x^{v-1} in B_0. The resultant functions

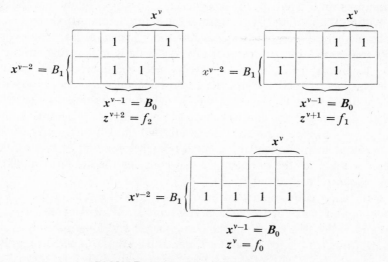

FIGURE 11.30   Translation functions.

are tabulated in Equations 11.29, 11.30, and 11.31. Note, from step 3 of the sequence, that $z^{\nu+1}$ is initially stored in B_1 and $z^{\nu+2}$ in B_0.

$$f_0 = B_1 \tag{11.29}$$

$$f_1 = (x^{\nu} \wedge \bar{B}_1) \vee (x^{\nu} \wedge B_0) \vee (\bar{x}^{\nu} \wedge \bar{B}_0 \wedge B_1) \tag{11.30}$$

$$f_2 = (\bar{x}^{\nu} \wedge B_0) \vee (B_0 \wedge B_1) \vee (x^{\nu} \wedge \bar{B}_0 \wedge \bar{B}_1) \tag{11.31}$$

The reset mechanism has been incorporated into the three-step sequence. Whenever the input $r = 1$, control is returned to step 1 to begin the input of a new character. The reader will notice that the single output, z, is specified separately on each line of the control sequence. This is necessary because the output is not a fixed function of some data register or input line, as was the case for the circuit of Fig. 11.27. Rather, the function is different for different steps in the sequence, i.e., it is also a function of the control signals. It is common in large systems for the output to be taken directly from a data register to be a fixed function of several data registers. In such cases, the outputs need be listed only once in the control sequence. ∎

The above example makes evident a very important difference between the control sequence method of design and the state diagram method presented in Chapter 10. Setting up the state diagram requires only an understanding of the required sequencing of inputs and outputs; no preliminary guesses as to the numbers of state variables or to the internal structure are required. By contrast, the control sequence approach essentially starts with a proposed basic structure. This method thus requires considerable intuitive understanding of possible circuit structures on the part of the designer, and it is very likely that the resultant design will not be minimal in the formal sense of Chapter 10. Once a circuit has been obtained by the control sequence method, it can, at least theoretically, be formally minimized, but the number of states usually makes this impractical.

Once a control sequence has been obtained, the creative portion of design effort is complete. Translation of the control sequence to a logic block diagram requires only a strict compliance with a few simple rules. In fact, this portion of the design process can be reduced to a computer program. At this point in the learning process, however, the reader will find it instructive to follow in detail the implementation of the control sequence for one or two examples. In the latter sections of the chapter, we will consider the design complete once a control sequence has been obtained.

We will find it convenient to develop separately the *control* portion and the *data storage* and *logic* portions of a circuit realization for the control sequence of Example 11.5. We first construct the data storage and logic portion as shown in Fig. 11.31. Note that Steps 1 and 2 are identical so that one line from the control unit will cause both of these steps to be executed. A second control line will cause the execution of Step 3. These lines are clearly marked as inputs at the left

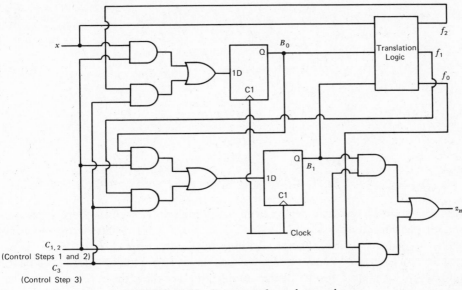

FIGURE 11.31   Data unit for code translator.

of Fig. 11.31. Note in the control sequence that the contents of both flip-flops are changed at every step. Thus, the clock line may be connected to the $C1$ input of flip-flops B_0 and B_1 without any clock control gating.

The circuit output z will be the output of a multiplexer network gate if it is a function of the control signals. The control step signals and the output values corresponding to that step serve as the input to one of the AND-gates of the two-level multiplexer network generating z. Only the control signal corresponding to the step being executed is 1, so the output value corresponding to this step is routed through the network to the output z. A similar two-level multiplexer network is placed at the input of each flip-flop so that the new flip-flop values corresponding to the current step are routed to the appropriate D inputs. An AND-gate will appear in the input network to a flip-flop for each control step, which actually causes that flip-flop to be updated. In this case, there are only two first-level AND-gates, one corresponding to Steps 1 and 2 and the other corresponding to Step 3. Since these two signals cover all three control steps, we have a check on our earlier observation that clock control was not required. To avoid unnecessarily complicating Fig. 11.31, the Step-3 functions, f_0, f_1, and f_2, are represented by a single block.

Let us now turn our attention to constructing a circuit that will generate the control signals $C_{1,2}$ and C_3. The first step is to obtain a state diagram from the control sequence of Example 11.5. Each control step will correspond to a state,

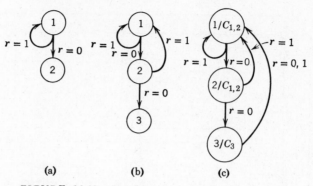

FIGURE 11.32   Developing a control state diagram.

with next-state arrows determined from the branch portion of the step. We begin by representing control Step 1 as state 1 with the appropriate arrows to next states, as shown in Fig. 11.32a. Adding next-state arrows from state 2, as specified by Step 2 of the control sequence, yields Fig. 11.32b. Adding the next-state arrow for state 3 results in the completed state diagram of Fig. 11.32c. The control circuit is a Moore circuit, providing a level output during the duration of each control step, to gate the appropriate signals to the flip-flops at clock time. On the state diagram, the control signals corresponding to each step are shown as outputs for each state, as shown in Fig. 11.32c. Note that only one control signal can be 1 at a time.

Once a state diagram has been obtained, there are two possible ways to obtain circuit for the corresponding control unit. One is to minimize the corresponding state table and proceed to a circuit design by the formal techniques of Chapter 13.

Example 11.6

Determine a minimal realization of the control circuit corresponding to Fig. 11.32c.

Solution The reader can easily verify that Fig. 11.32c, is a minimal-state sequential circuit. The state assignment depicted in Fig. 11.33a was obtained using Rules I and II of Chapter 10. From this, we can easily obtain the D flip-flop input equations given by Equation 11.32 and the output equations given by Equation 11.33.

$$Y_1^{v+1} = \bar{r} \cdot \bar{y}_2 \qquad Y_2^{v+1} = \bar{r} \cdot \bar{y}_2 \cdot y_1 \tag{11.32}$$

$$C_{1,2} = \bar{y}_2 \qquad C_3 = y_2 \tag{11.33}$$

An implementation of these expressions is given in Fig. 11.33.   ■

The data logic and storage unit of Fig. 11.31, together with the minimal control sequencer of Fig. 11.33, is probably as efficient a realization as can be found

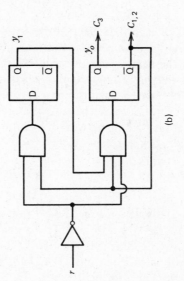

(b)

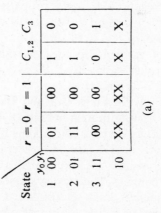

(a)

FIGURE 11.33   Minimal control sequencer for code translator.

for the code translator. A minimal-state table, determined by the formal techniques of Chapter 10, has 12 states, which agrees with the

$$(3 \text{ control states}) \times (4 \text{ data states})$$

used here. The fact that the number of states agrees does not mean that the complete circuits obtained by the two methods would be identical, but they would be of the same order of complexity.

Although the preceding technique will always produce the minimal control circuit, under some circumstances a preferable realization can be obtained through a "brute-force" technique of assigning a separate flip-flop to each state and using a branching network to route control to the proper next step. A branching network corresponding to

$$\rightarrow (r, \bar{r})/(1, 2) \qquad\qquad (11.34)$$

which represents the branch portion of Step 1 in the control sequence of Example 11.5, is given in Fig. 11.34. Prior to the clock pulse that executes Step 1, the output of the flip-flop is 1. This signal serves as one of the inputs to both AND-gates in the branching network. The other input of each is one of the branch functions of Equation 11.36. In general, there will be one gate for each branch function in a branch expression. Since one and only one of the branch functions can be 1, only the corresponding output of the branch network can be 1. In Fig. 11.34a, the output labeled Step 1 will be 1 if $r = 1$ and the flip-flop output $Q = 1$. If $r = 0$ and $Q = 1$, the output labeled Step 2 will be 1. This latter signal will be routed to the D input of the Step-2 flip-flop. If $r = 0$, the clock pulse that executes Step 1 will propagate the signal into the Step-2 flip-flop, thus passing control to Step 2. Where the output of more than one branch network can be routed to the input of a control flip-flop, OR-gates must be provided at the input of control flip-flops, as illustrated in Fig. 11.34b. If any of the illustrated branch signals from Steps 1, 2, or 3 are 1 when a clock pulse arrives, the Step-1 flip-flop is set to 1.

The flip-flops for Steps 1, 2, and 3 are all shown in Fig. 11.34c, together with the interconnecting branch networks. Also shown are the output signals C_3 and $C_{1,2}$. The latter signal is formed easily by connecting the outputs of flip-flops 1 and 2 through an OR-gate. We will refer to the realization of the control sequential circuit for a control sequence as a *control sequencer* throughout the remainder of the chapter. The reader should note that with this second method the control sequencer may be obtained as a direct translation of the control sequence, thus avoiding the intermediate step of finding a state diagram. The state diagram is needed only if we try to obtain a minimal state and minimal flip-flop realization.

Since the minimal control sequencer obtained by the formal minimization

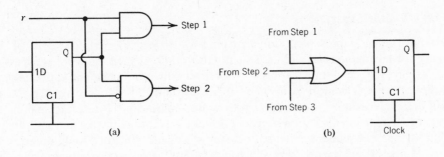

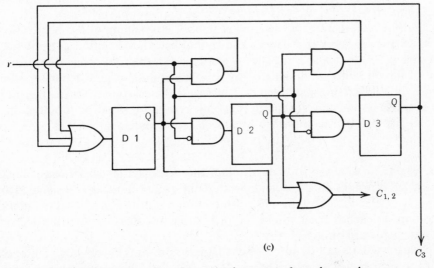

FIGURE 11.34 *Alternate control sequence for code translator.*

techniques was clearly more economical than that obtained by direct translation, the reader may wonder why the latter technique might be preferable. In some cases, even with the division of the circuit into data and control sections, the number of states in the control sequence may be so large as to make formal minimization impractical. Further, even though minimization produces a circuit with the minimum number of flip-flops, this does not necessarily mean that the overall circuit is minimal. Although not so for the above example, in many cases minimizing the number of states may require so much additional combinational logic that the overall cost is increased. Finally, even though it may be more expensive initially, the one-flip-flop-per-step approach may be chosen to ease the problems of testing, fault diagnosis, and maintenance. A sequencer configured as a physical implementation of the control flowchart can be most helpful in understanding the sequence.

11.8   Vector Operations

We have already presented sufficient notation to permit writing a control sequence for any sequential circuit. This notation is convenient where it is necessary to specify the next-state expression for every memory element separately. In larger systems, it is more convenient to treat sets of flip-flops as registers. As we will see, it is usually not necessary to write separate next-state expressions for individual flip-flops within a register. To take advantage of this fact and greatly simplify control sequences for large systems, we will introduce some notation for logical operations on registers and transfers into registers. To remain consistent with *Digital Systems* [1], we will use the vector notation of AHPL.

In the previous section, we denoted scalar variables or flip-flops in two ways, by boldface italic lower-case letters and by boldface italic capital letters with a subscript. We now define a register as a vector variable, i.e., a one-dimensional array of flip-flops, to be designated by a boldface italic capital letter (or short string of boldface italic capital letters.) Just as is the case in a software program, it is necessary somehow to specify the number of elements or flip-flops in a register. This can be accomplished by a declaration statement such as

$$\text{MEMORY: } A[3], B[8]$$

at the beginning of a control sequence indicates a sequential circuit consisting of at least 11 flip-flops, 3 in A and 8 in B, in addition to any designated by lower-case scalar variables. In most of the examples in this chapter, the number of flip-flops in each register will be set forth in the discussion so that the declaration statements are omitted for brevity.

We have used 1s and 0s to represent logical constants throughout the book. Constant vectors whose elements are 1s and 0s will also have a place in the language we are developing. Occasionally it will be convenient to transfer a constant into a register as illustrated for the 4-bit register A in Equation 11.35.

$$A \leftarrow (1, 0, 0, 1) \tag{11.35}$$

Sometimes we want to pick out individual flip-flops within a register and treat them independently. This can be accomplished easily by numbering the flip-flops in the register from *left to right*, starting with zero. We can identify a particular flip-flop in a register by affixing its number as a subscript of the symbol indicating the register. Thus, the leftmost flip-flop would be A_0, the rightmost bit A_3, as shown in Fig. 11.35.

FIGURE 11.35 Numbering of flip-flops in a register.

A long segment of a register may be specified by indicating the first and last flip-flops of the segment, separated by a colon. As an example the string AR_2, AR_3, AR_4, AR_5 could be denoted $AR_{2:5}$.

We may combine vectors into a single vector that may be treated for some operations as a single large vector by writing the individual vector symbols consecutively and separating them by commas. This process will be called *catenation*. If, for example, A is a 4-bit vector and B is a 3-bit vector, then C as given by Equation 11.36 is an 8-bit vector. A few of the individual element equivalencies are $C_4 = p$, $C_5 = B_0$, and $C_7 = B_2$.

$$C = A, p, B \qquad (11.36)$$

As suggested above, it is always possible to form a vector from a subset of the flip-flops of a register by subscripting and catenation. For example, if we wished to transfer the contents of three flip-flops of the 6-bit register D into the 3-bit register B, we could denote this operation as in Equation 11.37.

$$B \leftarrow D_1, D_4, D_5 \qquad (11.37)$$

or as in 11.38

$$B \leftarrow D_1, D_{4:5} \qquad (11.38)$$

If the vector notation just introduced is really to be of value in simplifying our program description of hardware, we must be able to express compactly Boolean operations involving registers. One convenient notation is the *reduction operation* over AND and OR, denoted $\wedge/A$ and $\vee/A$, respectively. The operation $\wedge/A$ is defined to mean the ANDing together of all the elements of A. That is,

$$\wedge/A = A_0 \wedge A_1 \wedge A_2 \wedge \cdots \wedge A_{\rho A - 1} \qquad (11.39)$$

Similarly,

$$\vee/A = A_0 \vee A_1 \vee A_2 \vee \cdots \vee A_{\rho A - 1} \qquad (11.40)$$

where ρA is the number of elements in A.

The three logical operators may also be applied to vectors on an element-by-element basis. For example,

$$\bar{A} = \bar{A}_0, \bar{A}_1, ..., \bar{A}_{\rho A} \qquad (11.41)$$

and

$$A \wedge B = A_0 \wedge B_0, A_1 \wedge B_1, ..., A_{\rho A - 1} \wedge B_{\rho B - 1} \qquad (11.42)$$

Clearly the latter operation is defined only if $\rho A = \rho B$. However, we will allow an $\wedge$ or an $\vee$ operation between a scalar and a vector and define it to mean an operation between the scalar and each element of the vector. For example,

$$a \wedge B = a \wedge B_0, a \wedge B_1, ..., a \wedge B_{\rho B - 1} \qquad (11.43)$$

Example 11.7

Compute the vectors to be placed in the register on the left of each of the following transfer statements:

$$A \leftarrow R_{5:6}, R_{0:4}$$

$$B \leftarrow R_{1:6}, 0$$

$$C \leftarrow (\wedge/R) \vee S$$

$$D \leftarrow \overline{R \wedge S}$$

Let $R = (1, 0, 0, 1, 0, 1, 1)$ and $S = (1, 1, 0, 0, 0, 0, 1)$

Solution

(a) $A = (1, 1, 1, 0, 0, 1, 0)$

(b) $B = (0, 0, 1, 0, 1, 1, 0)$

(c) $(\wedge/R) \vee S = (\wedge/1, 0, 0, 1, 0, 1, 1) \vee S$

$\qquad\qquad = (1 \wedge 0 \wedge 0 \wedge 1 \wedge 0 \wedge 1 \wedge 1) \vee S$

$\qquad\qquad = 0 \vee S = S$

(d) $\overline{R \wedge S} = \overline{(1, 0, 0, 0, 0, 0, 1)}$

$\qquad\qquad = (0, 1, 1, 1, 1, 1, 0)$ ∎

The *precedence* rule for operations in AHPL is similar to that for Boolean expressions.

1. Not.
2. Selection operators.
3. $\wedge$.
4. $\vee$ or $\oplus$.
5. Catenate.

Parentheses may be used where the precedence rules are insufficient and are sometimes used redundantly for added clarity.

11.9   Conditional Transfer

Two distinct items of information are required to specify a register transfer in an AHPL description:

1. A logical specification of the data to be transferred.
2. A set of conditions that indicates when and if the data are to be transferred into any of one or more data registers.

The first item provides for the routing of the proper data vector to the D inputs of the target memory elements. The second set of conditions causes clock pulses to be routed to the clock inputs of the selected target register at the time of a transfer. Construction of the data network as specified by item 1 and the clock network as specified by the conditions in item 2 can be pursued independently. The general form of a data transfer statement will be as shown in Equation 11.44.

$$\text{target} \leftarrow \text{data} \qquad\qquad (11.44)$$

As a specific example, assume we have three possible sources of data ($DATA1$, $DATA2$, $DATA3$), two possible targets (A, B), and five control variables (p, q, r, f, g) to select specific data sources and targets. The possible actions in this transfer may then be specified as shown in Fig. 11.36.

IF condition THEN data vector
IF p THEN $DATA1$
IF q THEN $DATA2$
IF r THEN $DATA3$

(a)   data switching

IF condition THEN register
IF f THEN A
IF g THEN B

(b)   target selection

FIGURE 11.36   Target and data conditions.

This figure indicates that the data to be transferred will be the vector $DATA1$ if $p = 1$ or will be $DATA2$ if $q = 1$ or $DATA3$ if $r = 1$. We assume that $(p \wedge q) \vee (p \wedge r) \vee (q \wedge r) = 0$. If $p = q = r = 0$, the data vector is a vector of 0s. If $f = 1$, the data are transferred into register, A. If $g = 1$, the data are transferred into B. If $f = g = 0$, no transfer takes place. If $f = g = 1$, the same data are transferred into both A and B. The implementation of Fig. 11.36 is given in Fig. 11.37 for 2-bit data vectors.

The network of Fig. 11.37 can be an implementation of a single step of a control sequence. Expression of the network in AHPL demands a more compact notation than the tabular form of Fig. 11.36. We use an asterisk to indicate a condition on either side of the transfer arrow.

$$S. \quad (A!B) * (f, g) \leftarrow (DATA1!DATA2!DATA3) * (p, q, r) \qquad (11.45)$$

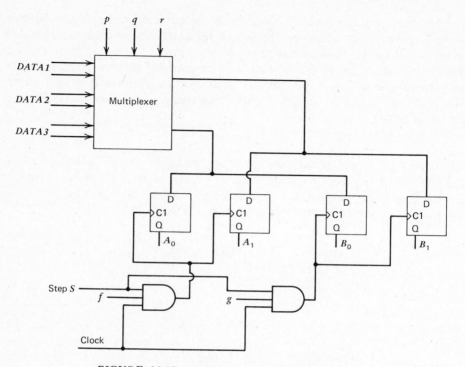

FIGURE 11.37   Realization of conditional transfer.

In the AHPL expression of Equation 11.45, the target registers and data vectors are separated by ! and the conditions that specify the clocking network and the data multiplexer are separated by commas. The target conditions and target registers are specified on the left of the arrow, and the data multiplexing network on the right. The transfer actually executed by Equation 11.45 depends on the conditions as given by Fig. 11.36.

We note that in addition to conditions f and g, the transfer specified by Equation 11.46 will take place only when control is at step S. This is enforced in Fig. 11.37 by the extra input labeled "Step S" on the two AND-gates that control the clocking of registers A and B.

In certain situations it is convenient to express a data multiplexing network independently of a transfer. Such a multiplexing network may be declared as a bus or an output. We use the symbol $=$ to denote a connection or data selection that is effective throughout the duration of a clock period. In Equation 11.46, for example, $ABUS$ is equal to $DATA1$ throughout any clock period for which step S is active and $p = 1$. Similarly $ABUS$ is $DATA2$ or $DATA3$ if q or r is 1.

$$S. \quad ABUS = (DATA1 \, ! \, DATA2 \, ! \, DATA3) * (p, q, r) \qquad (11.46)$$

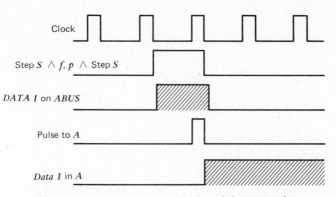

FIGURE 11.38   Timing of a bused data transfer.

The realization in Fig. 11.37 can also be described by the separate transfer and connection statements given in Equation 11.47.

$$S. \quad ABUS = (DATA1 \,!DATA2 \,!DATA3) * (p, q, r)$$

$$(A \,! B) * (f, g) \leftarrow ABUS \tag{11.47}$$

The timing of the execution of Equation 11.47 with $f = p = 1$ is given in Fig. 11.38. The separate declaration of a bus or multiplexer as required by Equation 11.47 is helpful in cases where the bus output fans out to a variety of targets that may be involved in separate transfer statements.

Example 11.8

The register configuration of Fig. 11.39a illustrates the principal data paths of a special-purpose memory unit that must accept a stream of data and addresses continuously but asynchronously placed in the 8-bit input register *INR*. The communications system providing data to *INR* will not be part of the design. We need only know the sequence and timing of data placed in *INR* and the relationship of three control lines, *ready*, *accept*, and *word*, to this system. The memory consists of 2^{16} 32-bit words, an associated 32-bit data register *DR* and a 16-bit address register *AR*. Each time a logical 1 is placed on line *write*, the word in *DR* is written into memory at the address currently located in *AR*. We will not be concerned with the internal implementation of the memory module. We will similarly ignore the read-from-memory capability, which we presume to exist using the same address and data registers.

The overall data transmission rate is sufficiently slow so that the memory interface for which an AHPL description is to be written will have ample time to deal with any 8-bit byte in *INR* before a subsequent byte is placed there. Each time a byte is placed in *INR*, a logical 1 will appear on line *ready* until the interface signals acceptance with a one-period level on line *accept*. The data and corresponding addresses will appear in the order given in Fig. 11.39b. The four data bytes that are to be catenated to form a 32-bit word appear first, beginning with the most significant 8 bits. The four data bytes are followed by the

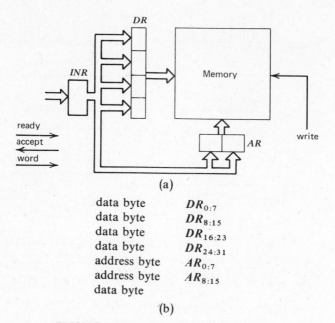

(a)

data byte	$DR_{0:7}$
data byte	$DR_{8:15}$
data byte	$DR_{16:23}$
data byte	$DR_{24:31}$
address byte	$AR_{0:7}$
address byte	$AR_{8:15}$
data byte	

(b)

FIGURE 11.39 *Communications memory.*

16-bit address in two consecutive bytes. Each address is followed by the first byte of the next data word, and so on. The *ready* signal corresponding to each first data byte will coincide with a logical 1 on the line *word*. Write a control sequence to describe operation of this control memory.

Solution   Prior to transferring a data byte from *INR*, it is necessary to check for a 1 on line *ready* and on line word since the data byte may be the first byte. This check and the subsequent transfer of a data byte are represented in the flowchart of Fig. 11.40. Also included is a mechanism for checking for the fourth data byte to alert the system that the next two bytes will form the address. We continue the sequence by checking for the *ready* = 1 signaling the presence of the first address byte in *INR*. This is followed by a similar sequence corresponding to the second address byte. With the complete address in *AR* we let *write* = 1 for one clock period prior to returning control to the beginning in anticipation of another data word.

Although the sequence of events for the interface is accurately reflected in Fig. 11.40, there remain some details that can only be specified in the AHPL description. First, it is convenient to check both *ready* and *word* in a single-branch statement.

1.

$$\rightarrow (\overline{ready}, \, ready \wedge word, \, ready \wedge \overline{word})/(1, 2, 3)$$

2. $CNT \leftarrow 0, 0$

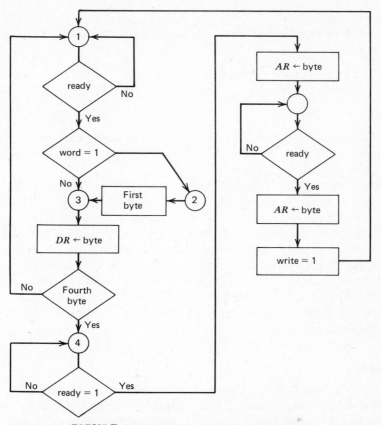

FIGURE 11.40 *Memory interface flowchart.*

In Step 2 a 2-bit register *CNT* is introduced. This counting register will be utilized to store the number of the byte to be transferred next into *DR*. If Step 2 is executed prior to Step 3, the latter step will cause *INR* to be transferred into the leftmost 8 bits of *DR*. Step 3 also increases the number in *CNT* by 1, informs the communications system that the byte in *INR* has been accepted, and effects a branch depending on the old value of *CNT*. If *CNT* contains (1, 1,), control proceeds to Step 4 to look for the first address byte. Otherwise control returns to Step 1 in anticipation of another data byte. As before Step 4 checks *ready*. If *ready* = 1, the first address byte is transferred into *AR* at Step 5. This procedure is repeated in Steps 6 and 7 for the second address byte. As indicated in the flowchart, the sequence is completed by the one period *write* = 1 in Step 8.

3. $(DR_{0:7} \, ! \, DR_{8:15} \, ! \, DR_{16:23} \, ! \, DR_{24:31}) * ((\overline{CNT}_0 \wedge \overline{CNT}_1),$
$(\overline{CNT}_0 \wedge CNT_1), (CNT_0 \wedge \overline{CNT}_1), (CNT_0 \wedge CNT_1)) \leftarrow INR;$
$CNT \leftarrow (CNT_0 \oplus CNT_1), \overline{CNT}_1; accept = 1$
$\rightarrow \overline{(CNT_0 \wedge CNT_1)}/(1)$

4. $\rightarrow \overline{(ready, ready)}/(4, 5)$
5. $AR_{0:7} \leftarrow INR$; $accept = 1$

6. $\rightarrow \overline{(ready, ready)}/(6, 7)$

7. $AR_{8:15} \leftarrow INR$; $accept = 1$

8. $write = 1$
    $\rightarrow (1)$ ■

Note in the above example that the outputs *write* and *accept* are specified only when they have a value other than 0. This follows the AHPL convention that *outputs are 0 unless specified otherwise*.

The above is not the only control sequence that will satisfy the design problem presented. For example, the similar tasks of selecting bytes for the data register and for the address register were handled in two distinct ways. A counter was provided to count through the 4 bytes to be loaded in *DR*. Two consecutive two-step sequences were employed to load the 2 address bytes in *AR*. The latter approach adds hardware to the control unit while avoiding a register analogous to *CNT* in the data unit. The larger the number of similar activities, the greater the likelihood that using a counter will be the best approach. Choosing the overall best approach from the array of alternatives is perhaps the most difficult step in the design process.

Problems

11.1 Construct the state diagram and state table of a sequential circuit with two pulse inputs, A and B, and a single level output, Z. Following a pulse on line B, the output Z is to be 1 provided that there have been an even number of pulses on line A, since the previous pulse on line B. Otherwise, a B pulse will reset the output to 0. The output will not change except on the arrival of a B pulse.

11.2. Determine the minimal state table of a sequential circuit, which has three pulse inputs x_1, x_2, and x_3, and a single level output Z. The output Z is to be 1 following a pulse x_1 until a total of at least four subsequent pulses have been observed on either of lines x_2 or x_3. Otherwise, the output of the circuit is to be 0. Any subsequent pulse on x_1 will return the output to 1 and cause the circuit to wait for four more pulses.

11.3. A sequential circuit is to have 3 pulse inputs A, B, and C and a single level output z. A pulse may appear on any one of the three lines at any time consistent with the pulse-mode assumption. The output is to be 1 if and only if the last 3 pulses in sequence have been AAA or ABA. Determine a state table for this sequential circuit.

11.4. A pulse-mode sequential circuit has four pulse inputs C, S_1, S_2, and N and two level outputs z_1 and z_2. The pulse-mode assumption is satisfied so that any two pulses are sufficiently separated in time to permit the circuit to react to each pulse independently. A pulse on line C will cause both z_1 and z_2 to go to 0. A pulse on line S_1 will cause z_1 to go to 1 without affecting z_2. A pulse on S_2 will cause z_2 to go to 1 without affecting z_1. A pulse on line N will complement the values of both z_1 and z_2. Determine a state diagram and a state table for this sequential circuit.

11.5. The state table of Fig. P11.5 has been developed for a pulse-input, pulse-output circuit. The indicated assignment of flip-flop states has already been

	Present	Next State, z		
$y_2\ y_1$	State	A	B	C
0 0	a	$c,1$	$d,1$	$d,1$
0 1	b	$d,0$	$d,1$	$c,0$
1 1	c	$d,0$	$b,0$	$a,1$
1 0	d	$c,1$	$a,0$	$b,0$

FIGURE P11.5

determined. Complete the design by determining minimal expressions for Z, J_{y2}, K_{y2}, J_{y1}, and K_{y1} for the following two hardware assumptions:
(a) *C1* input held at logical 1.
(b) J and K regarded as level inputs, pulses restricted to *C1* input.

11.6. Using J-K flip-flops, design a sequential circuit with three pulse inputs A, B, and C and a pulse output Z. The circuit is to yield an output pulse coinciding as nearly as possible with an input pulse if and only if the two previous input pulses have been on each of the other two lines. (For example, a pulse on line A will cause an output if the previous two pulses have been on lines B and C in either order.)

11.7. Determine circuit realizations of the state tables in Fig. 11.2 using J-K flip-flops.

11.8. A 20-μsec pulse delay line and a supply of NAND gates and J-K flip-flops are available for construction of the following circuit. Using these components, design a circuit with two pulse inputs, A and B, and a single pulse output, Z. The input pulses have widths of 0.1 μsec. For 20 μsec after each pulse on line A, each pulse on line B must generate an output pulse Z. Otherwise, the B pulses are ignored. Should one A pulse occur less than 20 μsec after another A pulse, then no further output pulses are to be generated until a third A pulse arrives that will not occur for at least 20 μsec. (*Hint:* Let the output of the delay line be a third input and construct a state table.)

11.9. A perfect 256-μsec pulse delay line is to be used as a memory for storing 256 bits of information in thirty-two 8-bit characters. Assume that all pulses are of 0.1 μsec duration. A circuit that will control the reading and writing of 8-bit characters to and from the memory is to be designed, using J-K flip-flops and NAND gates (negligible delay). The four inputs to the circuit are control lines, W and R, a data line, D, and a clock, C. The clock, which puts out 10^6 pulses per second, is to be used to synchronize the entry of data pulses on the delay line. Line D is a level input. Pulses will occur on lines W and R only intermediate between clock pulses. If a pulse appears on line W, the next eight bits appearing at the output of the delay line are to be replaced by pulses corresponding to the level of D at the time of the next eight clock pulses. In the absence of a write cycle, the circuit should cause data to be recirculated into the delay line. Following an R pulse, the next eight pulses appearing from memory are to constitute a circuit output. A W or an R pulse will never follow another W or R pulse by less than 8 μsec.

11.10. Use K-maps to derive Equations 11.20, 11.21, and 11.22.

11.11. Design in detail a counter that will count clock pulses modulo-16 if a level $L = 1$ or as a decade counter if $L = 0$.

11.12. Consider the flip-flop labeled c in Fig. P11.12 with its associated input logic as a control sequencer, and consider flip-flop a to be the only data

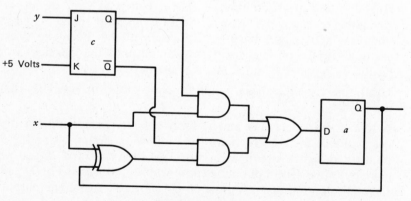

FIGURE P11.12

flip-flop in the sequential circuit. The control unit is a minimal realization and not a direct translation of the control sequence.

(a) Determine a state diagram of the control unit.
(b) Determine a complete control sequence for the sequential circuit.
(c) Construct the logic block diagram of a control sequencer that is a one-to-one translation of the above control sequence.

11.13. Familiarize yourself with the function of the above sequential circuit by studying the control sequence determined in Problem 11.12. Now synthesize the complete state diagram of a sequential circuit that will accomplish the same task. Determine a minimal realization of this state diagram using the techniques of Chapter 10, and compare its complexity with Fig. P11.12.

11.14. A straightforward realization of a design language step consisting of only a branch is given in Fig. P11.14. This circuit is effectively a 2-state sequen-

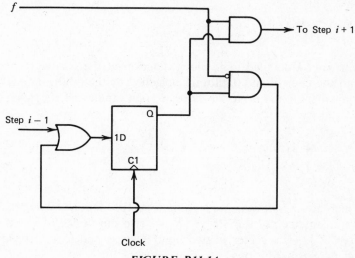

FIGURE P11.14

tial circuit. Use the techniques of Chapter 10 to realize this circuit in terms of

(a) A clocked R-S flip-flop
(b) A clocked J-K flip-flop

The inputs are the branch function f and a line from the step $i - 1$ flip-flop. The output leads to step $i + 1$.

11.15. Obtain an alternate realization of the 2-out-of-5 code checker described in
Problem 10.7 by first determining a control sequence. Base your approach
on a 3-bit counting register **BITS**, which will count the number of bits of
the 5-bit code word that have arrived thus far and a 2-bit counter **ONES**
that will count the number of these bits that have been 1s. Additional
memory elements may be added to the data and logic unit, if needed.
Consider the problem solved once a complete control sequence has been
obtained.

11.16. Construct a control unit realization corresponding to the following
AHPL sequence:

  1. $Z = X \vee A$
      $\rightarrow (a, \bar{a} \wedge b, \bar{a} \wedge \bar{b})/(2, 3, 4)$
  2. $A \leftarrow X$
      $\rightarrow (1)$
  3. $A \leftarrow X_{1:3}, X_0$
      $\rightarrow (4)$
  4. $A \leftarrow A_{1:3}, A_0$
      $\rightarrow (1)$

11.17. Construct a detailed logic block diagram of the hardware realization of
both the control and data units specified by the following AHPL description, where a and b are flip-flops, x is an input, and z a single output line

  1. $a \leftarrow x \vee b$
  2. $b \leftarrow x$
      $\rightarrow (a, \bar{a})/(1, 3)$
  3. $z = 1; b \leftarrow x \oplus b$
      $\rightarrow (1)$

11.18. Which of the following are legitimate AHPL action statements or parts of
action statements? Register A [2] has two flip-flops, B [3] and C [3] have
three, X [2] is a vector of two input lines, a and b are individual inputs,
and **BUS** is a 3-bit bus.

  (a) $b \leftarrow a$
  (b) $B \leftarrow ((A, a) ! (A, b)) * (C_0, \bar{C}_0)$
  (c) $(A_0, B_{0:1}) * a \leftarrow C$
  (d) $BUS \leftarrow C$
  (e) $B \leftarrow BUS + C$
  (f) $BUS = 0, 1, 0$
           $C \leftarrow BUS$
  (g) $B \leftarrow (B \oplus (A, a)) \oplus C$

(h) $a * B \leftarrow C$
(i) $B * a \leftarrow (X, b)$
(j) $\wedge / B, A \leftarrow C$

11.19. A digital system has an 8-bit vector, X, of input lines together with another input line *ready*. The system will have two output lines z and *ask*. Following each one period level placed on line *ask*, a new vector will become available on lines X. The availability of such a vector will be signaled by a one-period level on line *ready*. Each time an input vector is exactly the same as either of the two previous input vectors, the output z should be 1 for four clock periods prior to another signal on line *ask*. The line z should be 0 at all other times. Write an AHPL sequence describing this digital system. Define the data registers as required.

11.20. A random-sequence generator is driven by an external clock source. This generator provides a level output, z, which is constant between clock pulses, and may or may not change, on a random basis, when triggered by a clock pulse. A special-purpose computer is to be designed employing a 1-megahertz clock. This computer is to provide an output clock to drive the random process at a frequency of 1 kilohertz. The computer must also compute the number of level changes in the random process each second. The computer must also compute and display the average number of level changes per second over the first 2^8 seconds following the depression of its start button. Specify the necessary hardware, determine the control sequencer, and construct a block diagram of the control sequencer for this special-purpose computer. Accomplish division by shifting.

11.21. A stamp-vending machine dispenses 10¢, 15¢, and 25¢ packets of stamps and accepts nickels, dimes, and quarters. You may insert exact change or an excess amount and receive change. It is assumed you will not insert an excess amount if you have correct change, e.g., you might insert a quarter to pay for a 10¢ packet, but not a dime and a nickel. After inserting the coins, press one of three buttons to select the packet desired. If you have inserted enough money, you will receive the packet and any change due. If you have not inserted enough, your money will be refunded. The buttons are mechanically interlocked so that only one may be pressed at a time. The coins must be inserted one at a time into a single slot. A coin detector issues one of three pulses, N, D, or Q, for nickels, dimes, or quarters, respectively. The three buttons also emit pulses, X_1 for 10¢ packet, X_2 for 15¢ packet, X_3 for 25¢ packet. You are to design a pulse output circuit to control the dispensing of the packet and the change.

Output pulses should be emitted at the time of any X pulse according to the following code:

$$Z_1 = 10¢ \text{ packet} \qquad Z_2 = 15¢ \text{ packet} \qquad Z_3 = 25¢ \text{ packet}$$

$$Z_4 = 5¢ \text{ change} \qquad Z_5 = 10¢ \text{ change} \qquad Z_6 = 15¢ \text{ change}$$

$$Z_7 = \text{refund}$$

The circuit should reset following the X pulse.

Bibliography

1. Hill, F. J. and Peterson, G. R. *Digital Systems: Hardware Organization and Design*, 2nd Edition, Wiley, New York, 1978.
2. Chu, Y. "Why Do We Need Hardware Description Languages?" *Computer*, Dec. 1974, pp. 18–22.
3. Lipovski, G. J. "Hardware Description Languages," *Computer*, June 1977.

12

LOGICAL DESIGN USING MSI AND LSI PARTS

12.1 Introduction

In some quarters, it has been contended that computer-aided design and the use of *medium- and large-scale integrated circuits* (MSI and LSI) will invalidate traditional approaches to logical design and cause them to be replaced with new techniques. The authors do not fully agree with this point of view. It is true that the cost, uses, construction, and much of the design process for digital systems have been revolutionized. Many of the latter stages of the design process such as mask-making, printed-circuit-board layout, back-panel wiring, and testing have been automated. The initial, most creative, portions of the design process have been least affected.

It is also true that Boolean algebra, Karnaugh maps, and state tables are now less often used routinely as tools for minimization. These concepts are now models for understanding and communication. The larger and more complex digital systems become the greater the importance of a thorough understanding of sequential circuits becomes. Timing assumptions associated with the three types of sequential circuits are critical. Timing problems must be anticipated in advance. Design modifications are expensive. As we will see in Section 12.10, simulation becomes an important tool in verifying system timing in advance.

The design of combinational logic for implementation via custom LSI will not differ basically from the design for discrete component implementation. Where standard MSI parts are used, the designer's flexibility will be limited somewhat. Some sort of "macro-algebra" whose basic functions are realized by LSI parts is an intriguing idea, but none has been invented. Most special designs will require both individual gates and MSI parts.

A design language such as AHPL offers the best prospect for formalizing the process of design through interconnecting MSI parts. Nothing should be inferred from the term *language* beyond the usual in interpretation of the word. Neither AHPL nor any other hardware description language implies a built-in optimization procedure. This task is left to the engineer. AHPL offers a convenient, unambiguous means by which a design can be expressed and communicated to others. Where only a few integrated-circuit (IC) packages are used, optimization may focus on timing rather than circuit components.

In this chapter, we will develop AHPL as a tool for interconnecting IC parts. First, we require a clarification of terms used to identify these parts as well as a further development of some fundamentals of AHPL. These are the topics of the next two sections.

12.2   Definition of Terms

Integrated circuits were introduced in Chapter 5. Concisely, we define an IC as a circuit consisting of active and passive elements fabricated on a single semiconductor chip and mounted in an individual package. Technically, *chip* refers to the actual piece of silicon on which the circuit is fabricated, but the terms *chip* and *package* are often used interchangeably, e.g., when talking about chip-count or package-count. As noted in Chapter 5, digital ICs are often divided into categories of SSI, MSI, and LSI. The distinctions between these categories are usually stated in terms of the number of gates per package, less than 12 for SSI, 12 to 100 for MSI, more than 100 for LSI. From the standpoint of the designer, the mere statistic of gates per chip is of much less importance than the functional distinctions between circuits in the three categories.

The diagram of Fig. 12.1 will be helpful in clarifying the distinctions between these categories. In SSI we have the primitive logic elements, i.e., individual gates and flip-flops. They will usually be supplied with several elements in a package, but each unit is logically distinct, available for use independent of any other elements in the same package. In MSI we have functional units, sets of gates and flip-flops interconnected to perform definite functions, such as counting or decoding. While some advantages accrue from packing more elements into a single package, the main advantage of MSI relative to SSI derives from the fact

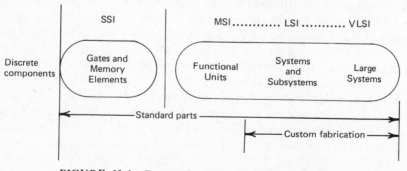

FIGURE 12.1   Forms of integrated circuit technology.

that the interconnections between elements are made on the chip. Consider the multiplexer of Fig. 8.23. The same multiplexing function could be realized with six SSI gate chips, at about the same cost for the chips themselves. With the SSI approach, however, we would have the cost of mounting six chips instead of one and the cost of all the wiring between the chips, resulting in a far more expensive realization. In LSI we have complete systems or subsystems, such as micro-processors and memories, typically made up of interconnections of functional units. A microprocessor, for example, includes registers, counters, decoders—elements that could be obtained individually in MSI form. Again, the big advan-tage is not so much in packing more elements on the chip as in putting all the interconnections on the chip. Recently a fourth category, very large-scale inte-gration (VLSI) has attracted a lot of interest. From the standpoint of the desig-ner, there is no significant functional distinction between LSI and VLSI—the systems are just bigger; for example, 16-bit microprocessors instead of 8-bit units. As suggested in Fig. 12.1, there is a sharp distinction between SSI and the other categories, while the distinctions among the other three categories are less distinct.

More important than drawing a fine line between MSI and LSI or VLSI is the distinction between standard parts and custom fabricated parts. As shown in Fig. 12.1, standard parts may be available in any category. Any digital circuit that a semiconductor manufacturer may consider to have profit potential will be manufactured and distributed as a standard IC part. Designs that subsequently turn out to be unprofitable are discarded. As certain designs become accepted, they are copied by other manufacturers.* Thus, the trend is toward ever greater standardization. Standard MSI parts include binary decoders, BCD decoders, decade counters, binary counters, shift registers, comparators, parity generators, etc. Sections of memory and read-only memory are the principal examples of standard LSI parts. The uniformity within these circuits makes for efficient fabrication and testing processes.

When a large number of copies of a given digital system are to be manu-factured, then custom fabrication of parts of the system as MSI or LSI may be considered. The designer should always consider the use of standard parts before resorting to custom fabrication. The trade-offs governing the choice of approach will be the topic Section 12.9. Once the designer has selected custom fabrication, some description of the design is submitted to a semiconductor manufacturer. Most vendors will accept logic block diagrams. The process is simplified for both customer and vendor when design information is transferred in the form of a

---

* This is not as sinister as it sounds. A customer is reluctant to adopt a part for a large-volume design unless the part can also be obtained from a *second source*. In second sourcing, the originator of a part licenses other manufacturers to produce the part.

hardware description language. Many vendors will ask for (but not demand) typical input output sequences for the various parts. These will assist the vendor in generating a test sequence for the chips.

For the designer, the jump from discrete components to ICs was an easy one. The language of switching theory had already accustomed the designer to thinking of logic circuits and memory elements as basic building blocks, without worrying about the components from which they were constructed. Mounting these logic elements in single packages was eagerly accepted once the processing capability was achieved by the manufacturer. Once a laboratory could make an IC, the natural improvements in processing led to the capability for manufacturing ever more complex digital circuits as IC parts. While welcoming this capability, the designer has not always been certain how best to utilize it.

As a first step in organizing the design process, we partition IC parts regardless of complexity into the following three classifications:

I.   Combinational logic unit.
II. Functional register.
III. Module.

Every IC part falls into one of these classifications, although the terms themselves may be used more generally. Any network composed strictly of combinational logic elements will be termed a *combinational logic unit*. An IC part that contains only combinational logic without memory may be referred to as a combinational logic unit or a Class I IC.

An IC part containing combinational logic and memory, but no internal sequential control, is a *functional register*, or a Class II IC. Among these are counters, shift registers, and partial arithmetic logic units, with accumulator flip-flops included. These parts may be treated as registers with one or more prewired transfers. Additional problems appear when more than one standard part must be catenated to form a register. Although a memory part is an array of registers, it may be treated as a functional register.

Any clocked digital IC part that cannot be classified as a combinational logic unit or functional register must be treated as a *module*. This term will also be used more broadly to refer to networks of IC parts featuring sequential control. A module may include a control unit that implements a sequence of AHPL steps. In contrast, a functional register executes a single AHPL step once each clock period in a manner determined by control signals connected from a source outside the functional register. A microprocessor is a module. In this book, we will view microprocessors only as elements to be interconnected within networks of IC parts. We will leave the subject of microprocessor programming to other books.

12.3 Interconnections

Whether a module, a functional register, or a combinational logic unit is a single IC part or not, it must be characterized by an identifiable set of input and output leads. As shown in Fig. 12.2a, combinational logic units may be nested inside modules, functional registers, and other combinational logic units. Functional registers may be nested within modeuls or other functional registers. Modules may not be nested within modules. In order to interconnect modules, we must include them within what we will loosely call a *structure* as shown in Figure 12.2b. We will not formalize the concept of a structure. For the syntax of this term, the reader is referred to Reference 1. Where it is necessary to use more than one module in this book, the descriptions will be shown together, and interconnections will be specified by identical names in the lists of input and output lines.

The inputs and outputs of a part must be declared in a hardware description. This is done in the declaration section. Adding some separators to mark the beginning and ending of individual system descriptions in a list, we have the formats given in Fig. 12.3.

From this point, we will develop the forms to be used in descriptions by a series of examples. For formal syntax, the reader is referred to References 1 and 2.

Systems of the type discussed in Chapter 11 are examples of modules. In that chapter we focused on the ⟨module body⟩ part of a module. For a complete description it is also necessary to declare memory elements, inputs, and outputs. To illustrate a complete AHPL description of a module, let us return to the communications memory of Example 11.8. Let the memory itself be a separate module, and let the module controlling the transfer of data be called MEMORY CONTROL. Since the register *INR* is not part of MEMORY CONTROL, let us

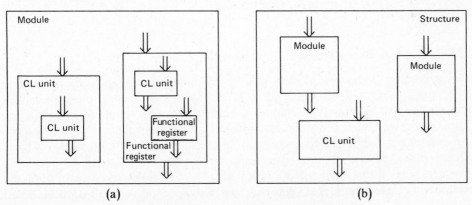

(a)   (b)

FIGURE 12.2   Nesting of Subnetworks.

MODULE: ⟨name⟩	CLUNIT: ⟨name⟩	FUNCTIONAL REG.: ⟨name⟩
⟨declarations⟩	⟨declarations⟩	⟨declarations⟩
⟨module body⟩	⟨unit body⟩	⟨body⟩
END	END	END

FIGURE 12.3 *Description formats.*

replace it by the input vector X [8]. Now we can write a formal module description of **MEMORY CONTROL** including declaration of inputs, outputs, and memory as follows:

MODULE: MEMORY CONTROL
    INPUTS: $X[8]$, *ready, word*
    MEMORY: $DR[32]$; $AR[16]$; $CNT[2]$
    OUTPUTS: AR; DR; *accept*; *write*
BODY
  1. $\rightarrow (\overline{ready}, ready \wedge word, ready \wedge \overline{word})/(1, 2, 3)$
  2. $CNT \leftarrow 0, 0$
  3. $(DR_{0:7}! \, DR_{8:15}! \, DR_{16:23}! \, DR_{24:31})*((\overline{CNT_0} \wedge \overline{CNT_1}),$
      $(\overline{CNT_0} \wedge CNT_1), (CNT_0 \wedge \overline{CNT_1}), (CNT_0 \wedge CNT_1) \leftarrow X;$
      $CNT \leftarrow (CNT_0 \oplus CNT_1), \overline{CNT_1}; accept = 1$
      $\rightarrow ((\overline{CNT_0 \wedge CNT_1}), (CNT_0 \wedge CNT_1))/(1, 4)$
  4. $\rightarrow (\overline{ready}, ready)/(4, 5)$
  5. $AR_{0:7} \leftarrow X; accept = 1$
  6. $\rightarrow (\overline{ready}, ready)/(6, 7)$
  7. $AR_{8:15} \leftarrow X; accept = 1$
  8. *write* $= 1$
      $\rightarrow (1)$
END

12.4 Logical Primitives and Combinational Logic Units

The preceding module description specifies all operations in terms of the Boolean primitives, AND, OR, Ex-OR, and NOT. As we saw in the last chapter, such a description can be directly translated into a realization in terms of gates realizing these primitives. Such a form of description is completely general since any logic operations may be described in this way, but it may not be a convenient form in many cases. Several logical operations are basic in the sense of occurring frequently in all sorts of systems, but are also quite complex in terms of the logic required to realize them. Having to describe such operations in full Boolean detail every time they occur can be very inconvenient.

As an example, the vector CNT in the preceding sequence is a two-bit counter, counting in the sequence

$$00 \to 01 \to 10 \to 11 \to 00 \to 01 \to \text{etc.}$$

In the sequence, the statement

$$CNT \leftarrow (CNT_0 \oplus CNT_1), \overline{CNT_1} \tag{12.1}$$

is a correct AHPL description of the increment operation, state in terms of the next value of the counter that must be transferred into the counter to implement the counting operation. This same form can be extended to counters of any number of bits, but the equations get very complicated for large counters. Since counting is such a common operation in digital systems, we provide for simplifying this notation by defining the *logical primitive* INC (increment). We use this logical primitive to represent the counting operation in AHPL statements of the form

$$CNT \leftarrow \text{INC}(CNT) \tag{12.2}$$

in which the register can have any name and can have any number of bits, as indicated by the declaration of the register.

Another operation that is frequently encountered is the decoding operation, discussed in Chapter 8. As an example of its use, suppose one step of a control sequence calls for using the contents of a 3-bit register B to select a particular bit from an 8-bit register A. That is, if $B = 0, 0, 0$, bit A_0 is desired, and so on. Assuming the selected bit is to be placed in a flip-flop, a detailed expression for the transfer statement is given by Equation 12.3.

$$r \leftarrow ((\bar{B}_0 \wedge \bar{B}_1 \wedge \bar{B}_2) \wedge A_0) \vee ((\bar{B}_0 \wedge \bar{B}_1 \wedge B_2) \wedge A_1) \vee ((\bar{B}_0 \wedge B_1 \wedge \bar{B}_2) \wedge A_2)$$
$$\vee ((\bar{B}_0 \wedge B_1 \wedge B_2) \wedge A_3) \vee ((B_0 \wedge \bar{B}_1 \wedge \bar{B}_2) \wedge A_4) \vee ((B_0 \wedge \bar{B}_1 \wedge B_2) \wedge A_5)$$
$$\vee ((B_0 \wedge B_1 \wedge \bar{B}_2) \wedge A_6 ((B_0 \wedge B_1 \wedge B_2) \wedge A_7) \tag{12.3}$$

Suppose instead that we assign a name to the 8-bit output vector of a three-input decoding network and use this name in place of the logical expression for the vector in the control sequence. That is, we will use the notation DCD (B) to refer to the output of the combinational logic network shown in Fig. 12.4. In the future, we shall use DCD (R) to represent the 2^n output leads of any n-to-2^n decoding network with R as the n-bit input vector. We may now use the compact expression 12.4 in place of 12.3

$$r \leftarrow \vee/(\text{DCD}(B) \wedge A) \tag{12.4}$$

The function DCD is a vector of minterms. An individual minterm i of a vector variable X may be written $\text{DCD}_i(X)$. We assume that the primitive $\text{DCD}(X)$ is available for X of any dimension.

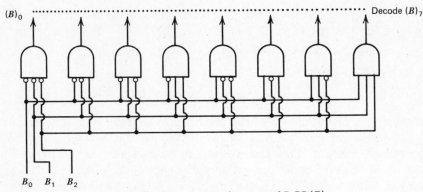

FIGURE 12.4   *One realization of DCD(B).*

Another basic operation for which it is convenient to define a logical primitive is binary addition. In Fig. 8.7 we showed a design for a binary adder. This figure is repeated as Fig. 12.5 with the inputs and outputs relabeled to conform to AHPL conventions. We could describe such a circuit in terms of Boolean expressions, but such expressions would be very complicated. Instead we define the logical primitive ADD to represent the binary addition of two n-bit vectors and a carry-in, producing an $(n + 1)$-bit sum.

Suppose we have a sequence in which we wish to add the contents of two registers, *MD*, and *AC*, with no input carry, placing the sum in *AC* and an overflow flip-flop, *If*. This operation can be described by the AHPL statement.

$$If, AC \Leftarrow \text{ADD}(AC, MD, 0) \tag{12.5}$$

We have shown circuits for decoders and adders to clarify the meaning of the logical primitives, but it is important to understand that the use of a logical

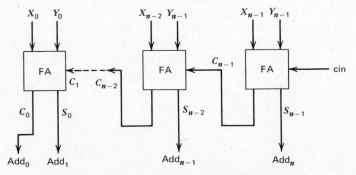

FIGURE 12.5   *One realization of a binary adder.*

primitive does not imply any particular circuit. When we use the primitive DCD in a sequence, it indicates that the operation of n-to-2^n-line binary decoding is to be carried out, but it does not imply any specific circuit. If we are using the sequence for simulation, i.e., as the input to a computer program that will simulate the operation of the system, then it is not necessary to specify the exact circuit. The use of DCD will simply signal the software to simulate the operation of decoding. If we want to translate the sequence into hardware, however, it is clearly necessary to specify the circuitry to be used.

Complex functions such as decoding and addition are commonly realized by combinational logic units in MSI form, which we will refer to as *CLUNITS*. If we wish to use a specific CLUNIT, we need a means to describe it in AHPL form, and a means to indicate its use in an AHPL sequence. As a first example, consider the SN74154 Decoder/Demultiplexer. Unlike an ordinary decoder, this device has two sets of inputs, 4-bit decoder inputs A and 2-bit data or enable inputs B. If B_0 and B_1 are both held low, the device functions as a conventional decoder of A, except that the output lines are inverted. If the B inputs are not held low, the device functions as a multiplexer, with the output selected by A following $B_0 \vee B_1$. The formal description of the behavior of this device in AHPL is as follows:

CLUNIT: IC74154(A; B)
  INPUTS: $A[4]$; $B[2]$
  OUTPUTS: $DEMUX[16]$
BODY
  $DEMUX = \overline{\text{DCD}(A) \wedge \overline{(B_0 \vee B_1)}}$
END

Note the use of the logic primitive DCD in the description. This does not imply the existence of separate decoding logic of the same general form of the circuit of Fig. 12.4; it simply shows that the circuit behavior is consistent with the use of a 4 to 16-line decoder followed by ANDing with the data variables and inversion at the outputs. Although a study of the actual circuit of the MSI part may be helpful in setting up the AHPL description of the part, the circuit details are unimportant in using the part. The situation is the same when using an MSI part as when using an individual gate; we need to know *what* the part does, but not *how* it does it.

Another point to be noted about this description is the naming of the inputs and outputs. In the same manner as variables in FORTRAN subroutines, the names serve as "dummy variables," representing the actual lines that will be connected when the part is used in a larger system. There is no requirement that the inputs be named A or B, or the outputs $DEMUX$. Now assume that we have a sequence describing a module in which we wish to use the 74154 to implement a 4 to 16-line decoding operation. First a declaration is added indicating that this device is being used.

$$\text{CLUNITS: DECODE4} < : \text{IC74154}$$

To indicate a specific application or *invocation* of the device in a sequence we may write a statement such as the following:

$$8. \ r \leftarrow \vee /(\text{DECODE4}(MC; 0, 0) \wedge AC) \qquad (12.6)$$

The declaration symbol, $<:$, specifies that any appearance of DECODE4 represents the output lines from a 74154 IC. The input lines are specified by the arguments used in the invocation. More than one application of DECODE4 with the same arguments will be implemented by one device with fan-out from the output lines. Referring to Equation 12.6 and the combinational logic unit description for the 74154, we see that MC provides the A inputs; and the B inputs will be held low, so that the device will function as a decoder. DECODE4 represents the 16-bit output $DEMUX$ in the description. The CLUNIT description must be included as a part of the system description along with the sequence describing the module in which it is used. In using CLUNITS in sequences we may use any name (such as DECODE4) that we like, except DCD, ADD, and INC, which are reserved for the logical primitives.

12.5 Functional Registers

While counting can be implemented using a register and separate increment logic, this is rarely done. Instead we use counters, MSI parts including register flip-flops and the necessary counting logic. Counters are examples of *functional registers*, which may be regarded as the sequential counterparts of CLUNITS. If we wish to use functional registers in the implementation of a module we have the same problems we have with CLUNITS. We must describe the behavioral characteristics of the part, we must be able to indicate that it is to be used in a module, and we must specify specific instances of its use.

Although functional registers may consist of more than one IC package, let us consider the general single-package functional register whose block diagram is depicted in Fig. 12.6. We separate the I/O pins into classes as shown. We will assume the class of pins including supply voltages and ground to be connected according to specification. These lines as well as the clock are not specifically treated in an AHPL description of the package.

Transfers may take place within the package synchronized by the clock as gated internally by signals on the control input lines. The targets of transfers are memory elements within the package. The origins of transfers may be these same memory elements or the data input lines. Unclocked transfers including setting,

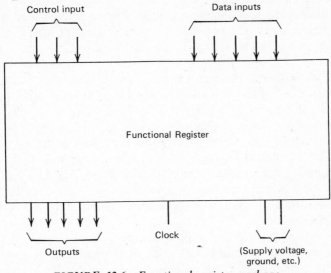

FIGURE 12.6   Functional register package.

clearing, and, in some cases, loading constants into memory elements may also be defined for functional registers. The clock line shown in Fig. 12.6 may be connected directly to the system clock or may be gated with a control level to invoke the package conditionally. The designer must take care to observe the rules articulated in Section 11.3 when specifying unclocked transfers.

As a first illustration of the description of a functional register and eventually its application, let us consider the SN 74163 four-bit counter. This counter consists of a vector of four flip-flops that may be incremented, cleared, or loaded with the values of a 4-bit data input vector. The AHPL description of the 74163 is given in Fig. 12.7.

FUNCTREG: IC74163(*notload, enp, notclear*; *DATA*)
  DATA INPUTS: *DATA*[4]
  CONTROL INPUTS: *notload, enp,ent,notclear*
  OUTPUTS: *COUNTER*[5]
  MEMORY: *D*[4]
BODY
  $D*((enp \land ent) \lor \overline{notload}) \lor \overline{notclear}) \leftarrow (INC(D)! \, DATA! \, 0, 0, 0, 0)$
      $*(ent \land enp \land notload \land notclear, \overline{notload} \land notclear, \overline{notclear});$
    $COUNTER = ent \land (\land /D),D$
END

FIGURE 12.7   Four-bit counter (SN74163).

A check of a manufacturer's data manual will reveal that the control lines represented by *notclear* and *notload* are active low or active 0 signals. As shown here our convention will be to preface the natural name of such signals with *not*.

As was the case with the CLUNIT discussed earlier, we are not concerned with the internal structure. Our concern is to describe the functioning of the part as compactly as possible. The result is a list of transfers that will take place in response to a clock pulse, with the nature of the transfers determined by the control signals. If $enp = ent = notload = $ and $notclear = 1$ at clock time, the counter will be incremented. If $notload = 0$ $notclear = 1$, the clock pulse will load the values on the $DATA$ lines into the counter, regardless of the values of enp and ent. If $notclear = 0$, the clock pulse will clear the counter to 0, 0, 0, 0, regardless of the value of the other signals. Note in the first line of Fig. 12.7 that the control variables are listed at the left of a semicolon, with the data variables at the right. This same ordering of variables will apply when the use of the counter is invoked in a control sequence.

As an example of the use of this part, assume a sequence in which we need a 4-bit counter to count the number of shifts of a data word in another register. We assign the name $SHFTCNT$ to this counter and indicate that it is to be implemented with a 74163 by the declaration

$$\text{FUNCTREG: } SHFTCNT <: \text{IC74163}$$

This declaration assigns the name $SHFTCNT$ to the outputs of a 74163 as specified in the sequence describing the part. These outputs may then be used as a source in any transfer statement. For example, the statement

$$MC \leftarrow SHFTCNT_{1:4}$$

will transfer the low-order 4 bits of the counter output to a 4-bit register MC.

Changing the value of the data in the counter will be specified by an *invoke* statement, requiring notation of the following form:

$$SHFTCNT(1, a, 1, 1; A) \leftarrow \qquad (12.7)$$

In this case, the counter is incremented if the variable $a = 1$. It will be noted from Fig. 12.7 that the functional register IC74163 will be incremented when $enp \wedge ent = 1$. For this particular operation on $SHFTCNT$, enp, $ent = a$, 1 so that $enp \wedge ent = a$. Thus the effect on the four memory elements within $SHFTCNT$ of the operation in Equation 12.7 is the same as would be that of Equation 12.8, if $SHFTCNT$ had been declared as a 4-bit vector of memory elements.

$$SHFTCNT*a \leftarrow \text{INC }(SHFTCNT) \qquad (12.8)$$

It will be noted that the 4-bit data vector A had no effect on this particular invocation. Any 4-bit vector could have been entered in Equation 12.7 in place of A.

As a further illustration, consider the AHPL module SYSTEM, which is partially described by Fig. 12.8a. Only declarations related to the particular 4-bit counter, *SHFTCNT*, and operations on this functional register are shown. Note that control and data inputs of *SHFTCNT* are listed only in the invocation statements and not the declaration. In Step 1

$$\textit{notload, enp, ent, notclear} = 1, 0, 1, 0 \tag{12.9}$$

calling for a clearing of the counter. In Step 4, **load** $= \bar{a}$ and **enp** $= \bar{a}$, so the values on the *ABUS* are loaded into the counter if $a = 1$, and the counter is incremented if $a = 0$. Steps 10 and 15 are unconditional load and increment statements, respectively.

The connections at the inputs to the *SHFTCNT* package are given in Fig. 12.8b. The data wiring is explicit with *DATA* always connected to the declared bus *ABUS*. The double lines represent 4-bit vectors implying that each AND or OR-gate shown in the data network represents *a vector of* 4 gates. The counter is loaded with the values on the *ABUS* by step 10 and by step 4 when $a = 1$. Since the **notload** input is active low, we have

$$\textit{notload} = \overline{(\text{step } 10) \vee ((\text{step } 4) \wedge a)} \tag{12.10}$$

as wired in Fig. 12.8b. We note that **ent** is 1 in all four steps so this line is connected to $+5$ volts. Thus incrementing the counter will be controlled by **enp**. We see from the control sequence that

$$\textit{enp} = (\text{step } 4) \wedge \bar{a} \vee (\text{step } 15) \tag{12.11}$$

This expression is wired in Fig. 12.8b.

Throughout this chapter we will assume that the designer can routinely construct the optimal wiring of a logic network by referring to a complete AHPL description. This translation process was illustrated by the above discussion if Fig. 12.8. Computer programs have been written to accomplish this translation process automatically, but this topic is beyond the scope of this book. For a discussion of an automatic hardware compiler, the reader is referred to Chapter 7 of Reference 2.

12.6 A Small Library of Standard Parts

A design automation system that might be based on AHPL would be expected to incorporate a library of AHPL descriptions of frequently used MSI parts. These descriptions would be used as references by the user in a design of mod-

MODULE: SYSTEM
  INPUTS: a; $A[4]$; $B[4]$
  ⋮
  FUNCTREG: $SHFTCNT <$: IC74163
  ⋮
  BUS: $ABUS[4]$
  ⋮
BODY
  1. $SHFTCNT$ $(1, 0, 1, 0;$ $ABUS) \leftarrow$
      ⋮
  4. $ABUS = A$; $SHFTCNT$ $(\bar{a}, \bar{a}, 1, 1;$ $ABUS) \leftarrow$
      ⋮
  10. $ABUS = B$; $SHFTCNT$ $(0, 0, 1, 1;$ $ABUS) \leftarrow$
      ⋮
  15. $SHFTCNT$ $(1, 1, 1, 1, ABUS) \leftarrow$
      ⋮

END

(a.)

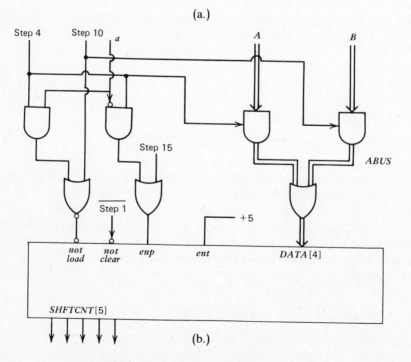

(b.)

FIGURE 12.8   *Wiring to functional register.*

ules and structures. The design automation system would usually also include a function level simulation program (see Section 12.10) that simulates the behavior of a module using the AHPL description. This simulation would incorporate the library descriptions of functional registers and CL units and would call them in much the same way that library subroutines are called by an ordinary programming language.

In order to pursue some design examples that will include the interconnection of MSI parts, it is necessary to have a small library of AHPL descriptions for reference. It would be pointless to try to develop a comprehensive library such as might be found in an industrial design automation system. Instead, we will focus on a short list of packages that will arise in the examples in the next section. The list will include a decoder/demultiplexer, a multiplexing bus, a 4-bit adder, a magnitude comparator, a 4-bit counter, a four-bit shift register, and a scratchpad memory. The CLUNITS are listed in Table 12.1, the FUNCTREGS in Table 12.2.

The first part in Table 12.1 is the decoder/demultiplexer previously discussed. The next unit, IC74153, forms two bits of a busing network with up to four data input vectors. The third unit in Table 12.1 uses the ADD primitive to describe a 4-bit adder that can be connected as part of a larger adder network. The last logic unit compares the magnitude of two 4-bit vectors A and B. If the binary magnitude of A is largest, $CMPR_0 = 1$. If B is largest, then $CMPR_2 = 1$. If $A = B$, then the outputs will assume the values of the input vector $C[3]$. $CMPR_1$ will be 1 only if $A = B \ AND \ C_1 = 1$.

The first functional register of Table 12.2 is the 4-bit counter described in Fig. 12.7. The second is a 4-bit shift register that can be loaded serially from the left or the right or in parallel from the values of $DATA[4]$, depending on the control inputs S_0 and S_1. If $S_0 = S_1 = 0$, no clocked transfer will take place. If *notclear* = 1 the four memory elements will be cleared asynchronously. The symbol ($\leftarrow\circ$) represents an asynchronous (unclocked) operation. In this case, when *notclear* goes low, the register is immediately cleared, regardless of the values of any other signals, just as with single flip-flops. Note that both the 74163 counter and the 74194 shift register are triggered by the rising edge of the clock just as are most edge-triggered D flip-flops. Designation of the triggering clock edge has not been incorporated in the language. A short discussion of clock wiring for standard parts will be provided in Section 12.7

The last functional register of Table 12.2 is a random-access memory chip storing 256 four-bit words. This is our first example of a two-dimensional array. Notice that the number of rows is given first in the memory declaration. The contents of the memory location addressed by input vector A is connected to the bus $DATA$, if the chip is enabled by higher-order address bits ($ce = 1$), $rw = 1$ (for read), and $od = 1$. An unclocked transfer from $DATA$ into the addressed

Table 12.1

IC74154.   4-line to 16-line decoder/demultiplexer

CLUNIT: IC74154(A; B)
  INPUTS: A[4]; B[2]
  OUTPUTS: $DEMUX$[16]
BODY
  $DEMUX = \overline{\mathrm{DCD}(A) \wedge (\overline{B_0 \vee B_1})}$
END

IC74153.   Dual 4-line to 1-line multiplexer

CLUNIT: IC74153(A; B; C; D; S; e)
  INPUTS: A[2]; B[2]; C[2]; D[2]; S[2]; e
  OUTPUTS: $TWOBUS$[2]
BODY
  $TWOBUS = ((\vee/(\mathrm{DCD}(S) \wedge (A_0, B_0, C_0, D_0)),$
  $\qquad\qquad (\vee/(\mathrm{DCD}(S) \wedge (A_1, B_1, C_1, D_1)))) \wedge e$
END

IC7483.   4-bit binary adder

CLUNIT: IC7483(A; B; *carry*)
  INPUTS: A[4]; B[4]; *carry*
  OUTPUTS: $ADDFOUR$[5]
BODY
  $ADDFOUR = \mathrm{ADD}(A; B; carry)$
END

IC7485.   4-bit magnitude comparator

CLUNIT: IC7485(A; B; C)
  INPUTS: A[4]; B[4]; C[3]
  OUTPUTS: $CMPR$[3]
BODY
  $D = A \wedge B$;
  $E = \overline{A} \wedge B$;
  $F = \overline{D \vee E}$;
  $CMPR_0 = D_0 \vee (F_0 \wedge D_1) \vee ((\wedge/F_{0:1}) \wedge D_2) \vee ((\wedge/F_{0:2}) \wedge D_3) \vee (\wedge/F) \wedge C_0$;
  $CMPR_1 = (\wedge/F) \wedge C_1$;
  $CMPR_2 = E_0 \vee (F_0 \wedge E_1) \vee ((\wedge/F_{0:1}) \wedge E_2) \vee ((\wedge/F_{0:2}) \wedge E_3) \vee (\wedge/F) \wedge C_2$
END

Table 12.2

IC74163.   Synchronous 4-bit counter

FUNCTREG: IC74163(*notload, enp, ent, notclear; DATA*)
  DATA INPUTS: *DATA*[4]
  CONTROL INPUTS: *notload, enp, ent, notclear*
  OUTPUTS: *COUNTER*[5]
  MEMORY: *D*[4]
BODY
  $D*((enp \wedge ent) \vee \overline{notload} \vee \overline{notclear}) \leftarrow (\text{INC}(D)! \, DATA ! \, 0, 0, 0, 0)$
    $*(ent \wedge enp \wedge notload \wedge notclear, \overline{notload} \wedge notclear, \overline{notclear});$
    $COUNTER = ent \wedge (\wedge/D), D$
END

IC74194.   *4-bit bidirectional universal shift register*

FUNCTREG: IC74194(*S, notclear; l, r, DATA*)
  DATA INPUTS: *l, r, DATA*[4]
  CONTROL UNITS: *S*[2], *notclear*
  OUTPUTS: *SHFTREG*[4]
  MEMORY: *D*[4]
BODY
  $D*((S_0 \vee S_1) \wedge notclear) \leftarrow ((D_{1:3}, \, l)! \, (r, D_{0:2})! \, DATA)*\text{DCD}_{1:3}(S);$
  $D*notclear \leftarrow \circ \, 0, 0, 0, 0;$
  $SHFTREG = D$
END

IC2111.   1024-bit static RAM

FUNCTREG: IC2111(*ce, od, rw; A, DATA*)
  DATA INPUTS: *A*[8]
  BUS: *DATA*[4]
  CONTROL INPUTS: *ce; od; rw*
  MEMORY: M[256,4]
BODY
  $DATA = \text{M}*(\text{DCD}(A) \wedge ce \wedge od \wedge rw);$
  $\text{M}*(\text{DCD}(A) \wedge ce \wedge \overline{rw}) \leftarrow \circ DATA$
END

---

* Triggered by rising edge of clock.

word takes place if $ce = 1$ and $rw = 0$. The notation for selecting a row of memory for a write or read is the same as the conditional transfer and conditional connection defined in Section 11.9, if "!" is interpreted to mean row catenation. COMBUS denotes a bus that can be used for input and output. The number of bits stored in this particular memory part is small in comparison to what is currently available, but functional register descriptions of more complex static memory parts would be quite similar.

12.7 Rising or Falling Edge-Triggering

The reader may have noticed that while we assumed in Chapter 11 that memory elements would be triggered by the falling edge of the clock, edge-triggered *D* flip-flops and many MSI parts respond to the rising edge of the clock. Of primary importance is the fact that *every memory in a clocked system must respond to the same edge of the system clock whether it be rising or falling.* We prefer to use the falling edge of the clock so that control gating of the clock pulse itself can be accomplished by a simple AND-gate. This is illustrated by (*control* ∧ *clock*) in Fig. 12.9a.

Clearly the rising edge of $\overline{clock}$ occurs at the same time as the falling edge of *clock*, as shown in Fig. 12.9a. The rising edge of $\overline{control \wedge clock}$ also occurs at this same point in time. Thus, memory elements responding to any of the five labeled edges will be effectively synchronized by the falling edge of *clock*. Typical approaches to implementing triggering synchronized to the falling edge of a clock are illustrated in Fig. 12.9b. Those rising-edge-triggered devices that are unconditionally clocked, such as functional registers and control flip-flops, are

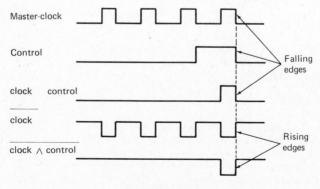

FIGURE 12.9a

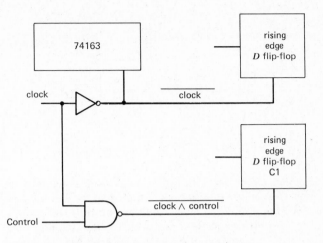

FIGURE 12.9b   Synchronized clocking.

connected to the clock through an inverter. Where gating of the clock by a control level is required, we merely gate the clock using a NAND-gate rather than an AND-gate.

12.8   Some Design Examples

In this section, we will consider a series of design examples that will be solved using AHPL. In each case, we will attempt to make as much use as possible of MSI parts. The examples were chosen to take advantage of the combinational logic units and functional registers tabulated in Tables 12.1 and 12.2.

Example 12.1

A serial-to-parallel interface unit is to be designed. This device will accept 8 consecutive serial bits and make them available as a vector on lines *DATAOUT*. A control input, *cntrl*, and a control output, *ready*, are provided to establish the beginning of an 8-bit byte. A block diagram of the module is shown in Fig. 12.10.

After a sequence of two consecutive 1s appears on the line *cntrl*, the unit should start shifting data from the *data* line into an 8-bit shift register. If the bit on *cntrl* immediately following the sequence of 11 is a 0, the data are to be shifted left. If the bit is a 1, the data should be shifted right. After 8 bits of data have been received, the circuit should resume checking for another sequence of 11 on the control line. Data should not be shifted while the unit is waiting for the sequence of 11 on the control line. The line, *ready*, should stay high while the unit is waiting for the 11 sequence. It should go low as soon as the sequence is detected and should go high only after the 8 bits of data have been shifted.

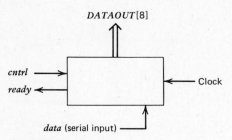

FIGURE 12.10   Serial parallel interface.

Solution   The design will be based on two 74194 four-bit shift registers as well as a 74163 counter to count the number of serial bits accepted. Since it is only necessary to increment the counter from 0 to 7, the most significant bit of D will never be affected by the counting process. We may therefore use this bit to store the direction of shifting. The following is an AHPL description of the module. The LABEL statement is used in this case to name constants making the invocation notation for loading and incrementing the counter more descriptive.

MODULE: SPINTERFACE
   INPUTS: *data*; *cntrl*.
   OUTPUTS: *DATAOUT*[8]; *ready*.
   FUNCTREG: *CNT* <: IC74163;
   A <: IC74194; *B* <: IC74194
   LABELS: *INCR* = 1, 1, 1, 1; *LOAD* = 0, 0, 0, 1.
BODY
  1. *ready* = 1; → $(\overline{cntrl})/(1)$
  2. *ready* = 1; → $(cntrl)/(1)$
  3. CNT (*LOAD*; *cntrl*, 0, 0, 0) ←
  4. $A(CNT_1, \overline{CNT_1}, 1; B_0, data, XXXX)$ ←
      $B(CNT_1, CNT_1, 1; data. A_3, XXXX)$ ←
      $CNT(INCR; XXXX)$ ←
      → $(\wedge/CNT_{2:4}, \ \wedge/CNT_{2:4})/(1, 4)$
END SEQUENCE
DATAOUT = *A*, *B*
END

The first two steps check the line *cntrl* for two successive 1s. When these 1s are received control goes to Step 3 where the received direction of shift is loaded into the most significant bit of the counter while the remaining 3 bits are cleared to 0 by the same LOAD operation. Step 4 is then executed eight times to shift the eight bits into A, B. The last shift is accomplished when $CNT_{2:4} = 1, 1, 1$ by the same clock edge that returns the module to Step 1. An inspection of Table 12.2 will reveal that the shift will be left if $CNT_1 = 0$,

$$A, B \leftarrow A_{1:3}, B, data \qquad\qquad (12.12)$$

and right if $CNT_1 = 1$,

$$A, B \leftarrow data, A, B_{0:2} \qquad\qquad (12.13)$$

for eight successive clock periods. The three occurrences of XXXX in Step 4 indicate that the remaining data inputs to the functional registers are don't cares. The statements between END SEQUENCE and END are in effect during every clock period.

A logic block diagram of the data portion of the module SPINTERFACE is given in Fig. 12.11. Specific IC pin connections are not shown. In addition to the three IC parts, only one NAND-gate is shown as part of the data unit and this gate could easily have

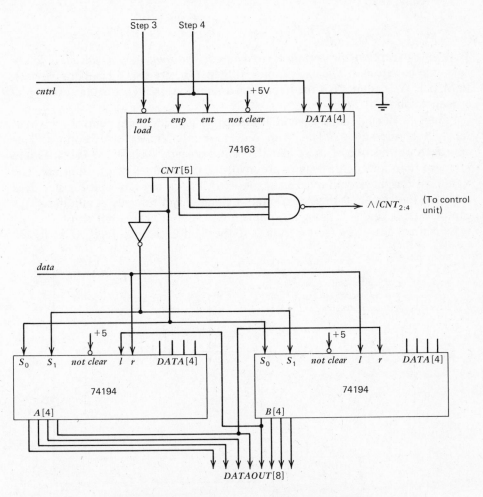

FIGURE 12.11   *Data portion* SPINTERFACE.

been considered part of the control unit. Not shown is a necessary inverter for the master clock line according to the discussion of Section 12.7. ∎

The realization of a control unit for a module such as that of Example 12.1 may be approached in several ways, two of which were discussed in Section 11.6. We note from the following branching portion of the control sequence of the module that this control sequence is a special case.

1. $\rightarrow \overline{(cntrl)}/(1)$
2. $\rightarrow (cntrl)/(1)$
3. $\rightarrow (4)$
4. $\rightarrow (\wedge/CNT_{2:4}, \overline{\wedge/CNT_{2:4}})/(1, 4)$

With one exception *control always remains in the same step or advances to the next step in sequence.* The exception is at Step 2, where control can branch back to Step 1. We assume that the steps form a loop so that Step 1 is the next step in sequence following Step 4.

Figure 12.12 suggests a general approach to implementing control units that satisfy the rule stated above with only a small number of exceptions. This approach utilizes one or more loadable shift registers, such as the 74194. Where no more than four control states are involved, one shift register is sufficient. The rightmost output is connected to input, r, to form a circular shift register. It is intended that the shift register contain a single 1 and three 0s at all times. This can be established by loading a suitable initial state into the shift register.

As control advances, the 1 circulates to the right in the shift register. As shown

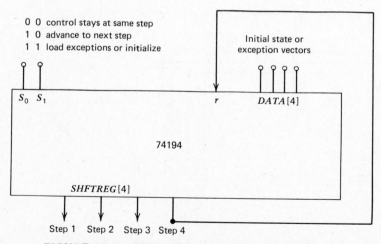

FIGURE 12.12   Shift register realization of control unit.

in Fig. 12.12, this is accomplished in the 74194 for each clock period in which S_0, $S_1 = 1, 0$. If S_0, $S_1 = 0, 0$, control remains at the current step. If S_0, $S_1 = 1$, 1, the control unit is set to the vector appearing on the four lines labeled *DATA*. Ths last option may be used to load an initial state, or to cause the control unit to branch to an exceptional next state (other than the current state or next state in sequence). If there is more than one initial state or exceptional state, a ROM-like set of OR-gates will be required at the *DATA* inputs. As we will see in the following example, an adaptation of the Karnaugh map may be used to determine the values of S_0 and S_1.

Example 12.2

Obtain a realization of the control unit of Example 12.1 using a single 74194 shift register.

Solution From the control sequence given above, we note that the only exceptional next control state occurs at Step 2, where control can branch to Step 1. Step 1 is also the initial state so the vector 1, 0, 0, 0 can be permanently connected to *DATA*, as shown in Fig. 12.13b.

To obtain the logic expressions for the control inputs $S[2]$ to the 74194, we refer to the Karnaugh-like map of Fig. 12.13a. Note that there are five rows, one for each control step and the first row, which is labeled *controlreset*. The control unit must be realized in such a way that a 1 on line *controlreset* will always cause control to be returned to the initial state. If a reset is not to take place, we assume that the system is in one of the four numbered control states.

The *controlreset* row of the map is 11 in each column, since a reset must be accomplished regardless of input. For the row corresponding to Step 1, the entries are 10 for *cntrl* = 1 to cause a right shift. If *cntrl* is 0, the module should remain in Step 1. This can be accomplished either by $S = 00$ (no change) or $S = 11$ (load 1, 0, 0, 0) so both entries are indicated. For control Step 2, the entries are 11 (load), if *cntrl* = 0, and 10 (right shift), if *cntrl* = 1. For Step 3, all entries are 10 for a right shift. In Step 4, we have 00 (no change) if $\wedge /CNT_{2:4} = 0$ and *1X* specifying either a load or right shift, if $\wedge /CNT_{2:4} = 1$.

To implement the first row of Fig. 12.13a requires that the expressions for S_0 and S_1 both include the term *controlreset*. Only the individual rows of Fig. 12.13 may be treated as precisely as Karnaugh maps. Once a control reset has been exercised, one, and only one, of the four steps may be assumed to be active at any time. If we choose 11 for the optional entries at Step 1, we see that $S_0 = 1$ for every condition except Step 4 when $\wedge CNT_{2:4} = 0$. On this basis we may express S_0 as

$$S_0 = controlreset \vee \overline{(\text{step4} \wedge (\overline{\wedge /CNT_{2:4}}))} \tag{12.14}$$

With this selection of the optional entries, S_1 must be 1 for Step 1 and Step 2 if *cntrl* = 0.

$$S_1 = controlreset \vee (\text{step1} \wedge \overline{cntrl}) \vee (\text{step2} \wedge \overline{cntrl})$$

The realizations of Equations 12.14 and 12.15 are given in Fig. 12.13b.

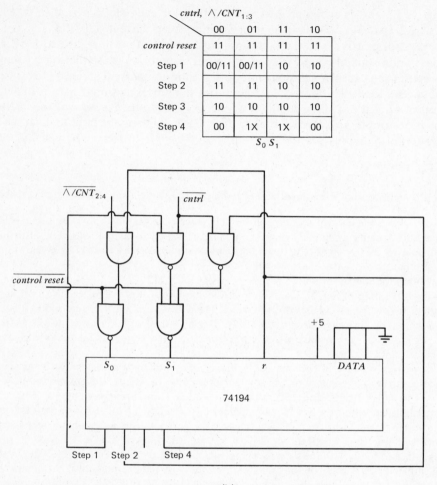

(b)

FIGURE 12.13   *Design of control unit for* SPINTERFACE.   ■

AHPL may be used as a mechanism for specifying the interconnection of MSI parts even if the entire network is combinational. The following example makes use of several of the combinational parts defined in Table 12.1.

Example 12.3

Use MSI parts to design a combinational logic network that will add the largest of three 4-bit numbers W, X, and Y to a fourth number D. The network will have five output bits.

Solution   The approach we will take is suggested by Fig. 12.14. We will first determine the values of the bus control lines *a*, *b*, and *c* in terms of comparator outputs. We propose three 7485 (see Table 12.1) comparators to compare the three possible pairs of vectors $(X; Y)$, $(X; W)$, and $(Y; W)$. Each comparator will have three outputs and the control values *a*, *b*, and *c* will be functions of these 9 output values. We note that the outputs of a comparator chip, e.g., the comparison of *W* and *X* must satisfy the following relations where the vectors *W* and *X* represent the corresponding binary numbers.

$$(W > X) \vee (W = X) \vee (W < X) = 1 \tag{12.16}$$

$$(W > X) \wedge (W = X) = (W > X) \wedge (W < X) = (W = X) \wedge (W < X) = 0 \tag{12.17}$$

The nonindependence of the comparator outputs as given by Equations 12.16 and 12.17 indicates immediately that the lines *a*, *b*, and *c* are not functions of all 9 comparator outputs. Let us agree to put *W* on the bus whenever this number is at least as large as both *X* and *Y*. Thus *a* does not depend on the relation between *X* and *Y*. In fact

$$a = (\overline{W < X}) \wedge (\overline{W < Y}) \tag{12.18}$$

Similarly let us put *X* on the bus if $W < X$ but $X \geq Y$. A Karnaugh map for *b* is shown in Fig. 12.15a. Note that all relations between *X* and the other two vectors are represented. From the map we confirm that

$$b = (W < X) \wedge ((X = Y) \vee (X > Y))$$

$$= (W < X) \wedge (\overline{X < Y}) \tag{12.19}$$

Given the preference already established for *W* and *X*, we place *Y* on the bus only if $W < Y$ and $X < Y$. Therefore

$$c = (W < Y) \wedge (X < Y) \tag{12.20}$$

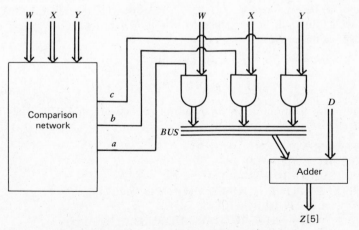

FIGURE 12.14   *Comparison-addition network.*

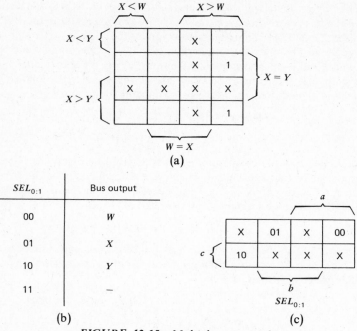

(a)

$SEL_{0:1}$	Bus output
00	W
01	X
10	Y
11	—

(b)

(c)

FIGURE 12.15   *Multiplexer control.*

One step remains since we must implement the bus shown using two copies of the combinational logic unit IC74153. We note that the two control inputs are internally decoded requiring the encoding of inputs *a*, *b*, and *c* on two wires. Let us label the 2-bit control vector to be generated as $SEL_{0:1}$. We shall connect the vectors *W*, *X*, and *Y* so that the bus outputs will be as tabulated in Fig. 12.15b. In Fig. 12.15c we obtain the following expression for *SEL*.

$$SEL_0 = c = (Y > W) \wedge (Y > X) \tag{12.17}$$

$$SEL_1 = b = (X > W) \wedge \overline{(X < Y)} \tag{12.18}$$

We are now in a position to express details of the network in AHPL. The unit we call COMPARADD is described as follows.

```
CLUNIT: COMPARADD(W, X, Y, D)
  INPUTS: W[4]; X[4]; Y[4]; D[4]
  OUTPUTS: CMPADD[5]
  CLUNITS: CMPONE <: IC7485; CMPTWO <: IC7485;
           CMPTHREE <: IC7485
           ADDBUS1 <: IC74153; ADDBUS2 <: IC74153;
           ADDER <: IC7483
```

BODY
$$SEL_0 = \text{CMPONE}_2(W; Y; 0, 0, 1) \wedge \text{CMPTWO}_2(X; Y; 0, 0, 1)$$
$$SEL_1 = \text{CMPTHREE}_2(W; X; 0, 0, 1) \wedge \overline{\text{CMPTWO}_2(X; Y; 0, 0, 1)}$$
$$CMPDD = \text{ADDER}(D; \text{ADDBUS1}(W_{0:1}; X_{0:1}; Y_{0:1}; 0, 0; SEL; 1),$$
$$\text{ADDBUS2}(W_{2:3}; X_{2:3}; Y_{2:3}; 0, 0; SEL; 1); 0)$$
END

∎

The inclusion of a logic block diagram of the network described above was originally contemplated. In this case, we found the diagram bulky and confusing. The AHPL description already given is the most complete and efficient representation of the network. It, together with the input/output declarations for the MSI parts of Table 12.1 (with pin designations added), provides sufficient information to permit wiring the network. Construction of a logic block diagram and verification of this statement is left as a problem for the reader.

12.9 Catenation of Functional Registers

In order to illustrate further the value of our register-transfer language and to improve the reader's facility with its use, we will consider the design of two more complex systems. Both of these systems will be required to handle so much information that it would not be practical to apply the classic sequential circuit-design methods. First we will design a very simple controller for a tester for combinational logic circuits fabricated in integrated circuit form.

This controller will require two 8-bit data registers together with an 8-bit counter. The data registers could be composed to individual D flop flops (typically available as two per IC package). Since the functional registers, 74194, are in our library we assume ready availability. Using two each of these parts for the 8-bit data registers could reduce the part count. In Example 12.1 we catenated 4-bit shift registers in a way that is not particularly convenient for larger systems, particularly those that may use several registers of the same length. In Table 12.3 we see a functional register description of an 8-bit counter and 8-bit data register without shift capability composed of the 4-bit functional registers of Table 12.2.

The first functional register in Table 12.3 merely catenates two shift registers and eliminates the shifting capability. The 8-bit counter is more interesting, since the number of outputs of the 4-bit counter is one more than the number of internal memory elements. This fact makes it almost essential to define a new 8-bit counter rather than catenating two 4-bit counters within a module description.

The intended applications of *EIGHTCOUNT* will require only that it be incremented and synchronously cleared to 0 so only two control inputs are

Table 12.3

I.   FUNCTREG: *EIGHTREG (load; DATA)*
      DATA INPUTS: $DATA[8]$
      CONTROL INPUTS: *load*
      OUTPUTS: *EIGHTREG*[8]
      FUNCTREG: $A <: $IC74194; $B <: $IC74194
  BODY
    $A(load, load, 1; 0, 0, DATA_{0:3}) \leftarrow$
    $B(load, load, 1; 0, 0, DATA_{4:8}) \leftarrow$
    $EIGHTREG = A, B$
  END

II.   FUNCTREG: *EIGHTCOUNT (clear, inc)*
      CONTROL INPUTS: *clear, inc*
      OUTPUTS: *EIGHTCOUNT*[8]
      FUNCTREG: $A <: $IC74163; $B <: $IC74163
  BODY
    $A(1, inc, inc, \overline{clear}; XX) \leftarrow$
    $B(1, A_0, inc, \overline{clear}; XX) \leftarrow$
    $EIGHTCOUNT = A_{1:4}, B_{1:4}$
  END

---

provided. The *clear* input is connected, through an inverter to the corresponding inputs of the 4-bit counters in the second functional register description of Fig. 12.3. The 4-bit counters have two inputs, *enp*, *ent*, both of which must be 1 to cause incrementing. Note that in A, the least significant four bits of *EIGHT-COUNT*, both *enp* and *ent* are connected to the single line *inc*. Therefore, the four bits of A incremented each time a clock pulse is gated into the part and $inc = 1$. The most significant four bits will only be incremented if

$$enp \wedge ent = inc \wedge A_0 = 1$$

This must be the case of the two parts are two functions together as a single 8-bit counter.

A logic block diagram of the *EIGHTCOUNT* is depicted in Fig. 12.16. This multipart functional register may be wired as a unit within a system just as the case with single parts. Within each instance the internal wiring shown above must be included.

Example 12.4

Design a controller to be used in conjunction with a digital magnetic tape cassette recorder to test combinational logic circuits with up to eight inputs and up to eight outputs. Usually the circuits to be tested will be fabricated in IC form with only the input and output leads available to the tester. The system is only required to perform fault

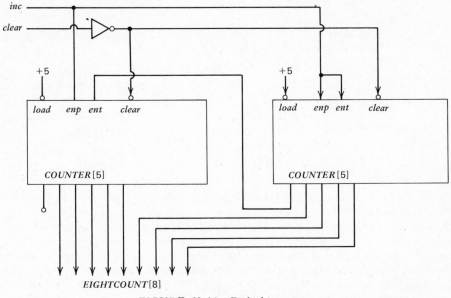

FIGURE 12.16 *Eight-bit counter.*

detection. That is, once it has been determined that some output does not take on its correct value for a given input vector, the circuit under test is rejected and the test sequence is started over again for the next copy of the circuit.

Pairs of 8-bit characters, consisting of an input vector followed by the corresponding correct output vector are stored on tape. The controller must be capable of sequencing through at least 256 such pairs. The system will not be required to operate at a particularly fast clock rate, so that the tape cassette unit and the controller may be assumed to be driven synchronously by the same clock. During each clock period in which the tape device receives a signal on the control line *read*, it will read an 8-bit character from the tape and place it in its output register *TAPE*. Following a clock period in which the tape device receives a signal on line *rewind*, the tape is rewound. At the completion of a rewind operation or when the system is first turned on, the tape reader responds with a one clock period signal on line *ready*.

The controller must provide the appropriate control signals to the tape unit. It must place each input vector in a register connected to the inputs of the circuit under test and compare the eight outputs with the correct output vector. Whenever a discrepancy between the actual output and the correct output is detected, an output signal z is to be turned on and remain on until the system is reset. A manually generated signal on a line labeled *go* will appear for one clock period after a new device to be tested has been placed in the test jig.

Solution The memory elements, both vector and scalar, required in the data and logic unit for the *test controller* are depicted in Fig. 12.17. There are three 8-bit registers. The

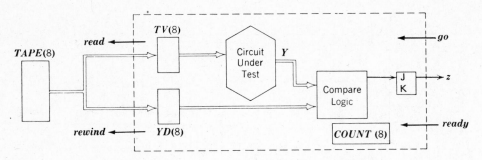

FIGURE 12.17 *Data and logic unit for test controller.*

register *TV* will be connected to the eight input leads of the circuit under test. The register *YD* is used to store the correct values expected on the eight output lines. The third vector *COUNT* will be used to count modulo-256 the number of test vectors applied to each circuit under test. When the count reaches zero, a signal on line *rewind* will cause the tape cassette to be rewound and will alert the operator that the circuit under test is good and that a new copy can be connected. For convenience, the lines connected to the outputs of the circuit under test will be labeled as vector *Y*. The flip-flop *z* (connected to the similarly labeled output line) is set to 1 whenever a fault is detected. The remaining signal lines from the controller to the cassette unit are not shown since they originate in the control sequence. Only the data and logic unit is shown in Fig. 12.17.

We begin with a declaration section for the module, TESTER, as given below. We use two copies of the 8-bit register and one copy of the 8-bit counter of Table 12.3. The vectors *TV* and *Y* to and from the circuit under test are declared as an input and output respectively.

> MODULE: TESTER
>   INPUTS: *TAPE*[8]; *go*; *ready*; *Y*[8]
>   MEMORY: *z*
>   OUTPUTS: *z*; *read*; *rewind*; *TV*[8]
>   FUNCTIONAL REGISTERS: *YD* <: *EIGHTREG*
>   *TV* <: *EIGHTREG*; *COUNT* <: *EIGHTCOUNT*
>   LABELS: *INCR* = 1, 1; *CLEAR* = 0, 0
> BODY

The actual control sequence begins at a point prior to the start of testing of a new device. The only transfer associated with the first two steps resets the error output *z* to 0. These steps are branches that hold control at these two steps until a *ready* signal appears to indicate that the tape cassette is rewound, and a *go* signal indicates that a new circuit is ready for test.

1. $\rightarrow (ready, \overline{ready})/(2, 1)$
2. $z \leftarrow 0$
    $\rightarrow (go, \overline{go})/(3, 2)$

Each individual test begins at Step 3. First, a *read* = 1 causes a test input vector to be placed in *TAPE*. This vector is then transferred to *TV*, exciting the test circuit. Next, the corresponding expected output vector is transferred to *YD* through *TAPE*.

3. *read* = 1
4. $TV(1; TAPE) \leftarrow$
5. *read* = 1
6. $YD(1; TAPE) \leftarrow$

At Step 7, *COUNT* is incremented, and at the same time the actual outputs are compared to the correct values. If there is disagreement in any one output value, $(\vee / Y \oplus YD)$ will be 1 and control will continue to Step 10. At Step 8, the error output flip-flop z is set to 1, the cassette is instruction to rewind, the counter is reset to zero, and then the control returns to Step 1 to wait for the next circuit.

7. $COUNT\ (INCR) \leftarrow$
     $\rightarrow (\vee /(Y \oplus YD),\ \overline{\vee /Y \oplus YD})/(8, 9)$
8. $z \leftarrow 1;\ rewind = 1;\ COUNT(CLEAR) \leftarrow$
     $\rightarrow 1$

If the test was successful, control proceeds to Step 9 where *COUNT* is checked to determine if all 256 test vectors have been applied. If there are more tests to be applied, control then returns to Step 3 to summon the next test input. At the completion of a test sequence, the cassette is instructed to rewind at Step 10 and control returns to Step 1.

9. $\rightarrow (\overline{\vee /COUNT},\ \vee /COUNT)/(10, 3)$
10. *rewind* = 1; $\rightarrow (1)$
END

The control sequencer circuit is shown in Fig. 12.18. Just as is often the case with circuits designed by the conventional methods, starting the operation requires a reset signal that is not included in the initial specifications. In this case, the reset signal would SET flip-flop #1 and CLEAR all the others. Setting flip-flop #1 (FF-1) enables the gates receiving the *ready* signal. As long as *ready* remains 0, succeeding clock pulses keep FF-1 set. When *ready* goes to 1, the next clock pulse clears FF-1 and sets FF-2. This generates control signal C_2, which goes to the C or K input of flip-flop z, so that it will be cleared by the next clock pulse. C_2 also enables the gates receiving *go*, causing FF-2 to clear and FF-3 to set when *go* = 1. Note that output line from each flip-flop is labeled with the transfer it is to enable, as well as with the control signal number. In fact, the control signal number is often omitted, since the number of the flip-flop makes it self-evident.

Steps 3 and 5 are examples of cases where outputs are generated solely by the control sequencer, as discussed earlier, with the *read* line to be set to 1 during these steps. As shown in Fig. 12.18, this is accomplished simply by ORing C_3 and C_5 together and connecting the output of the OR-gate to the *read* line. Thus, *read* = 1 during Steps 3 and 5, and is 0 during all other steps. The *rewind* signal is generated by a similar connection at Steps 8 and 10. The reader is strongly urged to go through this diagram carefully in conjunction with the control sequence to be sure that the correspondence between the sequence and each element of the circuit is thoroughly understood. ∎

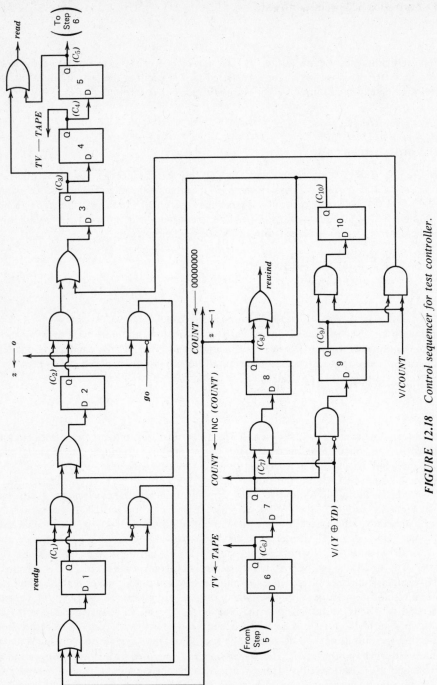

FIGURE 12.18 *Control sequencer for test controller.*

12.10 Use of Memory

This section will include an example that will require more storage for information than encountered until now. We will meet this new requirement by introducing a matrix notation to represent an array of registers. Suppose we have a large number of registers, each having the same number of bits. Rather than assigning a separate symbol to each such register, it would seem convenient to use the same symbol for all of the registers in the set and to distinguish between individual registers numerically. Organized in this way, the set of registers can be referred to as a memory. Since we are already using subscripts to designate individual bits within a register, we must resort to superscripts to identify registers within a memory. For example, an m-word memory $\mathbf{M}$ will consist of registers $\mathbf{M}^0$, $\mathbf{M}^1$, ..., $\mathbf{M}^{m-1}$. The mechanism for retrieving information from a memory will vary with the physical implementation.

If we take the symbol "!" to mean catenation of rows to form a two-dimensional array, we may use the asterisk to express the selection of an individual row as a function of a decoder output in both the read and write operations. Read may be expressed by

$$DATAOUT = \mathbf{M} * DCD(ADDRESS) \tag{12.21}$$

$DATAOUT$ will be equal to the row corresponding to the row number stored by $ADDRESS$. Write may be expressed by

$$\mathbf{M} * (DCD(ADDRESS) \wedge enable) \leftarrow \circ DATAIN \tag{12.22}$$

In this case $DATAIN$ will replace the word of $\mathbf{M}$ specified by the vector in $ADDRESS$ during any clock period in which $enable = 1$. Note that the transfer is unclocked.

The approach described above is consistent with the description of the memory part in Table 12.2. We will use these parts in the following example with no further concern about internal implementation.

Example 12.5

A simple machine-tool controller is to be designed using a random-access memory (RAM) to store sequences of tool positions. Sixteen-bit numbers are used to specify the tool position in three dimensions in a way that need not concern us here. There are four possible sequences of positions, any of which may be requested by an operator at any time. Each sequence is 256 words long. The system hardware must include a 16-bit register, PR, for storing the current tool position. The tool electronics continually monitor this register through three digital-to-analog converters.

Communications with the operator are provided by a start-stop flip-flop, ss, and a 2-bit sequence selector switch, SEQ, in which the bit combinations 00, 01, 10, and 11 correspond to sequences A, B, C, or D, respectively. A level on ss will cause the controller to

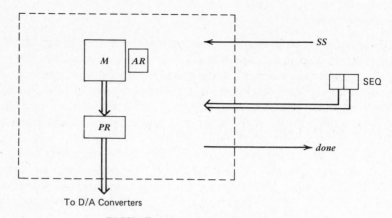

To D/A Converters

FIGURE 12.19　Controller registers.

begin reading out the sequence specified by **SEQ**. If the operator resets *ss*, the controller will terminate the sequence and clear **PR** to all 0s. If the controller completes a sequence, i.e., reads out all 256 words of the sequence in order, it will reset **PR** and issue a one-period level on line **done**. The operator may change the settings of **SEQ** at any time, except that a synchronizing mechanism is provided to prevent changes during a clock pulse.

It is to be noted that we are handling input in a different manner than in previous examples. Here input changes may occur at any time, but no action is taken on them until an appropriate point in the sequence.

Solution　In all, the memory consists of 1024 sixteen-bit words. We will use 16 of the 256×4 ARAM packages to realize this memory. Within the notational capability of AHPL presented thus far it is most convenient to compose 256×16 bit functional register from four of the ARAM parts. Four of these 16-bit word length functional registers could then be row catenated to form the complete memory. The following is a description of a multipackage functional register A16BITRAM.

FUNCTREG: A16BITRAM (*enable, read*; *ADDRESS*[8], *DATABUS*[16])
　DATA INPUTS: *ADDRESS*[8]
　COMBUS: *DATABUS*[16]
　CONTROL INPUTS: *enable*; *read*
　FUNCTREGS: ARAM_0 <: IC2111; ARAM_1 <: IC2111;
　　　　　　　ARAM_2 <: IC2111; ARAM_3 <: IC2111
BODY
　DATABUS = *DATA*_0, *DATA*_1, *DATA*_2, *DATA*_3;
　*A*_0 = *ADDRESS*; *A*_1 = *ADDRESS*; *A*_2 = *ADDRESS*
　*A*_3 = *ADDRESS*; *en*_0 = *en*_1 = *en*_2 = *en*_3 = *enable*;
　　　　　*rw*_0 = *rw*_1 = *rw*_2 = *rw*_3 = *read*;
　　*od*_0 = *od*_1 = *od*_2 = *od*_3 = *read* ∧ *enable*
END

The above description merely connects the proper input/output lines of the four declared IC2111 RAMs. These copies are not separately named but merely carry the name ARAM with a numeric extension of the form "_ i". We assume that all input and output lines from a copy _ i of ARAM carry the corresponding estension number. Note that the 16-bit data bus is wired directly to a catenation of the four data buses from the individual 4-bit RAMs. The address inputs to the individual chips are all connected to the same address input lines of the functional register. The same is true of the read/write and chip-enable lines. The output data control line is not used as an input to A16BITRAM. Instead the individual *od* lines are connected to *read ∧ enable*.

It is convenient that each sequence of tool positions is 256 words long. We may, therefore, use an 8-bit counter as the address register to identify individual words with in the sequence. The two bits *SEQ* may be used to select one of four 256-word × 16-bit RAMs. These bits may be regarded as the most significant two bits of the 10-bit address vector for a 1024 word memory, i.e., $AR[10] = SEQ, COUNT[8]$. The bits *SEQ* will be decoded and the resulting decoder outputs connected to the chip enable inputs of four copies of A16BITRAM. The eight bit counter is realized by a copy of the functional register *EIGHTCOUNT* as specified in the declaration portion of the test controller module given below.

>   MODULE: CONTROLLER
>   MEMORY: $PR[16]$
>   FUNCTIONAL REGISTERS: A16BITRAM_0;
>   A16BITRAM_1; A16BITRAM_2: A16BITRAM_3;
>   COUNT <: *EIGHTCOUNT*;
>   INPUTS: *ss*; *write*; $SEQ[2]$; $DATAIN[16]$
>   OUTPUTS: $PR[16]$; *done*
>   LABELS: $AR[10] = SEQ, COUNT$

The following is the control sequence portion of the controller. Note first the section between the keywords END SEQUENCE and END. The connections listed there are effective at all times. In this case the outputs of the 8-bit counter are connected to the address inputs of each of the A16BITRAMs. The *DATABUS*s of these memories are indicated as connected in bidirectional fashion so that any of the four memory segments enabled by a read operation will place a vector on the common *DATABUS* and vectors from this bus can be written into any of the segments. Here the LABEL statement is used to give a more descriptive name to a catenation of two registers.

>   BODY
>   **1.**   →$(ss, write, \overline{ss \lor write})/(2, 6, 1)$
>   **2.** $COUNT (1, 0) ←$ " clear "
>   **3.** $read = 1$; $enable\_0 = \overline{SEQ}_0 \land \overline{SEQ}_1$; $enable\_1 = \overline{SEQ}_0 \land SEQ_1$;
> $enable\_2 = SEQ_0 \land \overline{SEQ}_1$; $enable\_3 = SEQ_0 \land SEQ_1$;
> $PR ← DATABUS$
>   **4.** $COUNT(0, 1) ←$ " increment "
>   →$(ss \land \land /COUNT, ss \lor \land /COUNT)/(3, 5)$
>   **5.** $PR ← 16 \top 0$; $COUNT(1, 0) ←$ " clear "

(*continued*)

6. $DATABUS = DATAIN$; $read = 0$; $enable\_0 = \overline{SEQ}_0 \wedge \overline{SEQ}_1$;
  $enable\_1 = \overline{SEQ}_0 \wedge SEQ_1$; $enable\_2 = SEQ_0 \wedge \overline{SEQ}_1$;
  $enable\_3 = SEQ_0 \wedge SEQ_1$

7. $\rightarrow(write, \overline{write})/(7, 1)$

END SEQUENCE
  $ADDRESS\_0 = COUNT$; $ADDRESS\_1 = COUNT$;
  $ADDRESS\_2 = COUNT$; $ADDRESS\_3 = COUNT$;
  $DATABUS\_0 = DATABUS\_1 = DATABUS\_2$
  $= DATABUS\_3 = DATABUS$
END

In the above control sequence, the module remains at Step 1 to wait for a 1 on line *ss* or line *write*. If a 1 appears on *ss*, the 8-bit counter is cleared at Step 2. At Step 3 a word is read from memory and placed in *PR*. Note that only one of the four A16BITRAMs is enabled depending on the bits in *SEQ*. As mentioned, the eight bits used to select a particular word within one of the four memories are provided by *COUNT*. *COUNT* is incremented at Step 4 and then Step 3 is repeated if the counter did not contain all 1s and *ss* remained 1. Step 5 terminates the sequence by returning the tool to its rest position and placing a 1 period level on line *done*. Steps 6 and 7 are provided so that new tool sequences can be entered in the memory. ∎

12.11   Economics of Integrated Circuits

Often there is a cost-speed trade-off in the choice of technology with which to implement a given digital system. In this section, we will assume that speed is not the critical factor. We will consider situations where an adequate realization of a circuit can be obtained either by using individual gates and memory elements (ICs only), by using standard MSI parts and individual gates, or by custom LSI fabrication. As is often the case, the choice between the three approaches is to be made on economic considerations alone.

Let us first develop an expression for the cost involved in the manufacture of *n* MSI or LSI parts. A considerable cost is associated with layout, mask generation, simulation, and test sequence generation. These costs, which we will term *design costs*, are fixed regardless of the numbers of parts produced. These costs will vary with the complexity of the part. An approximate expression for design costs is given by Equation 12.23, where *g* is the number of gates in the part. *A* and *B* are constants that

$$\text{Design costs} = A + Bg \qquad (12.23)$$

will vary with the particular technology and will decrease with time. As of this writing, total design costs for a part are measured in thousands of dollars.

A per unit cost is associated with each part produced. This cost covers material, overhead, production labor, and testing costs. Thus, the total cost of *n* parts is given by Equation 12.24. The term $C + Dg$ is measured in pennies even for a fairly complex part.

$$\text{Cost of } n \text{ parts} = A + Bg + (C + Dg)n \qquad (12.24)$$

For a custom LSI part where n may be a few hundred or less, the design cost may be the dominant term. For a standard MSI part, where production may reach the tens of thousands, the design cost can often be neglected. Thus, the cost of a standard MSI part may take the form of Equation 12.25.

$$\text{Cost of standard part} = C + Dg \qquad (12.25)$$

Example 12.6

A logic designer has developed a complex combinational logic network composed of 85 gates. Ultimately a demand for m of these networks is projected. Two possible approaches to realizing the design are under consideration. One involves the use of a standard 90-gate MSI part that is very similar to the desired network and costs K dollars per part. A special 30-gate network connected around the usable portion of the MSI part must be used to obtain the desired realization. The possible custom fabrication of the network in a single package is also under consideration. Determine the most economical approach to obtaining m copies of the network. Assume that the cost of individual gates is \$0.05 per gate.

Solution   The component cost for m copies of the hybrid network is given by Equation 12.26.

$$\text{Hybrid component cost} = (K + 30 \cdot (0.05)) \cdot m = (K + 1.5) \cdot m \qquad (12.26)$$

An additional cost for wiring must be assessed for the hybrid approach. *This is a significant factor, in that wiring costs will often exceed the cost of parts.* In this case, let us assume a wiring cost of W dollars per network. Thus, the total cost for the m hybrid networks is given by Equation 12.27.

$$\text{Hybrid cost} = (K + W + 1.5)m \qquad (12.27)$$

For a fixed number of gates (85), Equation 12.24 for the cost of m custom parts reduces to the form of Equation 12.28.

$$\text{Cost of } m \text{ customs parts} = p + Qm \qquad (12.28)$$

This expression is plotted in Fig. 12.20. To provide a comparison, Equations 12.26 and 12.29 are plotted in the same figure. Clearly the custom approach is to be preferred for very large m and the hybrid approach for small m. The crossover point will occur for a smaller m due to wiring costs associated with the hybrid approach but not needed if custom packages are employed.

Suppose, for example, that a particular semiconductor manufacturer quotes a price of \$2000 plus \$2.00 per part for the custom parts. Suppose this same manufacturer will supply the standard parts for \$2.00 per part. That is, $K = $ \$2.00. If we assume a cost of

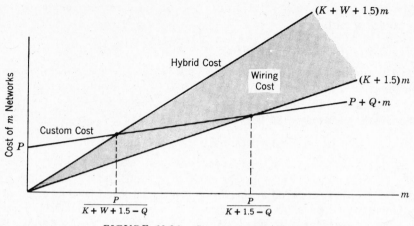

FIGURE 12.20   *Custom vs. hybrid MSI.*

$1.50 more to wire the hybrid network than to wire the custom part to a circuit board, the custom approach is to be preferred for all values of *m* satisfying the inequality.

$$2000 + 2m < 5m$$

$$m > 667 \tag{12.29}$$

If the similar standard part were not available, the custom approach would probably be chosen for a much smaller *m*. ∎

The above example represents a rather simple situation, but it will serve as a point of departure for more realistic cases. Typically, a digital system will be partitioned into several subnetworks. Standard MSI parts may be available approximating some but not all of the subnetworks. The partitioning itself may depend on whether or not custom circuits are economically attractive.

Often a third design alternative exists in addition to a custom part or a network of MSI components. The realization of relatively complex systems in which speed is not the primary design consideration can often be based on a microprocessor. Various special-purpose parts have been developed for use in conjunction with microprocessors. Thus where a microprocessor is applicable, this approach will usually result in the lowest possible parts count.

As suggested above, the total cost of a network will include design costs as well as per unit manufacturing costs. To compare the cost of two system designs requires an expression of the form of Equation 12.30, where *D* represents design costs, *P* represents the per unit cost including both wiring and parts costs, and *n* is the number of systems expected to be produced.

$$\text{Cost of } n \text{ systems} = D + P \cdot n \tag{12.30}$$

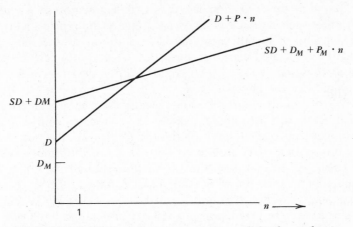

FIGURE 12.21 *Cost of microprocessor* vs. *all-hardware design.*

For microprocessor-based systems, a third term SD representing software developments cost must be added to form Equation 12.31.

$$\text{Cost of } n \text{ microprocessor-based systems} = SD + D_M + P_M \cdot n \quad (12.31)$$

If we assume that Equations 12.30 and 12.31 represent the cost of two design approaches to the same system, we may use these expressions to obtain a cost comparison between a microprocessor-based approach and an all-hardware approach. Since the microprocessor-based system will have the smallest parts count we would expect P_M to be less than P. It might be tempting to dismiss the software design costs, SD, as small in comparison to D and D_M. This is usually not the case. Programming a microprocessor to communicate with peripheral parts can be tedious and, therefore, expensive. Very often SD will be larger than D as well as DM. If this is the case, a comparison of the cost of n copies of the two systems would appear as shown in Fig. 12.21. As shown, the all-hardware system could very well be the most cost effective, if only a few copies are to be produced.

Figure 12.21 is most likely to be applicable where the microprocessor is used to accomplish a relatively simple function. If the full computational capability of the microprocessor is required, P might be so much larger than P_M that the cost of a single all-hardware system would be greater.

12.12   Simulation

Whenever an implementation of a digital system includes a custom MSI or LSI part, the system must "*work*" when it is first connected. It is not feasible to make a series of design modifications on the circuit while a prototype is under test on the bench. Each such design change would require the replacement of at least one costly diffusion or metallization mask. In addition, a time delay of days or weeks would be associated with the fabrication of each new part. Consequently, a trial-and-error design on the bench could take the designer weeks and bankrupt the company in the process.

Since design errors are inevitable, some means must be provided for checking out a system before it is actually fabricated. The answer, of course, is *simulation*. How one might approach the simulation of a sequential circuit will be our concern in this section. Very little need be said about the simulation of purely combinational logic. In that case, simulation would usually be only the computation of functional values for all possible input combinations.

There are many approaches to the simulation of sequential circuits. The approach chosen will be a function of the level of detail of circuit performance that must be monitored. As a general rule, the greater the level of detail provided or the greater the precision of the simulation, the greater the computer time consumed. This computer time can be significant when one considers the simulation of a circuit containing 100 or even 1000 gates. Some of the alternative approaches are tabulated in Fig. 12.22a. Figure 12.22b contains a hypothetical plot of the useful information that might be obtained from each type of simulation versus the amount of computation time consumed. No scale is presented because the plot merely presents an intuitive estimate based on some experience with these simulations.

The most accurate simulation would clearly be based on some circuit analysis program such as *SPICE* (point *A* of Fig. 12.22b). By accurately simulating the internal function of individual circuit elements, the precision of the output of all

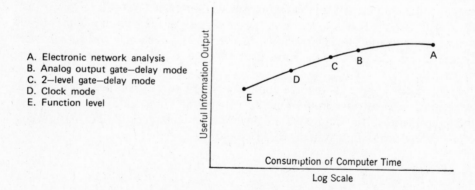

A. Electronic network analysis
B. Analog output gate–delay mode
C. 2–level gate–delay mode
D. Clock mode
E. Function level

FIGURE 12.22   *Alternative simulations.*

gates as functions of time could not be questioned. Unfortunately, the cost of running such programs is prohibitive for large logic networks.

At the opposite end (points D and E) of the curve in Fig. 12.22b is the clock-mode simulation. This type of simulation is the least costly and should be used wherever it will provide the insight required. Point D represents a clock-mode simulation based on a network level description, an interconnection list of gates and memory elements. The *function level* simulation, point E is based on clocked execution of statements in a hardware description language representation of a system. A function level simulation of AHPL is discussed in Reference 4. The use of a function level simulation in the design process is becoming increasingly common as the number of circuit elements in a single part continued to increase.

A graphic comparison of the computer time required by clock-mode and level-mode simulations is given in Fig. 12.23. By definition, a clock-mode simulation implies that one iteration through the network will be accomplished each clock period. That is, output values and next-state values are computed from the combinational logic networks and new values stored in the memory elements once each clock period. Many more iterations are required in a clock period if a clocked system is simulated in terms of delays in individual gates. Several iterations must take place between each input change to allow signals to propagate through multilevel gating. The clock itself must be regarded as a circuit input, so several iterations must be accomplished during the duration of a clock pulse. Some sophisticated simulations allow a variable time interval between iterations. Thus, fewer iterations per unit time are required during the latter part of the interval between clock pulses when the circuit is relatively stable. Various other time-saving techniques are possible.

More network elements must be considered in a gate-delay simulation. A flip-flop that can be treated as a unit in a clock-mode simulation must be treated as eight or ten separate gates in a gate-delay simulation. Thus, each iteration might be more time-consuming. Overall, one should expect at least a 10-to-1

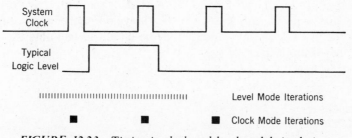

FIGURE 12.23   Timing in clock and level model simulations.

(conceivably 100-to-1) increase in computer time when resorting to a gate-delay as opposed to a clock-mode simulation.

Various approaches to a gate-delay simulation are possible. An accurate gate model would permit representation of a continuum of voltage levels between the 0 and 1 logic values. Such an approach is represented by point B in Fig. 12.22. Alternatively, gate outputs may be restricted to the logic levels 0 and 1 (point C). This has been the most commonly used simulation method for the verification of designs prior to making masks for MSI parts. A variety of approaches are possible for modeling gate delays. One of these will be considered at the end of Chapter 15.

Gate-delay simulations are very often *event driven*. That is, a gate-delay calculation for a given gate will take place during a particular iteration only if one of the inputs to that gate has changed value during the previous iteration. This approach will result in a saving in computation time in comparison to computing delays for every gate during every iteration. When compared to clock-mode simulation, the *event-driven gate-delay* simulation will represent a lesser but still significant difference in computation time.

12.13   Test Sequence Generation

The testing phase for discrete components or small-scale integrated circuits (individual gates) was usually regarded as a peripheral activity, seemingly added as an afterthought. The attitude toward the testing of MSI and LSI parts is necessarily much different. In Fig. 12.24, we see a representation of what might be regarded as a typical MSI part. The 100 gates in the network may be combinational or sequential. One of the inputs x_i may represent a clock, or the circuit may be designed in the level mode. As is commonly the case, we have shown more inputs than outputs.

The black box diagram of Fig. 12.24 may seem unrevealing, but it bears a meaningful resemblance to the dual in-line package as it is presented for test. In particular, all measurements must be taken on the four output leads, z_1, z_2, z_3, and z_4. Thus, if there is a failure in the network, its effect must be forced to appear on one of these lines. That is, an output of the faulty network must be caused to differ from what is known to be the value of the same output of the good network. All signals introduced into the circuit to facilitate these must be placed on the 10 input lines shown.

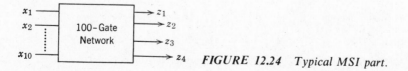

FIGURE 12.24   Typical MSI part.

Let us assume first that the network of Fig. 12.24 is strictly combinational logic, that is, four functions of ten variables. The most straightforward way to test such a circuit would be to check the outputs as all 2^{10} possible combinations of inputs were applied. Defining a test as a separate application of input signals and measurement of outputs, we find that this method would require 1024 tests to verify the functioning of the network of Fig. 12.24.

Now let us conjecture that we have somehow determined that no more than 500 failures can occur in the network of Fig. 12.24. Since the network is combinational, it may be possible to apply a set of inputs that will cause the effect of any given failure to appear at the output. In this case, no more than 500 tests, one for each failure, would be required. Very often, one test will cause an erroneous output in the event of any one of several different failures. Thus, the number of tests required might be considerably less than 500. Various approaches have been developed for finding an approximately minimal set of tests for complete testing of a combinational logic network [9, 10]. Perhaps the best known is the *D*-algorithm developed by Roth. Space will not permit a detailed presentation of these methods. The reader is referred to Reference 6. We will merely present a short example to suggest what may be involved.

Example 12.7

Suppose input 2 of the leftmost AND-gate in Fig. 12.25 is stuck at zero (SA0) as shown. Thus, the output of this AND-gate will always be 0. The output of this gate in the good network will be 1 only if $A = B = 1$. Thus A and B must be 1 in any test for this fault. Assuming $A = B = 1$, we place an X at the output of the leftmost AND-gate to indicate that this value in the good network will differ from the value in faulty network shown. It is now necessary to choose values of C and D which will cause this value conflict to be propagated to the network output. We may calculate gate outputs using

$$0 \wedge X = 0$$

$$0 \vee X = 1 \wedge X = X$$

$$1 \vee X = 1$$

Thus we see that the NOR-gate output will be X if and only if $C = 0$. From here, we see that the circuit output will be X if and only if $D = 1$.

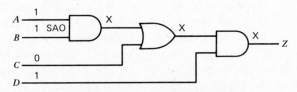

FIGURE 12.25   Test determination.

A test has now been determined for the fault under consideration. We note, however, that the same test will detect the following additional faults: leftmost AND-gate output SA0 Input 1 to NOR-gate SA0, NOR-gate output SA1, and input 1 of rightmost AND-gate SA1.   ∎

In the above example, only circuit failures that could be modeled as stuck inputs and outputs of individual gates were considered. This is the common practice. It is also common practice to assume in developing a set of tests that no more than one failure will occur at one time. This is referred to as the *single-fault assumption*. It may be justified in part by the following observations.

1. An arbitrary multiple fault is much less likely to occur than a particular single fault.
2. Massive multiple faults resulting from processing problems will be easily detected by most any set of tests.

Perhaps the most compelling justification of the single-fault assumption is that it is simply the only way to proceed. In a circuit which could have 500 single faults, 2^{500} combinations of multiple faults would be possible. Clearly the enumeration of multiple faults is impractical.

So far, we have discussed only combinational logic networks. The necessity of relying on the single-fault assumption is even stronger for sequential networks. The option of trying all possible combinations of input values does not exist in the latter case. The outputs are a function of the state of the network as well as the input values. A long sequence of inputs may be required to drive a sequential circuit to a particular state.

As yet no algorithmic procedure has been devised for determining optimal test sequences for sequential circuits. In the general case, this is a very difficult problem. It is the authors' opinion that the best approach is an interactive simulation. Input sequences may be suggested by the designer based on knowledge of the circuit function. The good network as well as all single-fault networks are then simulated (to some extent in parallel) for these sequences. Very likely the first sequence suggested will not cause the effect of every fault to be driven to the output during the simulation. The designer then uses the results of this simulation together with knowledge of the network to suggest a continuation of the input sequence. The simulation is continued, and the process is repeated until a test sequence that will detect all faults is obtained.

The simulation may be either clock or gate-delay mode. A clock-mode simulation can make use of the 3-value logic to handle indeterminate states that can arise in the simulation of faulty memory elements. The initial state of a network is indeterminate at the beginning of the testing process. This situation can be simulated by assigning initial X values to the memory element outputs.

The reader is referred to the literature [5, 8] for further details of these simulations. Our purpose here has been merely to stress the importance of test sequence generation and stimulate the reader's interest in the topic.

12.14 Summary

The increasing availability of economical IC realizations of complex functions will mean that the typical designer will be asked to design ever more sophisticated systems. Usually these systems will involve more elements than can be conveniently handled by state table methods. In this chapter, we have presented an approach that can be used in systems of any size. There is nothing inherent in our language approach that will assure a minimal realization. Still, it offers many advantages over intuitive methods that might be devised by the designer for individual jobs. Most important perhaps is its value in communications about a system between individuals and within an organization.

In this chapter, we have considered relatively small systems just beyond the reach of state techniques. In Reference 2, we extend our approach to larger systems, in particular the complete digital computer.

Problems

12.1. The SN74180 9-bit odd/even parity checker is to be added to the library of CLUNIT descriptions. Write a complete AHPL description of this part.

12.2. Repeat Problem 12.1 for the SN74147 encoder.

12.3. Repeat Problem 12.1 for the SN74181 4-bit arithmetic/logic unit.

12.4. The SN74160 decade counter is to be added to the library of functional registers. Write a complete AHPL description of this part.

12.5. Repeat Problem 12.4 for the SN74298 quadruple 2-input multiplexer with storage.

12.6. The block diagram in Fig. P12.6 shows a module COMPARESHIFT that was designed using AHPL. Write the AHPL description of this module.

12.7. A parallel-to-serial data converter is to be designed. When an input line *word* goes high, an 8-bit data word is to be loaded from lines $X[8]$. When the word has been loaded, the converter will raise the line *take* and shift the data out on line *serdat*, one bit at a time, least-significant digit first. When all 8 bits have been shifted, *take* should be dropped and the converter should wait for *word* to go up again. Design a module to perform this function, using 74194s and 74163s in the data unit, 74194s in the control

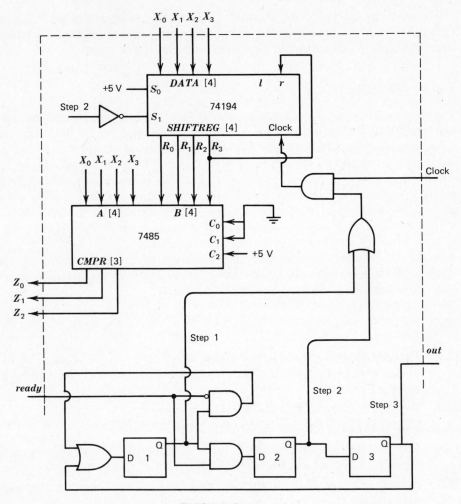

FIGURE P12.6

unit, and such discrete gates as may be required. Show the AHPL description and a complete block diagram of both the data and control sections.

12.8. A serial sequence generator is to be designed having the same inputs and outputs as the converter of Problem 12.7, except that only 4 bits of input appear on lines $X[4]$. The system will wait for line *word* to go high and then load the data from the X lines. It will then wait for *word* to go down, at which time it will raise *take* and shift the data out on *serdat* for eight clock periods in the order X_0, X_1, X_2, X_3, X_3, X_2, X_1, X_0. It will then drop *take* and wait for *word* to go high again. Whenever *take* $= 0$, *serdat* $= 0$. Write the AHPL description of this unit, assuming the use of one 74194 and one 74163 in the data section and draw the block diagram of the data section.

12.9. Design the control unit for the module of Problem 12.8:
(a) Using discrete flip-flops in the manner of Chapter 11;
(b) Using a 74194 shift register in place of the flip-flops.
Show the complete block diagrams of both control units.

12.10. Repeat Problems 12.8 and 12.9, except now assume an additional input, *control*. If *control* $= 0$, the unit is to function as described above. If *control* $= 1$, data is to be shifted out in the order X_3, X_2, X_1, X_0, X_0, X_1, X_2, X_3.

12.11. Construct a complete logic block diagram of the COMPARADD network derived in Example 12.3.

12.12. A BCD adder is to be designed. It will have 9 input bits, 8 bits for two BCD digits and 1 bit for an input carry; it will have 5 output bits, 4 bits representing the BCD sum digit, and 1 bit for the output carry. Write a complete AHPL description of the BCD adder as a CLUNIT, using the 7483 binary adder.

12.13. An available MSI part has four inputs and one output. The output will be 1 only if an even number of the inputs are 1 (zero is considered an even number). Write a CLUNIT description of this part. Use this part together with exclusive-OR gates in the design of another CLUNIT that will check for even parity over 7 bits, i.e., will produce a 1 out if the parity is not even.

12.14. A functional register is to be designed to implement a modulo-100 decimal counter. It will have 8 output bits, representing two decimal digits in BCD. It will count in decimal, i.e., 00 to 09, 10 to 19, etc., to 99, and then return to 00. Write an AHPL description of this unit, using 74163 counters and other logic as required. Show the block diagram of the complete unit.

12.15. Determine (a) a one flip-flop per control state and (b) a minimal-state realization of the control unit for Example 12.2. Compare the total parts count (using 7400 ICs) to that of Fig. 12.13.

12.16. Construct a state table for the control sequencer of Example 12.4 (Fig. 12.18) and develop a minimal-state version of this controller, using the formal techniques of Chapter 10. Compare the overall hardware requirements of the two versions.

12.17. Repeat Problem 12.9 for the module of Example 12.5.

12.18. Example 13.4 describes the design of an electronic ignition lock using classic sequential design techniques similar to those described in Chapter 10. We wish to write the AHPL control sequence for an improved version of this lock circuit. In this version the driver will press a start button and a randomly selected sequence of four letters will be displayed, after which the driver must punch the buttons in the same order. The driver will then press the start button again to obtain another sequence to be copied, continuing in this manner until three successive sequences have been copied correctly. There will be an input x providing a random sequence of 1s and 0s. Each time the line *start* (from the start button) goes up, 8 bits from x should be shifted into a register. (You may assume that changes in x are synchronized to the clock.) The four possible letters are A, B, C, and D, which will be coded 00, 01, 10, and 11, respectively. Each 8-bit word thus represents a random set of four letters. The four 2-bit codes must first be applied in sequence to the output lines $Z[2]$ that drive the display unit. The display unit will control the timing of the display. Each time a character is placed on Z, the lock unit should raise line *ready* and hold the character on the lines until the display unit raises the *next* line, indicating that it is ready for the next character.

When the four characters have been displayed, the unit will wait for the driver to enter the four characters from the keyboard over lines $Y[2]$, in the same code as above. Each time a key is pressed a line *char* will go up to indicate stable data on Y. If the sequence is entered correctly, the unit will record the fact and wait for the next sequence to start. When three consecutive sequences have been entered correctly, the unit will raise the line *on* to turn on the ignition. Figure P12.18 summarizes the various

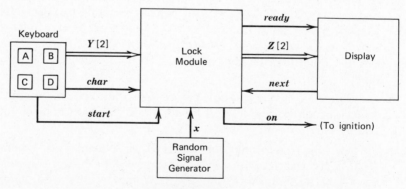

FIGURE P12.18

interconnecting lines. Write an AHPL sequence for the ignition lock module, using such MSI parts and discrete logic as you think appropriate. Show a block diagram of the module.

12.19. Determine a control sequence to realize the code translator of Problem 10.22. Compare the resulting realization with the minimal form determined by classic techniques.

12.20. A random-signal generator has a single input and a single output. The output is constant at 0 or 1 between signals on the input. Whenever a pulse appears on the input the output may or may not change (on a random basis) to the opposite value. A driver module is to be designed using a 1-MHz clock. The driver is to pulse the random-signal generator at a 1 KHz rate. The driver will also compute the average number of changes in the random signal per second, updating the average every 64 seconds. This average is to be sent out in binary on lines $Z[10]$. Write an AHPL description of the driver module. Use MSI parts wherever possible. Accomplish division by shifting.

12.21. An AHPL description is to be written for a module B with registers and communication lines as shown in Fig. P12.21. The function of module B is to facilitate transfer of information from system A to system C. System A might be a computer, and system C might be some sort of peripheral memory, such as a magnetic tape unit. In that case, system B will perform part of the function of a magnetic tape controller.

The output from system A will be a 32-bit vector, which we will refer to as a word. System C will accept only 8-bit vectors or bytes as input. System B then must sequentially disassemble the 32-bit word into four 8-bit bytes. The principal hardware elements in system B are a 32-bit register **LR** and an 8-bit register **SR**. Also included are an input

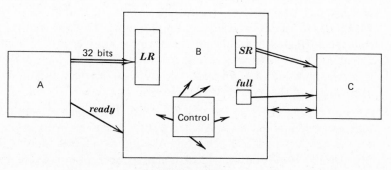

FIGURE P12.21

labeled *ready*, associated with *LR*, and a single flip-flop labeled *full*, associated with *SR*.

We assume that the clock source is provided by system C so that systems B and C are synchronized. During any clock period for which the flip-flop *full* is 1, system C will read a byte from register *SR*. In the clock period immediately after it places a byte in *SR*, system B will set *full* to 1. During the next clock period, system C will utilize the byte in *SR* and simultaneously clear *full* to 0. System B may place another byte in *SR* during the same clock period. Since systems B and C are synchronized, it is possible to read realiably from *SR* and place a new byte in *SR* during the same clock period. The flip-flops in *SR* must be master-slave for this scheme to work. In effect, we have assumed that system C can react in one clock period to simplify the design of system B.

Whenever a word is placed in *LR,* the flip-flop, *ready,* will be held at 1 by system A. After the fourth byte has been read from *LR*, system B will clear *ready* to zero.

12.22. Carry out the design of the conveyor belt control system of Problem 13.15 by the control sequence method. Then repeat the design, except now assume that as many as three additional packages may pass through the weighing machine before a weighed package reaches the x_4 photocell. Note the comparative ease with which this modification can be incorporated with the control sequence approach. By contrast, the traditional state table approach would require a complete redesign from start to finish.

12.23. Develop expressions similar to Equations 12.26 and 12.27 and a graph similar to Fig. 12.21 that could be used to compare the cost of n microprocessor-based designs with n custom-part realizations of the same functions.

12.24. Assume that the overall costs of manufacturing a finished digital product are made up of fixed costs (essentially independent of the approach used), wiring costs, and IC component costs. Ignore the fixed costs and assume that the wiring costs are proportional to the number of IC's used. Rework Example 12.6 assuming a wiring cost of $0.50 per IC package. Determine a crossover value of K in terms of total cost between the individual gate approach and the hybrid MSI approach.

12.25. Assume wiring costs as given in Problem 12.24. Suppose that a custom part can be supplied to satisfy the design discussed in Example 12.6 at a cost of $2000 plus $2.00 per copy. Assume $K = $2.00. Derive an inequality, similar to Equation 12.29, relating the total cost of the hybrid approach to the cost of custom parts.

12.26. Devise a minimal set of tests that will test for all single faults that can occur in the circuit of Fig. 12.25. Consider all stuck inputs and stuck outputs.

Bibliography

1. Hill, F. J. and Z. Navabi. "Extending Second Generation AHPL Software to Accommodate AHPL III," *Proceedings of the Fourth International Symposium on Computer Hardware Description Languages,* Palo Alto, Calif., Oct. 1979, pp. 47–53.
2. Hill, F. J. and G. R. Peterson, *Digital Systems: Hardware Organization and Design,* 2nd edition, Wiley, New York, 1978.
3. Swanson, R. E. "Extensions of AHPL and Optimization of the AHPL Compiler for MSI Design." Ph.D. Dissertation, University of Arizona, 1978.
4. Hill, F. J. and Z. Navabi. "Efficient Simulation of AHPL," *Proceedings of Sixteenth Design Automation Conference,* June 25–27, 1979.
5. Schorr, H. "Computer Aided Digital System Design and Analysis Using a Register Transfer Language," *IEEE Transactions on Electronic Computers,* Dec. 1964, pp. 730–737.
6. Duley, J. R. and D. L. Dietmeyer. "A Digital System Design Language (DDL)," *IEEE Transactions on Computers,* Vol. C-17, Sept. 1968, pp. 850–861.
7. Bell, G. and A. Newell. *Computer Structures: Readings and Examples,* McGraw-Hill, New York, 1971.
9. Borrione, D. "A Language for Simulation of Computer Architecture," *Proceedings of the 1975 International Symposium on Hardware Description Languages,* New York, Sept. 1975, pp. 143–152.
9. Piloty, R. "Segmentation Constructs for RTS III," *Proceedings of 1975 International Symposium on Hardware Description Languages,* New York, Sept. 1975, pp. 115–124.
10. Evans, D. J. "Accurate Simulation of Flip-flops and Timing Characteristics," *Proceedings of 15th Design Automation Conference,* June 1978.
11. Szygenda, S. A. and E. W. Thompson. "Digital Logic Simulation in a Time-Based Table-Driven Environment, Part 1, Design Verification," *Computer,* March 1975, pp. 24–36.
12. Chappell, S. G., P. R. Menon, J. F. Pellegrin, and A. M. Schowe. "Functional Simulation in the LAMP System," *13th Design Automation Conference Proceedings,* 1976, pp. 42–47.
13. Chang, H. Y., E. G. Manning and G. Metze. *Fault Diagnosis of Digital Systems,* Wiley-Interscience, New York, 1970.
14. Roth, J. P., W. G. Bouricius, and P. R. Schneider. "Programmed Algorithms to Compute Tests to Detect and Distinguish between Failures in Logic Circuits," *IEEE Transactions on Electronic Computers,* EC-16, 567–579, Oct. 1967.
15. Seshu, S. "An Improved Diagnosis Program," *IEEE Transactions on Electronic Computers,* EC-14, 76–79, Jan. 1965.
16. Brever, M. A. and A. D. Friedman. *Diagnosis and Reliable Design of Digital Systems,* Computer Science Press, Woodland Hills, Calif., 1976.

13

INCOMPLETELY SPECIFIED SEQUENTIAL CIRCUITS

13.1 Introduction

Throughout Chapter 10, we assumed that the output and next state were specified for every possible combination of input and present state. Thus, no don't-care conditions arose due to the circuit specifications, except somewhat indirectly when the number of internal states required was not a power of two. While such don't-cares are very useful in simplifying excitation functions, they are more a result of the form of realization chosen than of the actual circuit specifications. Because of this special character, such don't-cares, due to extra internal states, are sometimes known as *incidental* don't-cares [3].

In Chapter 11, we encountered another type of don't-care, resulting from the fact that certain *sequences* of inputs may never occur due to external constraints. In the sequence checker of Example 11.1, no pulse could occur more than once in a given three-pulse sequence so that a sequence *ABA*, for example, could not occur. In combinational circuits, if certain input conditions could never occur, the result was don't-care entries in the function maps. Similarly, in sequential circuits, input sequences that cannot occur will introduce *don't-care* entries in the state tables.

Just as don't-cares offered the possibility of simple realizations for combinational circuits, it is reasonable to expect that don't-cares in state tables may make possible further simplification than might otherwise be the case. For combinational circuits, the designer in effect specifies the don't-cares as 1s or 0s in whatever manner provides the simplest realization, and the appropriate choice is generally quite obvious. Sequential circuits being generally more complex, however, the problems of assigning don't-cares are correspondingly much more difficult.

First, there are several types of don't-cares in sequential circuits. The most common is that described above, resulting from input sequences that cannot occur, in which case both the next state and the output are unspecified. The justification is the same as for combinational don't-cares; if the event will never occur, then we "don't care" what the circuit would do in response to it. Second, we may have circuits that receive a continuous series of inputs, to which they respond with appropriate sequences of states, but the outputs are sampled only at specified times, not continuously. When the outputs are not being sampled, they may be unspecified even though the next states are specified. Third, we have situations where a given input is the *last* event of a sequence in the sense that it will always be followed by a general reset signal, which can reset the circuit from any state. For a Mealy circuit, the output will then be specified, but the next state will not.

In the case of combinational don't-cares, there are only two ways to assign the don't cares (0 or 1), but for don't-cares in a state table, there are generally many choices. Where two rows of a state table are identical in every column when *both* are specified, the obvious step is to assign the don't-cares in such a manner as to make the two rows identical, in which case they are equivalent by inspection. This is what was done in Example 11.1 to combine states q_3 and q_4 (see Fig. 11.12). Note that when a given column is unspecified in *both* rows, it may be left unspecified, resulting in don't-cares in the resultant transition table.

In Example 11.1, only two rows could be combined in this manner, and the resultant 4-state circuit was obviously minimal. In most cases, however, there are a number of ways that the don't-cares might be assigned, and it may not be obvious which choice will lead to the minimal state table. If the number of choices is not too large, we might assign all don't-cares in all possible ways and minimize the resultant completely specified state tables (by the methods of Chapter 10) to see which is minimal.

Example 13.1

A clock-mode sequential circuit has an input x and an output z. Starting in a reset state, it stays in reset as long as $x = 0$ and the output is unspecified. An input of $x = 1$ will move the circuit from reset state, R. This will be followed by a string of 0 inputs of arbitrary length, producing $z = 1$ outputs at the time of the second, fourth, sixth, etc., 0-inputs. When the circuit is not in state R, a 1-input may occur at any time and will return the circuit to state R with an output of $z = 0$. Develop a minimal state table.

Solution   A state diagram corresponding to the above operation and the corresponding state table are shown in Fig. 13.1. There is only one unspecified entry, the output for $x = 0$ and state R.

At first glance, it might seem that the safe way to proceed would be to form two state tables by first specifying the single don't care as a 0 and then as 1. These tables could then be minimized, finally choosing the one with the fewest states. This results in the two completely specified state tables of Fig. 13.2. For Fig. 13.2a, we see that only states R and 1 have identical outputs, but their equivalence would imply equivalence of R and 2. Thus, the table of Fig. 13.2a is minimal; and by similar argument the table of Fig. 13.2b is also minimal.

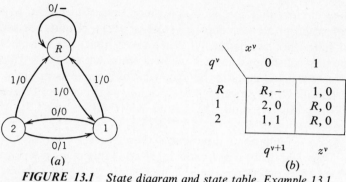

FIGURE 13.1 *State diagram and state table, Example 13.1.*

It appears that Example 13.1 cannot be realized with fewer than three states. But now consider the two-state circuit specified by the state diagram and table of Fig. 13.3, in which states A and B are equivalent to states 1 and 2 of Fig. 13.1, and the function of state R is "shared" between them. First consider the circuits in either states 1 or 2, or

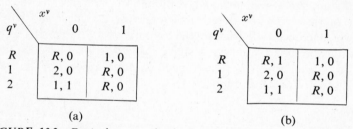

FIGURE 13.2 *Equivalent completely specified state tables, Example 13.1.*

states A or B, with a string of 0 inputs applied. Both circuits will generate as outputs an alternating string of 0s and 1s. For a string of 1s applied, both circuits will produce a string of 0s. Next, consider the circuit in state 2 or state B, with a 1 applied, followed by a string of 0s. The 2-state circuit will produce a 0 followed by an alternating string of 0s and

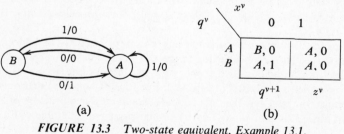

FIGURE 13.3 *Two-state equivalent, Example 13.1.*

1s. This behavior satisfies the specifications for the 3-state circuit since response to a string of 0s beginning in state R is a string of don't-care outputs. A similar analysis applies to a 1 followed by a string of 0s from states 1 or A. Thus, the circuits described by these two state diagrams are equivalent in the sense of generating the same required response to any sequence of inputs. ∎

The above example illustrates that arbitrary specification of don't-care entries in rows of a state table that are not identical may prevent minimization of the state table. As in Example 11.1, such simple and obvious specification may result in a table that is obviously minimal or so close to minimal that more complex and time-consuming efforts at minimization may not be justified. However, if absolute minimization is deemed important, more powerful techniques must be developed. For this purpose, we will need some new definitions and theorems.

13.2 Compatibility

We have seen that in some cases incompletely specified states can be combined to reduce the total number of states in a circuit. However, we cannot say that such states are equivalent because the formal definition of equivalence requires that both outputs and next states be defined for all inputs. Instead, we will say that incompletely specified states that can be combined are *compatible* with one another. When two states are identical wherever both are specified, as in the partial state table of Fig. 13.4, they can be made equivalent by assigning the unspecified entries the same values in both states.

States that are identical wherever both are specified are thus clearly compatible, but in the last section we saw a case where states that were not identical could be combined. In such cases we cannot simply specify the states so as to make them equivalent. To define compatibility we must go back to the basic concept of status having the same function in a circuit if they produce the same outputs in response to the same inputs. Since the circuit behavior is not specified for all possible input sequences, we cannot require identical outputs for all possible sequences, as was the case for equivalent states.

q^v	A	B	C	K
3	2, –	3, –	4, 0	1, –
4	2, 1	–, –	–, 0	1, 0

$$q^{v+1}, z^v$$

FIGURE 13.4

Definition 13.1 Let a sequence of inputs be applied to an incompletely specified circuit S^* in initial state q. The sequence is said to be *applicable* to q if no unspecified next-state entries are encountered, except possibly as a result of the last input of the sequence.

Definition 13.2 Two states, p and q, of a circuit S are said to be *compatible* if and only if

$$\lambda_S[\delta(p, X_1 X_2 \cdots X_K), X_{K+1}] = \lambda_S[\delta(q, X_1 X_2 \cdots X_K), X_{K+1}] \qquad (13.1)$$

whenever both of these outputs are specified, for every input sequence $X_1 X_2 \cdots X_{K+1}$ applicable to both states. The expression $\delta(q, X_1 X_2 \cdots X_K)$ denotes the state of the circuit after the sequence of inputs $X_1 X_2 \cdots X_K$ is applied to the circuit, originally in state q.

Suppose the output sequences resulting from some applicable input sequence are computed for a circuit initially in compatible states p and q, leaving blank those outputs that are not specified. Each of these output sequences can be computed only as far as the next states are specified. Definition 13.2 says that at no point will specified outputs in these two sequences differ.

Theorem 10.1 was used as the basis for eliminating pairs of states that cannot be placed in the same equivalence classes. Theorem 13.1 will be used similarly to eliminate pairs of states that are not compatible. The proofs of most theorems in this chapter will be omitted. They are similar to but in general more tedious than the proofs presented in Chapter 10.

THEOREM 13.1 If two states q_i and q_j of circuit S are compatible, then the following conditions must be satisfied for every input X.

1. $\lambda(q_i, X) = \lambda(q_j, X)$ whenever both are specified.

2. $\delta(q_i, X)$ and $\delta(q_j, X)$ are compatible whenever both are specified.

The processing of the implication table proceeds in exactly the same fashion as for the completely specified case. We first cross out any squares corresponding to pairs of states for which the *specified* outputs differ, i.e., for which condition 1 of Theorem 13.1 is not satisfied. In the remaining squares, we enter the numbers of pairs of states whose compatibility is implied. For example, if $\delta(q_i, X) = q_2$ and $\delta(q_j, X) = q_3$ in a certain state table, we will enter 23 in square i, j of the implication table. We enter check marks in squares corresponding to pairs of states that imply only their own compatibility. Once the initial implication table has been obtained, the elimination of further incompatible pairs of states by

---

* Strictly speaking, once constructed, any deterministic (nonrandom) circuit is completely specified. In Definition 13.1, we refer to the analytic determination of outputs and next states from the state table.

successive "passes" through the table proceeds in exactly the same manner as the elimination of nonequivalent states in completely specified circuits.

Example 13.2

A pulse input-pulse output sequence checker has four inputs, A, B, C, K, and two outputs Z_1, Z_2. At anytime the circuit may be in either a *check* or a *noncheck* mode. In the noncheck mode these four pulses may arrive in any order but only a K pulse will move the circuit into the check mode. Following this K pulse, the pulses A, B, C will arrive exactly once, in any order, followed by another K pulse. At the time of the K pulse following the sequence of A, B, C pulses, an output pulse will appear on line Z_2 if the pulses arrived in the order A, B, C and a pulse will appear on Z_1 if the order was C, B, A. If the A, B, C pulses arrive in any other order, there will be no output pulse. The same K pulse that can generate an output will return the circuit to the noncheck mode, requiring another K pulse to return it to the check mode. The circuit outputs will be observed only at the time of a K pulse. Therefore the output must be 00 when a K pulse drives the circuit to the check mode.

Solution A state diagram and state table are shown in Fig. 13.5. State q_1 constitutes the noncheck mode, from which a K pulse will take the circuit to q_3 the initial state of the check mode. Since only two of the A, B, or C pulses are needed to identify a sequence, the second pulse leads to one of three possible states q_7, q_8, q_9, corresponding to the three output possibilities. The output-producing K pulse sends the circuit back to q_1. It is seen that there are many don't-cares, because of the restriction to exactly one A pulse, one B pulse, and one C pulse per sequence and the fact that the output is specified only for the input K.

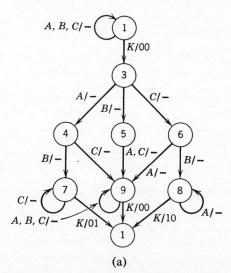

(a)

FIGURE 13.5 *State diagram and state table, Example 13.2.*

q^v	Input			
	A	B	C	K
1	1, –	1, –	1, –	3, 00
3	4, –	5, –	6, –	–, –
4	–, –	7, –	9, –	–, –
5	9, –	9, –	9, –	–, –
6	9, –	8, –	–, –	–, –
7	–, –	–, –	7, –	1, 01
8	8, –	–, –	–, –	1, 10
9	9, –	9, –	9, –	1, 00

$$q^{v+1}, Z_1 Z_2$$

(b)

FIGURE 13.5 (cont.)

The initial and final implication tables are shown in Fig. 13.6. The squares that contain checkmarks correspond to pairs whose compatibility is not dependent on the compatibility of any other states. The remaining squares, not crossed out in the final implication table, correspond to sets that are compatible because the implied states satisfy condition 2 of Theorem 13.1. ∎

The reader will recall that, in Chapter 10, equivalent pairs determined by the implication tables were combined to form equivalence classes. A similar process can be applied to compatible pairs. Although not formally proved, it should be clear that *pairs of states corresponding to squares not closed out in the final implication table are compatible pairs.*

Definition 13.3 A set of states, S_i, of a circuit S will be said to form a *compatibility class* if every pair of states in S_i is compatible.

Definition 13.4 A *maximal compatible* is a compatibility class, which will not remain a compatibility class if any state not already in the class is added to the class. A single state that is compatible with no other state is a maximal compatible.

Example 13.2 (continued)

Determine the set of all maximal compatibles for the sequence checker.

Solution The process is exactly that utilized to determine equivalence classes, as shown in Fig. 13.7. There are no compatibles indicated in columns 7 and 8. In column 6, we find that q_6 and q_7 form a compatible pair. Next we find q_5 compatible with q_6 and q_9, but here we note a situation we have not seen before. For completely specified circuits, if a state is equivalent to two other distinct states, then these states are equivalent to each

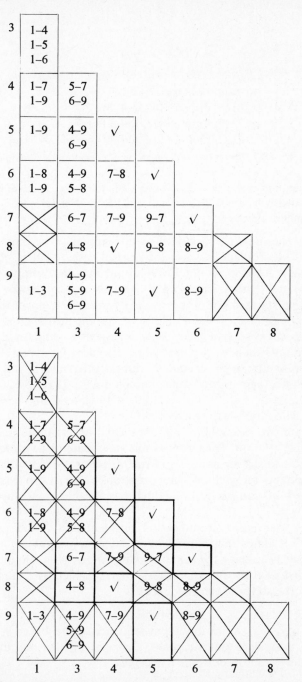

FIGURE 13.6 *Initial and final implication tables, Example 13.2.*

8 –
7 –
6 (6, 7)
5 (6, 7) (5, 6) (5, 9)
4 (6, 7) (5, 6) (5, 9) (4, 5) (4, 8)
3 (6, 7) (5, 6) (5, 9) (4, 5) (4, 8) (3, 7) (3, 8)
1 (6, 7) (5, 6) (5, 9) (4, 5) (4, 8) (3, 7) (3, 8) (1)

FIGURE 13.7   *Determination of maximal compatibles, Example 13.2.*

other. But here we find that q_6 and q_9 are not compatible, so instead of having a class (5, 6, 9), we have two compatible pairs, (5, 6) and (5, 9). The completion of the process follows in the same manner.   ■

In this case, the maximal compatibles all happen to be pairs, but that is a peculiarity of this problem and not the general case. What is significant is the fact that several states appear in more than one maximal compatible, as was the case with (5, 6) and (5, 9) as discussed above. It is easy to see why these pairs are compatible while states 6 and 9 are not from an inspection of the circled next-state entries in column *B* of Fig. 13.5*b*. Note that the next state for state 5 in this column is unspecified so that (at least with respect to column *B*) (56) and (59) are compatible regardless of the next states specified for 6 and 9. However, the next states for 6 and 9 are 8 and 9, respectively, and 8 and 9 are in turn incompatible since their outputs differ in column *K*. Therefore, 6 and 9 are not compatible. Similarly, state 3 is compatible with states 7 and 8, while these two states are not compatible with each other.

The above are two examples of a more formal observation that compatibility is *not* a transitive relation. Thus, compatible states cannot be grouped neatly into disjoint classes, and the formal basis of the procedures of Chapter 10 slips away.

This fact accounts for such puzzling situations as we saw in Example 13.1, and also makes the Huffman partition method inapplicable to incompletely specified circuits.

Nevertheless, our basic simplification problem is still the same: to find a circuit with fewer states that will produce the same output sequence. Since the simplified circuit need have the same output as the original circuit only for those outputs specified, slightly different criteria than specified by Definition 10.1 must be satisfied by the states of the two circuits.

Definition 13.5   A state p_j of a state table T is said to *cover* a state q_i of a state table S, written $p_j \geq q_i$, if, for any input sequence applicable to q_i and applied to both T and S initially in states p_j and q_i, respectively, the output sequences are identical whenever the output of S is *specified*.

The distinction between covering and compatibility is illustrated in Fig. 13.8. Note that, at least for the single input sequence 0112031, q_1 and q_2 are compa-

Input Sequence	0	1	1	2	0	3	1
Initial States	Output Sequences						
q_1	X	1	X	X	0	1	1
q_2	0	1	1	X	0	X	X
q_3	X	1	X	1	0	0	0
p	0	1	1	0	0	1	1

FIGURE 13.8

tible. In addition, the state p covers *both* q_1 and q_2 for the input sequence shown, and (q_1, p) and $q_2, p)$ are compatible pairs. Neither q_1 nor q_2, however, covers p. Thus, we observe that requiring a state r to cover a state s is more restrictive than requiring the two states to be compatible.

We further note that q_2 and q_3 are compatible, while q_1 and q_3 are not compatible. Thus, as we have observed from Example 11.1, compatibility is not a transitive property.

Definition 13.6 A state table T is said to cover a state table S if for every state q_i in S there exists some state p_j in T that covers q_i.

A circuit T that covers S necessarily performs the function of S. Given S, then, our goal is to find such a circuit, T, having a minimum number of states. Toward this goal, we present the following theorems.

THEOREM 13.2 If an internal state p of T covers both internal states q_j and q_i of S, then states q_i and q_j must be compatible.

Proof: If p covers q_i and q_j, then for any applicable input sequence $X_1 X_2 \cdots X_{K+1}$ for which

$$\lambda_S[\delta(q_j, X_1 \cdots X_K), X_{K+1}] \quad \text{and} \quad \lambda_S[\delta(q_i, X_1 \cdots X_2), X_{K+1}]$$

are both specified, we must have

$$\lambda_T[\delta(p, X_1 \cdots X_K), X_{K+1}] = \lambda_S[\delta(q_j, X_1 \cdots X_K), X_{K+1}] \tag{13.2}$$

and

$$\lambda_T[\delta(p, X_1 \cdots X_K), X_{K+1}] = \lambda_S[\delta(q_i, X_1 \cdots X_K), X_{K+1}] \tag{13.3}$$

Therefore,

$$\lambda_S[\delta(q_j, X_1 \cdots X_K), X_{K+1}] = \lambda_S[\delta(q_i, X_1 \cdots X_K), X_{K+1}] \tag{13.4}$$

and q_i and q_j are compatible.

The following is an immediate corollary of Theorem 13.2 and the definition of a compatibility class, Definition 13.3.

COROLLARY 13.3   If a state p_i of T covers a set of states, S_i, in S, these states must form a compatibility class.

For completely specified machines, it was only necessary to include each state of S in one equivalence class. A machine T with a state corresponding to each such equivalence class would then be equivalent to S. In the incompletely specified case, however, it may be necessary to include a state in more than one compatibility class.

Definition 13.7   A collection of compatibility classes is said to be *closed* if, for any class $\{q_1, q_2, \ldots, q_m\}$ in the collection and for every input X_i, all specified next states, $\delta(q, X_i)$, $\delta(q_2, X_i)$, ..., $\delta(q_m, X_i)$, fall into a single class in the collection.

We are now in a position to state the critical theorem of this section.

THEOREM 13.4   Let the n states of an incompletely specified sequential circuit S form a collection of m compatibility classes such that each of the n states is a member of at least one of the m compatibility classes. Then the circuit S may be covered by a circuit T, having exactly m states, $p_1, p_2, \ldots, p_m$, such that each compatibility class of S is covered by one of the states of T, if and only if the collection of m compatibility classes of S is closed.

From this theorem, it can be seen that our basic problem is to find a minimal closed collection of compatibility classes that includes all the states of the original circuit. It can be shown that the set of all maximal compatibles is a closed collection. Therefore, this set of maximal compatibles could be used to form the basis for an equivalent circuit. However, the resulting circuit will seldom be minimal, since there will usually be considerable overlap. Indeed, the number of maximal compatibles will often be greater than the number of states in the original circuit.

It is obvious that we will wish to delete redundant classes wherever possible. However, we must be careful not to violate the closure conditions by removing a class that is necessary to satisfy the implications of a class that is retained. Since any subset of a compatibility class is also a compatibility class, we may delete sets from classes in order to reduce closure requirements.

To facilitate discussion, we will refer to a closed collection that contains each state of a machine S in at least one class as a *cover collection*. Once we find a cover collection, the construction of a machine covering S is straightforward.

Example 13.2 (continued)

All the compatibility classes determined in Fig. 13.7 are listed in Fig. 13.9 together with any implied classes. We first note that state q_9 appears in only one compatibility class, so this class (5, 9), along with the state that did not combine at all, q_1, will be included in the

Comp. Classes	(6, 7)	(5, 6)	(5, 9)	(4, 5)	(4, 8)	(3, 7)	(3, 8)	(1)
Impl. Classes	√	√	√	√	√	(6, 7)	(4, 8)	√

FIGURE 13.9 *Compatibility classes and implied classes, Example 13.2.*

cover collection. The selection of these classes leaves states q_3, q_4, q_6, q_7, and q_8 to be covered. Four of these states can be covered by including classes (4, 8) and (6, 7) in the cover collection, which can be done without implying the inclusion of any other classes. It remains to cover state 3. This can be accomplished by including any of the three classes (3, 7), (3, 8), and (3) in the cover collection. We note that (3, 7) implies (6, 7) and (3, 8) implies (4, 8). Both of these implied classes have already been included, so any of the following can be used to form the final cover collection.

$$(1) \quad (5, 9) \quad (4, 8) \quad (6, 7) \quad (3, 7)$$
$$(1) \quad (5, 9) \quad (4, 8) \quad (6, 7) \quad (3, 8)$$
$$(1) \quad (5, 9) \quad (4, 8) \quad (6, 7) \quad (3)$$

The third of these closed collections is probably preferable since it will result in more don't-cares in the final state table.

Once a closed collection of compatibility classes has been obtained, the output entries in the table are specified as required to satisfy Definition 13.5. That is, for each state p_i of T and each input, the output must agree with any outputs which are specified for the states covered by p_i. Theorem 13.4 assures us that there will be no conflicts.

Similarly, each next-state entry for present state p_i must be the state covering the class that includes all of the next-state entries for the states covered by p_i. A minimal state table covering our machine is given in Fig. 13.10. As shown, states a, b, c, d, and e are assigned to cover classes (1), (3), (4, 8), (5, 9), and (6, 7), respectively. The outputs and next-state entries are obtained in much the same way as for completely specified circuits.

Comp. Class	q^ν	Inputs			
		A	B	C	K
(1)	a	$a, -$	$a, -$	$a, -$	$b, 00$
(3)	b	$c, -$	$d, -$	$e, -$	$-, -$
(4, 8)	c	$c, -$	$e, -$	$d, -$	$a, 10$
(5, 9)	d	$d, -$	$ⓓ, -$	$d, -$	$a, 00$
(6, 7)	e	$d, -$	$c, -$	$e, -$	$a, 01$

FIGURE 13.10 *Final minimal state table, Example 13.2.*

Consider the circled next-state entry for present state d and input pulse B. In Fig. 13.5, $\delta(5, B)$ is unspecified while $\delta(9, B) = 9$. Since state 9 is covered by class $(5, 9)$, we enter the corresponding state d as $\delta(d, B)$. The remaining entries in Fig. 13.10 may be obtained in a similar manner.   ∎

Let us next apply our modified minimization techniques to the 3-state table of Example 13.1 to see that they lead to the 2-state cover circuit which we determined by trial-and-error methods.

Example 13.1 (continued)

Minimise the 3-state table of Fig. 13.1b (repeated in Fig. 13.11a) by obtaining a closed collection of compatibility classes.

Solution   The simple implication table of Fig. 13.11b leads to the list of maximal compatibles and implied classes of Fig. 13.11c.

As might be expected since they form a collection of maximal compatibles, the two classes provide a cover collection. The two classes imply each other so that both must be used to satisfy closure. Thus, it is the closure requirement that prevents us from arbitrarily combining q_R with one or the other of q_1 or q_2, as was originally suggested. It may seem strange that compatible classes that overlap can be used at the same time, but a comparison of the final state table of Fig. 13.11d with the original (Fig. 13.1b) shows that q_A satisfies all specifications of q_1 and q_R, and q_B does the same for q_2 and q_R.

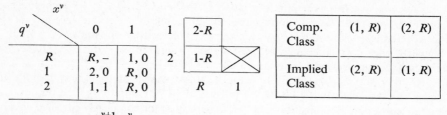

(a)   (b)   (c)

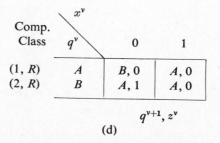

(d)

FIGURE 13.11   *Minimization of state table of Example 13.1.*   ∎

In both of the above examples, the closure problems were trivial, and the choice of closed covers was fairly obvious. In some cases where there are a large number of interacting closure requirements, it may be desirable to reduce the size of some of the compatibility classes. It is apparent that any subset of a maximal compatible is also a compatibility class, and deleting one or more states from a maximal compatible may reduce the closure requirements. One such situation illustrated by the following example.

Example 13.3

Find a minimal-state table equivalent to the state table of Fig. 13.12a.

Solution The final version of the implication table is shown in Fig. 13.12b, and the determination of maximal compatibles is shown in Fig. 13.12c.

The maximal compatibles are listed in Fig. 13.12d, with the implied pairs. The reader may easily verify that the smallest closed cover obtainable from this listing is the set

$$[(abc), (ae), (bd), (de)] \qquad\qquad (13.6)$$

p^v	$X = 00$	$X = 01$	$X = 11$	$X = 10$
a	$b,-$	$e,0$	$e,0$	$b,-$
b	$c,0$	$d,-$	$-,-$	$-,-$
c	$-,-$	$e,-$	$a,0$	$b,0$
d	$a,-$	$b,1$	$-,-$	$-,-$
e	$b,1$	$-,-$	$-,-$	$d,1$

p^{v+1}, z

(a)

(b)

d	(de)
c	(de)
b	(bc), (bd), (de)
a	(ae), (abc), (bd), (de)

(c)

Comp. Classes	ae	abc	bd	de
Impl. Pairs	bd	bc ae de	ac	ab

(d)

	q^{v+1}				z^v			
	X^v				X^v			
q^v	00	01	11	10	00	01	11	10
(ab)	1	2	3	3	1/2	0	0	0	$-$
(bc)	2	2	3	1	1/2	0	$-$	0	0
(de)	3	1	1/2	$-$	3	1	1	$-$	1

(e)

FIGURE 13.12 *Minimizing state table, Example 13.3.*

If we delete state c from the class (abc), we remove the implication of class (ae), thus obtaining the three-state closed cover

$$[(ab), (bc), (de)] \tag{13.7}$$

Similarly, deleting state b from (abc) leads to the three-state cover

$$[(ac), (ae), (bd)] \tag{13.8}$$

The minimal state table corresponding to the cover of Equation 13.7 is shown in Fig. 13.12e. In setting up this table we find that the overlapping classes create a situation we have not seen before. Consider the next-state entries for state q_1, representing the class (ab). In the first three columns of the original state table, the next-state entries are (bc), (de), and (e), corresponding to states q_2, q_3, and q_3, respectively. In the fourth column, the next-state entry is (b), which appears in both q_1 and q_2. Thus the next-state entry can be either q_1 or q_2, as indicated by the entry 1/2. Such entries may be considered partial don't-cares and usually provide extra flexibility in design. The other optional entries in the table arise in the same manner.   ∎

13.3   A Complete Example*

The examples in the previous section illustrate some of the problems that can occur due to closure requirements, but the state table was given arbitrarily rather than derived from a description of some desired circuit function. The reader may wonder if such situations are likely to occur in "real-life" problems. Let us now consider a complex example that will lead to all three types of don't-cares and a difficult covering problem.

Example 13.4

An electronic ignition lock is to be designed for the purpose of preventing drunk drivers from starting their cars. The basic idea is to flash an arbitrary sequence of letters on a screen very briefly and then require the driver to punch a set of buttons in the same order. Tests have shown that better discrimination between drivers who are drunk and those who simply have poor retention of visual images is obtained by requiring the task to be performed in several short sequences rather than one long sequence. On this basis, the following scheme is proposed. There will be four buttons, labeled A, B, C, D, and three "legal" sequences, as listed in Fig. 13.13a. When the system is turned on, Sequence 1 will be displayed, after which the driver must punch the buttons in that order. If the driver is successful, Sequence 2 will then be displayed, and if it is successfully repeated, Sequence 3 will be displayed. Only if all three sequences are correctly followed will the ignition be turned on. If the driver makes a mistake at any point, the keyboard will be mechanically locked to prevent further operation and may be released only by removing the ignition key and starting over.

---

* This section may be omitted without affecting the continuity of the remaining chapters.

Seq. No.	Sequence
1	A D B C
2	A C B D
3	B D A C

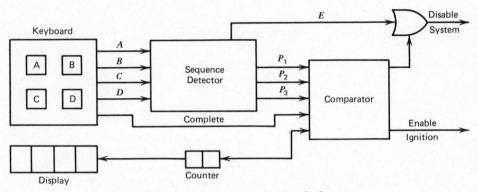

FIGURE 13.13   *Electronic ignition lock system.*

The basic block diagram of the system is shown in Fig. 13.13b.

The counter contains the number of the sequence being used, always starting with 1 when turned on. The sequence detector checks the sequence. If an error is made in any sequence, the error signal E is issued to disable the system. If a sequence is correctly completed, a pulse is issued on one of the lines P_1, P_2, or P_3, corresponding to sequence 1, 2, 3, respectively. This information is compared by the comparator with the number of the sequence that was displayed. If they agree for sequence 1 or 2, the counter is stepped to the next number to start the next sequence.

After agreement on Sequence 3, the ignition is turned on. If the pulse received by the comparator does not match the contents of the counter, a disabling error signal is issued by the comparator. The keyboard is so constructed that each button may be pushed only once for a given sequence. A 1-level will be supplied to the comparator by the keyboard on a line labeled *Complete* after the fourth button has been depressed. The comparator will ignore any pulse inputs occuring before *Complete* = 1. Develop a minimal flow table for the pulse-input, pulse-output sequence detector.

Solution   A state diagram for the sequence detector is developed in Fig. 13.14a. The three expected input sequences lead in 4 steps from the reset state q_0 back to state q_0. Inputs that do not fall in any of the three sequences are shown as generating output pulse E and leading to a figurative don't-care state. The P_1, P_2, and P_3 outputs for the three expected input sequences are don't-cares until the fourth pulse appears. We refine Fig. 13.14a slightly to form the state table of Fig. 13.14b by separating the error output E from the

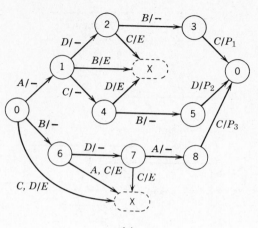

(a)

q	A	B	C	D
0	1, 0, −	6, 0, −	−, 1, −	−, 1, −
1	−, −, −	−, 1, −	4, 0, −	2, 0, −
2	−, −, −	3, 0, −	−, 1, −	−, −, −
3	−, −, −	−, −, −	0, 0, P_1	−, −, −,
4	−, −, −	5, 0, −	−, −, −	−, 1, −
5	−, −, −	−, −, −	−, −, −	0, 0, P_2
6	−, 1, −	−, −, −	−, 1, −	7, 0, −
7	8, 0, −	−, −, −	−, 1, −	−, −, −
8	−, −, −	−, −, −	0, 0, P_3	−, −, −

$$q^{v+1}, E, P$$

(b)

FIGURE 13.14 *State diagram and table, sequence detector, Example 13.4.*

other three pulse outputs. Starting in state q_0, either A or B can be the start of a correct sequence, leading to states q_1 or q_6, respectively. The error output, E, is 0 since no error has occurred, and the other outputs are don't-cares, as previously noted. C or D is "illegal" as a first input, resulting in an E pulse. Since this output will lock the keyboard, preventing further inputs, the next state is a don't-care. This is an example of the third type of don't-care described at the beginning of the chapter.

In state q_1, correct inputs of C or D lead to q_2 or q_4, and an incorrect input of B produces an error output. An A input cannot occur due to the restriction that no button can be pushed more than once, so all entries are don't-cares. Operation proceeds in this general manner until the fourth input, at which time a pulse output indicates the number

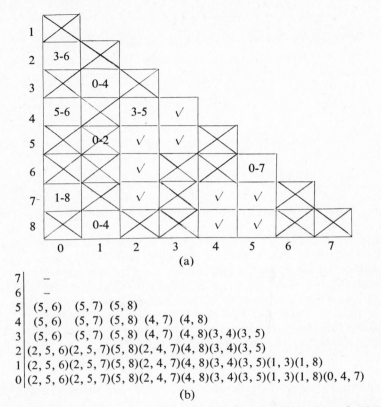

FIGURE 13.15 · Implication table and maximal compatibles, Example 13.4.

of the sequence just completed and the circuit returns to state q_0 to await the next sequence.

The final implication table and the formation of the maximal compatibles are shown in Fig. 13.15. All states appear in at least one compatible pair, so no single states need be included in the list. The maximal compatibles and their implied classes are shown in Fig. 13.16. We first note that only two 3-state classes do not overlap, so a minimum of

Maxi-mal Compa-tibles	(2, 5, 6)	(2, 5, 7)	(2, 4, 7)	(0, 4, 7)	(5, 8)	(4, 8)	(3, 4)	(3, 5)	(1, 3)	(1, 8)
Implied Classes	(0, 7)	$\sqrt{}$	(3, 5)	(5, 6) (1, 8)	$\sqrt{}$	$\sqrt{}$	$\sqrt{}$	$\sqrt{}$	(0, 4)	(0, 4)

FIGURE 13.16   Maximal compatibles and implied classes, Example 13.4.

Maximal Compatibles	x^v q^v	A	B	C	D
(0, 4, 7)	a	c, 0, −	b, 0, −	−, 1, −	−, 1, −
(2, 5, 6)	b	−, 1, −	d, 0, −	−, 1, −	a, 0, 2
(1, 8)	c	−, −, −	−, 1, −	a, 0,	b, 0, −
(3)	d	−, −, −	−, −, −	a, 0,	−, −, −

$$q^{v+1}, E,$$

FIGURE 13.17   *Minimal state table, Example 13.4.*

four classes will be required for a cover. If we can find a 4-state cover, we will know it is minimal.

As a first step, we look for any states that appear in only one maximal compatible. A maximal compatible or some subset of a maximal compatible that contains such states must be included in the cover collection. In this case, q_0 appears only in the class (0, 4, 7), so we tentatively place (0, 4, 7) in the cover collection. This class implies (5, 6) and (1, 8). The (5, 6) implication is satisfied by (2, 5, 6), which in turn implies (0, 7), which is satisfied by (0, 4, 7). Similarly, (1, 8) implies (0, 4), which is satisfied by (0, 4, 7). All three classes can be used, since they imply each other, thus satisfying closure. Further, they cover all states except q_3, the addition of which yields

$$(0, 4, 7), (2, 5, 6), (1, 8), (3)$$

as a 4-state minimal cover. The corresponding minimal state table is shown in Fig. 13.17.   ∎

In the above example, we note that a number of states combine without implication. In Example 13.1, we saw that a minimal circuit could not always be obtained after an arbitrary assignment of don't-cares, but in that case there were implications involved and the difficulty arose from the need to satisfy the closure requirement. But if there are no implications and thus no closure requirements, can we safely combine states by inspection? Let us try it in Example 13.4 and see.

In that example, we see that (2, 5, 7) is a compatible without implication, so let us combine them as state 2 by inspection before setting up the implication table. The reduced state table and corresponding implication table after this simplification are shown in Fig. 13.18. Now the only compatible classes are (3, 4) and (4, 8), which cover only three of the seven states, so the other four must be included individually, resulting in a 6-state cover. This may seem surprising, since a 4-state cover is known to exist, and combining (2, 5, 7) did not set up any implications.

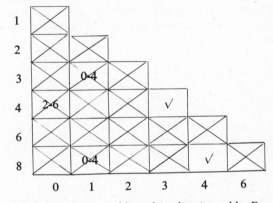

FIGURE 13.18 *Reduced state table and implication table, Example 13.4.*

A partial explanation is seen in the fact that the combination of states 2, 5, 7 forced 2 and 5 to assume the specifications of 7, which is incompatible with 6. This in turn prevents the combination of $(0, 4, 7)$, which in turn eliminates $(1, 8)$, etc. Thus, the problem is not just closure but the basic intransitive nature of compatibility. The conclusion is that combination by inspection of any states other than those identical in every entry, including don't-cares, may prevent state table minimization.

Example 13.4 (continued)

Complete the design of sequence detector for the electronic ignition lock.

Solution The minimal state table is shown in Fig. 13.19, with the next state and output sections separated.

There are so many don't-cares in the next-state and P functions that the choice of assignment is unlikely to make much difference in the final circuits for the state variables

q^v	$(y_2y_1)^v$	A	B	C	D	A	B	C	D	A	B	C	D
a	00	c	b	–	–	0	0	1	1	–	–	–	–
b	01	–	d	–	a	1	0	1	0	–	–	–	010
c	11	–	–	a	b	–	1	0	0	–	–	100	–
d	10	–	–	a	–	–	–	–	0	–	–	001	–

q^{v+1} E^v $(P_3, P_2, P_1)^v$

FIGURE 13.19   *Final state table, Example 13.4.*

and P outputs. However, the E function has only a few don't-cares, so all three possible assignments were tried for E, with Assignment 1 resulting in the simplest equations. The corresponding state variable assignments are indicated in the second column of Fig. 13.19.

$(y_2y_1)^v$	A	B	C	D
00	11	01	× ×	× ×
01	× ×	10	× ×	00
11	× ×	× ×	00	01
10	× ×	× ×	00	× ×

$(y_2y_1)^{v+1}$

$(y_2y_1)^v$	A	B	C	D
00	1	0	×	×
01	×	1	×	0
11	×	×	0	0
10	×	×	0	×

J_{y_2}

$(y_2y_1)^v$	A	B	C	D
00	0	×	×	×
01	×	0	×	×
11	×	×	1	1
10	×	×	1	×

K_{y_2}

$(y_2y_1)^v$	A	B	C	D
00	1	1	×	×
01	×	0	×	0
11	×	×	0	×
10	×	×	0	×

J_{y_1}

$(y_2y_1)^v$	A	B	C	D
00	0	0	×	×
01	×	1	×	1
11	×	×	1	0
10	×	×	×	×

K_{y_1}

FIGURE 13.20   *Transition table and excitation maps, Example 13.4.*

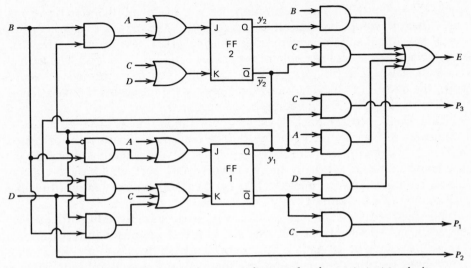

FIGURE 13.21   *Final circuit, sequence detector for electronic ignition lock.*

The transition table and excitation maps corresponding to Fig. 13.19 are shown in Fig. 13.20. The excitation equations are read from these, and the output equations are read directly from Fig. 13.19, as given below. The final circuit is shown in Fig. 13.21. We assume that the clock input is tied to logical 1 so that the *J* and *K* inputs are pulse inputs.

$$\left.\begin{array}{l} J_{y_2} = A + By_1 \\ K_{y_2} = C + D \end{array}\right\} \tag{13.13}$$

$$\left.\begin{array}{l} J_{y_1} = A + B\bar{y}_1 \\ K_{y_1} = By_1 + C + D\bar{y}_2 \end{array}\right\} \tag{13.14}$$

$$E = Ay_1 + By_2 + C\bar{y}_2 + D\bar{y}_1 \tag{13.15}$$

$$\left.\begin{array}{l} P_1 = C\bar{y}_1 \\ P_2 = D \\ P_3 = Cy_1 \end{array}\right\} \tag{13.16}$$ ∎

The reader may wish to question our overall approach to the design of the drunk detector. The reader might visualize a more efficient realization through considering the entire design as a unit rather than designing the sequence checker as a separate device. The state table resulting from this approach would likely have fewer don't-cares and, even though larger, might be easier to minimize. The restriction that each button could be pushed only once might seem unnecessary and even undesirable since it makes it easier to punch the correct sequence accidentally. This restriction led to many of the don't-cares in Fig. 13.14b. It would be difficult to argue with an assertion that don't-cares should be ignored and the system designed using AHPL. The drunk detector has been included as a final example of the minimization of incompletely specified flow tables. Whether the savings resulting from such minimization would be worth the cost in engineering time would depend on the mode of realization of the circuit and the number of copies to be produced.

Incompletely specified clock-mode and pulse-mode circuits arise infrequently. Don't-cares will be the rule in the level-mode sequential circuits to be discussed in the next chapter. A level-mode design will usually be of a fundamental device that might be replicated many times at the system level. Under these circumstances the expenditure of effort to find the most economical and most reliable realization might be justified.

When the need for minimization of incompletely specified tables does occur, the techniques in this chapter will generally suffice. Occasionally in very complex systems there may be so many overlapping compatibility classes and interacting implications that the determination of a minimal cover is very difficult. A more formal technique for handling these cases is presented in Appendix A, but the typical designer will rarely find need for more than has been presented here.

Problems

13.1. A single input line X, on which no more than two consecutive 1-bits or no more than two consecutive 0-bits may occur, is to serve as the input to two distinct sequential circuits. Both circuits are synchronized by the same clock.

(a) Determine the incompletely specified state table of a circuit with a single output, Z_1, such that $Z_1^{v+2} = X^v X^{v+1} + X^{v+2}$ (logical OR).

(b) Determine the incompletely specified state table of a circuit with a single output, Z_2, such that $Z_2^{v+2} = (X^v + X^{v+1}) \cdot X^{v+2}$. Do not use overlapping notation for the states of (a) and (b).

(c) Using the implication table, show that 3-state covers can be obtained for each of the circuits of parts (a) and (b).

(d) Obtain the state table of a 3-state circuit with four outputs, 00, 01, 10, and 11, that will simultaneously accomplish the function of both circuits.

13.2. (a) Determine the set of all compatible pairs of states of the incompletely specified sequential circuit given in Fig. P13.2.

(b) Determine the set of all maximal compatibles.

(c) Verify that compatibility of states is not an equivalance relation for this circuit.

	q^{v+1}, Z		
q^v	$X = 0$	$X = 1$	$X = 2$
q_1	$q_3,0$	$q_5,1$	
q_2	$q_3,0$	$q_5,$	
q_3	$q_2,0$	$q_3,0$	$q_1,$
q_4	$q_2,$	$q_3,$	$q_5,$
q_5		$q_5,0$	$q_1,$

FIGURE P13.2

13.3. A *strongly connected* sequential circuit, S, is a circuit such that, for every pair of states p and q of S, there exists a sequence of inputs $X_1 X_2 \cdots X_K$ such that $\delta(q, X_1 X_2 \cdots X_K) = p$.

Although the result may not serve a useful function, form a circuit simply by writing the two state tables of Problem 13.1 (a) and (b) as a single table. The outputs and initial states of each row of the new table remain the same as in the original tables.

(a) Show that the circuit formed as above is not strongly connected.

(b) Using the Paull-Unger implication table, obtain a 5-state equivalent of the above circuit, which is strongly connected.

13.4. Determine a minimal state equivalent of the sequential circuit described by the flow table in Fig. P13.4.

	q^{v+1}, Z			
q^v	$X = 0$	$X = 1$	$X = 2$	$X = 3$
A	$B,$	$C,0$		$D,$
B		$E,0$		
C	$D,$	$F,$	$C,$	
D	$E,$	$C,0$		$A,$
E		$F,0$		
F		$,1$	$D,$	$B,$

FIGURE P13.4

13.5. Determine a minimal state table for a circuit with a single input and a single output that is only sampled once every five clock times. When sampled, the output must indicate whether three or more of the last five input bits have been 1 (including the current bit). Also provided are a clock and a synchronizing input that resets the circuit to an initial state following each sampling instant. The latter need not be shown in the state table.

13.6. Determine the state table of a circuit that will accomplish the same function as the circuit of Problem 10.7. The only change is that input X_2 will be 1 only for the one-clock pulse immediately preceding the input of each coded character. The state table is now incompletely specified. Show that 14 states are still required in the minimal circuit.

13.7. One information input line X_1, a clock, and the outputs of a modulo-4 counter serve as the inputs to a sequential circuit. Binary-coded decimal (BCD) numbers appear sequentially in 4-bit characters (lowest-order bit first) on line X_1, synchronized with the output of the counter. That is, $c_2 c_1 = 00$ when the lowest-order bit arrives, and $c_2 c_1 = 01$, 10, and 11, respectively, as the next three bits arrive. Derive the incompletely specified state table for a circuit that will translate the BCD code to excess-3 code with a delay of three clock pulses.

13.8. Obtain a minimal-state table of a synchronous sequential circuit that which will perform the translation shown in Fig. P13.8 with only a two-period delay. The process may be assumed to be synchronized by a reset signal not described in the state table. As shown, only the first two bits of the output character are sampled, so the third is redundant. Only the four valid characters of Fig. P13.8 will occur as inputs.

X^{v+2}	X^{v+1}	X^v	Z^{v+4}	Z^{v+3}	Z^{v+2}
0	0	1	−	1	0
0	1	0	−	1	1
1	0	0	−	0	1
1	1	1	−	0	0

FIGURE P13.8

q^v	q^{v+1}, Z						
	$X = 1$	$X = 2$	$X = 3$	$X = 4$	$X = 5$	$X = 6$	$X = 7$
q_1	$q_1,0$	$q_3,1$	$q_5,$		$q_5,$	$q_5,1$	,
q_2	$q_1,0$	$q_3,1$	$q_5,1$				$q_8,0$
q_3		$q_4,$		$q_1,$	$q_1,0$		$q_5,$
q_4	$q_1,$	$q_4,$	$q_5,$		$q_1,$	$q_4,$	$q_5,1$
q_5	$q_5,0$		$q_3,0$	$q_4,1$	$q_1,0$	$q_5,$	
q_6	$q_5,1$	$q_4,0$	$q_3,1$	$q_1,0$	$q_1,$	$q_4,$	$q_5,1$
q_7	$q_1,0$	$q_2,$	,1	$q_6,1$	$q_7,1$	$q_8,0$	
q_8		$q_2,1$		$q_4,1$		$q_8,0$	$q_7,0$

FIGURE P13.9

13.9. Determine the maximal compatibles of the circuit described by the state table in Fig. P13.9. Determine a 4-state cover of this circuit.

13.10. Determine a minimal covering collection for the sequential circuit partially described by the implication table of Fig. P13.10. .

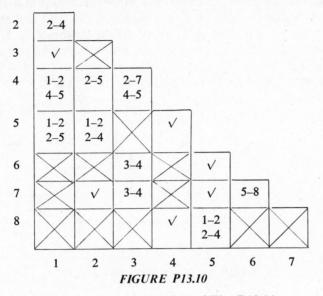

FIGURE P13.10

13.11. Repeat 13.10 for the implication table of Fig. P13.11.

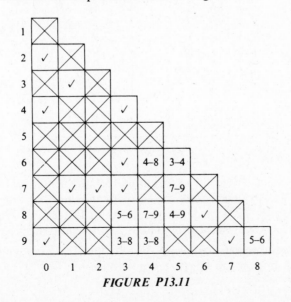

FIGURE P13.11

13.12. A clocked detector circuit is connected to a controller, as shown in Fig. P13.12. Only when $x_1 = 1$ are the outputs z_1, z_2 gated to the controller. x_1 and x_2 are level signals changing only between clock pulses. When x_1 goes to 1 for two successive clock periods, then x_2 for the next four clock periods will represent a decimal digit in the Gray code of Fig. P13.12b, bit b_0 first. At the time of the fourth bit (b_3), $z_1 = 1$ if the digit received as even, and $z_2 = 1$ if it was odd. At all other times that $x_1 = 1$, $z_1 z_2 = 00$. If x_1 remains 1 for the two clock periods following receipt of a digit, another digit will follow, and the process will repeat until x_1 goes to 0, which may occur only between digits. Design the ODD/EVEN detector using J-K flip-flops.

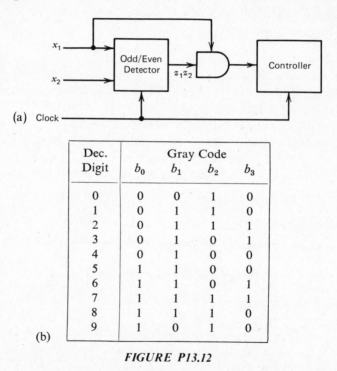

(a)

(b)

Dec. Digit	Gray Code			
	b_0	b_1	b_2	b_3
0	0	0	1	0
1	0	1	1	0
2	0	1	1	1
3	0	1	0	1
4	0	1	0	0
5	1	1	0	0
6	1	1	0	1
7	1	1	1	1
8	1	1	1	0
9	1	0	1	0

FIGURE P13.12

13.13. In the sequence detector of Example 13.4, the next state on the final input of the third sequence could be a don't-care since the state of the detector once the car has started is unimportant. Determine whether this additional don't-care would provide any further simplification of the design.

13.14. Redesign the sequence detector of Example 13.4, removing the restriction that each button can be pushed only once.

13.15. At the Monongahela Works of the International Croquet, Badminton, and Quoits Co. Ltd., packages of wickets come down a conveyor belt and pass through a weighing machine. As the packages enter the weighing machine, they pass a photocell that emits a pulse x_1 to activate the weighing machine, which then emits a pulse x_2 if the package is under-weight, a pulse x_3 if it is overweight, and no signal if the weight is correct. Farther down the line, each package passes another photocell, which emits a pulse x_4 to cause overweight or underweight packages to be diverted into one of two special areas to have the weight corrected. The placement of the special stations and the spacing of the packages is such that an additional package may pass through the weighing machine before the previously weighed package reaches the x_4 photocell. A sequential circuit is required to keep track of whether a given package should be diverted to the underweight area or the overweight area, or pass straight through. Design a circuit to accept the four x pulses as inputs and emit a z_1 pulse at the time of the x_4 input if the package is underweight, a z_2 pulse if it is overweight. Although the spacing of packages on the belt is not uniform, you may assume that spacing is sufficiently controlled so that input pulses cannot overlap. Assume the S-C flip-flops used have separate direct reset inputs, which should be connected to a special reset line to reset the circuit when the process is started.

13.16. Determine the AHPL control sequence of an improved version of the drunk detector described in Section 13.3. Remove the restriction that each button may be pushed only once. As before, require each driver to respond correctly to each of three sequences. Now, however, let these sequences be chosen at random from a prespecified set of eight of the possible sequences. A device with one random output x is available. At the time of each clock pulse, the probability is 0.5 that $x = 0$ and 0.5 that $x = 1$. The values of successive outputs are totally uncorrelated. A given sequence may be allowed to appear more than once in a set of three sequences. The switch panel may be assumed to include a synchronizing mechanism so that a signal will be active for only one clock period each time a button is pushed. Specify storage elements as required and determine a control sequence.

Bibliography

1. Grasselli, A. and F. Luccio. "A Method of minimizing the Number of Internal States in Incompletely Specified Sequential Networks," *IEEE Trans. on Electronic Computers*, **EC-14:** 3, 330–359 (June 1965).

2. Paull, M. C. and S. H. Unger. "Minimizing the Number of States in Incompletely Specified Sequential Switching Functions," *IRE Trans. on Electronic Computers*, **EC-8:** 3, 356–357 (Sept. 1959).

3. Hartmanis, J. and R. E. Stearns. *Algebraic Structure Theory of Sequential Machines*, Prentice-Hall, Englewood Cliffs, N.J., 1966.

4. Harrison, M. A. *Introduction to Switching and Automata Theory*, McGraw-Hill, New York, 1955.

5. Miller, R. E. *Switching Theory, Sequential Machines*, Vol. 2, Wiley, New York, 1965.

6. Moore, E. F. *Sequential Machines, Selected Papers*, Addison-Wesley, Reading, Mass., 1964.

7. Ginsburg, S. *Introduction to Mathematical Machine Theory*, Addison-Wesley, Reading, Mass., 1962.

8. Givone, D. D. *Introduction to Switching Circuit Theory*, McGraw-Hill, New York, 1970.

9. Kohavi, Z. *Switching and Finite Automata Theory*, McGraw-Hill, New York, 1970.

10. Sheng, C. L. *Introduction to Switching Logic*, Intext, Scranton, Pa., 1972.

14

LEVEL-MODE SEQUENTIAL CIRCUITS

14.1 Introduction

In Chapter 9, we saw that, if we remove the requirement for input pulses to trigger state transitions, we have the most versatile category of sequential circuits, *level-mode* circuits. The general form for this type of circuit is a combinational logic circuit with feedback. Feedback around otherwise combinational networks will usually result in sequential operation.* Figure 14.1 shows a circuit that we will now analyze in terms of the timing chart of Fig. 14.2 to see if it is indeed sequential.

As a starting condition, we assume both inputs and the output are at 0. Because of the feedback, the output is a function of itself and these assumptions are not necessarily consistent. In this case, if $Y = 0$, the output of NOR-gate 1 is $\bar{R}_y = 0$, so OR-gate 2 implements $Y = S + \bar{R}y = 0 + 0 = 0$, which checks.

At time t_1, S goes to 1. This causes the output, Y, of OR-gate 2 to go to 1 after a short propagation delay. In turn, $\bar{y}$ goes to 0 and $\bar{R}y$ to 1, again after short propagation delays. OR-gate 2 is now implementing $1 + 1 = 1$, so when S returns to 0 at time t_2, there is no further change. Indeed, the $\bar{R}y = 1$ term has "locked in" the output at 1, so that S has no further effect.

---

* Not all combinational circuits with feedback are sequential. See p. 141 of Reference 1.

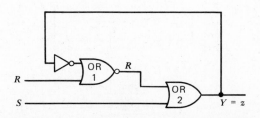

FIGURE 14.1 *Combinational circuit with feedback.*

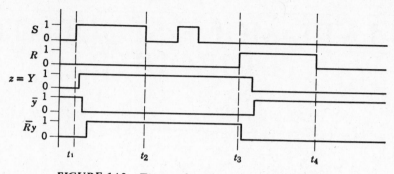

FIGURE 14.2   *Timing chart for circuit of Figure 14.1.*

Next, with $S = 0$, R goes to 1 at t_3. This drives $\bar{R}y$ to 0 and Y to 0 in turn. Then $\bar{y}$ goes to 1, locking in $\bar{R}y$ at 0. Thus, when R returns to zero at t_4, there is no further change and we are back at the starting condition.

Consider the condition of the circuit immediately before t_1 and immediately after t_2. At both times, the inputs are the same, $S = R = 0$, but the output is different. Thus, the output is not dependent solely on the inputs, and the circuit is sequential. Thus, we do not have to use flip-flops or other memory devices to produce sequential circuits. Indeed, reader who compares Fig. 14.1 with Fig. 9.3 will recognize that this circuit is the basic *S-R* flip-flop, redrawn to emphasize the form of the basic model. It might also be noted that the form of Fig. 9.3 seems to have two feedback variables, while Fig. 14.1 has only one. This illustrates one of the special characteristics of level-mode circuits—that it is somewhat arbitrary just which variables you identify as feedback variables, i.e., state variables. In clock-mode and pulse-mode circuits, the state variables are the

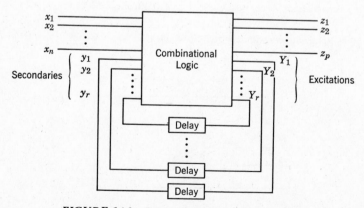

FIGURE 14.3   *Basic model, level-mode circuit.*

outputs of flip-flops, so there is no question as to their identity. In level-mode circuits, the feedback loops form continuous paths and which part of the loop you wish to identify with the state variables is open to choice. Fortunately, as we will see, this apparent ambiguity causes no problems in analysis or design.

The general model for the level-mode circuit is repeated in Fig. 14.3. Recall that the delay elements are not as a rule specific elements inserted in the feedback path for that purpose, but rather represent a "lumping" of the distributed delays in the combinational logic elements into single delay elements, one for each feedback variable. These delay elements may be considered as providing short-term memory. When an input changes, the delay enables the circuit to "remember" the present values of the variables, $y_1, y_2, \ldots, y_r$, long enough to develop new values of $Y_1, Y_2, \ldots, Y_r$, which in turn, after the delay, become the next-state values of the y's. Note the distinction between the y's and the Y's. In the steady-state condition, they are the same, but during transition they are not. The y's will be referred to as the *secondaries* and the Y's as the *excitations*. Both correspond to the state variables in the clocked and pulse circuits, and either may be referred to by that name when the distinction between them is not important.

In the timing chart of Fig. 14.2, we note that changes in the secondaries and excitations, i.e., changes in state, may occur in response to any change in the value of any input. Furthermore, in that analysis, we allowed only one variable to change at a time and allowed no further changes to occur until all variables had stabilized. When such constraints on the circuit imputs can be enforced, we will say that the circuit is operating in the *fundamental mode*. This assumption about input changes considerably simplifies our design problems. Although the fundamental mode is a method of operation, not a special type of circuit, we will refer to circuits designed with this assumption as fundamental-mode circuits. In the next sections, we will consider only fundamental-mode circuits. In section 14.9, we will consider the special problems introduced by elimination of the fundamental-mode restrictions.

14.2   Analysis of a Fundamental-Mode Circuit

We will now see that a fundamental-mode circuit can be described by transition and flow tables in essentially the same manner as clocked and pulse circuits. The transition table for a fundamental-mode circuit is a tabulation of the excitations and the outputs. For the circuit of Fig. 14.1, the equation for the excitation is

$$Y = S + \bar{R}y \qquad (14.1)$$

Since the excitation is the same as the output in this case, this is also the equation for z. The map corresponding to this equation is shown in Fig. 14.4.

Just as with the clocked and pulse circuits, we list the inputs across the top and the secondaries (state variables) down the side, one row for each of the 2^r combinations of secondaries. Since the excitations are the next values of the state variables in fundamental-mode circuits, no distinction will be made between excitation maps and transition tables. The excitation maps will differ from the transition tables only in that they will be maps of single-state variables, while the transition tables will show all state variables. For only one state variable, they are identical.

Since we must assume that both inputs will not change simultaneously from their unexcited or 0 states, don't-care conditions are in the $SR = 11$ column. This is reflected in the specification that both inputs of an S-R flip-flop must not be pulsed simultaneously. The circled entries represent *stable* states, that is, states for which $y = Y$. For example, for entry $SRy = 000$, $Y = 0$. Since Y represents the next value of y, this is a stable condition. This corresponds to the starting condition in the timing table of Fig. 14.2. Similarly, $SRy = 010$ is a stable entry since a 0 output will be unaffected if R goes to 1. The square $SRy = 100$ corresponds to the case where S goes to 1 while the flip-flop output is 0. As specified by Equation 14.1, the resultant value of Y is 1. Since this is not equal to the present value of y, it represents an unstable value and is not circled. After a delay, y changes to 1 and the circuit moves to the circled stable entry at $SRy = 101$. As S returns to 0, the circuit assumes the stable state indicated in square $SRy = 001$ without further charge in output. Finally, R going to 1 moves the circuit through unstable entry $SRy = 011$ to stable entry $SRy = 010$ and then to $SRy = 000$ when R returns to 0.

We have seen that more than one stable state may be found on the same row of a fundamental-mode transition table. It is convenient, in fact, to let each possible combination of inputs and secondary states to be numbered as a separate state. Doing this to the transition table of Fig. 14.4 results in the *flow* (state) table of Fig. 14.5. Note the distinction between this flow table and those of clocked- and pulse-mode circuits, for which the states are only a function of the

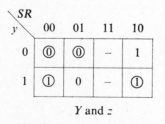

FIGURE 14.4   *Transition table for circuit of Fig. 14.1.*

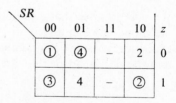

FIGURE 14.5   *Flow table for circuit of Fig. 14.1.*

state variables. Also note that the circled entries indicate the final destination when the state variables must change, i.e., the stable state into which the circuit will settle when all state variables have stabilized.

14.3 Synthesis of Flow Tables

We have just seen how an existing circuit can be analyzed in terms of transition and flow tables. In the design process, we must first go from a verbal statement of the desired circuit performance to a flow table, and thence to transition table and actual circuit. The likelihood of error in the process of translating the problem statements into a flow table is minimized by initially going to a *primitive* flow table.

Definition 14.1 A *primitive flow table* is a flow table in which only one stable state appears in each row.

The primitive flow table corresponding to the flow table of Fig. 14.5 is shown in Fig. 14.6. This table describes the same response as the reduced flow table of Fig. 14.6. Starting in stable state ① with $SR = 00$, S going to 1 takes us through 2 to ②. When S returns to 0, the transition is through 3 to ③. Again, changes in S will move the circuit back and forth between ② and ③. When R goes to 1, the transition is through 4 to ④, and R going to 0 takes the circuit through 1 to ①.

It is often convenient in fundamental-mode problems first to represent the problem as a timing chart rather than a state diagram. This procedure is followed in the following example.

SR	00	01	11	10	z
	①	4	—	2	0
	3	—	—	②	1
	③	4	—	2	1
	1	④	—	—	0

FIGURE 14.6 *Primitive flow table for circuit of Fig. 14.1.*

Example 14.1

Design a fundamental-mode circuit with two inputs, x_1 and x_2, and a single output, z. The circuit output is to be 0 whenever $x_1 = 0$. The first change in the input x_2, occurring while $x_1 = 1$ will cause the circuit output to go to 1. The output is then to remain 1 until x_1 returns to 0. Changes in inputs will always be separated sufficiently in time to permit

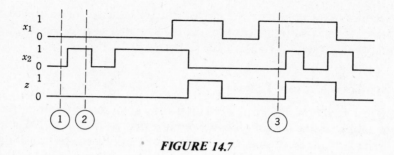

FIGURE 14.7

the circuit to stabilize before a second input transition takes place. Some typical input output sequences are illustrated in Fig. 14.7.

Solution The first step of the design process is to determine a primitive flow table, for which we must assume some convenient starting stable state. In this case it is most convenient to choose a state for which $x_2 x_1 = 00$. For this state, the output will be 0, regardless of which input was the last to change. This initial state is indicated as ① in Figs. 14.7 and 14.8. We also indicate the output corresponding to the stable state. For now, we indicate the outputs only for stable states on the flow table. The outputs for unstable states will be considered later.

Since both inputs are not allowed to change simultaneously, we enter a dash in the $x_2 x_1 = 11$ column of the first row of the flow table. This will eventually result in a don't-care condition in the circuit realization. Dash marks will similarly be entered in each row of the primitive flow table in that column differing in both inputs from the inputs for the stable state of that row.

If x_2 goes to 1 while x_1 is still 0, there will still be no circuit output, so we let the circuit go to a stable state ②, with $z = 0$. If x_1 goes to 1 while $x_2 = 0$, the circuit goes to stable state ③, also with $z = 0$. From state ②, x_2 returning to 0 will take the circuit back to ①, as will x_1 going to 0 from state ③. From state ②, x_1 going to 1 will take the circuit to a new state, ④, also with $z = 0$. We have so far determined the partial flow table of

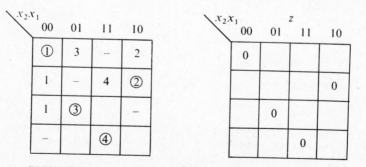

FIGURE 14.8 *Partial primitive flow table, Example 14.1.*

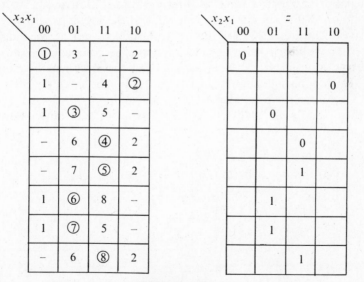

FIGURE 14.9

Fig. 14.8. The uncircled entries, or unstable states, merely represent the circuit destination following an input transition.

Both states ③ and ④ correspond to the last change having been from $x_1 = 0$ to $x_1 = 1$, so that a change in x_2 should cause the output to go to 1. Thus, from state ③, x_2 going to 1 takes us to ⑤, with $z = 1$. Similarly, from state ④, x_2 going to 0 takes us to ⑥, with $z = 1$. From state ④, x_1 going to 0 will return the circuit to the previous state, ②. These additions to the primitive flow table are shown in Fig. 14.9.

FIGURE 14.10

From either state ⑤ or ⑥, the output should remain 1 as long as x_1 remains 1, regardless of further changes in x_2. So we provide states ⑦ and ⑧, both with $z = 1$, such that the circuit will cycle between ⑤ and ⑦ or ⑥ and ⑧, for consecutive changes in x_2. A change of x_1 to 0 will take ⑥ or ⑦ to ①, and ⑤ or ⑧ to ②. The completed primitive flow table is shown in Fig. 14.10. ■

14.4 Minimization

It would, of course, be possible to implement Fig. 14.10 as is, utilizing eight internal states corresponding to the eight possible combinations of three secondaries. If the primitive flow table were identical to the state tables of Chapter 10 or 11, we might consider application of the various techniques developed in those chapters. However, if we compare the primitive flow table of Fig. 14.10 with a typical state table, we see a significant difference, namely the absence of a present-state column in the flow table.

This difference arises from the fact that a clocked circuit is stable only between clock pulses, and the state of the circuit at these times is independent of the present inputs. Therefore, all the stable states may be grouped in a single column, corresponding to no clock input. A fundamental-mode circuit will be stable any time all the inputs are stable, and the state will depend on the values of the inputs. Thus, a present-state entry may appear in any column of the flow table.

The distinctions may be further clarified by considering a partial-state table and a partial primitive-flow table, as shown in Figs. 14.11a and b. First, consider that the circuit of Fig. 14.11a is in state q_1, and the inputs are 10 when a clock pulse arrives. The transition is from q_1 to unstable 3 during the clock interval and thence to stable q_3 when the clock interval is complete, as shown by the dotted arrow in Fig. 14.11a. Next assume that the circuit of Fig. 14.11b is in stable ① with inputs 00. If the inputs go to 10, the circuit moves across to unstable 3 and then down to stable ③, as shown.

Since there is only one stable state in each row of the primitive-flow table, we

q^v	q^{v+1}				z				States				Output			
	00	01	11	10	00	01	11	10	00	01	11	10	00	01	11	10
1	1	2	=	3	0	–	–	–	①	2	–	3	0	–	–	–
2	5	2	4	–	–	0	–	–	5	②	4	↓	–	0	–	–
3	1	–	8	3	–	–	–	0	1	–	8	③	–	–	–	0

(a) State Table	(b) Primitive Flow Table

FIGURE 14.11 *Partial-state and primitive-flow tables.*

Present State	Next State 00	01	11	10	Output 00	01	11	10
①	①	2	–	→3	0	–	–	–
②	5	②	–4	–	–	0	–	–
③	1	–	8	③	–	–	–	0

(a)

Present State	Next State 00	01	11	10	Output 00	01	11	10
⑨	⑨	2	8	⑨	0	–	–	0
②	5	②	4	–	–	0	–	–

(b)

FIGURE 14.12   *Primitive flow table with present state column added.*

can regard these as the present states for the individual rows and group them in a present state column, as shown in Fig. 14.12a. Now the transition from ① to ③ may be regarded as a transition from a present to a next state. Thus, we see that there is little real distinction between the state table for a clocked circuit and the primitive-flow table for a fundamental-mode circuit. The only distinction is that for *every row of the flow table there must be one input combination for which the next state is the same as the present state.* This condition could be imposed on clocked circuits if the designer so desired. This would only limit flexibility, having no effect on the determination of a covering circuit.

We also see that states may be combined in the same manner on both state and flow tables. In Fig. 14.11a, states q_1 and q_3 are obviously compatible and could be combined. Similarly, states ① and ③ in Fig. 14.12a are compatible and could be combined. Should they be combined, they would be replaced by a state, say, ⑨, which would be stable in both columns 00 and 10, as indicated in Fig. 14.12b. That is, the next state would be the same as the present state in both columns.

We have established a precise analogy between the minimization problems for clocked circuits and fundamental-mode circuits. We may proceed then to utilize all of the techniques of Chapters 10 and 11 on Example 14.1.

Example 14.1 (Continued)

We now proceed to reduce the number of rows in the primitive flow table of the earlier section of Example 14.1. We repeat Fig. 14.10 as Fig. 14.13 in the present-state, next-state format justified by the above discussion.

The implication table corresponding to this flow table is shown in Fig. 14.14a. It is obtained in exactly the same way as for the clocked circuits. We enter an × for any pair with conflicting outputs, a ✓ for any pair compatible without implication, and the implied pairs in the remaining squares. We then eliminate pairs that imply incompatible pairs to produce the final implication table of Fig. 14.14b.

The maximal compatibles are easily determined to be

$$(12), (13), (24), (5678) \tag{14.2}$$

q^v	00	01	11	10	00	01	11	10
		x_2x_1				x_2x_1		
①	①	3	–	2	0			
②	1	–	4	②				0
③	1	③	5	–		0		
④	–	6	④	2			0	
⑤	–	7	⑤	2			1	
⑥	1	⑥	8	–		1		
⑦	1	⑦	5	–		1		
⑧	–	6	⑧	2			1	
		q^{v+1}				z		

FIGURE 14.13

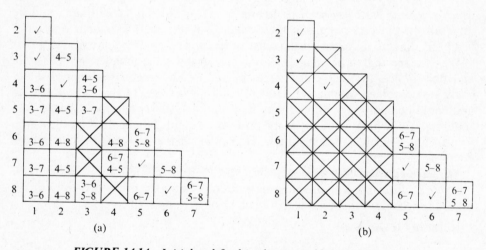

FIGURE 14.14 *Initial and final implication tables, Example 14.1.*

The smallest closed collection may be found by inspection to be

$$(13), (24), (5678) \tag{14.3}$$

The 3-state table corresponding to this closed collection is shown in Fig. 14.15. Note that the combining of states may result in more than one stable state in a single row. In the primitive-flow table, we see that, for present state ① and input $x_2 x_1 = 00$, the next state is also ①. Similarly, in state ③, for input $x_2 x_1 = 01$, the next state is also ③. Thus, when states ① and ③ are combined into state ⓐ, the next state must be the same as the

q^v	$x_2 x_1$				$x_2 x_1$			
	00	01	11	10	00	01	11	10
(13)	ⓐ	ⓐ	ⓐ	c	b	0	0	–	0
(24)	ⓑ	a	c	ⓑ	ⓑ	0	–	0	0
(5678)	ⓒ	a	ⓒ	ⓒ	b	–	1	1	–
	q^{v+1}				z			

FIGURE 14.15   Minimal-flow table, Example 14.1

present state, for both $x_2 x_1 = 00$ and $x_2 x_1 = 01$. The reasoning is similar in arriving at the pairs of stable states in the other rows of the minimal table.

In row ① of the primitive table, the entry for $x_2 x_1 = 11$ is a don't-care, as this would correspond to a forbidden double transition from the stable ① entry. In row ③, however, the entry in this column is an unstable 5, corresponding to the transition from ③ to ⑤. When states 1 and 3 combine into state ⓐ, the entry in this column must be an unstable c, corresponding to a transition to the class containing state ⑤. This does not mean that a double transition, from $x_2 x_1 = 00$ to $x_2 x_1 = 11$ is any more "legal" than it was efore. If the circuit is operated properly in the fundamental mode, the only transition through this unstable c will be from the stable ⓐ corresponding to $x_2 x_1 = 01$. The other cases where the circuit is to go to a new stable state are handled in a similar manner, as shown in Fig. 14.15.

Now let us reconsider our earlier assumption that the outputs are of no concern during the transitions. If an output variable is to change value as a result of a state change, then this variable is a don't-care during the transition. Assume an output variable is to change from 0 to 1 in a certain transition. If a 0 is entered as the value during the transition, then the change of variable will not take place until the end of the transition. If a 1 is entered, the change will take place at the start of the transition. Since it makes no difference exactly when in the transition the output change occurs, the entry is a don't-care. Referring to the final flow table of Fig. 14.15, we see that all transitions to or from state ⓒ involve a change in output. The outputs during these transitions thus remain unspecified.

If an output variable is *not* supposed to change as the result of a transition, the situation becomes a bit more complicated. It may be that the outputs are sampled by whatever other circuits may be involved only between transitions, in which case the transition outputs are don't-cares. But if the outputs during transitions may affect other circuits, we usually must specify them. Refer again to Fig. 14.15 and consider the transition from ⓐ to b to ⓑ, as the input goes from 00 to 10. If the output were a 1 for unstable b, as might occur if we leave it unspecified, a 1 pulse might appear on the output line during the transition. If the circuit driven by z were, for example, a pulse circuit, this pulse might cause some undesired transition. Thus, the output corresponding to unstable b must be specified as 0. Similarly, the output for unstable a in row ⓑ is specified as 0.

Several more problems must be considered before an adequate circuit can be constructed for Example 14.1. Before considering these difficulties, let us develop another fundamental mode circuit of some practical interest. ∎

Example 14.2

Certain forms of digital computer control units [11] make liberal use of a fundamental-mode sequential circuit called a control delay. This circuit has a control pulse input x, a clock input, C, and a control pulse output, z, as illustrated in Fig. 14.16a. Pulses on line x will always be separated by several clock periods. Whenever a pulse occurs on line x, it will overlap a clock pulse and be of approximately the same width as a clock pulse. That is, line x will only go to 1 after the clock as gone to 1 and will return to 0 only after the clock has returned to 0. For each input pulse, there is to be an output pulse on line z coinciding with the next clock pulse following the x pulse. Thus, each x pulse results in a z pulse delayed by approximately one clock period, as shown in Fig. 14.16b.

The input line x originates at the output of a similar control delay with perhaps additional gate and line delay in between. The control delay output will be a very slightly delayed version of the clock due to internal gate delay. Thus, we assume that no two inputs will change simultaneously so that the *fundamental-mode* constraint is satisfied. Develop a minimal-state table for this circuit.

Solution   State ①, representing a condition between clock pulses and at least two clock periods after the last pulse on line x, is a natural starting state. From here, the next input change can only be from 0 to 1 on line C. We let this change take the circuit to state ② as shown in Fig. 14.17a. From here, the next change may be the leading edge of a pulse on line x, in which case the circuit goes to state ③. If no x pulse occurs, the circuit returns to state 1 when C goes to 0. We have assumed that the clock will always return to 0 ahead of a coincident x pulse. Thus, the circuit must go from state ③ to state ④, as shown in Fig. 14.17. Next, x will return to 0, taking the circuit to state ⑤. We have assumed that another control pulse will not occur until at least one clock period after an output has been generated. Therefore, the circuit will go from state ⑤ to state ⑥ with the arrival of the next clock pulse and remain in that state for the duration of that clock pulse. The output $z = 1$ while the circuit is in state ⑥. A complete tabulation of outputs for all stable states is given in Fig. 14.17c. Following the output pulse, the circuit returns to state ①.

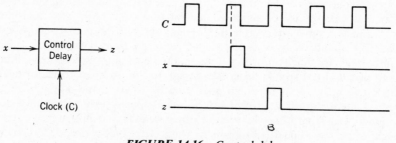

FIGURE 14.16   *Control delay.*

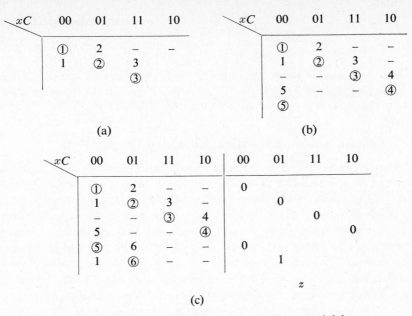

(a)

(b)

(c)

FIGURE 14.17 *Developing flow table for control delay.*

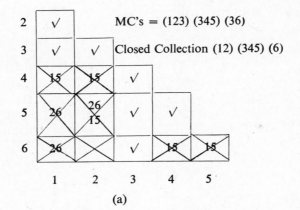

MC's = (123) (345) (36)

Closed Collection (12) (345) (6)

(a)

(b)

FIGURE 14.18 *Minimal-flow table for control delay.*

Using the implication table of Fig. 14.18a, the primitive-flow table is readily minimized to form the flow table of Fig. 14.18b. The outputs corresponding to unstable states are specified in this table to prevent unwanted output transitions.   ∎

14.5 Transition Tables, Excitation Maps, and Output Maps

The next step in the design is to assign specific combinations of the secondaries to the rows of the reduced flow table. Here we have all the state assignment problems of the clock- and pulse-mode circuits, plus some special problems peculiar to the fundamental mode. Let us proceed for now as though there were no new problems. As an example, let us construct a transition table corresponding to the flow table just developed for the control delay. We first assign the combinations of secondary values $y_2 y_1 = 00$ to $\textcircled{a}$, $y_2 y_1 = 01$ to $\textcircled{b}$, and $y_2 y_1 = 11$ to $\textcircled{c}$. Substituting these values for the stable states in Fig. 14.18b results in the partial transition table of Fig. 14.19a. This figure is a tabulation of the excitations, $Y_2 Y_1$, as a function of the inputs and the present values of the secondaries,

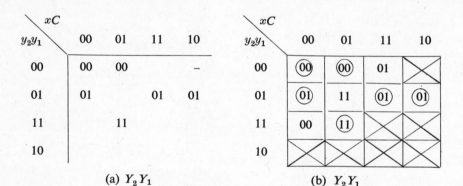

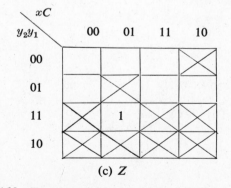

FIGURE 14.19   *Transition table and output map for control delay.*

$y_2 y_1$. The excitations for unstable states must be the same as for the correspond-ing stable states. For example, the inputs 11 cause a transition from state $@$ to state $\textcircled{b}$ through unstable state b. To cause this change to take place in the physical circuit, we enter $Y_2 Y_1 = 01$ in place of unstable state b. Thus, when xC takes the value 11, the excitations are logically specified as 01. After a circuit delay, the secondaries take on the values $y_2 y_1 = 01$. Now the secondaries and excitations are the same (01), and the circuit is stable.

Excitation values corresponding to all unstable states are entered in completed transition table of Fig. 14.19b. In succeeding problems, we will obtain the transi-tion table in one step by making no distinction between stable and unstable states in the assignment of excitation values. The output table of Fig. 14.19c is a direct translation of the minimal state table.

For convenience in reading off the excitation equations, we may separate the transition table into separate *excitation maps* for the excitation variable, as shown in Fig. 14.20a. From these maps and the output map, we read the equations

$$Y_2 = \bar{x}Cy_1 \tag{14.4}$$

$$Y_1 = x + \bar{y}_2 y_1 + Cy_1 \tag{14.5}$$

$$z = y_2 \tag{14.6}$$

The corresponding circuit is shown in Fig. 14.20b.

If this circuit were clock or pulse mode, we would also check the other two possible assignments to see which is minimal. However, as mentioned above, fundamental-mode circuits have some special problems, so let us look at the functioning of this circuit in some detail before worrying about an absolute optimum.

Let us assume that the circuit is in the stable state $xCy_2 y_1 = 0111$, corre-sponding to the stable $\textcircled{c}$ state in the second column, third row of the state table. The corresponding values at various points in the circuit are shown in Fig. 14.20b. Note that the 1 values for Y_2 and Y_1 are produced by gates $A1$ and $A3$, respectively. Now assume that C goes from 1 to 0, as shown. This should take the circuit through the unstable a entry in the third row to the stable $@$ in the first column, first row, with both state variables changing to 0.

Recalling that a NAND-gate will have a 1 out if any input is 0, we see that the $C = 0$ signal will drive the outputs of $A1$ and $A3$ to 1. Providing $A2$ does not change too fast, the 1 at the output of $A3$ will then drive Y_1 to 0 at the output of $A4$. This value of Y_1, fed back to the input of $A2$, will lock its output at 1, in turn stabilizing Y_1 at 0, as required.

But now assume that the gates have unequal delays, as shown in Fig. 14.20b. Then the change in C will first drive the output of $A1$ to 1, which will in turn

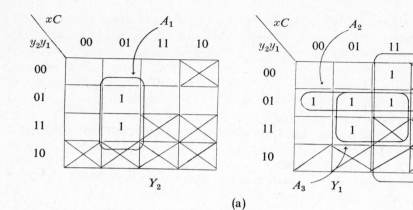

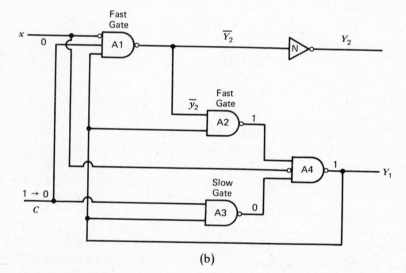

FIGURE 14.20   *Excitation maps and circuit for control delay.*

drive $A2$ to 0 *before* the output of $A3$ goes to 1. As a result, Y_1 will not change and the circuit will make an erroneous transition to $Y_2 Y_1 = 01$, the stable ⓑ state in the first column.

The difficulty just described occurred because both secondaries were required to change simultaneously following a single input change but did not change at the same time due to unequal circuit delays. Such a situation is termed a *race* because the nature of the transition may depend on which variable changes fastest. Sometimes races may be "fixed" by introducing extra delays, but this approach is expensive and doesn't always work. In the next section, we will investigate the nature of these races more carefully and show that they can be eliminated by a better state assignment.

14.6 Cycles and Races

It is possible for a circuit to assume more than one unstable state prior to reaching a new stable state. If for a given initial state and input transition such a sequence of unstable states is unique, then it is termed a *cycle*. For example, in the table and corresponding Y map of Fig. 14.21, the circuit will cycle through three unstable states during a transition from ① to ④. Note in the Y map that the next state for $y_2 y_1 x_2 x_1 = 0001$ is not $y_2 y_1 = 10$. Instead, the machine proceeds from 00 to 01 to 11 to 10. Each of the transitions involves a change in only one secondary. Thus, the circuit proceeds reliably through each of the unstable states to the final stable state. The cycle may be entered from stable states ② and ③ as well.

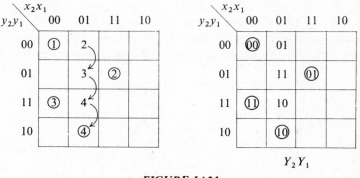

FIGURE 14.21

Where a change of more than one secondary is specified by the Y map, the resulting situation is termed a *race*. The race illustrated in Fig. 14.22 is a *non-critical race*. Suppose the inputs change to 01 while the machine is in state $y_2 y_1 = 00$. If y_2 changes first, the machine goes to unstable state 10. If y_1 changes first, the machine goes to 01. For either of these unstable states, the Y map shows the next state to be stable ②. Thus, no matter what the outcome of the race, the stable state ②, as designated by the Y map, is reached.

The situation encountered in the control delay, as discussed in the last section, is an example of a *critical* race, a situation in which there are two or more stable states in a given column, and the final state reached by the circuit depends on

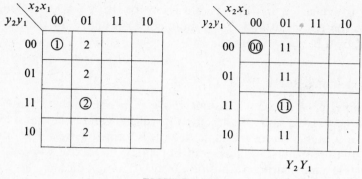

FIGURE 14.22

the order in which the variables change. In other words, the desired result might occur or might not occur depending on the actual circuit delays.

Another critical race is illustrated in Fig. 14.23. There the circuit is supposed to move from stable ② ($y_2 y_1 = 01$) in the $x = 1$ column to stable ③ ($y_2 y_1 = 10$) in the $x = 0$ column. If Y_1 changes first, the excitations will be $Y_2 Y_1 = 00$, leading to an erroneous transition to stable ①. If Y_2 changes first, the excitations will be $Y_2 Y_1 = 11$, leading to unstable 1 in the third row and thence to stable ①. Thus, no matter what the delays, the wrong transition will occur.

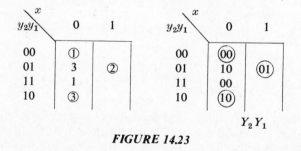

FIGURE 14.23

It might seem that the first type of critical race would be less troublesome since it can, in theory, be eliminated by proper adjustment of delays. In practice, however, constraining delays is difficult and expensive so that both types of critical races are to be avoided. In some problems, this may be accomplished by a judicious assignment of state variables. In others, it becomes necessary to use more than a minimal number of states.

Example 14.2 (continued)

We are now in a position to construct a more satisfactory version of the control delay. It is only necessary to eliminate the critical race in column 00 of Fig. 14.19b and replace it with a cycle through the spare state. A new transition table is given in Fig. 14.24. The only changes are in the lower two squares of column $xC = 00$. Thus, from stable state $\textcircled{11}$, the circuit will cycle as indicated by the arrows when C goes to 0. In terms of the excitation maps, the only change is an additional 1 in the third row. The new equation for Y_2 is:

$$Y_2 = y_2 y_1 + \bar{x} C y_1 \tag{14.7}$$

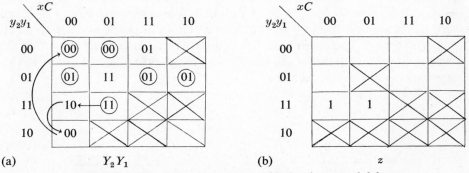

(a) $Y_2 Y_1$ (b) z

FIGURE 14.24   *Revised transition table for the control delay.*

There is no change in the equation for Y_1. To avoid a possible output pulse during the cycle, an additional 1 in the output map must be specified, as shown in Fig. 14.24b, but this does not change the equation for z. The revised circuit for the control delay shown in Fig. 14.25 will function properly as a control delay provided the input pulses satisfy the fundamental mode constraints as previously assumed.

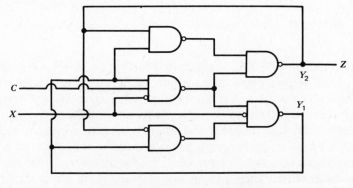

FIGURE 14.25   *Control delay.*

14.7   Race-Free Assignments

To eliminate critical races, we must assign the state variables in such a manner that transitions between states require the change of only one variable at a time. To this end, it is desirable that stable states between which transitions occur be given adjacent assignments, i.e., assignments differing in only one variable. To find such assignments, it is convenient to make a *transition diagram* on the Boolean hypercube. Consider the state table shown in Fig. 14.26a and the transition diagram in Fig. 14.26b. To set up the transition diagram we start by assigning a to 00. Noting that a has transitions to c, we give c an adjacent assignment, at 01, and indicate the transition by an arrow from a to c. Next, the transition from c back to a is indicated by an arrow from c to a. Since there is a transition from c to b, b is given an adjacent assignment at 11, leaving 10 for d. The diagram is then completed by filling in arrows for all the other transitions. In this case, all transitions are between adjacent states, so this assignment is free of critical races.

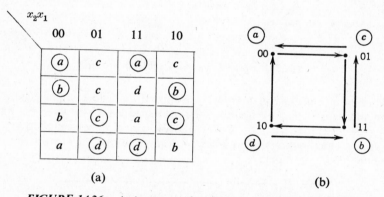

(a)   (b)

FIGURE 14.26   *Assignments of states to prevent critical races.*

A slightly different situation is illustrated in Fig. 14.27. Now d has a transition to c, resulting in a diagonal transition, i.e., a change of two variables. Note that this diagonal transition cannot be eliminated since d has transitions to three other states and cannot be adjacent to all of these with only two state variables. However, this diagonal transition is seen to represent a noncritical race and will cause no problems. Whenever there is only one stable state in a column, transitions into that column need not be considered in choosing a state assignment since critical races cannot occur unless there are at least two stable entries in a column. Thus, the same assignment will be valid for this state table.

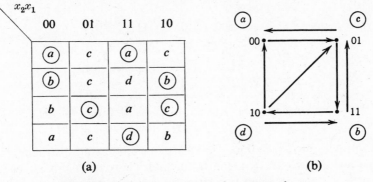

(a)   (b)

FIGURE 14.27   *Assignment with noncritical race.*

In Fig. 14.28, we see a more difficult situation. Here we have several critical races, as indicated by the diagonal transitions, and it is clear that no permutation of the assignment can eliminate all diagonal transitions. For example, the first row requires both Ⓑ and Ⓓ to be adjacent to Ⓐ, while the second row requires Ⓑ and Ⓓ to be adjacent. Clearly, not all these requirements can be satisfied.

In situations such as the above, the critical races can be eliminated only by the use of spare states. If the number of states in the original state table is not a power of two, spare states will automatically be available. This was the case in the control delay, where there were only three states and the spare state was used for a cycle to avoid the critical race. If the original number of states is a power of two or if a satisfactory assignment is not found using available spare states, spare states must be created by adding extra state variables.

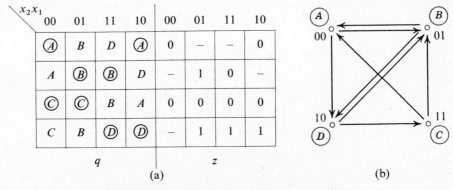

(a)   (b)

FIGURE 14.28   *State assignment with critical races.*

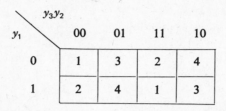

y_1	00	01	11	10
0	1	3	2	4
1	2	4	1	3

(header: y_3y_2)

FIGURE 14.29 *Universal assignment for 4-state tables.*

There are two basic techniques for using the spare states, *shared-row* assignments and *multiple-row* assignment. In the multiple-row assignment, as the name implies, each state is assigned two or more combinations of state variables, i.e., two or more rows in the transition table. Figure 14.29 shows a universal multiple-row assignment for four-state circuits. The extra rows are created by using three state variables. The four states are arbitrarily numbered 1 to 4, and their assignments are indicated by their location on the K-map of the state variables. There are two state variable combinations assigned to each state, with the two assignments for each state being logical complements.

The structure of this assignment is shown more clearly in Fig. 14.30. The two combinations assigned to each state are on diagonally opposite corners of the cube. Both assigned combinations for a state are adjacent to one of the combinations for each of the other states. This is illustrated by the arrows in Fig. 14.30, indicating adjacent transitions from either ① state to any of the other states.

The development of the transition table for the universal assignment for the state table of Fig. 14.28 is shown in Fig. 14.31. We start by setting up an augmented state table (Fig. 14.31a) with the eight-state variable combinations listed and the corresponding state assignment. Since state numbers are arbitrary, we let $1 \rightarrow A$, $2 \rightarrow B$, etc. Although there is no logical difference, we denote the two representations of each state as A_1, A_2, B_1, B_2, etc., for ease of reference. For each row assigned to a given state, the stable entries are the same as in the

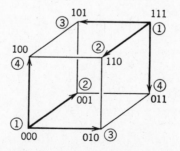

FIGURE 14.30 *Universal 4-state assignment on hypercube.*

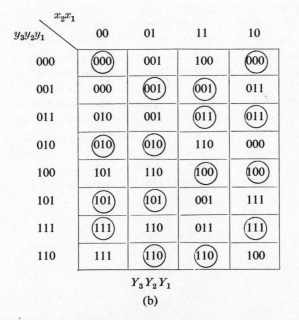

		x_2x_1							
$y_3y_2y_1$		00	01	11	10	00	01	11	10
A_1	000	A_1	B_1	D_2	A_1	0	–	–	0
B_1	001	A_1	B_1	B_1	D_1	–	1	0	–
D_1	011	C_1	B_1	D_1	D_1	–	1	1	1
C_1	010	C_1	C_1	B_2	A_1	0	0	0	0
D_2	100	C_2	B_2	D_2	D_2	–	1	1	1
C_2	101	C_2	C_2	B_1	A_2	0	0	0	0
A_2	111	A_2	B_2	D_1	A_2	0	–	–	0
B_2	110	A_2	B_2	B_2	D_2	–	1	0	–

States   z

(a)

$y_3y_2y_1$	x_2x_1 00	01	11	10
000	000	001	100	000
001	000	001	001	011
011	010	001	011	011
010	010	010	110	000
100	101	110	100	100
101	101	101	001	111
111	111	110	011	111
110	111	110	110	100

$Y_3 Y_2 Y_1$

(b)

FIGURE 14.31   *Development of transition table for state table of Fig. 14.28.*

original state table. For example, state $\textcircled{A}$ is stable in the first and fourth columns, so rows $\textcircled{A_1}$ and $\textcircled{A_2}$ are stable in the same columns. Similarly the outputs are the same for each row assigned a given state. For state $\textcircled{A}$, the output row is 0, $-$, $-$, 0, so the output rows for $\textcircled{A_1}$ and $\textcircled{A_2}$ are both 0, $-$, $-$, 0.

The transitional entries are filled in to provide adjacent transitions. State $\textcircled{A}$ makes transitions to $\textcircled{B}$ and $\textcircled{D}$. $\textcircled{A_1}$ is adjacent to $\textcircled{B_1}$ and $\textcircled{D_2}$, and $\textcircled{A_2}$ is adjacent to $\textcircled{B_2}$ and $\textcircled{D_1}$, so the transitional entries are filled in accordingly. We can now see why the external behavior of the circuit is not changed by the use of multiple states. Consider the circuit to be in state $\textcircled{A}$, with a stable output of 0 and either $x_2 x_1 = 00$ or $x_2 x_1 = 10$. It makes no difference whether the circuit is actually in $\textcircled{A_1}$ or in $\textcircled{A_2}$. From state $\textcircled{A}$, it should go to $\textcircled{B}$ or $\textcircled{D}$. From $\textcircled{A_1}$, it will go to $\textcircled{B_1}$ or $\textcircled{D_2}$; from $\textcircled{A_2}$, it will go to $\textcircled{B_2}$ or $\textcircled{D_1}$. Either way, the new stable state will correspond to $\textcircled{B}$ or $\textcircled{D}$ of the original state table. The same argument holds for any other state transitions. The final transition table, Fig. 14.31b, is obtained by replacing the state entries with the assigned combinations of state variables.

The multiple-row assignment method always requires extra state variables since the number of rows must be at least twice the number of states. Adding state variables will increase the cost, so *whenever extra rows are available because the number of states is not a power of two, the shared-row approach is generally preferable*. In this method, the extra rows are not assigned to any specific state but instead are *shared* as required to convert any critical races into cycles of adjacent transitions. For 3-state tables, any assignment such as the one shown in Fig. 14.32 will lead to a critical-race-free shared-row realization. Any transitions between 1 and 2 or 1 and 3 are adjacent; transitions between 2 and 3 cycle through the spare row. We have already used this assignment for the control-delay example.

For 4-row tables, there will be no spare rows to share unless we add a state variable. But we have already seen a universal 4-row assignment with three state variables, so either this approach or the shared-row approach can be used for such tables. Almost all 5-state machines can also be handled using three state variables. A universal assignment for n states is an assignment that will work for every n state flow and any numbering of the states. Universal assignments for 5

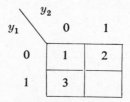

FIGURE 14.32   *Universal shared-row assignment for 3-state tables.*

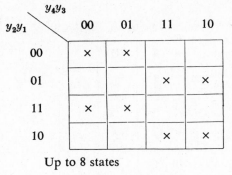

Up to 8 states

(a)

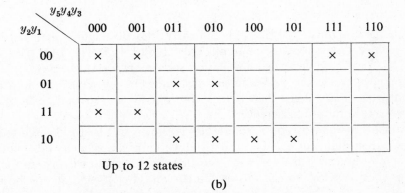

Up to 12 states

(b)

FIGURE 14.33 Universal shared-row assignments.

to 8 states and 9 to 12 states are shown in Fig. 14.33 [12]. Since there is only one row per state and the numbering is arbitrary, we do not indicate any state numbers in Fig. 14.33 but simply indicate by × the rows to which states should be assigned. For 5 to 8 states, simply assign one ×ed row from Fig. 14.33a to each state, and similarly for 9 to 12 states for Fig. 14.33b.

Although universal assignments require an extra variable to allow for all possible cases, shared-row assignments that do not require extra variables can often be found for specific tables, particularly if there are several spare states available without adding extra variables. A useful concept in finding such assignments is the *destination set*. In any column of a state table, a destination set consists of any stable state and any rows that make transitions into that state. For example, in the state table of Fig. 14.34, in the first column, the stable states are ①, ③, ⑤; and row 4 leads to ①, row 2 to ⑤. The destination sets for

$x_2 x_1$

	00	01	11	10	00	01	11	10
1	①	3	4	①	0	–	0	0
2	5	4	②	②	–	1	1	0
3	③	③	2	1	1	1	1	–
4	1	④	④	④	–	1	0	1
5	⑤	⑤	⑤	4	0	1	1	1

States

FIGURE 14.34   Example 5-state table.

this column are thus $(1, 4)(2, 5)(3)$. Similarly, the destination states for the second, third, and fourth columns are $(1, 3)(2, 4)(5)$, $(1, 4)(2, 3)(5)$, and $(1, 3)(2)(4, 5)$, respectively. The destination sets are important because, to avoid critical races, the members of each set must either be adjacent or so located in relation to the spare states that cyclic transitions for all sets in a given column may be made without interference.

A standard assignment for 5-state tables is shown in Fig. 14.35a. We term it *standard* rather than *universal* because it will work for almost every table but will fail for a few rare cases, as we will see. For this case we arbitrarily assign the five states as shown in Fig. 14.35b on the K-map, and, Fig. 14.35c on the Boolean hypercube. In the first column of the state table, the destination sets are $(1, 4)(2, 5)(3)$, requiring transitions from 4 to 1 and from 2 to 5. We see in Fig. 14.35c that this is impossible with this assignment as the two cycles must use the same spare rows enroute to different destinations. Rows may be *shared* only in the sense that they may be used for different cycles in different columns. In any single column, a transitional row may be part of a cycle to only one destination.

This problem occurs only in columns having $(2, 2, 1)$ destination sets, i.e., columns having 2, 2, and 1 members in three destination sets. It can be eliminated by permuting the assignment so that the two 2-member sets do not occupy diagonally opposite vertices of the hypercube. In terms of the assignment map, the 2-member sets must not occupy the connected rows as shown in Fig. 14.35d. There are 15 possible $(2, 2, 1)$ column configurations and, as long as all 15 do not occur in the same table, it is possible to avoid assigning any $(2, 2, 1)$ sets in the "forbidden" pattern. This assignment will thus fail only in the unlikely event of a state table with 15 or more columns, including all 15 possible $(2, 2, 1)$ columns [12].

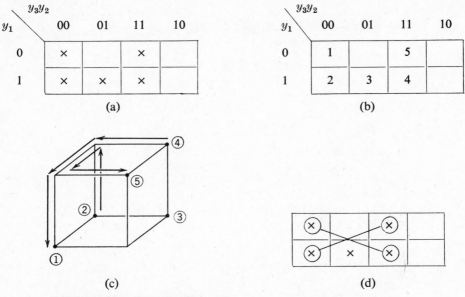

FIGURE 14.35 *Standard shared-row assignment for 5-state tables.*

The easiest way to use this assignment is to construct any arbitrary $(2, 2, 1)$ group of destination sets that does *not* occur in the destination sets of the state table and assign them to the "forbidden" pattern. This guarantees that none of the $(2, 2, 1)$ columns which are in the state table can be assigned to that pattern.

For the table of Fig. 14.34, we see that the $(2, 2, 1)$ pattern $(1, 2)(3, 5)(4)$, for example, does not appear among the destination sets, so we assign $(1, 2)$ and $(3, 5)$ to the diagonal locations, as shown in Fig. 14.36a. For ease of reference, the spare rows are labeled α, β, γ. To set up the augmented state table (Fig. 14.34b), we start by filling in the stable states and the corresponding outputs, just as for a multiple-row assignment.

The transitional entries are filled in with reference to the map of the assignment (Fig. 14.36a) to determine if the transition is direct or whether a cycle through spare states must be provided. State ① makes transitions to ③ in column 01 and to 4 in column 11. The transition to ③ is direct, but the transition to ④ uses the cycle ① $\rightarrow \alpha \rightarrow 4 \rightarrow$ ④, as seen in the third column of Fig. 14.36b. Transitions ② $\rightarrow$ ④ and ② $\rightarrow$ ⑤ are both direct. Transition ③ $\rightarrow$ ① is direct, but ③ $\rightarrow \gamma \rightarrow 2 \rightarrow$ ② cycles through γ. The other entries are filled in similarly. We note here an advantage of the shared-row method. It is rare that all the spare row entries are needed, resulting in many don't-cares. By contrast, the multiple-row method does not provide any don't-cares other than those in the original state table.

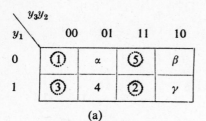

(a)

$y_3y_2y_1$	x_2x_1 00	01	11	10	00	01	11	10
① 000	①	3	α	①	0	–	0	0
③ 001	③	③	γ	1	1	1	1	–
④ 011	α	④	④	④	1	1	0	1
α 010	1	–	4	4	–	–	0	–
β 100	–	–	–	–	–	–	–	–
γ 101	–	–	2	–	–	–	1	–
② 111	5	4	②	②	0	1	1	0
⑤ 110	⑤	⑤	⑤	α	0	1	1	–

(b)

$y_3y_2y_1$	x_2x_1 00	01	11	10
000	⟨000⟩	001	010	⟨000⟩
001	⟨001⟩	⟨001⟩	101	000
011	010	⟨011⟩	⟨011⟩	⟨011⟩
010	000	× × ×	011	011
100	× × ×	× × ×	× × ×	× × ×
101	× × ×	× × ×	111	× × ×
111	110	011	⟨111⟩	⟨111⟩
110	⟨110⟩	⟨110⟩	⟨110⟩	010

$Y_3 Y_2 Y_1$

(c)

FIGURE 14.36   *Valid 5-state assignment for Fig. 14.34.*

Up to now, very little has been said with regard to the assignment of outputs corresponding to unstable states. Some rules must be followed, and we will introduce them now. The output entries for the transitional states must be chosen to eliminate any spurious pulses. If the output is to be the same before

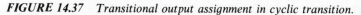

States: ④ → α → 1 → ①   ④ → α → 1 → ①
Output: 1   –   –   0   1   0   1   0
              (a)   (b)

States: ④ → α → 1 → ①   ④ → α → 1 → ①
Output: 1   1   –   0   1   –   0   0
              (c)   (d)

FIGURE 14.37 *Transitional output assignment in cyclic transition.*

and after the transition, it must saty constant throughout the cycle. If the output is to change, it must change only once during a cycle. Therefore, it can be a don't-care for only one transitional state during the cycle. Consider the transition from ④ in the second column to ① in the first column, in Fig. 14.36b, a transition that cycles through x. If we let both transitional outputs be don't-cares, as shown in Fig. 14.37a, the result, depending on groupings of don't-cares, might be a momentary 1 pulse, as showr in Fig. 14.37b. To prevent this, either the first transitional state must be specified as 1, or the second as 0, as shown in Figs. 14.37c and 14.37d, respectively. Then choice between the latter two should be made on the basis of which seems likely to result in the simplest output equations.

The transition table follows directly from the augmented state table by replacing each state by the corresponding state variable combination, as shown in Fig. 14.36c.

There are no generally applicable 3-variable assignments for 6 and 7 state tables, but suitable assignments can often be found through careful analysis of the destination sets. This may be illustrated by the following example.

Example 14.3

Obtain a critical-race-free state assignment for the table of Fig. 14.38.

x_1x_2			
00	01	11	10
3	6	①	①
②	3	1	②
③	③	③	5
3	④	④	1
2	⑤	6	5
⑥	⑥	⑥	2

FIGURE 14.38 *Example 6-state table.*

Solution   The destination sets are $(1, 3, 4)(2, 5)(6)$, $(1, 6)(2, 3)(4)(5)$, $(1, 2)(3)(4)(5, 6)$, and $(1, 4)(2, 6)(3, 5)$. First we note a destination set with more than two members, $(1, 3, 4)$, in the first row. In such cases, it is often possible to cycle within the destination set, e.g., ① → 4 → ③ or ④ → 1 → ③. In view of the flexibility offered by such cycles, it is usually best to consider the 2-member destination sets first. In this case, we note that state 2 is a member of four such sets $(2, 5)$, $(2, 3)$, $(1, 2)$, and $(2, 6)$. In a 3-variable assignment, a single state can be adjacent to only 3 other states, so these four destination states require that ② be adjacent to one of the spare states, which we shall designate α. Of the four states paired with ② in destination sets, ⑤ and ⑥ are not part of the $(1,.3, 4)$ set, so we place them adjacent to the $(2, \alpha)$ pair, giving the initial assignment shown in Fig. 14.39a. The remaining state paired with ②, ① and ③ must be adjacent to ② or α, so we locate them as shown in Fig. 14.39b. Finally, noting that state ④ can cycle to ③ through state ①, we place 1 and the remaining spare state, β, as shown in Fig. 14.39c. We then check

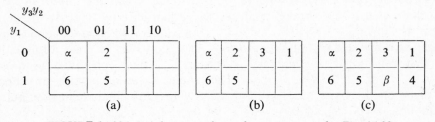

FIGURE 14.39   *Development of race-free assignment for Fig. 14.38.*

the destination sets for all columns to see that all transitions can be made without critical races with the assignment of Fig. 14.39c. The development of the augmented state table and transition table is shown in Fig. 14.40.

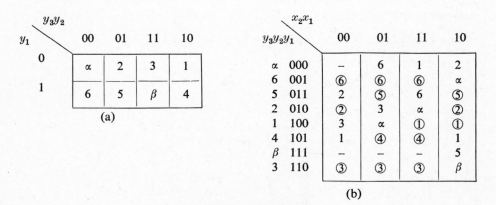

FIGURE 14.40   (a) (b)

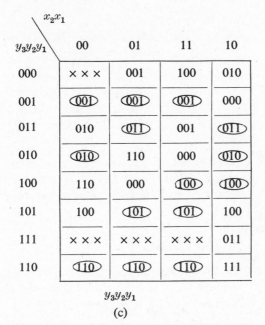

$$y_3y_2y_1$$

(c)

FIGURE 14.40   *Assignment, augmented state table, and transition table for Fig. 14.38.*

■

Shared row assignments ultilizing cycles tend to be more economical than multiple-row assignments, but we often pay for this economy in slower operation. An assignment such as the universal 4-state assignment of Fig. 14.29 is known as a *single transition time* (STT) assignment, since each transition involves the change of only a single variable. By contrast, consider the standard 5-state assignment as shown in Fig. 14.41, and assume that a transition is required from ⑤ → β → γ → ②. This requires that y_2 first change from 1 to 0, then y_1 from 0 to 1, and finally y_3 from 1 to 0, thus requiring three transition times to complete the change. Not every state table will call for this particular transition (Fig. 14.34 does not); but since it can occur, this assignment is characterized as a *three transition time* (3TT) assignment. The standard 8-state and 12-state assignments of Figs. 14.33a and 14.33b are 3TT and 5TT assignments, respectively.

Since speed of operation is often very important, the designer may wish to find assignments that are faster than the standard shared-row assignments. Unfortunately, for tables with more than four states, faster assignments are either very difficult to find or require an excessive number of state variables, or both. In most cases, the designer will probably do best to use one of the assignment

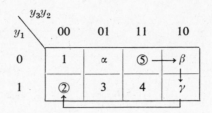

FIGURE 14.41   *Five-state assignment with triple transition.*

techniques discussed in this section. For those cases when speed is critical to the extent that a considerable expenditure in design time and extra hardware is warranted, the reader is referred to Reference 13 for a number of techniques for finding fast assignments.

14.8   Hazards in Sequential Circuits

In Fig. 14.42, we show the flow table and transition table for a sequential circuit. The circuit is free from critical races and so, presumably, will function properly. The minimal equations for the state variables are

$$Y_2 = y_2 x_2 + y_2 \bar{y}_1 x_1 + \bar{y}_1 x_2 \bar{x}_1 + y_2 y_1 \bar{x}_1 \tag{14.8}$$

and

$$Y_1 = \bar{y}_2 \bar{x}_2 y_1 + y_2 y_1 x_1 + y_2 x_2 x_1 + y_1 x_2 \bar{x}_1 \tag{14.9}$$

The logic circuit for Y_1 is shown in Fig. 14.43. The logic for Y_2 is only indicated symbolically since it is not pertinent to the present discussion.

Let us assume that the circuit is in the stable state ⓒ at $y_2 y_1 x_2 x_1 = 1111$. Under this condition, both gates A2 and A3 will develop 1 at their outputs,

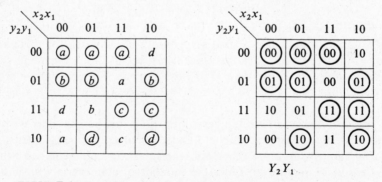

FIGURE 14.42   *Flow table and transition table for hazard example.*

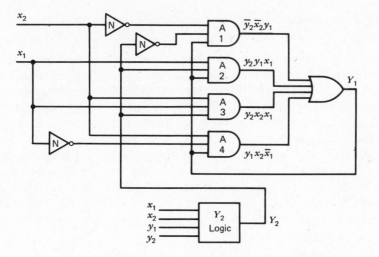

FIGURE 14.43   *Circuit illustrating static hazard.*

producing $Y_1 = 1$ at the output of the OR-gate. Now let x_1 change to 0. We see that the circuit should move to the stable state ⓒ at $y_2 y_1 x_2 x_1 = 1110$, that is, neither Y_1 or Y_2 should change. But let us consider the circuit action in detail. When x_1 goes to 0, the outputs of gates A2 and A3 will go to 0, and the output of gate A4 will go to 1 due to $\bar{x}_1$ going to 1. However, because of the delay in the inverter, $\bar{x}_1$ will not go to 1 as fast as x_1 goes to 0, so that there will probably be a short interval when the outputs of all AND-gates will be 0. The output of the OR-gate will thus momentarily go to 0. Since y_2 does not change, the circuit will momentarily see the state variable input $y_2 y_1 = 10$ and may make an erroneous transition to stable state ⓓ at $y_2 y_1 x_2 x_1 = 1010$.

Next, let us assume that same starting state, $y_2 y_1 x_2 x_1 = 1111$, and now let x_2 go to 0. Now the circuit should go through unstable b to stable ⓑ at $y_2 y_1 x_2 x_1 = 0101$. Again, Y_1 should not change. As before, gates A2 and A3 are 1 initially. When x_2 goes to 0, gate A3 goes to 0 and, at the same time, the transition in Y_2 is started. After delay through the Y_2 logic, y_2 goes to 0, taking A2 to 0, and A1 goes to 1 due to $\bar{y}_2$ going to 1. However, as before, the delay through the inverter will cause all gates to be 0 momentarily, in turn causing Y_1 to go to 0 momentarily. The circuit may thus make an erroneous transition to stable a at $y_2 y_1 x_2 x_1 = 0001$.

The above are examples of malfunctions caused by *static hazards*.

Definition 14.2   Let there be a state transition in a fundamental mode circuit such that one or more of the secondary variables is required to remain constant. If it is possible, due to unequal delays in the circuit, for one or more of the constant state variables to change momentarily, then a *state hazard* is said to exist.

Note that a static hazard is a characteristic of the combinational logic. In the example given earlier, if y_1 were a normal input rather than a feedback variable, Y_1 would still go to 0, but only momentarily. In normal combinational design or in clocked or pulse circuits, static hazards are not of concern since momentarily erroneous signals are not generally troublesome. However, if a momentarily incorrect signal is fed back to the input, as in the above example, a permanent error can occur. A static hazard will not always cause a malfunction, depending on the various delays, but reliable design requires that the hazard be eliminated.

The hazard can be seen quite clearly on the K-map of the Y_1 circuit, Fig. 14.44a. In the first case, the change in x_1 moves the circuit out of the cubes $y_2 y_1 x_1$ and $y_2 x_2 x_1$ and into the cube $y_1 x_2 \bar{x}_1$. In the second case, the change in

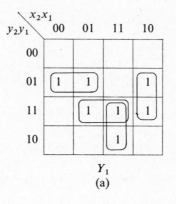

Y_1

(a)

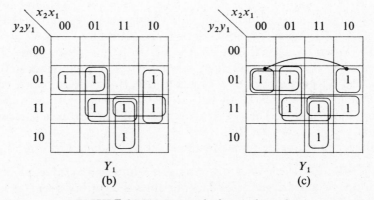

FIGURE 14.44   Removal of static hazards.

x_2 moves the circuit out of the cube $y_2 y_1 x_1$ and into the cube $\bar{y}_2 \bar{x}_2 y_1$. As we have seen, the circuit moves out of the starting cubes faster than it moves into the final cubes. Each cube represents one product input to the output OR-gate. One of the product-terms must be at 1 to provide a 1 output. Whenever the circuit must move from one cube to another, there is a possibility of a momentary interval when neither corresponding product is at 1. On this basis, the remedy is fairly obvious. Any pair of 1s between which a transition may take place must be in a single cube. Thus, the two hazards in this example can be removed by adding two more cubes, as shown in Fig. 14.44b.

We note that there is another static hazard between the 1s in $y_2 y_1 x_2 x_1 = 0100$ and $y_2 y_1 x_2 x_1 = 0110$, so that these two 1s should also be in a single cube. On the other hand, the two 1s in $y_2 y_1 x_2 x_1 = 0110$ and $y_2 y_1 x_2 x_1 = 1110$ need not be a single cube. From the flow table, we see that both of these 1s correspond to stable states, and vertical transitions between stable states are not possible in fundamental-mode circuits. On this basis, the minimal hazard-free realization is that shown in Fig. 14.44c. The resultant circuit will have two more AND-gates than the original, but it will be hazard-free.

The situation illustrated here, where the output is supposed to remain 1, is known as a *static hazard in the 1s*. It should be obvious that the same thing could happen when the output is supposed to remain 0, in which case we have a *static hazard in the 0s*. It is also possible to have a hazard when the output is supposed to change. When certain patterns of delay occur, the circuit may go momentarily to the new value and then back to initial value before making the permanent transition, that is, $0 \to 1 \to 0 \to 1$ instead of $0 \to 1$. This situation is known as a *dynamic hazard*. Like the static hazard the dynamic hazard is seldom of real concern in combinational circuits but may cause malfunctions in fundamental-mode sequential circuits.

The detection of all possible types of hazards in a given circuit is very complicated and has been the subject of extensive investigation [3]. However, as designers, all we need is some rule which is sufficient to ensure freedom from hazards. The following theorem provides this rule.

THEOREM 14.1   *A second-order SOP circuit that is free of all static hazards in the 1s will be free of all static and dynamic hazards.*

The proof of this theorem is too involved to be considered here. Note that this is not a necessary condition, i.e., hazard-free circuits of higher order are possible. However, the conditions are much more complex for higher-order circuits. For purposes of designing hazard-free fundamental-mode circuits, we need only restrict ourselves to second-order SOPs, ensuring that any pair of adjacent 1s on the excitation maps, corresponding to a legitimate input transition, are included in a single product-cube. For POS design, the same rules may be applied to design on the 0s.

Example 14.3

Develop a hazard-free realization for the excitation map shown in Fig. 14.45a. The minimal equation, representing the groupings shown in Fig. 14.45a, is given by

$$Y_1 = \bar{x}_1 y_2 + x_1 x_2 y_1 + x_1 \bar{x}_2 \bar{y}_2 \qquad (14.10)$$

We see that there are two hazards, between 5 and 13 and between 11 and 15. The revised groupings of Fig. 14.45b eliminate these hazards and, therefore, all static and dynamic hazards. The resulting Boolean expression,

$$Y_1 = \bar{x}_1 y_2 + x_2 y_1 y_2 + x_1 y_1 \bar{y}_2 + x_1 \bar{x}_2 \bar{y}_2 + x_1 x_2 y_1 \qquad (14.11)$$

is not minimal, but it is hazard-free.

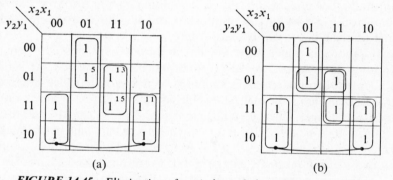

FIGURE 14.45 *Elimination of static hazards by redundant products.*   ∎

Like races, the static and dynamic hazards are due to unequal delays in various signal paths. Still another type of hazard arises when delays result in one excitation changing before the circuitry generating another excitation is even aware of the input change. In Fig. 14.46, we show the transition table, excitation maps, and output map for a sequential circuit designed to count modulo-4 the number of changes in input x.

The excitation equations with static hazards eliminated are given by Equations 14.12 and 14.13.

$$Y_2 = y_1 \bar{x} + y_2 x + y_1 y_2 \qquad (14.12)$$

$$Y_1 = y_1 \bar{x} + \bar{y}_2 x + \bar{y}_2 y_1 \qquad (14.13)$$

Since we have carefully avoided all critical races and all static and dynamic hazards, the resulting circuit should work as intended. Or will it?

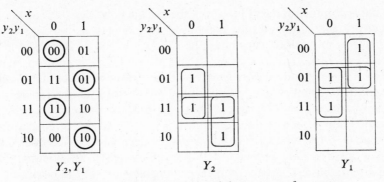

FIGURE 14.46 *Transition table and design maps for counter.*

Consider the realization of these equations as shown in Fig. 14.47. Assume the circuit is in the stable state ⑪, for which the signals on the various lines are as shown. Now let x go to 1, which should change Y_1 and take the circuit to stable ⑩. The input $x = 1$ ($\bar{x} = 0$) drives AND-3 to 0 and thus Y_1 to 0. The change in Y_1 in turn takes the output of AND-1 to 0. If the circuit is functioning properly, AND-2 will change to 1 at the same time AND-3 changes to 0, so that the subsequent change of AND-1 will not affect Y_2. But suppose that the delay in AND-2 is very long so that its output is still 0 when the output of AND-1 goes to 0. Then OR-1 will see all 0 inputs and Y_2 will go to 0. Instead of seeing itself

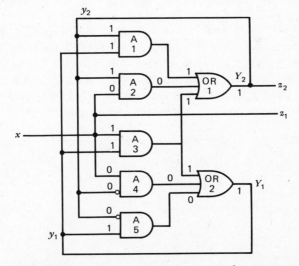

FIGURE 14.47 *Essential hazard in input-change counter.*

at the stable combination $xy_2 y_1 = 110$, the circuit now sees itself at the unstable condition $xy_2 y_1 = 100$ and thus makes an erroneous transition to stable ⓪①.

At first, this might look very much like the static hazard discussed in connection with Fig. 14.42, but in that case there was only one excitation involved, so that the problem could be corrected by adding gates in the circuitry generating that excitation variable. Here the problem involves an interaction between two excitation variables. The variable Y_1 has changed and propagated back through the circuitry for Y_2 before that circuitry has completed responding to the initial input change. This type of hazard, which is peculiar to sequential circuits, is known as an *essential hazard*. As the name suggests, an essential hazard is created by the basic logical structure, or specification, of the circuit and is not affected in any way by the particular form of realization used. The hazard can be detected on the original state table.

Definition 14.3 An *essential hazard* exists in a state table if there is a stable state from which three consecutive changes in a single input variable will take the circuit to a different state than the first change alone [4].

This behavior will occur if any two adjacent columns of the state table exhibit either of the two patterns shown in Fig. 14.48a and 14.48b. In both cases, if the circuit starts in ①, a single change of variable takes the circuit to ② and two more changes take it to ④. The reader should note that the numbering of the states and ordering of the rows in these two tables is quite arbitrary; the patterns may appear quite different in a larger flow table. For example, the transition indicated in the table of Fig. 14.48c, ③ → ⑤ → ⑥ → ②, corresponds to the pattern of Fig. 14.48b.

The reader will note that for the hazard described in connection with Fig. 14.47 to occur the delay through AND-2 would have to be longer than the cumulative delay along the path AND-3, OR-2, AND-1. The reader may feel that such a combination of delays would be very unlikely, but it is not. In typical IC gates, the standard production tolerances on delay specification allow a range of three-or-four-to-one between minimum and maximum delays times.

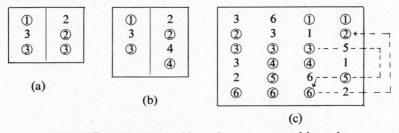

FIGURE 14.48 *Flow tables indicating essential hazards.*

Essential hazards can be found in many circuits of considerable practical importance including almost all counters. Since the effect of these essential hazards cannot be eliminated by modifying logic, they can be handled only by carefully controlling the number of gates in each feedback loop. Very often realizations can be worked out with pairs of cross-coupled gates making up loops that must respond very rapidly. Gates on a single chip will usually have less variance in delay times than typical gates on different chips. Thus, essential hazards are less likely to cause trouble in a level-mode circuit implemented in a single IC package.

Occasionally it will be necessary to insert a pair of inverters in a feedback loop so that the secondaries do not change until the input change has propagated to all parts of the circuit. The deliberate insertion of delay is not at all desirable, since speed is generally of the essence, but there is often no other remedy for the essential hazard.

The reader may ask why hazards and races do not occur in clock- or pulse-mode circuits. Recall from the discussion of Chapter 9 that flip-flops are designed so as to ensure that no resultant changes in the flip-flop excitations would be seen until after the end of the clock pulse. Furthermore, the interval between clock pulses was required to be long enough to allow all changes to stabilize. Taken in this light, we may regard clocking and the use of flip-flops as means of introducing controlled delays to eliminate the timing problems associated with level-mode circuits.

14.9   General Level-Mode Circuits

We now must consider the special problems occurring when the restrictions are removed that only one input may change at a time and that no further changes may occur until the circuit has stabilized. Essentially this means that we will now allow simultaneous changes in two or more inputs. While no two events are ever precisely simultaneous, if one change occurs and then a second follows while the circuit is still responding to the first, the changes are simultaneous in the sense that the circuit is trying to respond to both at the same time. In a sense, the speed of response of the circuit defines the finest division by which we can measure time. If the separation of two changes is less than this finest division, then the changes are simultaneous within this frame of reference.

Whenever two circuits inputs come from two unrelated and unsynchronized sources, simultaneous input changes can occur. There are many practical situations where we must deal with signals from unrelated sources. As an example, consider a computer-controlled manufacturing process. The inputs will be derived from sensors monitoring the process at a variety of points. From our knowledge of the process, we can determine the signal rates of individual variables and select our logic circuitry accordingly. But there is no way we can guarantee that two or more signals may not change at the same time. In such circumstances, the best solution may be to synchronize the inputs to an internal clock, but this will also be an unrelated source, so the same problem of simultaneous signal changes still exists.

Let us begin by considering what might go wrong in a specific multiple-input transition. Note that relaxing the fundamental-mode restrictions does not necessarily mean that every possible change may occur regardless of the present stable state. In many practical situations, there may be only one or two points within the flow table from which multiple-input changes might occur. For each stable state in the flow table, the designer must consider whether a multiple change might occur and, if so, analyze carefully the possible functioning of the circuit, as in the following.

Note that the table of Fig. 14.49 has no critical races. Although no assignment is indicated, all transitions will be adjacent with the rows arranged as shown. Suppose that the only simultaneous input change that might be expected in Fig. 14.49a is a change in $x_2 x_1$ from 00 to 11 when in stable state $\textcircled{a}$, as indicated.

Because of random delays in logic paths, not all gates will see the changes in both inputs at the same time. Thus, at the outset, there may be gates that see the inputs in any one of the four combinations of values. This is depicted graphically by darkening the first row of Fig. 14.41. Some delay combination may, for example, cause a critical gate to see only the change of x_1 from 0 to 1. Thus, the circuit might go, momentarily at least, to state $\textcircled{b}$, and different points in the circuit might take on values corresponding to the second row of the table. Now suppose there is another gate that still sees the inputs as 00 but sees the state variables as corresponding to the second row. This gate might thus send the

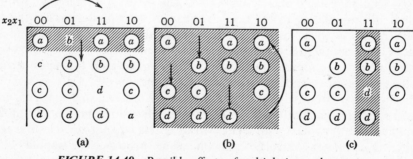

FIGURE 14.49 *Possible effects of multiple input changes.*

circuit to state ⓒ, making it possible for some points in the circuit to take on the values in the third row of the table. Then a gate seeing the inputs correctly as 11 might take it to ⓓ, where it might stabilize, or go on to ⓐ if some other gate sees the inputs as 10, as shown in Fig. 14.49b. Thus, depending on delays, the circuit might momentarily pass through any point in the state table as suggested by the shading of the entire table in Fig. 14.49b. Any arbitrary combination of input and secondary values could conceivably be assumed by some section of some realization of the flow table at some point in time shortly after the input change. On the other hand, many of the combinations of values might be remote if not impossible in any particular realization. Eventually, the effect of the dual-input change will have propagated throughout the circuit. Once this has occurred, the circuit will be restricted to states in column 11, as depicted in Fig. 14.41c. Since there are 3 stable states in this column, a malfunction would seem to be a real possibility.

The situation illustrated above is very similar to the essential hazard in that not all gates receive full information about the input change at the same time. The essential hazard is a special case in that difficulty can arise even with only a single input change.

Let us now consider tactics by which we might limit the chaos suggested by the maximally pessimistic analysis above. First, if there is only one stable state in a column, there can be no malfunction. Sooner or later all changes propagate sufficiently to bring the circuit into the proper column; and if there is only one stable state in that column (noncritical race), the circuit must terminate there. Unfortunately, this is not a condition over which we generally have any control, as the number of stable states in a column is determined by the basic function the circuit is to perform.

If the multiple change is into a column with two or more stable states, reliable operation will be ensured if the following two rules are satisfied.

Rule 1   There must be enough delay in each feedback loop so that all input changes will have propagated through at least the first level of gating before any secondary changes have propagated around a loop.

Rule 2.   If a partial input change, i.e., a change in a subset of the inputs involved in a multiple change, can cause the circuit to leave a row in the state table, this behavior must be consistent with the intended behavior associated with the multiple change.

If sufficient delay is included to satisfy Rule 1, then the range of movement due to an input change can be limited to two rows of the flow tables. To do this, we must specify a transition table so that no races critical or noncritical are initiated in a row where a multiple input change can occur. This situation is illustrated in

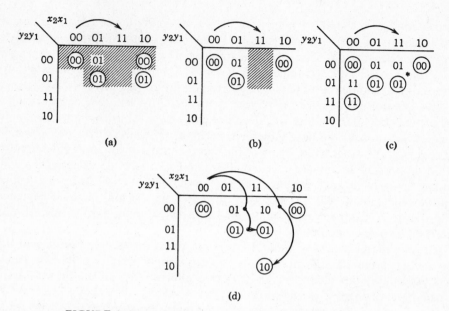

FIGURE 14.50 *Limiting the range of multiple input changes.*

Fig. 14.50. If the x_2 change is seen before the x_1 change $(00 \rightarrow 10)$, there is no change of state, so the circuit stays on the proper row until the x_1 change is seen. If the x_1 change is seen first $(00 \rightarrow 01)$, y_1 may change, so that part of the circuit may see the state as on the second row. Because of the delay in feedback, no gate will see the y_1 change until it sees the inputs as 11. The circuit is thus initially limited to the shaded area of Fig. 14.50a, and then, when all gates see the 11 input, to the shaded area in Fig. 14.50b.

The feedback delay prescribed by Rule 1 thus ensures that the circuit moves into the proper column, but what happens then will depend on the satisfaction of Rule 2. Figure 14.50c illustrates one manner in which this rule might be satisfied. Note that both unstable states in row 00 specify a transition to stable state ⓪①. Thus, no matter how the dual-input change might initially be interpreted, the circuit will always stabilize at the point marked with the asterisk, *provided Rule 1 is satisfied*. We have just illustrated the most common way in which Rule 2 can be satisfied. That is, the intended destination for any multiple-input change must be the same as the final destination reached by an equivalent sequence of single-input changes. In this case, the intended destination of the change $00 \rightarrow 11$ is the same as the final destination of the sequential change $00 \rightarrow 01 \rightarrow 11$.

It is important to note that both rules must be satisfied. Figure 14.50d shows a version of the table for which Rule 2 is not satisfied and for which the circuit is about equally likely to wind up in either ⓪① or ①⓪ even though Rule 1 may be satisfied. In terms of the design process, the designer must consider Rule 2 at the time the state table is set up. The designer will then ensure that Rule 1 is satisfied only after the complete circuit has been designed by selecting gates to achieve the proper delay relationships.

Let us now consider a complete design example.

Example 14.5

Because of gate delays, line delays, and communication between units with separate clocks, a control pulse to a section of a computer system may arrive out of synchronism with the clock for that section. In order to avoid unpredictable operation in the clocked circuits of the computer, such a control pulse must be routed through some type of pulse synchronizer. This pulse synchronizer will have two inputs, the computer clock line C and the control pulse line P. Pulses on line P may arrive at random times relative to clock pulses. The inputs are pulses, but because they may violate the pulse-mode restriction of one input pulse at a time, we must consider them as consecutive changes in levels and apply level-mode techniques.

Corresponding to every pulse on line P, the synchronizer is to emit a pulse on output line Z, synchronized with a clock pulse. Pulses on line P are of approximately the same duration as clock pulses but will occur less frequently, i.e., always separated by several clock periods. This is an example of a situation in which some, but not all, multiple-input changes may occur.

Solution   As with fundamental-mode design, it is possible to start with a primitive-flow table, but in this case we find it convenient to proceed directly to a table with more than one stable state per row. Let state ① be a quiescent stable state following an output pulse, i.e., a situation in which no further output should occur until another input appears on P. As illustrated in Fig. 14.51a, the circuit merely moves back and forth between the stable states labeled ① as the clock goes off and on in the absence of an input control pulse. When a pulse appears on line P, the circuit goes to stable state ②, as indicated in Fig. 14.51b. Whether the ultimate circuit senses single changes or multiple changes on row ①, it will eventually stabilize in row ② when a clock pulse arrives, given satisfaction for Rule 1.

Note that the output, though not shown, must remain 0 for all stable states shown in Fig. 14.51b. It might seem that a 1 output could be specified for ② with input 11, since

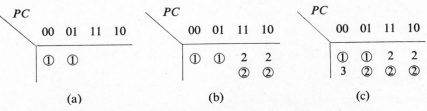

FIGURE 14.51   *Developing a flow table for the pulse synchronizer.*

this corresponds to the direct change $00 \rightarrow 11$ and would apparently produce a full-width output pulse. But to do so would violate our rule that we must not specify different behavior for a simultaneous change than for any sequence of single changes, since we do not wish an output if either P or C changes alone. In turn, we must specify a stable ② with 0 output for input 01, or else we would have a partial output when the pulses overlapped and P ended first. The circuit can move out of row ② only when both inputs return to 0, as shown in Fig. 14.51c.

The net effect of the above specification is to ensure that no output will occur until after the circuit has sensed both inputs at 0 following a pulse on P. In the worst case, as shown in Fig. 14.52, this means that an output pulse may be delayed relative to a control pulse by more than one clock period. Given the minimum interval of several clock periods between control pulses, this delay presents no problem and is necessary to achieve reliable operation in response to multiple-input changes.

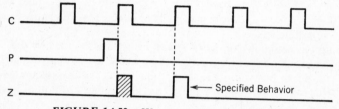

FIGURE 14.52   *Worst-case synchronizer delay.*

Once the circuit has reached stable state ③, the next clock pulse will take it to state ④, generating an output of 1 for the duration of the clock pulse. The assumed interval between control pulse inputs precludes the possibility of multiple-input changes on rows ③ and ④ and permits the don't-care entries shown in Fig. 14.53. In the output table, the values for z corresponding to unstable states have been specified where necessary.

The only possible multiple-input change that has not been discussed (and found to satisfy Rules 1 and 2) is identified by an arrow in Fig. 14.53. If the circuit initially only senses the change in P (from 1 to 0), the circuit will leave row ② via unstable 3 and move on to stable state ④ when the complete input change is sensed. This will result in the shaded output pulse in Fig. 14.52. If the circuit actually interpreted the changes from

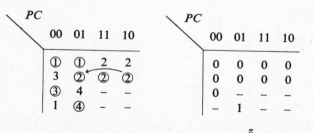

FIGURE 14.53   *Flow table for synchronizer.*

$PC = 10$ to $PC = 01$ as taking place simultaneously, the circuit will wait a full clock period to put out the unshaded pulse in Fig. 14.52. Clearly the two possible forms of circuit behavior are quite different; but does it matter? In both cases, the output is a single pulse synchronized with a clock pulse and of clock-pulse duration. From the problem statement, we conclude that it makes no difference which type of circuit behavior actually takes place. This is an interesting special case of the satisfaction of Rule 2. The circuit "does something different," but not in any significant sense.   ■

We now turn our attention to obtaining a circuit realization. States are assigned without difficulty yielding the transition table of Fig. 14.54a. Individual Karnaugh maps are included to illustrate the minimal multiple-output realization for Y_2 and Y_1 as given by Equations 14.14 and 14.15. Note the inclusion of the term $y_2 y_1$ to avoid a static hazard in Y_2. Equation 14.16 for output z is taken directly from Fig. 14.53.

$$Y_2 = \bar{P}\bar{C}y_1 + y_1 y_2 + y_2 C \tag{14.14}$$

$$Y_1 = \bar{P}\bar{C}y_1 + \bar{y}_2 y_1 + P \tag{14.15}$$

$$z = y_2 \bar{y}_1 \tag{14.16}$$

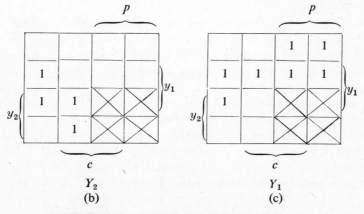

(a)

(b)   (c)

FIGURE 14.54   *Transition table for synchronizer.*

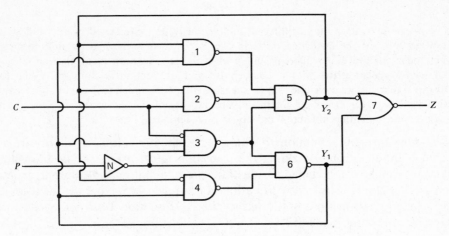

FIGURE 14.55   *Pulse synchronizer.*

A logic block diagram implementation of the excitation equations is given in Fig. 14.55. Although outside the scope of this book, there will always be some possibility of malfunction of this circuit caused by a narrow pulse from gate 3 driving gates 1 and 5 simultaneously into the linear region. Several clock periods may be required [14] for this condition to disappear.

Problems

14.1. Determine the primitive-flow table of a fundamental-mode sequential circuit with two inputs, x_2 and x_1. The single output, z, is to be 1 only when $x_2 x_1 = 10$ provided that this is the fourth of a sequence of input combinations 00 01 11 10. Otherwise the output is to be 0. Include a tabulation of outputs assuring that no spurious 1 outputs will occur during transitions between two states with 0 outputs. Both inputs will not change simultaneously.

14.2. A controller is to be designed for an automatic parking lot gate of the type where the driver inserts a coin in a slot and the gate then opens long enough to let the car enter. When the driver inserts the coin, a coin detector will emit a pulse P, which will cause the gate control signal z to go to 1 to open the gate. As the car passes through the gate, it will pass by a photocell that will cause a signal x to go to 1. When the car is past the gate, x will go back to 0, at which time z should go to 0, closing the gate. To prevent persons from defeating the system by blocking off the photocell to keep the gate open, there will be a treadle switch in the pavement immediately in front of the gate that will emit a pulse T whenever car wheels pass over it. The car passing through will trip this switch twice. If it is triggered a third time before x has gone back to 0, z should go to 0 to close the gate. Assume that the switch is located relative to the photocell so that x will go to 1 before the front wheels hit the switch, and the rear wheels will hit the switch before x goes back to 0. Determine a primitive-flow table of a circuit to develop the control signal z.

14.3. A circuit has a single-input line on which pulses of various widths will occur at random. Construct the primitive-flow table of this circuit with a single output, z, on which a pulse will occur coinciding with every fourth input pulse. Because this circuit is part of a random-process generation scheme, the power on state is of no interest.

14.4. A sequential circuit has two level inputs and two level outputs. The outputs are coded to count in the straight binary code, modulo-4, as shown in Fig. P14.4. The count is to increase by one for each 0-to-1 transition of either input occurring when the other input is at the 0 level. The two inputs will not change at the same time. Determine a primitive-flow table of a circuit to meet this specification.

z_2z_1	Count	Next Count
00	0	1
01	1	2
10	2	3
11	3	0

FIGURE P14.4

14.5. Determine a minimal-state equivalent of the primitive-flow table given in Fig. P14.5. Suppose that the don't care outputs were all specified such

	x_2x_1				z x_2x_1			
00	01	11	10	00	01	11	10
ⓐ	f	i	k	0	–	0	–
ⓑ	f	g	j	0	–	0	–
ⓒ	d	h	j	0	–	0	–
c	ⓓ	i	k	–	1	–	1
a	ⓔ	i	l	–	1	–	–
a	ⓕ	h	j	–	1	–	1
a	e	ⓖ	k	0	–	0	–
c	f	ⓗ	k	0	–	0	–
a	d	ⓘ	j	0	–	0	–
c	f	i	ⓙ	–	1	–	1
a	d	h	ⓚ	–	1	–	1
b	d	h	ⓛ	0	–	0	0

FIGURE P14.5

that unstable states always have the same output as similarly lettered stable states. Thus, the state table becomes completely specified. Repeat the minimization after this modification and compare the two solutions. Would this approach in general prevent the determination of a minimal table? Why not?

14.6. Determine a minimal-row equivalent of the flow table of Fig. P14.6.

	q $x_2 x_1$				z $x_2 x_1$		
00	01	11	10	00	01	11	10
①	2	–	5	0	0	–	0
–	②	3	–	–	0	0	–
–	4	③	9	–	–	0	0
1	④	–	–	–	1	–	–
–	–	6	⑤	–	–	0	0
–	7	⑥	8	–	0	0	0
1	⑦	–	–	0	0	–	–
1	–	–	⑧	0	–	–	0
1	–	–	⑨	0	–	–	0

FIGURE P14.6

14.7. Determine minimal-state equivalents of the primitive-flow tables determined in:
(a) Problem 14.1.
(b) Problem 14.2.

14.8. Assign secondaries to the minimal-flow table of Fig. P14.8. Construct a transition table that avoids critical races. Obtain Boolean expressions for each excitation variable.

	q $x_2 x_1$				$z_2 z_1$ $x_2 x_1$		
00	01	11	10	00	01	11	10
①	2	①	3	01	—	01	—
1	②	②	②	—	00	00	00
1	③	2	③	—	10	—	10

FIGURE P14.8

14.9. Assign the eight combinations of values of three secondaries to the six rows of the flow table in Fig. P14.9. Determine a transition table that avoids critical races. Do not minimize the flow table.

	q x_2x_1				z x_2x_1			
	00	01	11	10	00	01	11	10
	①	2	–	5	0			
	②	②	4	6	1	1		
	–	2	④	6			1	
	1	2	–	⑤				0
	1	⑥	8	⑥	0			1
	–	6	⑧	5			0	0

FIGURE P14.9

14.10. Obtain a minimal-flow table covering the table of Fig. P14.9. Assign secondary states to rows of the minimal-state table using only two secondaries. If possible, obtain a transition table for this assignment that avoids critical races.

14.11. Make a secondary assignment for the flow table of Fig. P14.11. Construct a transition table without critical races.

	q x_2x_1				z x_2x_1			
	00	01	11	10	00	01	11	10
	ⓐ	ⓐ	d	b	0	1		
	a	ⓑ	c	ⓑ		0		0
	ⓒ	a	ⓒ	d	1		1	
	c	b	ⓓ	ⓓ			1	1

FIGURE P14.11

14.12. For the circuit and secondary assignment of Fig. P14.12, obtain expressions for Y_2, Y_1, and z that are free from all static and dynamic hazards.

y_2y_1	q x_2x_1				z x_2x_1			
	00	01	11	10	00	01	11	10
00	ⓐ	ⓐ	ⓐ	b	1	1	0	0
01	a	ⓑ	c	ⓑ	1	1	1	0
11	ⓒ	ⓒ	ⓒ	d	0	0	1	1
10	c	–	a	ⓓ	0	–	0	1

FIGURE P14.12

14.13. For the flow table shown in Fig. P14.13, determine a state assignment free of critical races, using three state variables.

x_1x_2			
0 0	0 1	1 1	1 0
(A), 0	(A), 0	E, 0	D, 0
D, 0	E, 0	(B), 0	(B), 0
A, 0	(C), 0	F, 0	(C), 0
(D), 0	G, 0	(D), 0	(D), 0
D, 0	(E), 0	(E), 0	F, 0
(F), 0	A, 0	(F), 0	(F), 0
(G), 1	(G), 0	B, 0	C, 0

q, z

FIGURE P14.13

14.14. For the flow tables shown in Fig. P14.14, determine assignments free of critical races. Assume the tables are minimal.

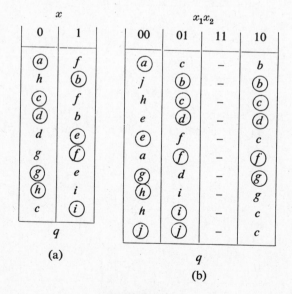

x	
0	1
(a)	f
h	(b)
(c)	f
(d)	b
d	(e)
g	(f)
(g)	e
(h)	i
c	(i)

q

(a)

x_1x_2			
00	01	11	10
(a)	c	–	b
j	(b)	–	(b)
h	(c)	–	(c)
e	(d)	–	(d)
(e)	f	–	c
a	(f)	–	(f)
(g)	d	–	(g)
(h)	i	–	g
h	(i)	–	c
(j)	(j)	–	c

q

(b)

FIGURE P14.14

14.15. For the flow tables of Fig. P14.15, determine assignments free of critical races and determine excitation equations free of combinational hazards.

x_1x_2

00	01	11	10
ⓐ,0	ⓐ,0	d,0	e,0
ⓑ,0	ⓑ,0	d,0	ⓑ,0
b,0	ⓒ,0	ⓒ,0	e,0
e,–	ⓓ,0	e,0	b,0
ⓔ,0	ⓔ,0	c,0	e,0

q,z

(a)

x_1x_2

00	01	11	10
①,0	2,–	①,1	5,–
1,–	②,0	②,0	4,–
③,0	5,–	4,–	③,1
3,–	5,–	④,0	④,0
⑤,0	⑤,0	1,–	⑤,1

q,z

(b)

FIGURE P14.15

14.16. Write hazard-free excitation and output expressions for the following flow tables. Make certain that there are no transient output pulses.
(a) Figure P14.9.
(b) Figure P14.11.

14.17. Inspect the flow tables of Fig. P14.17 for essential hazards. If any exist, indicate them by specifying the single and triple transitions fulfilling the condition of Definition 14.3. For example, for the flow table of Fig. 14.48b, the essential hazard would be indicated as

$$① \to 2 \to ② \to 3 \to ③ \to ○$$

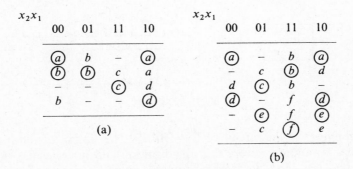

FIGURE P14.17

14.18. A troublesome problem associated with a switch is contact bounce. There are generally many switches and relays in the console and input output sections of computers. It is very difficult and expensive to construct switches and relays so that the contacts do not bounce. This problem is illustrated in Fig. P14.18. At t_1, the switch moves from A to B. It makes

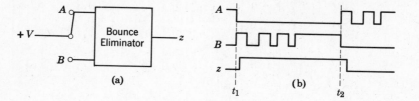

FIGURE P14.18

initial contact at B and then bounces several times, producing a series of several short pulses before settling permanently at B. We may assume it does not bounce back far enough to contact A. At t_2, the switch is moved back to A, with similar results.

Design a "bounce eliminator" to produce a z output as shown following B but eliminating the multiple transitions.

14.19. Another common feature of computer consoles is a step control, which enables an operator or service engineer to step the computer through a program one cycle at a time. For this purpose, we require a circuit where a push button can be used to parcel out the clock pulses, one clock pulse per activation of the push button. The timing diagram is shown in Fig. P14.19. We assume that the push button signal has had the bounce

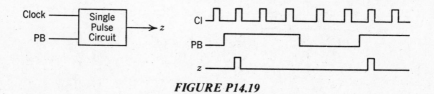

FIGURE P14.19

eliminated by the circuit of Fig. P14.18. The timing of the push-button signal is random, except that the duration and interval may be assumed to be very long compared to clock duration and interval. Design a fundamental-mode circuit that will accomplish the above function.

14.20. A fundamental-mode circuit has two inputs and two outputs. The two outputs should indicate which of the inputs has changed most recently and whether the number of times it has changed is odd or even.
   (a)   How many secondaries will be required?
   (b)   Obtain a minimal-flow table of such a circuit.
   (c)   Obtain a race-free transition table.
   (d)   Design a circuit to realize the transition table of c that is free of static and dynamic hazards.
   (e)   If any essential hazards exist, indicate them on the flow table.

14.21. A fundamental-mode circuit is to be designed that will identify a particular sequence of inputs and provide an output to trigger a combination lock. The inputs to the circuit are three switches labeled X_3, X_2, and X_1. The output of the circuit is to be 0 unless the input switches are in the position $X_3 X_2 X_1 = 010$ where this position occurs at the conclusion of a sequence of input positions $X_3 X_2 X_1 = 101\ 111\ 011\ 010$. Note that two switches are not required to be switched simultaneously.

Using only NAND-gates, design a fundamental-mode circuit that will have an output of 1 only at the conclusion of the above-described sequence of switch positions. A new correct sequence may begin every time the switches are set to $X_3 X_2 X_1 = 101$. Be sure that the circuit is free from hazards and critical races.

14.22. Complete the design of the circuit of Example 14.1, starting with the final flow table of Fig. 14.15. Make an assignment free of critical races, determine hazard-free equations, and draw the final circuit.

14.23. Complete the design of the controller for the parking lot gate, previously considered in Problems 14.2 and 14.7b.

14.24. Let us now consider a more elaborate version of the parking lot gate of Problem 14.2. Assume that there are two lanes entering a parking garage and that two gates, as described in Problem 14.2, are arranged side by side. After these gates, the two lanes merge into a single-lane ramp to the parking floors. The operation of the gates will be just as described in Problem 14.2 except that we wish to control the spacing of cars as they go up the ramp. For this purpose, we locate another treadle switch a short distance up the ramp. After a car passes through either of the gates, its front wheels should pass over the ramp treadle before either gate can open again. Let the output pulse of the ramp treadle be R. If there are two cars at the gates at the time of an R pulse, they should be released in alternate order. If two cars arrive at the gates at the same instant, they may be released in either order, but take care in your design to ensure that both cannot be released at the same time. Design a circuit to meet these requirements.

14.25. A detector circuit is to be designed for a reaction tester. At random intervals, a pulse X_1 will be emitted by a pulse generator. This X_1 pulse will turn on a red light for the duration of the pulse. When the person being tested sees the light come on, the person is to push a button that emits a pulse X_2, which may be assumed to be shorter in duration than the X_1 pulse. The output of the detector, Z, is to be 0 unless a complete X_2 pulse occurs during the X_1 pulse, in which case the output is to go to 1 and remain there until a pulse R appears on a reset line. Design a detector to meet these requirements. Include the reset line in your design.

14.26. On television quiz show, three contestants are asked a question, all at the same time. A contestant who thinks he or she knows the answer pushes a button. Whoever pushes a button first gets to answer the question. Design a detector circuit to turn on a light in front of the contestant who pushed a button first. Let the signals from the three push buttons be X_1, X_2, X_3 and let the outputs controlling the corresponding lights be Z_1, Z_2, Z_3. Include provision for a reset line controlled by the host. Since the contestants are racing each other, two or more signals will occasionally coincide. The circuit may resolve these "ties" in a random fashion, but the circuit must be designed so that no two lights can ever come on at the same time.

14.27. Repeat Problem 14.26, except now assume that we wish an indication of the order in which the contestants pushed their buttons. There will be three lights in front of each contestant, numbered 1, 2, 3. The appropriate light should come on in front of each contestant, indicating whether the person was first, second, or third to push a button.

Bibliography

1. Miller, R. E. *Switching Theory*, Vol. 1, Wiley, New York, 1965.
2. McCluskey, E. J. "Fundamental and Pulse Mode Sequential Circuits," *Proc. IFIP Congress, 1962*, North Holland Publ. Co., Amsterdam, 1963.
3. McCluskey, E. J. *Introduction to the Theory of Switching Circuits*, McGraw-Hill, New York, 1965.
4. Unger, S. H. "Hazards and Delays in Asynchronous Sequential Switching Circuits," *IRE Trans. on Circuit Theory*, **CT-6**, 12(1959).
5. Huffman, D. A. "The Design and Use of Hazard-Free Switching Networks," *J. ACM*, **4**, 47 (1957).
6. Huffman, D. A. "The Synthesis of Sequential Switching Circuits," *J. Franklin Institute*, **257**, 161–190, 275–303 (March—April 1954).

7. Huffman, D. A. "A Study of the Memory Requirements of Sequential Switching Circuits," *Technical Report No. 293*, Research Laboratory of Electronics, M.I.T., April 1955.

8. Caldwell, S. H. *Switching Circuits and Logical Design*, Wiley, New York, 1958.

9. Eichelberger, E. B. "Hazard Detection in Combinational and Sequential Switching Circuits," *IBM J. Res. and Dev.*, **9**: 2 (1965).

10. Maley, G. A. and J. Earle, *The Logic Design of Transistor Digital Computers*, Prentice-Hall, Englewood Cliffs, N.J., 1963.

11. Hill, F. J. and G. R. Peterson. *Digital Systems: Hardware Organization and Design*, Wiley, New York, 1973.

12. Saucier, G. A. "Encoding of Asynchronous Sequential Networks," *IEEE Trans. on Computers*, **EC 16**: 3, (1967).

13. Unger, S. H. *Asynchronous Sequential Switching Circuits*, Wiley, New York 1969.

14. Chaney, T.J. and C.E. Molnar, "Anamalous Behavior of Synchronizer and Arbiter Circuits," *IEEE Trans. on Computers, EC 22*: 4, (1973).

15

A SECOND LOOK AT FLIP-FLOPS AND TIMING

15.1 Introduction

Chapter 14 was devoted to the development of techniques for describing and designing level-mode and fundamental-mode sequential circuits. It is the opinion of the authors that a thorough understanding of the special problems of level-mode analysis will not only be of value in the design of fundamental-mode circuits but will also provide insight into the use of these circuits at higher levels of design. Edge-triggered and master-slave flip-flops are the level-mode components most commonly used in digital system design. This may surprise the reader who is accustomed to thinking of a flip-flop as a clock-mode circuit. Once designed to satisfy the clock-mode criteria of one state change per clock pulse, a flip-flop is a clock-mode circuit. While it is being designed to satisfy this criterion, we must regard the flip-flop as level mode. The clock line must be treated as any other level input.

It is not likely that the commonly available master-slave J-K flip-flop was originally designed using the techniques of Chapter 14. Nevertheless we will find that a flow-table formulation will graphically display the capabilities and limitations of this device. Before discussion of the master-slave memory element, we will first develop a flow table for the edge-triggered D flip-flop. With some resourcefulness, the excitation equations that actually describe the standard TTL device can be derived from a level-mode flow table. It is entirely possible that this circuit was first designed in this manner.

Between the treatment of the two types of flip-flops, we will approach once again the topic of *clock skew*, a problem that arises when clock pulses reach different parts of a system at different times because of unequal propagation delays. We will examine in detail the effect of this phenomenon and consider its implications as we examine alternatives in the design of the J-K flip-flop. When we reach the point of determining a realization of the transition table, we will introduce a new approach for eliminating static and dynamic hazards.

The remainder of the book will not depend on the material of this chapter, and many readers may wish to skip this chapter. Others may find it satisfying to see that formal level-mode procedures can lead directly to time-honored designs. The student who is likely to become involved in the design of high-speed clock-mode circuits will find that this material will provide needed insight into the capabilities and limitations of available flip-flops. A thorough understanding of memory-element timing will allow the reader to anticipate problems that might otherwise appear after a circuit has been constructed.

15.2   Edge-Triggered *D* Flip-flop

Perhaps the most widely used of all clocked memory elements is the rising-edge-triggered *D* flip-flop, which was first discussed in Chapter 9 (Fig. 15.1a). This device was originally developed as part of the TTL SSI logic family. The same circuit with the preset and clear lines removed and the three feedback loops cut is shown in Fig. 15.1b.

Since 3 pairs of gates are immediately cross-coupled in Fig. 15.1a, it is evident that at least three loops will have to be cut to reduce the network to combinational logic. In Fig. 15.1b y_1, y_2, and y_3 may be regarded as secondary inputs while Y_1, Y_2, and Y_3 are excitations. An inspection of Fig. 15.1b will reveal that it is indeed a combinational logic network consisting of two level NAND-NAND realizations of Y_1, Y_2, and Y_3. The three first-level gates are shared in a multiple-output realization.

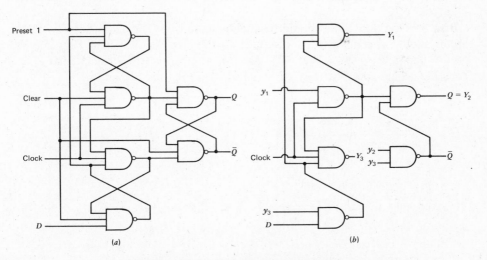

FIGURE 15.1   *Edge-triggered D flip-flop.*

The expressions for the three state variables are given by

$$Y_3 = y_1 C + y_3 D + \bar{C} \tag{15.1}$$

$$Y_1 = y_1 C + y_3 D \tag{15.2}$$

$$Y_2 = y_1 C + y_3 y_2 \tag{15.3}$$

So far we have merely analyzed the rising-edge-triggered D flip-flop. Let us attempt the synthesis of the same device beginning with the derivation of a level-mode flow table. *The specifications of a rising-edge D flip-flop call for D to be stable at the time of the rising edge of the clock.* There is, however, no constraint at the time of the falling edge, so the circuit cannot be considered fundamental mode. The fact that we need worry about only one clock edge will greatly simplify the design.

In Fig. 15.2a we begin at stable state ① (marked with an asterisk) for which D, the clock input, C, and the output, Z, are all 0. No change in circuit output may take place until we first have $D = 1$ and $C = 0$, so that the next clock transition will be a rising edge. Thus the circuit remains in stable state ① until $C = 0$ and $D = 1$. The inputs may change to these values at approximately the same time so that the transition to stable state ② in Fig. 15.2b is not a fundamental-mode transition. This is not a problem, since the only exit from the first row of the table leads directly to stable state ②.

From stable state ② in Fig. 15.2b only single-input transitions can take place. If D returns to 0, the output does not change and the circuit returns to state ①. In the event of a rising clock edge, the circuit goes to stable state ③ for which $Z = 1$.

Row 3 of the flow table should be analogous to row ① except that $Z = 1$. The circuit should remain in this row until $C = D = 0$. As shown in Fig. 15.3, the circuit then goes to stable state ④, which is analogous to ② except that $Z = 1$. Thus the design specifications are satisfied by a four-row flow table.

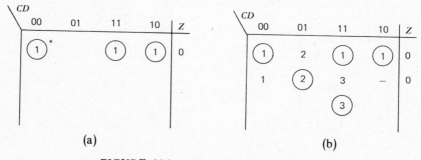

FIGURE 15.2   *Flow table development.*

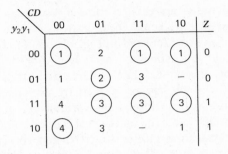

FIGURE 15.3 *Flow table for edge-triggered flip-flop.*

The state of assignment Fig. 15.3 includes only single secondary transitions and can be implemented directly. The resulting circuit, while featuring only two excitations, would require more gates than the six-gate version of Fig. 15.1. We leave the verification of this fact as a problem for the reader.

Since Fig. 15.1b features 3 state variables, one might ask what justification there is for considering circuits with more than the minimum number of excitations. It is possible in the clock mode as well as the level mode that the most economical realization may not have a minimal number of state variables, although such cases are very much the exception. Since adding extra state variables is a trial-and-error process, it should probably be considered only if a very

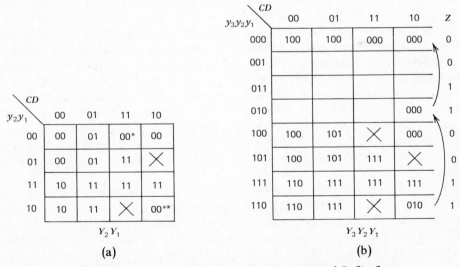

FIGURE 15.4 *Transition tables for edge-triggered D flip-flop.*

large production volume is anticipated for the resulting design. This was certainly the case for the edge-triggered D flip-flop.

Using the state assignment depicted in Fig. 15.3 results in the transition table of Fig. 15.4a. Note that the realization of Y_1 would be simplified if the entry in the square marked by one asterisk in Fig. 15.4a were 01. Similarly Y_2 would be simplified if the entry in the square marked by two asterisks were 10. These simplifications will occur if a third state variable Y_3 is added as shown in Fig. 15.4b. Thus row 000 shares the function of stable state ① with row 100. For $CD = 10$ the circuit cycles from 110 to 010 to 000 instead of going directly to 100 as in Fig. 15.4a. The entry for $CD = 11$ and $y_3 y_2 y_1 = 100$ can now be a don't-care, since this square could be entered only if D could change simultaneously with the rising edge of the clock.

A minimal multiple-output realization of Fig. 15.4 is given by Equations 15.1, 15.2, and 15.3 with

$$Z = Y_2 \tag{15.4}$$

15.3 Edge-Triggering and Clock Skew

The edge-triggered D flip-flop is a particularly convenient digital system component, since the circuit output is affected by the input only at the time of a clock transition. Input sensing and output changes take place at a single point in time. The successful functioning of a system made up of edge-triggered elements requires that the triggering edge reach all memory elements at approximately the same time. Sometimes a particular realization of a system will introduce delay in the clock line and cause this clock-mode assumption to be violated. Consider the clock-mode sequential circuit of Fig. 15.5. Each flip-flop shown will be assumed to be rising-edge-triggered, so that any state change will take place immediately as the rising edge of a clock pulse reaches its c-k input terminal. In effect, the circuit is a simple 2-bit shift register. Assume that, because of some peculiarity in wiring the clock distribution lines, a delay has appeared between the clock line entering flip-flop 1 and the clock line entering flip-flop 2. This may have been caused by a long wire or by a pair of inverters inserted to help handle heavy loading on the clock line. In high-speed circuits, wire length can become quite critical.

The initial values of the flip-flops are shown in Fig. 15.5b to be $y_1 = y_2 = 1$, and the input is 0. If the circuit were working properly, the next clock pulse would place a 0 in flip-flop 1, and the second pulse would propagate the 0 on to y_2. In Fig. 15.5b, we observe the problem caused by the delay in the clock line. Note that y_1 quite properly goes to 0 upon the rising edge of clock pulse 1. This value is propagated directly to the input of flip-flop 2 before the clock pulse,

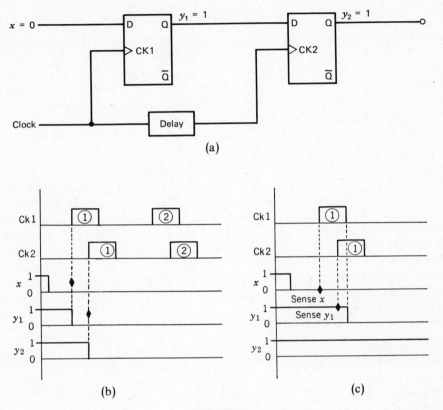

FIGURE 15.5

$CK2$, reaches this device. Thus the positive transition on $CK2$ occurs when $D_2 = 0$, so that y_2 goes to 0. Thus, a state change has taken place at clock pulse 1 that should not have occurred until pulse 2.

Where typical edge-triggered or master-slave flip-flops are used, great care must be taken to avoid the above behavior. One must ensure that the clock inputs to all flip-flops are synchronized. Another approach might be to use a somewhat more complicated flip-flop. Suppose that the data input was sensed at the leading edge of the clock pulse but the outputs did not change until the trailing edge, as shown in Fig. 15.5c. In that case the *original* value of y_1 would be sensed at the input of flip-flop 2 by pulse 1 and the circuit would perform properly provided the clock-pulse width exceeded the clock skew. As we will see in the next section, adding this edge-sensing capability to the master-slave flip-flop considerably complicates its design.

15.4   A Flow Table for a *J-K* Master-Slave Flip-flop

In this section, we will obtain a flow table for a master-slave flip-flop. Departing slightly from our basic procedure of the previous chapter, we will introduce some constraints that will tend to channel the design toward an expected result. Whether one would expect these constraints as part of a predesign specification is arguable, but each can be defended in terms of good design practice.

First we assume that the concept of a master element and a slave element has meaning apart from the final realization. To enforce this property, we assume two outputs, the master output z_1 and the slave output z_2. Once the device is constructed, only the slave output will be available outside of the IC package. We begin the flow table by defining a stable state ⓐ corresponding to $z_2 = z_1 = 0$. As long as the clock C remains 0, the circuit will stay in state ⓐ as indicated in Fig. 15.6. When the clock goes to 1, the output z_1 of the master element should be set to 1 provided that $J = 1$. Because of the defined function of the *J-K* flip-flop, the value of K is immaterial. Therefore, we enter unstable state b in the last two columns of the first row of Fig. 15.6. This will cause the circuit to go to stable state ⓑ, for which the outputs are defined as $z_2 z_1 = 01$. We want no further activity until the clock returns to 0, hence the four ⓑ entries on the $C = 1$ half of the second row in Fig. 15.6.

```
CJK
        000    001    011    010 │ 100    101    111    110 │ z₂z₁

        ⓐ      ⓐ      ⓐ      ⓐ  │                 b      b  │ 0 0
                                │ ⓑ      ⓑ      ⓑ      ⓑ  │ 0 1
```

FIGURE 15.6   *Developing J-K master-slave flow table.*

We can now choose between two approaches to the completion of the flow table in Fig. 15.6. One will lead to a standard master-slave flip-flop and the second will lead to a more complicated edge-sensing master-slave. For now we follow the design for the standard master-slave flip-flop by filling out the first row with stable state ⓐ entries. Thus, a 1 will be propagated into the master element if J changes to 1 while $C = 1$. Once the master element has been set to 1, no further activity will take place until $C \rightarrow 0$, regardless of changes in J and K. When $C \rightarrow 0$, the circuit goes to stable state ⓒ, for which $z_2 z_1 = 11$, as shown in Fig. 15.7. We complete the table by providing for a return to state ⓐ after the next occurrence of $C = K = 1$. This requires a fourth stable state ⓓ corresponding to $z_2 z_1 = 10$, as given in Fig. 15.7.

Since the output transition in the master-slave flip-flop occurs on the falling

CJK	000	001	011	010	100	101	111	110	$z_2 z_1$
	(a)	(a)	(a)	(a)	a	a	b	b	0 0
	c	c	c	c	(b)	(b)	(b)	(b)	0 1
	(c)	(c)	(c)	(c)					1 1

(a)

CJK	000	001	011	010	100	101	111	110	$z_2 z_1$
	(a)	(a)	(a)	(a)	a	a	b	b	0 0
	c	c	c	c	(b)	(b)	(b)	(b)	0 1
	(c)	(c)	(c)	(c)	c	d	d	(c)	1 1
	a	a	a	a	(d)	(d)	(d)	(d)	1 0

(b)

FIGURE 15.7 *Flow table for standard master-slave J-K flip-flop.*

edge of the clock pulse, it might seem that this could be considered to be a falling-edge-triggered flip-flop, but this is not the case. In a true edge-triggered flip-flop such as the *D* flip-flop of Section 15.2 (which could equally well be designed for falling-edge triggering), the nature of the state transition is completely determined by the value of the data input at the time of the triggering edge. But this is not the case for the master-slave flip-flop.

Consider the input signals given in Fig. 15.8 to be applied to the master-slave flip-flop described by the flow table of Fig. 15.7 with the output, Z_2, initially 0. Note in Fig. 15.8a that *J* goes to 1 while the clock is 1 causing the circuit to go to state ⓑ and the output to 1 with the falling edge of the clock. In Fig. 15.8b *J* returns to 0 while the clock is on. In this case the circuit remains in state ⓑ, and the output again goes to 1 with the falling edge of the clock. We note that the inputs for Fig. 15.8a and 15.8b differ from each other on both edges of the clock, yet the circuit responded in the same way. Clearly the transition is not controlled by the input on either edge. The flip-flop is enabled for the entire duration of the clock pulse, and consistent response can be obtained only by holding the data inputs stable for this entire period. This implies that input and state changes must propagate through every level of combinational logic while the clock is 0. This could be a serious drawback in some cases, such as when using a square-wave clock.

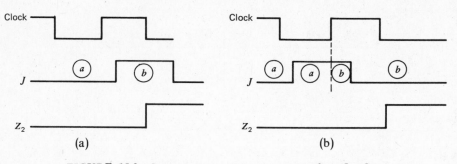

FIGURE 15.8 *Inconsistent response in master-slave flip-flop.*

As we saw in Fig. 15.5, the use of a flip-flop that senses the leading edge and changes outputs on the trailing edge can be a considerable advantage in dealing with clock skew. Example 15.1 develops the flow table of an edge-sensing master-slave flip-flop. Unfortunately, such devices are not readily available.

Example 15.1

Obtain the flow table for an edge-sensing master-slave flip-flop.

Solution   Note that two blank spaces remain in the first row of Fig. 15.6. If the flip-flop is to be an edge-sensing master-slave, it is important to distinguish between $J \rightarrow 1$ followed by $C \rightarrow 1$ and $C \rightarrow 1$ followed by $J \rightarrow 1$. In the latter case, the circuit must not be allowed to go to state b. To ensure this, it will be necessary to define another state for the circuit to assume as soon as $C \rightarrow 1$ while $J = 0$. This will require entering the corresponding unstable states in the blank spaces of the first row of Fig. 15.6.

We continue from Fig. 15.6 by entering unstable state e in the blank spaces of the first row. This leads to the partial flow table of Fig. 15.9a. Remembering the discussion of Section 14.9, we see that the stable state reached following a multiple input change on the first row would be unpredictable. To avoid multiple changes, *we must impose the restriction that both J and K be stable when C goes to 1 and that the circuit will stabilize following the leading edge of the clock pulse before a subsequent change in J or K will take place.* If we assume that the maximum possible clock skew is known, it will always be possible to enforce this restriction, although sometimes a reduction in clock frequency may be required. With this restriction, the first row of Fig. 15.9a is essentially fundamental mode.

State e serves as a waiting state to assure that changes in J or K while the clock is 1 will not affect the master element. When the clock returns to 0, the circuit returns to state a. From state b, the circuit will continue to c as before when C returns to 0. The lower three rows of the completed flow table as shown in Fig. 15.9b are analogous to the upper three rows. State f has been added to hold the output at 1 if a signal on line K occurs while the clock is 1. A three-state-variable realization of Fig. 15.9 can be constructed. We will leave this as a homework problem for the reader.

CJK									$z_2 z_1$
	000	001	011	010	100	101	111	110	
	(a)	(a)	(a)	(a)	e	e	b	b	00
	a	a	a	a	(e)	(e)	(e)	(e)	00
					(b)	(b)	(b)	(b)	01

(a)

CJK									$z_2 z_1$
	000	001	011	010	100	101	111	110	
	(a)	(a)	(a)	(a)	e	e	b	b	00
	a	a	a	a	(e)	(e)	(e)	(e)	00
	c	c	c	c	(b)	(b)	(b)	(b)	01
	(c)	(c)	(c)	(c)	f	d	d	f	11
	c	c	c	c	(f)	(f)	(f)	(f)	11
	a	a	a	a	(d)	(d)	(d)	(d)	10

(b)

FIGURE 15.9   *Flow table for edge-sensing J-K master-slave flip-flop.*   ∎

15.5   Another Approach to Level-Mode Realization

Before trying to obtain a realization of the standard J-K flip-flop let us introduce a new model on which level-mode realizations can be based. Where the standard realization would require the addition of several gates to eliminate static and dynamic hazards, a realization based on the model of Fig. 15.10 might be preferable. The realization of only one excitation variable, Y_i, is shown, but each excitation will be realized by a similar network. The inputs may be any of the circuit inputs or secondaries.

The excitation, Y_i, is the output of a pair of a cross-coupled NAND-gates, looking very much like an R-S flip-flop. As shown in the figure, we denote the outputs of the input NAND-gates as $\bar{f}_1 \cdots \bar{f}_k$ and $\bar{g}_1$ to $\bar{g}_m$, (k and m have no necessary relation to the number of variables), and then define

$$\overline{S_{y_i}} = \bar{f}_1 \cdot \bar{f}_2 \cdots \bar{f}_k \tag{15.5a}$$

or

$$S_{y_i} = f_1 + f_2 + \cdots + f_k \tag{15.5b}$$

and

$$R_{y_i} = \bar{g}_1, \bar{g}_2 \cdots \bar{g}_m \tag{15.6a}$$

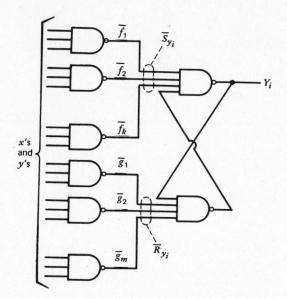

FIGURE 15.10   *Alternate level-mode realization.*

or

$$R_{y_i} = g_1 + g_2 + \cdots + g_m \qquad (15.6\text{b})$$

With these definitions, we see that the output of the lower of the cross-coupled gates is given by

$$\overline{\overline{R_{y_i} \cdot y_i}}$$

so that

$$Y_i = \overline{\overline{S_{y_i}} \cdot \overline{\overline{R_{y_i} \cdot y_i}}}$$

$$= S_{y_i} + \overline{R_{y_i}} \cdot y_i \qquad (15.7)$$

Equation 15.7 we recognize as the next-state equation of the *R-S* flip-flop, so that the proposed modification consists of storing the state variables in unclocked latches in a manner reminiscent of clock-mode design except that there are no clock pulses involved.

Now consider a level-mode circuit with a given state variable equal to 1 and an input change such that this state variable is not supposed to change. If the 1 is produced by one gate before the input change and a different gate after, it is possible for the state variable to go to 0 momentarily during the change, a phenomenon known as a static combinational hazard. In the standard level-mode circuit, since this momentary change is fed back directly, it can cause an erroneous transition.

In the alternate design of Fig. 15.10, since the value of Y_i is stored in a latch, there is no need for any signal controlled by the inputs to remain at 1 to hold Y_i; all that is required is that R_{y_i} remain at 0 whenever Y_i is to be 1. That is, no matter what combinational hazards may exist in S_{y_i}, no transitions of Y_i from 1 to 0 can occur unless R_{y_i} goes to 1. With AND-OR design, we "create" 1s with each product function (the f's and g's in Fig. 15.10) so there is no possibility of a combinational hazard creating an erroneous 1 in the S_{y_i} or R_{y_i} functions. By a similar argument, static hazards in the 0s cannot occur since S_{y_i} will be 0 whenever Y_i is to be 0.

Not only does this design method eliminate static hazards, it also may produce simpler designs. Because of the latches, 1s are *required* in the S_{y_i} or R_{y_i} functions only for transitions where the state variables are to change. Where Y_i is to remain 0 or 1, R_{y_i} or S_{y_i}, respectively, are don't-cares. Thus, if the number of transitions involving changes of state variables is small, the alternate design may be simpler than the classic design of Chapter 14. Unfortunately, this alternate method does not eliminate problems due to races or essential hazards. Both of these phenomena are inherent in the flow table structure and independent of the mode of circuit realization.

15.6   Realizing the Master-Slave *J-K* Flip-flop

A glance at the flow table of Fig. 15.7b indicates that no problem will be caused by either of the multiple-input changes that can occur. The natural state assignment of $z_2 = y_2$ and $z_1 = y_1$ will yield a transition table free from cycles and races. It remains to determine an economical realization free from static and dynamic hazards, in which the essential hazard associated with switching the clock on and then off is carefully controlled.

A transition table for the above state assignment is given in Fig. 15.11. Using this table as a K-map for both Y_2 and Y_1, we determine the following straightforward realization of the flip-flop.

$$Y_2 = y_1 \bar{C} + y_2 C + {}^*y_2 y_1 \tag{15.8}$$

$$Y_1 = y_1 \bar{C} + \bar{y}_2 y_1 + y_1 \bar{K} + \bar{y}_2 C J \tag{15.9}$$

The term $y_2 y_1$ as marked by the asterisk is included to eliminate a static hazard. A circuit implementation of these expressions will require 8 gates, including one to cover the $y_1 \bar{C}$ term in both expressions, three inverters, and a total of 23 gates inputs.

Although the above implementation can be made to function satisfactorily, let us now try the approach discussed in the preceding section. The transition table of Fig. 15.11 can be directly translated using Equation 15.7 to form the four S

| | CJK | | | | | | | |
y_2y_1	000	001	011	010	100	101	111	110
00	(00)	(00)	(00)	(00)	(00)	(00)	01	01
01	11	11	11	11	(01)	(01)	(01)	(01)
11	(11)	(11)	(11)	(11)	(11)	10	10	(11)
10	00	00	00	00	(10)	(10)	(10)	(10)

$$Y_2Y_1$$

FIGURE 15.11　*Transition table for master-slave flip-flop.*

and Karnaugh maps of Fig. 15.12. From these we obtain the following simple set and clear equations.

$$S_{y_1} = C \cdot J \cdot \bar{y}_2 \qquad R_{y_1} = C \cdot K \cdot y_2 \qquad (15.10)$$

$$S_{y_2} = \bar{C} y_1 \qquad\qquad R_{y_2} = \bar{C} \cdot \bar{y}_1 \qquad (15.11)$$

Consistent with the standard flip-flop notation, in which the outputs of the cross-coupled gates are the Q and $\bar{Q}$ outputs, we note that $\bar{y}_2$ and $\bar{y}_1$ are directly available from the lower outputs of the cross-coupled pairs. With this observation, we construct the master slave flip-flop as shown in Fig. 15.13. In this case, the alternate method does lead to a simpler design than that corresponding to Equations 15.8 and 15.9, requiring two fewer inverters and 4 fewer inputs. Let us emphasize again, however, that there is no way of knowing in advance which method will lead to the simpler design.

This flip-flop is very similar to the standard *J-K* flip-flop in TTL. As discussed,

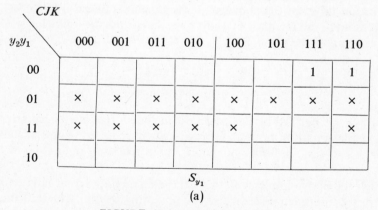

FIGURE 15.12　*Excitation maps.*

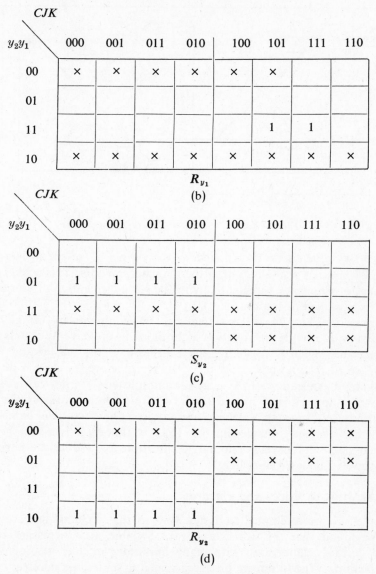

FIGURE 15.12 (continued)

this device is race-free and all static and dynamic hazards have been eliminated. However, an essential hazard associated with turning the clock on and off remains in the circuit. Suppose, for example, that the inverter marked by the asterisk in Fig. 15.13 is slow. In that case, a 0-to-1 transition of the clock might

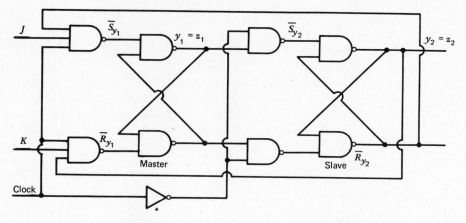

FIGURE 15.13   *Master-slave J-K flip-flop.*

cause a new value to be propagated to the output y_1 of the master element before a 0 appeared at the output of the indicated inverter. This could cause the new value to propagate immediately into the slave element, in direct contradiction to the specified operation.

As shown in Fig. 9.24, this problem can be eliminated by using the input NAND-gates to provide the inversion of the clock to the slave. With this arrangement the same signal is triggering the transition in the master and disabling the slave, so there is no possibility of y_1 changing before the slave is disabled.

Obtaining a realization of the edge-sensing master-slave device characterized by the 6-row flow table of Fig. 15.9 will be left as a homework problem for the reader. Because three state variables are required, the circuit will be more complicated than the one obtained in Fig. 15.13. Although it is a superior device, most manufacturers have not found justification for adding the edge-sensing master-slave to their product line.

15.7   Analysis of Races and Hazards

In Chapter 14, we attempted to design systems that included no races or hazards. We found that this was not always possible. It is necessary to design essential hazards deliberately into many important circuits in order to make them function. In addition, we observed that it is sometimes necessary to allow for simultaneous changes in the values of more than one input. Problems that could arise in this connection were termed function hazards. The proper functioning of circuits containing function hazards or essential hazards will in some way be dependent on gate delay.

After completing the design in Section 15.6, we noticed an essential hazard in the design of the *J-K* flip-flop. Although we might have picked it up earlier, the fact that we did not is not atypical. Often some type of race or hazard will turn up in a design in a way not anticipated by the designer might be unsure how serious a problem it might be. It would be desirable to have some computational technique that could be used to identify cases of uncertain behavior and evaluate the likelihood that they will cause trouble. This technique should be applicable to the final design.

One approach would be an outright simulation of the circuit as designed. Simulation was discussed in Chapter 13. Eichelberger [1] has developed a computational technique that will usually obtain the desired information more explicitly and with considerably less computer running time than a full circuit simulation. Eichelberger's method utilizes a 3-value logical calculus, which we will take time to introduce at this point.

In Chapter 4, we observed that a Boolean algebra must have a number of elements given by 2^n for some integer n. Thus, the three-value calculus cannot be described by a Boolean algebra. Of the postulates and theorems of Sections 4.1 and 4.4, all but Postulate VI and Theorem 4.10 hold for the three-value calculus, but this precludes almost any algebraic simplification. For our purposes, we require only that the three-value computational system be closed under the operations $+$, $\cdot$, and NOT.

In addition to 0 and 1, the third element will be denoted $\times$. This element may be regarded as an indeterminate value or unknown value in Eichelberger's method. That is, when the output of a gate is $\times$, it is not known whether the value should be 1 or 0. We define the complement of $\times$ very simply as $\bar{\times} = \times$, thus violating the postulates of a Boolean algebra. We assure ourselves of a closed system by defining the mappings AND and OR as given in Figs. 15.14b and 15.14c. An intuitive feel for Fig. 15.14 can be achieved by thinking of $\times$ as an unknown value. Clearly, if A is unknown, $\bar{A}$ is unknown. $0 + \times$ and $\times + \times$ will always be determined by the unknown value $\times$. $1 + \times = 1$ regardless of the value of $\times$. The situation is analogous for the AND operation.

A	$\bar{A}$
0	1
$\times$	$\times$
1	0

(a)

A \ B	0	$\times$	1
0	0	$\times$	1
$\times$	$\times$	$\times$	1
1	1	1	1

(b) $A + B$

A \ B	0	$\times$	1
0	0	0	0
$\times$	0	$\times$	$\times$
1	0	$\times$	1

(c) $A \cdot B$

FIGURE 15.14   Three-value logic defined.

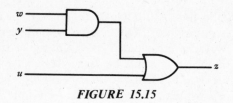

FIGURE 15.15

Example 15.2

The value of y in the simple circuit of Fig. 15.15 is unknown while w and u will always be known as 0 or as 1. For what combination of values of w and u will the output, z, depend on the unknown value of y?

Solution If we let $y = \times$, we have $z = w \cdot \times + u$. From Fig. 15.4, we see that $z = \times$ if and only if $w = 1$ and $u = 0$. ■

In any level-mode circuit, including those designed according to the model described in Section 15.4, Boolean expressions can be obtained expressing all excitations in terms of the inputs and secondaries. Eichelberger's method utilizes these expressions to compute the circuit response to changing inputs as specified by the following two-step procedure. The technique, which utilizes the 3-value calculus, will identify the existence of any possibility of a sequential circuit reaching an erroneous state due to the existence of a critical race or hazard. The computation must be performed corresponding to each input change where problems are suspected.

Step 1 With the changing input variables assigned the values $\times$ and all other inputs and secondaries as originally specified, evaluate the excitation functions to determine if one or more have changed from 1 or 0 to $\times$. If so, change the corresponding secondary values to $\times$ and repeat the process until no additional changes in the excitation functions are determined.

Following Step 1, some of the secondaries may have stabilized at $\times$. The question arises as to whether these secondaries will assume the desired values after the input change is completed in Step 2.

Step 2 With the changing inputs assigned their new values (0 or 1) and all other variables equal to their values at the end of Step 1, evaluate all excitations. If one or more of the excitations changes from $\times$ to 0 or 1, change the corresponding secondaries accordingly; and recompute the excitations. Repeat the process until no further changes are noted.

The results after Steps 1 and 2 may be evaluated using Theorem 15.1.

THEOREM 15.1 If an excitation is 1 (0) following Steps 1 and 2 for a given input change and a given initial state, then this excitation must stabilize at 1 (0) for the same initial state and input change in the physical circuit regardless of gate delays.

If an excitation remains equal to $\times$ following Step 2, then the circuit is delay dependent. This procedure will reveal as suspect any circuit containing any race or hazard. Conclusions drawn from this procedure are more conservative than those that might be provided by an actual 3-value simulation. If the procedure identifies a hazard that was unnoticed by the designer, it may be possible to revise the circuit design. Where the design cannot be modified, as with an essential hazard, for example, it may be necessary to resort to an actual simulation. The following is a very simple example of the procedure.

Example 15.3

From Fig. 15.3 we see that the elementary D flip-flop of Fig. 15.16a has been implemented without eliminating a static hazard. Carry out Steps 1 and 2 of the above procedure for the circuit of Fig. 15.16a, and repeat the process for a modified design with the static

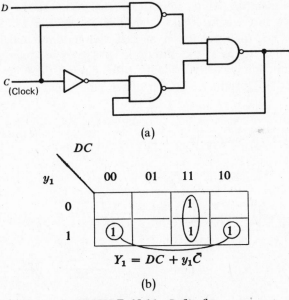

(a)

$$Y_1 = DC + y_1\bar{C}$$

(b)

FIGURE 15.16 D flip-flop.

hazard eliminated. Consider only the input change where initially $D = C = y_1 = 1$, and C changes to 0. From Fig. 15.16b, the reader can observe that the static hazard may well affect the result of this input change.

	D	C	Y_1		D	C	Y_2
Initial	1	1	1		1	1	1
Step 1	1	×	×		1	×	1
Step 2	1	0	×		1	0	1

$$Y_1 = DC + Y_1\bar{C} \qquad Y_1 = DC + Y_1\bar{C} + Y_1 D$$

(a)   (b)

FIGURE 15.17

Solution The Step 1 and 2 calculations for the 0 flip-flop described are given in Fig. 15.17a. The values listed for Y_1 are computed from the expression listed in the figure. When C assumes an × in Step 1, both DC and $y_1\bar{C}$ are ×, so Y_1 takes on this value. When C goes to zero, $Y_1 = y_1\bar{C} = \times$.

The addition of the term $y_1 D$ as given in Fig. 15.17b eliminates the static hazard. Thus, Y_1 remains 1 as C goes to × and then to 0. In more complicated circuits, a critical race or hazard which would be identified by this procedure may not be so easily spotted on a Karnaugh map. ∎

As a second example, let us investigate the effect of the essential hazard observed to exist in the master-slave *J-K* flip-flop of Fig. 15.13. In a complex circuit such as this, the application of Eichelberger's method is simplified by cutting the feedback loops at the outputs of the gates producing the excitation. Although it is not immediately evident in this circuit, there are only two state variables, y_1 and y_2, produced by the upper gates of the cross-coupled pairs.

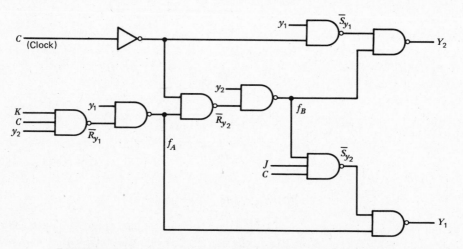

FIGURE 15.18   *J-K master-slave flip-flop with feedback loops cut.*

Cutting the loops at these points results in the combinational logic network of Fig. 15.18. The outputs are the excitations Y_2 and Y_1 and the corresponding secondary inputs are y_2 and y_1. For convenience, two reference points in the network are labeled f_A and f_B. The reader will notice that these points correspond to the outputs of the lower gates of the two cross-coupled pairs. In order to apply Eichelberger's procedure, we must determine excitation expressions for Y_2 and Y_1. First we determine expressions for f_A and f_B by referring to the figure.

$$f_A = \overline{y_1 \cdot \overline{K} \cdot \overline{C} \cdot y_2}$$
$$= \bar{y}_1 + K \cdot C \cdot y_2 \tag{15.12}$$
$$f_B = \bar{y}_2 + \bar{C} f_A$$
$$= \bar{y}_2 + \bar{C}\bar{y}_1 + \bar{C}KCy_2 = \bar{y}_2 + \bar{C}\bar{y}_1 \tag{15.13}$$

Now we can use these expressions in developing expressions for Y_2 and Y_1.

$$Y_2 = \bar{f}_B + \bar{C}y_1$$
$$= \overline{\bar{y}_2 + \bar{C} \cdot \bar{y}_1} + \bar{C}y_1$$
$$Y_2 = y_2 \cdot (C + y_1) + \bar{C}y_1$$
$$= y_2 C + y_1 \bar{C} + y_1 y_2 \tag{15.14}$$
$$Y_1 = \bar{f}_A + JCf_B$$
$$= \overline{\bar{y}_1 + KCy_2} + JC\bar{y}_2$$
$$= y_1 \bar{K} + y_1 \bar{C} + y_1 \bar{y}_2 + JC\bar{y}_2 \tag{15.15}$$

The reader may note that Equations 15.14 and 15.15 are the same as the hazard-free expression for Y_2 and Y_1 (Equation 15.8 and 15.9) obtained by the conventional methods. This should not be taken as an indication that the two methods lead to identical designs; indeed, we have already noted that the circuits are different. What it does reflect is that both circuits have the same function and can, with appropriate analysis and manipulation, be represented by identical Boolean forms.

The reader will note Boolean algebra Postulate VI, $\bar{a} \cdot a = 0$, was used liberally in the development of Equations 15.14 and 15.15. As indicated earlier, this postulate is not valid for the three-value calculus. The effect of its application is to limit the proliferation of $\times$ values. Let us first apply Eichelberger's procedure to Equations 15.14 and 15.15. If both y_1 and y_2 stabilize at 0 or 1 following Step 2, it will be necessary to repeat the computation on the original circuit in Fig. 15.18. We apply Step 1 to initial conditions $y_1 = y_2 = 0$, $J = 1$, and $K = C = 0$.

C y_2 y_1	$y_1\bar{K} + y_1\bar{C} + y_1\bar{y}_2 + JC\bar{y}_2 = Y_1$		$y_2C + y_1\bar{C} + y_1y_2 = Y_2$	
0 0 0		0		
Step 1 × 0 0	0 + 0 + 0 + ×	×	0 + 0 + 0	0
× 0 ×	× + × · $\bar{\text{x}}$ + × + ×	×	0 + × · $\bar{\text{x}}$ + 0	×
× × ×	× + × + × + ×	×	× + × + ×	×
Step 2 1 × ×	× + 0 + × + ×	×	× + 0 + ×	×

FIGURE 15.19   *Input change calculation.*

The Step 1 and Step 2 computations are tabulated in Fig. 15.19 for C changing from 0 to × to 1. As an aid to following the computation the values of each product term of the Boolean expressions are given at each step. Note in this case that three Step-1 iterations are required before both y_1 and y_2 are stable at ×. Both of these values remain × after Step 2, thus verifying the essential hazard.

As we have said, the result of this test is necessarily conservative. The circuit may perform satisfactorily anyway. In this case, it would be necessary to resort to a simulation to obtain a better estimate of the reliability of the circuit. In other cases, a design change might be based on the results of this computation.

15.8   Level-mode Simulation

As mentioned in Chapter 12, heavily used gate-delay simulations are usually event driven to reduce computation time. These programs are complex and will not be discussed here. The following paragraphs suggest an approach to level-mode or gate-delay simulation, which is straightfroward and easy to implement.

In simulating a level-mode circuit, it is convenient to consider the output of each gate (not just the secondaries) as state variables. Physically, the block labeled *connection matrix* in Fig. 15.20a is nothing more than a network of wires. In a computer program, this function is simulated by a shuffling of values from the array-storing gate outputs to the gate-input storage array. This shuffle can be organized in a variety of ways. We leave this problem to the individual programmer.

The model for a single gate is given in Fig. 15.20b. Implementing this model requires three steps during each iteration through the network. The first step involves computing the logical function specified for a particular gate for the current gate inputs. The second step models the gate delay, and the third step determines an updated value of the gate output. The delay and rise time of any practical logic circuits are influenced by a number of time constants. In the simplest possible gate model we will consider here, rise time is neglected altogether. That is, the gate output may jump from 0 to 1 or 1 to 0 in a single

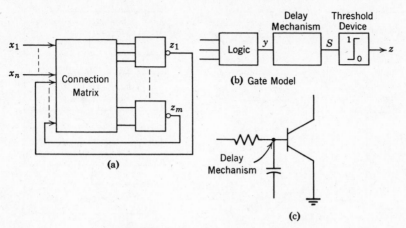

FIGURE 15.20   *Level-mode model.*

iteration. We imagine delay to be controlled by a single time constant, depicted at the base of a transistor in Fig. 15.16. A discrete approximation of this nonlinear integration is a saturating up-down counting mechanism as specified in Fig. 15.21.

S_{min} is a small negative integer, and S_{max} is a small positive integer. The combined effect of the delay counter and output threshold may be explained as follows. If the logical function, y, of the gate inputs is 0 for iteration, v, the count, S, is reduced. If $y = 1$, S is increased. As specified in Fig. 15.21a, S cannot exceed S_{max} or assume a value less than S_{min}. In Fig. 15.21b, we see that the new output will be 1 if $S^{v+1} \geq 0$. If the $S^{v+1} < 0$, the output will be 0. Suppose, for example, that S is initially saturated at S_{min} and $z = 0$. The output will change to 1 only after $y = 1$ for S_{min} successive iterations. The situation is similar where the delay counter is initially saturated at S_{max}. Thus, S_{min} and S_{max} represent nominal gate delay values. A gate modeled by Fig. 15.21 will exhibit most of the important properties of an actual circuit. Very narrow pulses, for example, can disappear in

Y	S^v	S^{v+1}
0	S_{min}	S_{min}
0	$\alpha > S_{min}$	$\alpha - 1$
1	$\beta < S_{max}$	$\beta + 1$
1	S_{max}	S_{max}

S^{v+1}	z^{v+1}
$S^{v+1} < 0$	0
$S^{v+1} \geq 0$	1

(a) Delay counter   (b) Output threshold

FIGURE 15.21

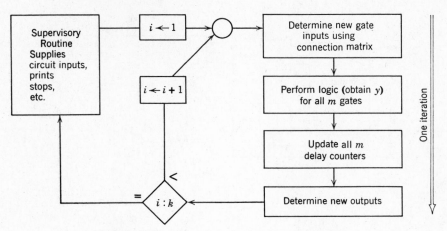

FIGURE 15.22   *Flowchart for level-mode simulation.*

a network modeled in this way. The principal advantage of the counting scheme is efficiency of programming. It can be implemented on small fixed-point machines, no multiplication or division is required, and the counts can be stored as 8-bit bytes.

Each iteration through the complete network is characterized by the flow chart of Fig. 15.22. The important point to notice in Fig. 15.22 is that each step is accomplished for all gates before the next step is begun. Thus, all gate inputs remain unchanged until the output of every gate is updated. Gates may, therefore, be considered in any order irrespective of network connections. After some number K iterations, the program enters a supervisory routine, which permits changes in the circuit inputs $x_1 \cdots x_n$ and performs various housekeeping operations. We will not concern ourselves with the details of this routine.

As pointed out earlier, many approaches to level-mode simulation are possible. We have only attempted to show what might be done using one very economical gate model.

Example 15.4

Illustrate the sum, S, and the output, z, of the AND gate in Fig. 15.23 for several iterations with inputs as shown. Assume the delay model of Fig. 15.22. Let $S_{max} = 2$ and $S_{min} = -3$.

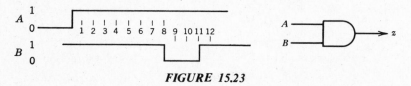

FIGURE 15.23

Solution We assume that S is initially equal to $S_{\min}$. The value of S begins to change when the input A goes to one at iteration 1. The output changes to 1 after iteration 3 as shown in Fig. 15.24, while S continues to increase until saturating at iteration 5. We note that the new output will not affect any other gate until iteration 4.

Iteration	1	2	3	4	5	6	7	8	9	10	11	12	13	14	15
S	-3	-2	-1	0	1	2	2	2	1	0	-1	0	1	2	2	2

FIGURE 15.24

At iteration 8, the narrow pulse on line B begins to affect S. We note that the output pulse is narrower than the input pulse. This can happen for pulses narrower than two gate delays. ∎

15.9 Summary

In this chapter, we have utilized the understanding of fundamental mode circuits gained in Chapter 14 to explore in detail the function of a master-slave flip-flop. It is unlikely that the reader will ever be called upon to design such a device. You are much more likely to have occasion to use them in situations where the validity of the clock-mode assumption may be called into question. You should now be able to perceive when this might be the case and to take whatever action might be appropriate.

Problems

15.1. Obtain an alternate flow table for the rising-edge-triggered D flip-flop in which the circuit changes state on any falling edge of the clock. For that clock transition, it must always go to the same next stable state. Depending on D, the circuit may go to two different states at the time of the rising edge of the clock. Obtain the most economical possible realization of this flow table and compare the result to Fig. 15.1a.

15.2 Derive the flow table for a D-type flip-flop that operates on both the positive and negative transitions of the clock. The circuit has two inputs D and C and a single output z. The value of D at the time C changes from 0 to 1 becomes the flip-flop output z. The output then remains unchanged until the next clock transition. The output is again updated when C

changes from 1 to 0 and so on. Also, obtain an output table for the circuit. Assume fundamental-mode operation.

Discuss the usefulness of this type of memory element as a controlled clock device as discussed in Chapter 11.

15.3. Obtain the flow table of a rising-edge-triggered J-K flip-flop.

15.4. Develop the level-mode flow table for a master-slave T flip-flop. Recall that the T flip-flop is regarded as a pulse mode circuit. The only input is the pulse line, T.

15.5. Obtain a level-mode flow table for an edge-sensing master-slave D flip-flop. That is, the value of D present at the time of the leading edge of a clock pulse will become the flip-flop output at the time of the trailing edge of the pulse.

15.6. Use the technique of Section 15.5 to obtain a hazard-free realization of the transition table given in Fig. 14.42.

15.7. Repeat problem 15.6 for the transition table of Fig. 15.54.

15.8. Use the approach of Section 15.5 to obtain a realization of the flow table given in Fig. 15.3.

15.9. Obtain a realization of the edge-sensing master-slave J-K flip-flop given in Fig. 15.9. Make sure the circuit is free from races and static and dynamic hazards. Are there essential hazards in this flow-table? If so, identify them in the state table and find the critical gates in the circuit realization. Discuss the probability of circuit malfunction due to each of these essential hazards.

15.10. Tabulate the functional values of the following Boolean expressions for all combinations of inputs using the three value logic of Section 15.7.

$$\text{(a)} \quad A \oplus B \qquad \text{(b)} \quad \bar{A} + B$$

15.11. Use Eichelberger's method to identify the hazard in the circuit of Fig. 14.43. Recall that the hazard will have an effect as circuit values change from $y_2 y_1 x_2 x_1 = 1111$ to $y_2 y_1 x_2 x_1 = 1110$.

15.12. Use the technique of Section 15.8 manually to step through a simulation of the circuit in Fig. P15.12. Let $S_{max} = 2$ and $S_{min} = -3$ for all gates.

(a) Assume that the circuit is initially stable with $y_1 = 1$ and $D = C = 0$. Begin the simulation as C goes to 1. After 6 iterations, C returns to 0. Continue the simulation for a total of 15 iterations.

(b) Assume that the circuit is initially stable, with $D = C = y_1 = 1$. Let C go to 0 at the beginning of the simulation. Carry out the simulation for 15 iterations.

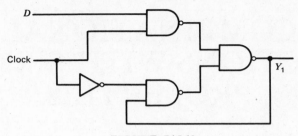

FIGURE P15.12

15.13. A single inverter is to be simulated using the technique of Section 15.8. Let $S_{max} = 3$ and $S_{min} = -4$. Suppose the gate is initially stable with an input of logical 0. Manually calculate 15 iterations of the simulation beginning at the onset on a narrow input pulse.
   (a)   pulse duration = 3 iterations
   (b)   pulse duration = 5 iterations
   (c)   pulse duration = 9 iterations

Bibliography

1. Eichelberger, E. B. "Hazard Detection in Combinational and Sequential Switching Circuits," *IBM Journal of Research and Development*, **90** (March 1965).
2. Harrison, Michael, *Introduction to Switching and Automata Theory*, McGraw-Hill, New York, 1965.
3. *Signetics Data Manual*, Signetics Corporation, Sunnyvale Calif., 1976, p. 87.
4. Evans, D. J. "Accurate Simulation of flip-flop Timing," *Proceedings of Design Automation Conference*, no. 15, p. 398, June 1978.

16

COMBINATIONAL FUNCTIONS WITH SPECIAL PROPERTIES

16.1   Introduction

In Chapters 6, 7, and 8, we dealt with general techniques applicable to the design of combinational circuits of all types, with emphasis on two-level, or second-order, design. In this chapter, we return to combinational design to consider some special classes of logic functions that do not lend themselves readily to second-order design and some special techniques for dealing with such functions.

One particular type of function that is not economically realized in second-order form is that in which the function value is determined only by the *number* of literals that take the value 1. Such functions are called symmetric functions and will be discussed in detail in the next section. Section 16.5 deals with the general problem of breaking Boolean functions into higher-order representations. These representations are called decompositions. We will see that symmetric functions are readily decomposable functions.

In Section 16.7, we turn to another special type of combinational circuit, the *iterative* circuit, made up of a cascade of identical subcircuits. We will see that the design of such circuits has much in common with sequential circuit design. The concluding sections of this chapter develop the concepts of lattices, unate functions, threshold logic, and, indeed, many current topics of switching theory research.

16.2   Symmetric Functions

In Chapter 8, we encountered the expression

$$f(c, x, y) = c\bar{x}\bar{y} + \bar{c}x\bar{y} + \bar{c}\bar{x}y + cxy \tag{16.1}$$

in the generation of a sum bit in binary addition (see Fig. 8.2a). This function is an example of a *symmetric function*.

As we learned, symmetric functions may often be most economically realized with more than two levels of logic. Notice that the value of Equation 16.1 is 1 whenever an odd number of the variables are 1. The n-variable odd parity check function is the extension of this pattern to n variables. An efficient multilevel NAND-gate realization of this latter symmetric function was also given in Chapter 8.

In general, a symmetric function may be defined as follows.

Definition 16.1 A switching function $f(x_1^{j_1}, x_2^{j_2}, \ldots, x_n^{j_n})$ is *symmetric* with respect to the literals $x_1^{j_1}, x_2^{j_2}, \ldots, x_n^{j_n}$, if and only if it remains unchanged after any permutation of these literals, where the j_i may take on only the values 0 and 1, and

$$x_i^{j_i} = x_i \quad \text{if } j_i = 0 \qquad \text{and} \qquad x_i^{j_i} = \bar{x}_i \quad \text{if } j_i = 1 \tag{16.2}$$

Consider, for example, Equation 16.1. Since it contains only three variables, we may check exhaustively all six permutations of the variables to determine if the expression remains unchanged. In Fig. 16.1, we see Equation 16.1 as it appears after each permutation of $\{c, x, y\}$. It should be clear to the reader that the function is unchanged in each case. The function

$$x\bar{y}z + \bar{x}yz + \bar{x}\bar{y}\bar{z} \tag{16.3}$$

is also symmetric, but in this case the literals of symmetry are x, y, and $\bar{z}$ rather than x, y, and z.

This becomes clear if we let

$$q = \bar{z}$$

Thus, Equation 16.3 becomes

$$x\bar{y}\bar{q} + \bar{x}y\bar{q} + \bar{x}\bar{y}q$$

which is clearly unchanged by a permutation of the variables.

The reader may notice that functions symmetric with respect to some complemented literals and some uncomplemented literals may not be easily identified.

Permutation	Expression
1 Identity: $c \to c, x \to x, y \to y$	$c\bar{x}\bar{y} + \bar{c}x\bar{y} + \bar{c}\bar{x}y + cxy$
2 $c \to x, x \to y, y \to c$	$x\bar{y}\bar{c} + \bar{x}y\bar{c} + \bar{x}\bar{y}c + xyc$
3 $c \to y, y \to x, x \to c$	$y\bar{c}\bar{x} + \bar{y}c\bar{x} + \bar{y}\bar{c}x + ycx$
4 $c \to c, x \to y, y \to x$	$c\bar{y}\bar{x} + \bar{c}y\bar{x} + \bar{c}\bar{y}x + cyx$
5 $c \to x, x \to c, y \to y$	$x\bar{c}\bar{y} + \bar{x}c\bar{y} + \bar{x}\bar{c}y + xcy$
6 $c \to y, x \to x, y \to c$	$y\bar{x}\bar{c} + \bar{y}x\bar{c} + \bar{y}\bar{x}c + yxc$

FIGURE 16.1 *Permutations of the variables of Equation 16.1.*

Symmetric functions have some very interesting properties that are expressed in the following set of theorems.

As was mentioned previously, the value of a symmetric function depends only on the number of literals of symmetry which are 1. This is set down formally in the following key theorem, which was originally stated by Shannon [6].

THEOREM 16.1 A necessary and sufficient condition that a switching function of n variables be symmetric is that it may be specified by a set of integers $\{a_k\}$ where $0 \le a_k \le n$, such that if exactly $a_m (m = 1, 2, \ldots, k)$ of the literals have the value 1, then the function has the value 1, and not otherwise.

Proof:
    A. (Necessity) Suppose that a function is 1 when the first a_m literals are 1 but is 0 when some other a_m literals are 1. Suppose each of the a_m literals of the first set is permuted to one of the literals of the second set. Clearly, a different function will result. Thus, the function would not be symmetric.
    B. (Sufficiency) Suppose that a function is 1 if and only if exactly a_m of the literals are 1. Following any permutation of the literals, the function will still be 1 when any a_m of the literals are 1. Similarly for any a_m in $\{a_k\}$, so the function is symmetric.

Q.E.D.

We may now express any symmetric function in the form

$$S_{\{a_k\}}^n(x_1^{j_1}, x_2^{j_2}, \ldots, x_n^{j_n}) \tag{16.4}$$

Thus, Equation 16.1 may be written as

$$S_{1, 3}^3(c, x, y) \tag{16.5}$$

and Equation 16.3 as

$$S_1^3(x, y, \bar{z}) \tag{16.6}$$

Similarly, the n-variable even parity check function may be written

$$S_{0, 2, 4, \ldots, n}^n(x_1, x_2, \ldots, x_n) \tag{16.7}$$

if n is even. If n is odd, the last subscript is $n - 1$.

16.3 Boolean Combinations of Symmetric Functions

Theorem 16.1 not only permits a convenient notation for symmetric functions, but it also provides for a considerable simplification of the algebraic operations on symmetric functions.

THEOREM 16.2

$$S_{a_1, \ldots, a_j, b_1, \ldots, b_m}^n(x_1^{j_1}, x_2^{j_2}, \ldots, x_n^{j_n}) + S_{a_1, \ldots, a_j, c_1, \ldots, c_k}^n(x_1^{j_1}, x_2^{j_2}, \ldots, x_n^{j_n})$$
$$= S_{a_1, \ldots, a_i, b_1, \ldots, b_m, c_1, \ldots, c_k}^n(x_1^{j_1}, x_2^{j_2}, \ldots, x_n^{j_n}) \tag{16.8}$$

Proof: The proof follows immediately from Theorem 16.1. That is, a function that is 1 when $a_1, \ldots, a_j, b_1, \ldots, b_{m-1}$ or b_m of the literals are 1 or when $a_1, \ldots, a_j, c_1, \ldots, c_{k-1}$ or c_k of the same literals are 1 is 1 when $a_1, \ldots, a_j, b_1, \ldots, b_m, c_1, \ldots, c_{k-1}$ or c_k of the literals are 1.

THEOREM 16.3

$$[S_{a_1, \ldots, a_i, b_1, \ldots, b_m}^n (x_1^{j_1}, x_2^{j_2}, \ldots, x_n^{j_n})] \cdot [S_{a_1, \ldots, a_i, c_1, \ldots, c_k}^n (x_1^{j_1}, x_2^{j_2}, \ldots, x_n^{j_n})]$$
$$= S_{a_1, \ldots, a_i}^n (x_1^{j_1}, x_2^{j_2}, \ldots, x_n^{j_n}) \quad (16.9)$$

Proof Similar to that of Theorem 16.2.

Example 16.1

$$S_{1, 2, 4}^5 (V, W, X, Y, Z) + S_{2, 3, 4}^5 (V, W, X, Y, Z) = S_{1, 2, 3, 4}^5 (V, W, X, Y, Z)$$

while

$$[S_{1, 2, 4}^5 (V, W, X, Y, Z)] \cdot [S_{2, 3, 4}^5 (V, W, X, Y, Z)] = S_{2, 4}^5 (V, W, X, Y, Z)$$

It is possible to state another theorem that follows directly from Theorem 16.1. Since the general statement of the theorem is rather awkward, we will instead illustrate it by example. ∎

Example 16.2

$$\overline{S_{1, 2, 3, 6}^6 (x_1, x_2, \ldots, x_6)} = S_{0, 4, 5}^6 (x_1, x_2, \ldots, x_6) \quad (16.10)$$

Note that $S_{0, 4, 5}^6$ is 1 when 0, 4, and 5 of the literals $x_1, x_2, \ldots, x_6$ are 1. The function $S_{1, 2, 3, 6}^6$ is never 1 when $S_{0, 4, 5}^6$ is 1, but the former function is 1 for every number of 1-laterals for which $S_{0, 4, 5} = 0$. The complement of any symmetric function of any number of variables may be obtained in this way. ∎

The theorem illustrated by Example 16.2 concerns the complement of a symmetric function. The next theorem concerns the function obtained by complementing the variables of a symmetric function.

THEOREM 16.4

$$S_{a_1, a_2, \ldots, a_k}^n (x_1^{j_1}, x_2^{j_2}, \ldots, x_n^{j_n}) = S_{n-a_1, \ldots, n-a_k}^n (x_1^{1-j_1}, x_2^{1-j_2}, \ldots, x_n^{1-j_n}) \quad (16.11)$$

Proof If a_i of the literals $x_1^{j_1}, x_2^{j_2}, \ldots, x_n^{j_n}$ are 1, then $n - a_i$ of these literals are 0, and $n - a_i$ of the literals $x_1^{1-j_1}, x_2^{1-j_2}, \ldots, x_n^{1-j_n}$ are 1. Thus, the two sides of Equation 16.11 are equal for any assignment of variable values.

Example 16.3

$$S_{2, 3}^4 (A,B,C,D) = AB\bar{C}\bar{D} + A\bar{B}C\bar{D} + A\bar{B}\bar{C}D + \bar{A}BC\bar{D} + \bar{A}B\bar{C}D + \bar{A}\bar{B}CD$$
$$+ ABC\bar{D} + AB\bar{C}D + A\bar{B}CD + \bar{A}BCD$$
$$= S_2^4 (\bar{A},\bar{B},\bar{C},\bar{D}) + S_1^4 (\bar{A},\bar{B},\bar{C},\bar{D}) = S_{4-2, 4-3} (\bar{A},\bar{B},\bar{C},\bar{D}) \quad (16.12)$$

∎

THEOREM 16.5 Any symmetric function $S^n_{a_1, a_2, \ldots, a_k}(x_1^{j_1}, x_2^{j_2}, \ldots, x_n^{j_n})$ may be represented by the conjunction of a subset of the following *elementary symmetric functions*:

$$\sigma^n_0 = x_1^{1-j_1} \cdot x_2^{1-j_2} \cdots x_n^{1-j_n}$$

$$\sigma^n_1 = x_1^{j_1} \cdot x_2^{1-j_2} \cdots x_n^{1-j_n} + x_1^{1-j_1} \cdot x_2^{j_2} \cdot x_3^{1-j_3} \cdots x_n^{1-j_n}$$

$$+ \cdots + x_1^{1-j_1} \cdot x_2^{1-j_2} \cdots x_{n-1}^{1-j_{n-1}} \cdot x_n^{j_n}$$

$$\vdots$$

$$\sigma^n_n = x_1^{j_1} x_2^{j_2} \cdots x_n^{j_n}. \tag{16.13}$$

Proof Each elementary symmetric function, σ_i, is 1 if and only if exactly i of the literals are 1. Thus, by Theorems 16.1 and 16.2, any symmetric function may be formed as the conjunction of a subset of these functions.

Example 16.4

$$S^4_{0, 2, 4}(A, B, C, \bar{D}) = S^4_0(A, B, C, \bar{D}) + S^4_2(A, B, C, \bar{D}) + S^4_4(A, B, C, \bar{D})$$

$$= \bar{A}\bar{B}\bar{C}D + (AB\bar{C}D + A\bar{B}CD + A\bar{B}C\bar{D} + \bar{A}BCD + \bar{A}B\bar{C}\bar{D}$$

$$+ \bar{A}\bar{B}C\bar{D}) + ABC\bar{D}$$

$$= \sigma^4_0 + \sigma^4_2 + \sigma^4_4 \qquad \blacksquare$$

Example 16.5

A certain communication system utilizes 16 parallel lines to send four 4-bit characters at a time. Each of these 4-bit characters has a parity bit to establish even parity over the four bits of that character. Parity is to be checked at a certain relay point in the communications system, and retransmission of all four characters is to be demanded by the relay point if there is odd parity over two or more of the characters.

Write, in symmetric function notation, the expression for a combinational circuit whose output is 1 whenever retransmission is to be demanded. Let w_1, w_2, w_3, w_4; x_1, x_2, x_3, x_4; y_1, y_2, y_3, y_4; and z_1, z_2, z_3, z_4 represent the bits of the four characters.

Solution It is easily seen that the overall expression is not a symmetric function. Consider, for example, the following sets of bits, which might be received at the relay point.

$w_1\ w_2\ w_3\ w_4$	$x_1\ x_2\ x_3\ x_4$	$y_1\ y_2\ y_3\ y_4$	$z_1\ z_2\ z_3\ z_4$
1 1 1 0	1 1 1 0	1 1 0 0	0 0 0 0

$w_1\ w_2\ z_3\ w_4$	$x_1\ x_2\ z_4\ x_4$	$y_1\ y_2\ y_3\ y_4$	$z_1\ z_2\ w_3\ x_3$
1 1 0 0	1 1 0 0	1 1 0 0	0 0 1 1

For the first set of bits, the functions should be 1 since the first two characters have odd parity. The second set of bits has the same number of 1s, but the output should be 0 since all characters have correct parity.

It is, however, straightforward to write the parity check function over each set of four bits as a symmetric function. For example, odd parity over w_1, w_2, w_3, and w_4 is given by

$$P_w = S_{1,3}^4(w_1, w_2, w_3, w_4)$$

Similarly, the function that determines from the four parity checks whether two or more of these are 1 is also symmetric. Thus, overall function may be expressed as

$$f = S_{2,3,4}^4[S_{1,3}^4(w_1, w_2, w_3, w_4), S_{1,3}^4(x_1, x_2, x_3, x_4)$$
$$S_{1,3}^4(y_1, y_2, y_3, y_4), S_{1,3}^4(z_1, z_2, z_3, z_4)] \qquad \blacksquare$$

We have said nothing so far about the physical implementation of symmetric functions. In terms of NAND or NOR-gates, symmetric functions enjoy no advantage over other functions with respect to economic realization. In applications that will tolerate relays or the linear single threshold devices to be discussed later in the chapter, symmetric functions may be realized quite economically.

16.4 Higher-Order Forms

In Fig. 6.21, we saw an example of a function in which a higher-order realization was more economical than the best second-order realization. Nevertheless, we have so far concentrated on second-order circuits, primarily on the assumption that their ease of design and higher speed would outweigh any possible cost disadvantages. This assumption is quite generally justified, but there are certainly situations where the cost factor will require the use of higher-order circuits. Perhaps the simplest and most commonly applied method of finding higher-order forms of switching functions is factoring. Factoring generally involves starting with a second-order form, looking for common terms, and then applying the distributive laws to factor them out.

One particular class of functions for which factoring frequently pays off is the class of symmetric functions, discussed in the last section. Let us consider the even parity function of four variables, $S_{0,2,4}^4(A, B, C, D)$. The K-map of this function is shown in Fig. 16.2.

We see that this function, as is the case with all parity functions, results in a "checkerboard" pattern on the K-map. From the standpoint of a second-order realization, this pattern is highly undesirable. There is not a single possible combination of 0-cubes into larger cubes, so an SOP realization will require eight 4-input AND-gates and one 8-input OR-gate, for a total of nine gates with 40 inputs. We might note that this function is the worst possible case for second-order realization. Deleting any minterms would obviously simplify the function, and moving or adding any minterms will result in larger cubes, also simplifying

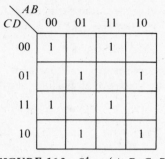

FIGURE 16.2 $S_{0,2,4}^4(A, B, C, D)$.

the function. From this observation, we can show that the most expensive possible nonredundant second-order realization of a function of n variables will have

$$N_g = 2^{n-1} + 1 \qquad (16.14)$$

gates with

$$N_i = (n + 1)2^{n-1} \qquad (16.15)$$

inputs.

We have already demonstrated in Chapter 8 a less expensive higher-order realization of the 8-variable parity check function. Let us now approach this problem from the point of view of factoring. To start, we first write the canonic expression for $S_{0,2,4}^4(A, B, C, D)$, which can be read directly from the map:

$$f(A, B, C, D) = \bar{A}\bar{B}\bar{C}\bar{D} + AB\bar{C}\bar{D} + \bar{A}B\bar{C}D + A\bar{B}\bar{C}D$$

$$+ \bar{A}\bar{B}CD + ABCD + \bar{A}BC\bar{D} + A\bar{B}C\bar{D} \qquad (16.16)$$

The map is of further assistance as we search for common factors in the various terms. We note, for example, that the product $\bar{C}\bar{D}$ is common to the two minterms in the first row, which can thus be factored to the form

$$\bar{A}\bar{B}\bar{C}\bar{D} + AB\bar{C}\bar{D} = \bar{C}\bar{D}(\bar{A}\bar{B} + AB)$$

Similarly, $\bar{C}D$ is common to the minterms in the second row, CD to the minterms in the third row, and $C\bar{D}$ to the fourth-row minterms. The function can then be factored to the form

$$f(A, B, C, D) = \bar{C}\bar{D}(\bar{A}\bar{B} + AB) + \bar{C}D(\bar{A}B + A\bar{B})$$

$$+ CD(\bar{A}\bar{B} + AB) + C\bar{D}(\bar{A}B + A\bar{B}) \qquad (16.17)$$

The four terms in Equation 16.17 may be combined in pairs to form

$$f(A, B, C, D) = (\bar{C}\bar{D} + CD)(\bar{A}\bar{B} + AB) + (\bar{C}D + C\bar{D})(\bar{A}B + A\bar{B})$$

$$\qquad (16.18)$$

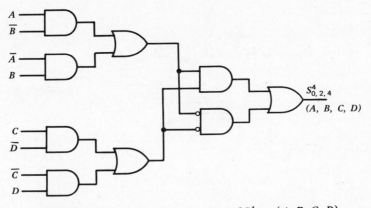

FIGURE 16.3   *Factored realization of* $S^4_{0,\,2,\,4}(A,\,B,\,C,\,D)$.

Finally, noting the relationship of the factors in Equation 16.18 to the exclusive-OR, we have the form

$$f(A,\,B,\,C,\,D) = \overline{(C \oplus D)}\overline{(A \oplus B)} + (C \oplus D)(A \oplus B) \qquad (16.19)$$

This equation leads to the realization of Fig. 16.3, requiring nine gates with 18 inputs, plus two inverters. We have thus eliminated 22 inputs at the cost of two inverters, generally a quite worthwhile saving.

If the circuit is to be realized with NAND logic, we can eliminate the inverters at the cost of two more inputs, by a technique known as "bundling" [3]. This technique is applicable whenever a signal derived from a NAND (or NOR) circuit is to be inverted at the input to some other gate, as shown in Fig. 16.4a. This places two inversions in series, which cancel, so that the inputs to the first gate can be "bundled" together and passed on directly to the input of the second gate, as shown in Fig. 16.4b. Applying this technique to the circuit of the parity checker results in the circuit of Fig. 16.5 having nine gates with 20 inputs.

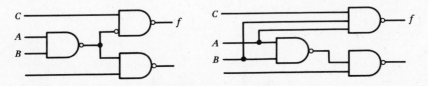

FIGURE 16.4   *Bundling.*

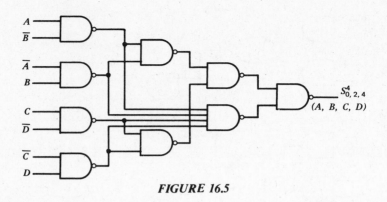

FIGURE 16.5

16.5 Simple Disjoint Decomposition

Simplification of circuits by factoring is a satisfactory technique in many cases, but it is largely a trial-and-error process, dependent on a good deal of experience and often just plain luck in spotting common factors in the function. It is evident that a more systematic method of finding simpler designs of higher order would be very useful. In essence, the process of simplification involves breaking a complex single function into a number of related simpler functions. The general name for this process is *functional decomposition*. The methods to be presented here, which we believe to be the most generally useful, are primarily the work of Ashenhurst [2] and Curtis [4].

The first type of decomposition we will consider is the simple disjoint decomposition.

Definition 16.2 Let $x_1, x_2, \ldots, x_n$ be a set of n switching variables and let A and B be disjoint subsets of X such that $A \cup B = X$. If $F[\varphi(A), B]$ takes on the same functional values as $f(X)$ wherever the latter are specified, then F and φ are said to form a *simple disjoint decomposition* of f. The set of variables, A, shall be

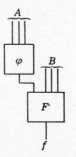

FIGURE 16.6   Simple disjoint decomposition.

known as the *bound* variables, the set, B, as the *free* variables. The partitioning of the variables shall be indicated by $B \mid A$. The basic form of circuit realization of the simple disjoint decomposition is shown in Fig. 16.6.

Example 16.5

Obtain a simple disjoint decomposition of

$$f_1(x_1, x_2, x_3, x_4) = x_1 \bar{x}_2 x_3 x_4 + x_1 x_2 x_3 \bar{x}_4 + \bar{x}_1 \bar{x}_2 \bar{x}_3 \bar{x}_4 + \bar{x}_1 x_2 \bar{x}_3 x_4 \quad (16.20)$$

Solution

$$f_1(x_1, x_2, x_3, x_4) = x_1 x_3(x_2 \bar{x}_4 + \bar{x}_2 x_4) + \bar{x}_1 \bar{x}_3(\bar{x}_2 \bar{x}_4 + x_2 x_4) \quad (16.21)$$

$$= x_1 x_3 \varphi(x_2, x_4) + \bar{x}_1 \bar{x}_3 \overline{\varphi(x_2, x_4)} \quad (16.22)$$

$$= F[\varphi(x_2, x_4), x_1, x_3] \quad (16.23)$$

where

$$\varphi = x_2 \bar{x}_4 + \bar{x}_2 x_4 \quad (16.24)$$

This corresponds to the partitioning $x_1, x_3 \mid x_2, x_4$. ∎

Nothing was said in the previous example as to how it was determined that a simple disjoint partition of the form $x_1, x_3 \mid x_2, x_4$ could actually be obtained.

Let us explore the possibility of a general method of determining whether or not a particular decomposition exists. Consider, for example, the function

$$f(x_1, x_2, x_3, x_4) = \sum m(4, 5, 6, 7, 8, 13, 14, 15) \quad (16.25)$$

which may be decomposable into the form $F[\varphi(x_1, x_4), x_2, x_3]$. As a first step, we can collect the minterms containing each of the four combinations of $x_2 x_3$ and then factor these terms out to give the form

$$f(x_1, x_2, x_3, x_4) = \bar{x}_2 \bar{x}_3 \alpha(x_1, x_4) + \bar{x}_2 x_3 \beta(x_1, x_4)$$

$$+ x_2 x_3 \lambda(x_1, x_4) + x_2 \bar{x}_3 \delta(x_1, x_4) \quad (16.26)$$

This is the step that produced Equation 16.21 above and Equation 16.17 in the parity function example. This factoring can be done for any function whether it is decomposable or not. Now we will make a special map of the function, called a *partition map* (Fig. 16.7). This is simply a K-map with the free variables running down the side and the bound variables across the top. Note that the minterm numbers are still those based on the original ordering of the variables specified in Equation 16.25.

Comparing the partition map with Equation 16.26, we can see that the first row must correspond to the function $\alpha(x_1, x_4)$, the second row to $\beta(x_1, x_4)$, the third row to $\lambda(x_1, x_4)$, and the fourth row to $\delta(x_1, x_4)$. Again, this is still completely general. Now recall that for a simple disjoint function, F must be a

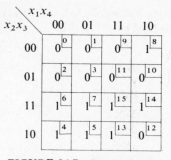

FIGURE 16.7 *Partition map.*

function of x_2, x_3 and a *single* function $\varphi(x_1, x_4)$. In the above, let the first row function α be the function φ. Then, for the condition of a *single* function to be met, the remaining functions must be either α, $\bar{\alpha}$, 0, or 1. This in turn means that the remaining rows must be either identical, complementary, all 0s, or all 1s. In this example, the conditions are met and the function can be read off the partition matrix as

$$f(x_1, x_2, x_3, x_4) = \bar{x}_2 \bar{x}_3 \varphi(x_1, x_4) + x_2 \bar{x}_3 \overline{\varphi(x_1, x_4)} + x_2 x_3$$

$$= F[\varphi(x_1, x_4), x_2, x_3] \tag{16.27}$$

where $\varphi(x_1, x_4) = x_1 \bar{x}_4$.

We now have a rule for detecting a simple disjoint decomposition on the partition map. However, this rule may be rather difficult to apply in large partition maps. As a further aid, let us define *column multiplicity* as the number of different column patterns of 1s and 0s. For example, in the partition of Fig. 16.7, there are two distinct column patterns,

$$\begin{bmatrix} 0 \\ 0 \\ 1 \\ 1 \end{bmatrix} \quad \text{and} \quad \begin{bmatrix} 1 \\ 0 \\ 1 \\ 0 \end{bmatrix}$$

The following theorem offers a simple method by which it can be determined if a partition map corresponds to a simple disjoint decomposition.

THEOREM 16.6 A partition map corresponds to a simple disjoint decomposition if and only if the column multiplicity is less than or equal to two.

Proof If a function is decomposable into row functions, the functional values of each row of the Karnaugh map must form one of the four functions, 0, 1, φ, or $\bar{\varphi}$. Rows corresponding to the functions 0 and 1 have the same entry in every column and thus have no effect on column multiplicity. If a column has a 1 in some φ-row, it must have a 0 in every $\bar{\varphi}$-row and a 1 in every φ-row. Similarly, a 0 in a φ-row implies a 0 in every

φ-row and a 1 in every $\bar{\varphi}$-row. Thus, there are only two distinct types of columns possible. If we assume a column multiplicity of two, we may similarly deduce the existence of a simple disjoint decomposition.   Q.E.D.

With the above, we have the basic rule needed to find decompositions. However, the amount of work is still rather extensive, since there are many possible partitions of the variables into bound and free variables. To reduce the amount of labor involved, we introduce *decomposition charts*, which are simply K-maps showing the position of the minterms for all possible combinations of bound and free variables. The decomposition charts for 4-variable functions are shown in Fig. 16.8. To use the charts, we simply circle the minterm numbers of the function on all charts and then look for column multiplicities of 2 or 1. Note that the 4×4 charts should be viewed sideways to reverse the order of bound and free variables. Thus, for example, the leftmost 4×4 chart will indicate decomposition of the form $F(\varphi(x_3, x_4), x_1, x_2)$ when viewed normally, and decompositions of the form $F(\varphi(x_1, x_2), x_3, x_4)$ when viewed sideways. The 2×8 charts should only be viewed normally, since decompositions with only one bound variable are trivial.

x_2		0				1		
x_3x_4	00	01	11	10	00	01	11	10
0	0	1	3	2	4	5	7	6
x_1 1	8	9	11	10	12	13	15	14

x_1		0				1		
x_3x_4	00	01	11	10	00	01	11	10
0	0	1	3	2	8	9	11	10
x_2 1	4	5	7	6	12	13	15	14

x_1		0				1		
x_2x_4	00	01	11	10	00	01	11	10
0	0	1	5	4	8	9	13	12
x_3 1	2	3	7	6	10	11	15	14

x_1		0				1		
x_2x_3	00	01	11	10	00	01	11	10
0	0	2	6	4	8	10	14	12
x_4 1	1	3	7	5	9	11	15	13

x_3x_4	00	01	11	10
00	0	1	3	2
01	4	5	7	6
x_1x_2 11	12	13	15	14
10	8	9	11	10

x_2x_4	00	01	11	10
00	0	1	5	4
01	2	3	7	6
x_1x_3 11	10	11	15	14
10	8	9	13	12

x_2x_3	00	01	11	10
00	0	2	6	4
01	1	3	7	5
x_1x_4 11	9	11	15	13
10	8	10	14	12

FIGURE 16.8   *Four-variable decomposition charts.*

Example 16.6

Use decomposition charts to find all simple disjoint decompositions of

$$f(x_1, x_2, x_3, x_4) = \sum m(4, 5, 6, 7, 8, 13, 14, 15) \tag{16.28}$$

Solution The decomposition charts for this function are shown in Fig. 16.9. The maps of Figs. 16.9b and 16.9e viewed normally indicate the partitionings $x_2 \,|\, x_1x_3x_4$ and $x_1x_2 \,|\, x_3x_4$. The maps of Figs. 16.9f and 16.9g viewed sideways indicate the partitionings $x_2x_4 \,|\, x_1x_3$ and $x_2x_3 \,|\, x_1x_4$. The resultant decompositions can then be read from the charts as:

$$\begin{aligned} f(x_2|x_1x_3x_4) &= \bar{x}_2\varphi(x_1, x_3, x_4) + x_2\overline{\varphi(x_1, x_3, x_4)} \\ &= \bar{x}_2(x_1\bar{x}_3\bar{x}_4) + x_2\overline{(x_1\bar{x}_3\bar{x}_4)} \end{aligned} \tag{16.29}$$

where $\varphi(x_1, x_3, x_4) = x_1\bar{x}_3\bar{x}_4$.

$$\begin{aligned} f(x_1x_2|x_3x_4) &= \bar{x}_1x_2 + x_1x_2\varphi(x_3, x_4) + x_1\bar{x}_2\overline{\varphi(x_3, x_4)} \\ &= \bar{x}_1x_2 + x_1x_2(x_3 + x_4) + x_1\bar{x}_2\overline{(x_3 + x_4)} \end{aligned} \tag{16.30}$$

where $\varphi(x_3, x_4) = x_3 + x_4$.

$$\begin{aligned} f(x_2x_4|x_1x_3) &= x_2x_4 + x_2\bar{x}_4\varphi(x_1, x_3) + \bar{x}_2\bar{x}_4\overline{\varphi(x_1, x_3)} \\ &= x_2x_4 + x_2\bar{x}_4(\bar{x}_1 + x_3) + \bar{x}_2\bar{x}_4\overline{(\bar{x}_1 + x_3)} \end{aligned} \tag{16.31}$$

x_2	0				1			
x_3x_4	00	01	11	10	00	01	11	10
x_1 0	0	1	3	2	④	⑤	⑦	⑥
1	⑧	9	11	10	12	⑬	⑮	⑭

(a)

x_1	0				1			
x_3x_4	00	01	11	10	00	01	11	10
x_2 0	0	1	3	2	⑧	9	11	10
1	④	⑤	⑦	⑥	12	⑬	⑮	⑭

(b)

x_1	0				1			
x_2x_4	00	01	11	10	00	01	11	10
x_3 0	0	1	⑤	④	⑧	9	⑬	12
1	2	3	⑦	⑥	10	11	⑮	⑭

(c)

x_1	0				1			
x_2x_3	00	01	11	10	00	01	11	10
x_4 0	0	2	⑥	④	⑧	10	⑭	12
1	1	3	⑦	⑤	9	11	⑮	⑬

(d)

x_3x_4	00	01	11	10
00	0	1	3	2
x_1x_2 01	④	⑤	⑦	⑥
11	12	⑬	⑮	⑭
10	⑧	9	11	10

(e)

x_2x_4	00	01	11	10
00	0	1	⑤	④
x_1x_3 01	2	3	⑦	⑥
11	10	11	⑮	⑭
10	⑧	9	⑬	12

(f)

x_2x_3	00	01	11	10
00	0	2	⑥	④
x_1x_4 01	1	3	⑦	⑤
11	9	11	⑮	⑬
10	⑧	10	⑭	12

(g)

FIGURE 16.9 *Decomposition charts, Example 16.5.*

where $\varphi(x_1, x_3) = \bar{x}_1 + x_3$.

$$f(x_2 x_3 \mid x_1 x_4) = x_2 x_3 + x_2 \bar{x}_3 \varphi(x_1, x_4) + \bar{x}_2 \bar{x}_3 \overline{\varphi(x_1, x_4)}$$

$$= x_2 x_3 + x_2 \bar{x}_3 (\bar{x}_1 + x_4) + \bar{x}_2 \bar{x}_3 \overline{(\bar{x}_1 + x_4)} \qquad (16.32)$$

where $\varphi(x_1, x_4) = \bar{x}_1 + x_4$. ∎

The same procedure can be applied to functions of more variables. The decomposition charts for 5 variables are shown in Fig. 16.10. The method can, in

$x_2 x_3$		00				01				11				10		
$x_4 x_5$	00	01	11	10	00	01	11	10	00	01	11	10	00	01	11	10
0	0	1	3	2	4	5	7	6	12	13	15	14	8	9	11	10
x_1 1	16	17	19	18	20	21	23	22	28	29	31	30	24	25	27	26

$x_1 x_3$		00				01				11				10		
$x_4 x_5$	00	01	11	10	00	01	11	10	00	01	11	10	00	01	11	10
0	0	1	3	2	4	5	7	6	20	21	23	22	16	17	19	18
x_2 1	8	9	11	10	12	13	15	14	28	29	31	30	24	25	27	26

$x_1 x_2$		00				01				11				10		
$x_4 x_5$	00	01	11	10	00	01	11	10	00	01	11	10	00	01	11	10
0	0	1	3	2	8	9	11	10	24	25	27	26	16	17	19	18
x_3 1	4	5	7	6	12	13	15	14	28	29	31	30	20	21	23	22

$x_1 x_2$		00				01				11				10		
$x_3 x_5$	00	01	11	10	00	01	11	10	00	01	11	10	00	01	11	10
0	0	1	5	4	8	9	13	12	24	25	29	28	16	17	21	20
x_4 1	2	3	7	6	10	11	15	14	26	27	31	30	18	19	23	22

$x_1 x_2$		00				01				11				10		
$x_3 x_4$	00	01	11	10	00	01	11	10	00	01	11	10	00	01	11	10
0	0	2	6	4	8	10	14	12	24	26	30	28	16	18	22	20
x_5 1	1	3	7	5	9	11	15	13	25	27	31	29	17	19	23	21

FIGURE 16.10 *Decomposition charts for 5-variable functions.*

x_3	0				1			
x_4x_5	00	01	11	10	00	01	11	10
x_1x_2 00	0	1	3	2	4	5	7	6
01	8	9	11	10	12	13	15	14
11	24	25	27	26	28	29	31	30
10	16	17	19	18	20	21	23	22

x_2	0				1			
x_4x_5	00	01	11	10	00	01	11	10
x_1x_3 00	0	1	3	2	8	9	11	10
01	4	5	7	6	12	13	15	14
11	20	21	23	22	28	29	31	30
10	16	17	19	18	24	25	27	26

x_2	0				1			
x_3x_5	00	01	11	10	00	01	11	10
x_1x_4 00	0	1	5	4	8	9	13	12
01	2	3	7	6	10	11	15	14
11	18	19	23	22	26	27	31	30
10	16	17	21	20	24	25	29	28

x_2	0				1			
x_3x_4	00	01	11	10	00	01	11	10
x_1x_5 00	0	2	6	4	8	10	14	12
01	1	3	7	5	9	11	15	13
11	17	19	23	21	25	27	31	29
10	16	18	22	20	24	26	30	28

x_1	0				1			
x_4x_5	00	01	11	10	00	01	11	10
x_2x_3 00	0	1	3	2	16	17	19	18
01	4	5	7	6	20	21	23	22
11	12	13	15	14	28	29	31	30
10	8	9	11	10	24	25	27	26

x_1	0				1			
x_3x_5	00	01	11	10	00	01	11	10
x_2x_4 00	0	1	5	4	16	17	21	20
01	2	3	7	6	18	19	23	22
11	10	11	15	14	26	27	31	30
10	8	9	13	12	24	25	29	28

x_1	0				1			
x_3x_4	00	01	11	10	00	01	11	10
x_2x_5 00	0	2	6	4	16	18	22	20
01	1	3	7	5	17	19	23	21
11	9	11	15	13	25	27	31	29
10	8	10	14	12	24	26	30	28

x_1	0				1			
x_2x_5	00	01	11	10	00	01	11	10
x_3x_4 00	0	1	9	8	16	17	25	24
01	2	3	11	10	18	19	27	26
11	6	7	15	14	22	23	31	30
10	4	5	13	12	20	21	29	28

x_1	0				1			
x_2x_4	00	01	11	10	00	01	11	10
x_3x_5 00	0	2	10	8	16	18	26	24
01	1	3	11	9	17	19	27	25
11	5	7	15	13	21	23	31	29
10	4	6	14	12	20	22	30	28

x_1	0				1			
x_2x_3	00	01	11	10	00	01	11	10
x_4x_5 00	0	4	12	8	16	20	28	24
01	1	5	13	9	17	21	29	25
11	3	7	15	11	19	23	31	27
10	2	6	14	10	18	22	30	26

FIGURE 16.10   (*continued*)

theory, be extended to any number of variables, but the number and size of the charts rapidly becomes impractical beyond 5 variables. Curtis [4] gives the charts for 6 variables but there are 29* of them, making the practicality of the method for 6 variables rather doubtful. However, the generality of the method provides the basis for computer mechanization of the process for large numbers of variables.

Functions with don't-cares can be handled very simply on the decomposition charts. Circle the 1s of the function in the usual manner and indicate the don't-cares by lines through the corresponding minterms. Then assign the don't-cares to achieve a column multiplicity of two or less.

Example 16.7

Determine a simple disjoint decomposition of

$$f(x_1, x_2, x_3, x_4, x_5) = \sum m(1, 2, 7, 9, 10, 18, 19, 25, 31) + d(0, 15, 20, 23, 26)$$

Solution The decomposition chart for the partition $x_3 x_4 | x_1 x_2 x_5$ is shown in Fig. 16.11a, with the minterms circled and the don't-cares lined out. We see that the chart

---

* Curtis also gives a scheme for reducing the number of charts to 16, but the patterns are more difficult to recognize on the condensed charts.

x_1		0				1		
$x_2 x_5$	00	01	11	10	00	01	11	10
$x_3 x_4$ 00	0̶	①	⑨	8	16	17	㉕	24
01	②	3	11	⑩	⑱	⑲	27	2̶6̶
11	6	⑦	1̶5̶	14	22	2̶3̶	㉛	30
10	4	5	13	12	2̶0̶	21	29	28

(a)

x_1		0				1		
$x_2 x_5$	00	01	11	10	00	01	11	10
$x_3 x_4$ 00	0̶	①	⑨	8	16	17	㉕	24
01	②	3	11	⑩	⑱	⑲	27	㉖
11	6	⑦	⑮	14	22	2̶3̶	㉛	30
10	4	5	13	12	2̶8̶	21	29	28

(b)

FIGURE 16.11 *Decomposition charts, Example 16.7.*

has a column multiplicity of four in both directions. If the don't-cares at 15 and 26 are assigned as 1s, a column multiplicity of two is achieved with the chart viewed in the normal direction, as shown in Fig. 16.11b. From this chart, interpreted as a K-map, we can read off the realization

$$f(x_3 x_4 | x_1 x_2 x_5) = (\bar{x}_3 \bar{x}_4 + x_3 x_4)\varphi(x_1, x_2, x_5) + \bar{x}_3 x_4 \overline{\varphi(x_1, x_2, x_5)} \quad (16.33)$$

where $\varphi(x_1, x_2, x_5) = x_2 x_5 + \bar{x}_1 x_5$. ∎

16.6   Complex Disjoint Decomposition

It is apparent that there must be many forms of decomposition other than simple disjoint. In this section, we consider decompositions resulting from the partitioning of the variables into three or more disjoint subsets.

Definition 16.3   Let $X = x_1, x_2, \ldots, x_n$ be a set of n switching variables and let $A_1, A_2, \ldots, A_m$ be disjoint subsets of X such that $A_1 \cup A_2 \cup \cdots \cup A_m = X$. Decompositions of the form

$$f(X) = F[\varphi_1(A_1), \varphi_2(A_2), \ldots, \varphi_m(A_m)]$$

or

$$f(X) = F[\varphi_1(A_1), \varphi_2(A_2), \ldots, \varphi_{m-1}(A_{m-1}), A_m] \quad (16.34)$$

shall be known as multiple disjoint decompositions. Decompositions of the form

$$f(X) = F(\lambda[\varphi(A), B], C) \quad (16.35)$$

or, in general, of the form

$$f(X) = F[\varphi_{m-1}(\varphi_{m-2}[\cdots (\varphi_1(\varphi_0(A_0), A_1], A_2) \cdots A_{m-2}], A_{m-1}), A_m] \quad (16.36)$$

shall be known as iterative disjoint decompositions. Finally, combinations of these forms, such as,

$$f(X) = F[\gamma[\varphi(A), B], \lambda(C), D] \quad (16.37)$$

shall be known as *complex disjoint* decompositions. In all cases, the decompositions are nontrivial only if all the bound sets contain at least 2 variables, since functions of a single variable are trivial.

Ashenhurst [2] has given a set of theorems relating the above types of decompositions to simple disjoint decompositions. These theorems enable us to extend the use of the decomposition chart to the determination of complex decompositions.

Rather then present the somewhat tedious proofs of these theorems, we will content ourselves with an example of the application of each theorem.

THEOREM 16.7 Let $f(X)$ be a function for which there exist two simple disjoint decompositions:

$$f(X) = F[\lambda(A, B), C]$$

$$= G[\varphi(A), B, C] \tag{16.38}$$

then there exists an iterative disjoint decomposition

$$f(X) = F[\rho[\varphi(A), B], C] \tag{16.39}$$

where

$$\rho[\varphi(A), B] = \lambda(A, B)$$

As an example of the application of this theorem, consider the function

$$f(x_1, x_2, x_3, x_4, x_5) = \sum m(5, 10, 11, 14, 17, 21, 26, 30)$$

Two decomposition charts of this function are shown in Fig. 16.12. The first indicates the existence of a decomposition

$$f(X) = F[\lambda(x_1, x_3, x_5), x_2, x_4] \tag{16.40}$$

while the second, viewed sideways, shows a decomposition

$$f(X) = G[\varphi(x_1, x_3), x_2, x_4, x_5] \tag{16.41}$$

Thus, the conditions of Theorem 16.7 are met with $A = (x_1 x_3)$, $B = (x_5)$, and $C = (x_2, x_4)$. From Fig. 16.12a, we read off the realization

$$f(x) = \bar{x}_2 \bar{x}_4 \lambda(x_1, x_3, x_5) + x_2 x_4 \overline{\lambda(x_1, x_3, x_5)} \tag{16.42}$$

where $\lambda(x_1, x_3, x_5) = x_3 x_5 + x_1 x_5$.

Thus, from Theorem 16.7, $\lambda(x_1, x_3, x_5)$ can be decomposed according to the partition $x_1 x_3 | x_5$. By inspection of the equation for λ, we see that the decomposition is given by

$$\rho[\varphi(A), B] = x_5(x_3 + x_1)$$

where $\varphi(A) = x_3 + x_1$.

This example is trivial in the sense that we did not need Theorem 16.7 to tell us that $\lambda(x_1, x_3, x_5)$ could be factored in the above manner. However, the theorem tells us that, no matter now large the number of variables, the charts will enable us to locate decompositions in which the subfunctions can in turn be further decomposed.

x_1		0				1		
x_3x_5	00	01	11	10	00	01	11	10
x_2x_4 00	0	1	⑤	4	16	⑰	㉑	20
01	2	3	7	6	18	19	23	22
11	⑩	⑪	15	⑭	㉖	27	31	㉚
10	8	9	13	12	24	25	29	28

(a)

x_2		0				1		
x_4x_5	00	01	11	10	00	01	11	10
x_1x_3 00	0	1	3	2	8	9	⑪	⑩
01	4	⑤	7	6	12	13	15	⑭
11	20	㉑	23	22	28	29	31	㉚
10	16	⑰	19	18	24	25	27	㉖

(b)

FIGURE 16.12 *Decomposition charts indicating iterative decomposition.*

THEOREM 16.8 Let $f(X)$ be a function for which there exist two simple disjoint decompositions

$$f(X) = F[\lambda(A), B] \tag{16.43}$$

$$= G[\varphi(B), A] \tag{16.44}$$

then there exists a multiple disjoint decomposition

$$f(X) = H[\lambda(A), \varphi(B)] \tag{16.45}$$

The conditions of Equations 16.43 and 16.44 will be indicated by a single decomposition chart, which satisfies the column multiplicity requirements when viewed both normally and sideways. An example is the parity function considered earlier. The decomposition chart corresponding to the map of Fig. 16.2 is shown in Fig. 16.13a. Clearly, we have a column multiplicity of two in both directions. Viewed normally, the chart indicates the decomposition

$$f(x_1 x_2 | x_3 x_4) = (\bar{x}_1 \bar{x}_2 + x_1 x_2)\overline{\varphi(x_3 x_4)} + (\bar{x}_1 x_2 + x_1 \bar{x}_2)\varphi(x_3, x_4) \tag{16.46}$$

where $\varphi(x_3 x_4) = x_3 \bar{x}_4 + \bar{x}_3 x_4$.

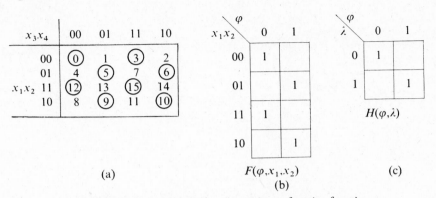

FIGURE 16.13   *Multiple decompositions of parity function.*

Viewed sideways, the chart indicates

$$f(x_3 x_4 \,|\, x_1 x_2) = (\bar{x}_3 \bar{x}_4 + x_3 x_4)\overline{\lambda(x_1, x_2)} + (\bar{x}_3 x_4 + x_3 \bar{x}_4)\lambda(x_1, x_2) \quad (16.47)$$

where $\lambda(x_1 x_2) = \bar{x}_1 x_2 + x_1 \bar{x}_2$.

The theorem indicates the existence of a multiple decomposition of the form $H[\varphi(x_3, x_4), \lambda(x_1, x_2)]$. To determine H, we make a K-map of $F[\varphi, x_1, x_2]$ as given by Equation 16.46 (Fig. 16.13b). The values of $\lambda(x_1, x_2)$ are indicated on this map, immediately enabling us to construct the map of $H(\varphi, \lambda)$ as shown in Fig. 16.13c. From this map, we read off

$$H(\varphi, \lambda) = \bar{\varphi}\bar{\lambda} + \varphi\lambda \quad (16.48)$$

where $\varphi = x_3 \oplus x_4$ and $\lambda = x_1 \oplus x_2$.

This will be recognized as the same result as obtained by factoring. Again the example is perhaps trivial, but the method extends to functions of any number of variables.

THEOREM 16.9   Let $f(X)$ be a function for which there exist two simple disjoint decompositions

$$f(X) = F[\lambda(A), B, C] \quad (16.49)$$

$$= G[\varphi(B), A, C] \quad (16.50)$$

then there exists a multiple disjoint decomposition

$$f(X) = H[\lambda(A), \varphi(B), C] \quad (16.51)$$

x_3		0				1		
$x_4 x_5$	00	01	11	10	00	01	11	10
$x_1 x_2$ 00	0	(1)	3	(2)	(4)	5	(7)	6
01	(8)	9	(11)	10	12	(13)	15	(14)
11	24	(25)	27	(26)	(28)	29	(31)	30
10	(16)	17	(19)	18	20	(21)	23	(22)

(a)

x_1		0				1		
$x_2 x_5$	00	01	11	10	00	01	11	10
$x_3 x_4$ 00	0	(1)	9	(8)	(16)	17	(25)	24
01	(2)	3	(11)	10	18	(19)	27	(26)
11	6	(7)	15	(14)	(22)	23	(31)	30
10	(4)	5	(13)	12	20	(21)	29	(28)

(b)

FIGURE 16.14   *Decomposition chart for $S_{1,3,5}^5(x_1, x_2, x_3, x_4, x_5)$.*

As an example, consider the parity function of five variables $S_{1,3,5}^5$ $(x_1 x_2 x_3 x_4 x_5)$, for which two decomposition charts are shown in Fig. 16.14. Chart (a) viewed sideways gives the decomposition

$$f(x_3 x_4 x_5 | x_1 x_2) = (\bar{x}_3 \bar{x}_4 \bar{x}_5 + \bar{x}_3 x_4 x_5 + x_3 \bar{x}_4 x_5 + x_3 x_4 \bar{x}_5)\lambda(x_1, x_2)$$
$$+ (\bar{x}_3 \bar{x}_4 x_5 + \bar{x}_3 x_4 \bar{x}_5 + x_3 \bar{x}_4 \bar{x}_5 + x_3 x_4 x_5)\overline{\lambda(x_1, x_2)} \quad (16.52)$$

where $\lambda(x_1, x_2) = \bar{x}_1 x_2 + x_1 \bar{x}_2$.

Chart (b), viewed sideways, gives the decomposition

$$f(x_1 x_2 x_5 | x_3 x_4) = (\bar{x}_1 \bar{x}_2 \bar{x}_5 + \bar{x}_1 x_2 x_5 + x_1 \bar{x}_2 x_5 + x_1 x_2 \bar{x}_5)\varphi(x_3, x_4)$$
$$+ (\bar{x}_1 \bar{x}_2 x_5 + \bar{x}_1 x_2 \bar{x}_5 + x_1 \bar{x}_2 \bar{x}_5 + x_1 x_2 x_5)\overline{\varphi(x_3, x_4)} \quad (16.53)$$

where $\varphi(x_3, x_4) = \bar{x}_3 x_4 + x_3 \bar{x}_4$.

The conditions of Theorem 16.9 are thus satisfied with $A = \{x_1, x_2\}$, $B = \{x_3, x_4\}$, and $C = \{x_5\}$. To determine the composite function, we construct a K-map of Equation 16.52, as shown in Fig. 16.15a, indicating the cor-

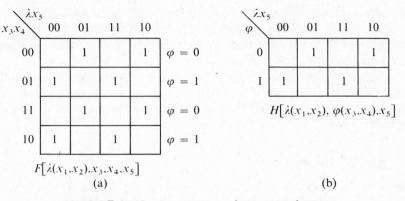

FIGURE 16.15 *Construction of composite function.*

responding values of $\varphi(x_3, x_4)$. From these, the map of $H[\lambda(x_1, x_2), \varphi(x_3 x_4), x_5]$ may be constructed as shown in Fig. 16.15b. From the map, we read off the function

$$H[\lambda, \varphi, x_5] = (\bar{\lambda}\bar{x}_5 + \lambda x_5)\varphi + (\bar{\lambda}x_5 + \lambda\bar{x}_5)\bar{\varphi} \qquad (16.54)$$

This completes the decomposition corresponding to Theorem 16.9, but we note that the final map of H (Fig. 16.14b) is also decomposable. Thus, we can define

$$\rho(\lambda, x_5) = \bar{\lambda}x_5 + \lambda\bar{x}_5$$
$$= \lambda \oplus x_5$$

Then the function becomes:

$$f(X) = \bar{\rho}\varphi + \rho\bar{\varphi} \qquad (16.55)$$

where $\rho = \lambda \oplus x_5$, $\varphi = x_3 \oplus x_4$, and $\lambda = x_1 \oplus x_2$.

It is apparent that functions might also be decomposed in a nondisjoint fashion, and Curtis [4] has developed the theory quite extensively. However, the methods involved are so complex as to make them of doubtful value. It is apparent that there is little point trying to decompose a function unless the possibility of significant economy exists. As we indicated earlier, many of the functions whose two-level realizations are most expensive are subject to disjoint decomposition.

The reader may have noted in the example illustrating Theorem 16.9 that the two decomposition charts (Fig. 16.14) are identical. Since the value of a symmetric function is dependent only on the *number* of variables taking on the value 1, it is apparent that this condition (identical decomposition charts) will always hold for symmetric functions. Thus, we may conclude that one or more of the theorems given here will be applicable to most (if not all) symmetric functions. This observation, in turn, suggests that the methods of disjoint decomposition will generally be adequate for the simplification of symmetric functions.

It is clear that the value of decomposition will increase as the number of variables increases. However, the design procedures for functions of more than six variables get so complex that is is generally more efficient to break the problem into smaller parts at the *specification* stage.

16.7   Iterative Networks

An iterative network is a highly repetitive form of a combinational logic network. This repetitive structure makes it possible to describe iterative networks utilizing techniques already developed for sequential circuits. We will limit our discussion to one-dimensional iterative networks represented by the cascade of identical cells given in Fig. 16.16a. A typical cell with appropriate input and output notation is given in Fig. 16.16b. Note the two distinct types of

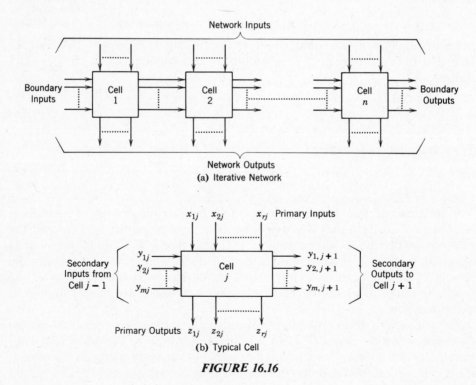

(a) Iterative Network

(b) Typical Cell

FIGURE 16.16

inputs, primary inputs from the outside world and secondary inputs from the previous cell in the cascade. Similarly, there are two types of outputs, primary, to the outside world, and secondary, to the next cell in the cascade. At the left of the cascade are a set of boundary inputs we will denote in the same manner as secondary inputs. In some cases, these inputs will be constant values. A set of boundary outputs emerges from the rightmost cell in the cascade. Although these outputs are to the outside world, they will be labeled in the same manner as secondary outputs. In a few cases, the boundary outputs will be the only outputs of the iterative network.

The next three examples will serve to illustrate the analysis and synthesis of a few practical iterative networks. We will find that it is possible to use all the techniques associated with the synthesis of clock-mode sequential circuits to synthesize iterative networks. As with sequential circuits, we will first analyze a network already laid out in iterative form. By first translating the network of Example 16.8 to a form very similar to a state table, we hope to make the reverse process more meaningful.

Example 16.8

Although the reader is accustomed to thinking of counters as self-contained sequential circuits, such need not be the case. Very often it is desirable to include a counting capability in a general-purpose arithmetic register in a computer. Going one step further to include the possible counting capability in a bank of several arithmetic registers provides a rationale for the network of Fig. 16.17.

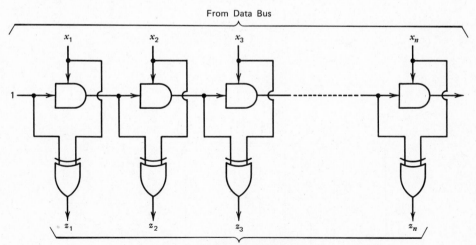

FIGURE 16.17 Increment network.

The inputs $x_1 \cdots x_n$ in Fig. 16.17 come from the output of a data bus. Depending on control inputs to the bus, this information vector might be the contents of any one of several data registers. The outputs of the iterative network $z_1, z_2, \ldots, z_n$ represent the new contents of the same register after incrementing. A control pulse will cause the new information to be triggered back into the same register. The arrangement in Fig. 16.17 makes it possible to share the incrementing logic over the entire bank of registers rather than including it separately in each register. Admittedly, in an MSI environment, the economic advantage would be important only if the number of registers in the bank was quite large. Another criticism of the network is the possible need for the effect of x_1 to propagate through $n - 1$ levels of AND-gates. A practical realization would no doubt use multi-input AND-gates. Finally, the AND-gate and exclusive-OR network for bit 1 and the AND-gate for bit n are unnecessary. Nevertheless we will consider the format of Fig. 16.17, which is an example of a simple iterative network. The reader will easily find that the number representing $x_1 \cdots x_n$ (x_1 is the least significant bit) is actually incremented by noting that a bit is complemented whenever all less significant bits are 1.

We base our analysis on the typical cell, cell j, given in Fig. 16.18a. The functional values of y_{j+1} and z_j are tabulated in Fig. 16.18b for all combinations of values of inputs x_j and $\bar{y}_i$. Note that this tabulation very closely resembles the *transition table* of clock-

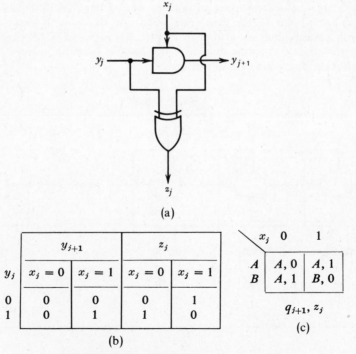

(a)

y_j	y_{j+1}		z_j	
	$x_j = 0$	$x_j = 1$	$x_j = 0$	$x_j = 1$
0	0	0	0	1
1	0	1	1	0

(b)

	x_j 0	1
A	$A, 0$	$A, 1$
B	$A, 1$	$B, 0$

q_{j+1}, z_j

(c)

FIGURE 16.18   *Development of state table.*

mode sequential circuits. The primary inputs are listed horizontally as are sequential-circuit inputs. The secondary inputs $y_{1_j} \cdots y_{m_j}$ (in this case only y_j) take the place of the present states of memory elements in the table. The secondary outputs take the place of the memory element next states. By defining $y = 0$ as state A and $y = 1$ as state B, we translate the transistion table to the *state table* form of Fig. 16.18c. ∎

As was the case with sequential circuits, the first step in the synthesis of an iterative network will be to reduce an English-language description to state-table form. From there, all of the techniques of Chapters 10 to 12 may be used to obtain an iterative network from a state table. The following two examples will serve to illustrate the synthesis process.

Example 16.9

A set of bits $x_1 \cdots x_n$ may be considered as received information in a communications system. A particular decoding scheme requires a check for odd parity over the first bit, the first two bits, the first three bits, etc. Design an iterative network that will accomplish this function.

Solution The reader will notice that the desired outputs are the symmetric functions S_1^2, $S_1^2, S_{1,3}^3, S_{1,3}^4$, etc. In general, sets of symmetric functions can quite easily be implemented as iterative networks.

A state table may be constructed directly by letting the *secondary input state* to block j be A if parity over primary inputs $x_1, x_2, \ldots, x_{j-1}$ is odd. If parity over these bits is even, the secondary input state will be B. The output z_j is to be one if there is odd parity over bits $x_1, x_2, \ldots, x_j$. Thus, $z_j = 1$ when $x_j = 0$ and the input state is A and when $x_j = 1$ and the input state is B. The secondary output will be A precisely when $z_j = 1$ and will be B when $z_j = 0$. Thus, we have the state table of Fig. 16.19.

x_j	0	1
B	$B, 0$	$A, 1$
A	$A, 1$	$B, 0$

$$q_{j+1}, z_j$$

FIGURE 16.19 *State table for iterative parity checker.* ∎

The next example will be the first to require more than two secondary input states. We will continue the design process in this case to develop the network realization of a single cell from the state table.

Example 16.10

Bits $x_1, x_2, \ldots, x_n$ are interrupt signals sent to a computer central processor from n peripheral equipments to request service from the processor. At any given time, the computer can communicate with no more than two of these peripherals. Peripheral 1 (corresponding to interrupt x_1) has the highest priority, with peripheral 2 second highest, etc. It is, therefore, desired to construct a network with outputs $z_1, z_2, \ldots, z_n$ such that no more than two of the outputs will be 1 at a given time. The 1 outputs will correspond to the two highest-priority inputs with active interrupts. (Interrupt i is active if $x_i = 1$.) If only one interrupt is active, only one output line will be 1 and no output will be 1 in the case of no active interrupts.

Solution We define three secondary input states for cell j. The secondary input state will be A if no interrupts of higher priority than peripheral j are active. The state is B if one higher priority interrupt is active and the state is C if two or more higher priority interrupts are active. Thus, the output $z_j = 1$ only if $x_j = 1$ and the input state is A or B. If $x_j = 0$, the secondary output state is the same as the input state. If $x_j = 1$, an input state B must generate an output state C and an input state A must generate an output state B. This information is tabulated in Fig. 16.20a.

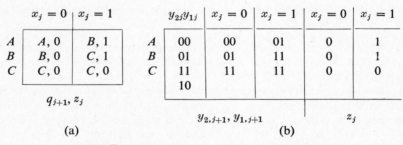

FIGURE 16.20   Tables for cell in priority network.

A convenient state assignment defines $y_2 y_1 = 00$ as A, $y_2 y_1 = 01$ as B and $y_2 y_1 = 11$ as C. This leads to the transition table of Fig. 16.20b. The equations for the secondary outputs or next state equations (16.56, 16.57) may be obtained directly by treating the transition table as a Karnaugh map.

$$y_{2, j+1} = y_{2j} + x_j \cdot y_{1j} \tag{16.56}$$

$$y_{1, j+1} = y_{1j} + x_j \tag{16.57}$$

$$z_j = x_j \cdot \bar{y}_{2j} \tag{16.58}$$

The primary output, z_j, is similarly obtained from the transition table. The network for a single cell of the priority network may be obtained by implementing these equations. ■

Two additional examples of iterative circuits of considerable interest are the adder and the magnitude comparator. (See Problems 16.11 and 16.12.) Both of these networks have two inputs per cell but are otherwise of the same order of difficulty as the above example. The comparator is a special case in that there are no primary outputs. The only outputs of the overall network are the secondary outputs for the right most cell. As indicated in Fig. 16.16a, these outputs are called boundary outputs.

Although iterative networks were first considered in the days when the primary switching element was the relay, they have taken an increased significance with the advent of LSI. In this context, iterative circuits are something of a research topic. The repetitive structure offers the advantage of easier optimization of chip layout. Variations of one- and two-dimensional iterative networks may provide the language through which complex functions can be systematically organized or single chips. Iterative networks have also been considered as a modeling technique in the generation of test sequences for LSI sequential circuits.

16.8   Ordering Relations

Another interesting special class of functions is the class of *unate* Boolean functions. These functions will be particularly important to our discussion of threshold logic. Before we proceed to define *unate* rigorously, consider the following two Boolean functions:

$$f_1 = x_1 x_2 x_3 + x_1 x_2 x_4 + \bar{x}_3 \bar{x}_4 \tag{16.59}$$

$$f_2 = x_1(x_2 + \bar{x}_3 \bar{x}_4) \tag{16.60}$$

Note that there is no possible way to express f_1 without, for example, both the literal x_3 and its complement $\bar{x}_3$ appearing in the expression. The function f_2 is expressed in such a way that, if a literal appears in the expression, then its complement does not appear. Functions which can be expressed this way are unate.

In order to formalize the definition of unate, we first consider the concept of an *ordering* relation. In Chapter 10, the concept of a relation was introduced, although only equivalence relations were illustrated. Certain relations that do not satisfy the symmetric law are also of interest.

Definition 16.4   A relation, $\leq$, is a *partial-ordering relation* if and only if it is reflexive and transitive, that is

$$x \leq x$$

and

$$x \leq y \quad \text{and} \quad y \leq z \quad \text{imply } x \leq z$$

and it is anti-symmetric, that is,

$$x \leq y \quad \text{and} \quad y \leq x \quad \text{imply } x = y \tag{16.61}$$

Example 16.11

The set of all real numbers is partially ordered. In addition, either or both of the following must be true where a and b are real numbers.

$$a \leq b \quad \text{or} \quad b \leq a \tag{16.62}$$

A partially ordered set of objects, any two of which satisfy Equation 16.62, is said to be *totally ordered*. ∎

Example 16.12

An example of a partially ordered set that is not totally ordered may be defined as follows. Let S consist of the four sets A, B, $A \cup B$, and $A \cap B$, where $A \neq B$ and neither A nor B is the null set. If we define the ordering relation to be set inclusion, we see in Fig. 16.21 that the definition of partial ordering is satisfied for any two elements, but that

$$A \nsubseteq B \quad \text{and} \quad B \nsubseteq A \tag{16.63}$$

Therefore, S is not totally ordered. ∎

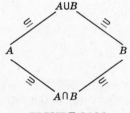

FIGURE 16.21

Definition 16.5   Let S be a set that is partially ordered by $\leq$. An element $u \in S$ is said to be an *upper bound* of the subset, X, $(X \subseteq S)$ if every $x \in X$ satisfies $x \leq u$. The element u is a least *upper bound* of X, if $u \leq v$ where v is any upper bound of X. A *lower bound* and a *greatest lower bound* are similarly defined.

Definition 16.6   A lattice is defined as a set, L, partially ordered by $\leq$, such that every two elements of L have a least upper bound and a greatest lower bound.

Example 16.13

Let S be the set of switching functions of 3 variables x_1, x_2, x_3. Let a relation be defined on S as

$$x \leq y \quad \text{if} \quad x + y = y$$

Show that S, ordered by $\leq$, forms a lattice.

Solution   Let a, b, and c be any three elements of S. Clearly, $\leq$ is reflexive. If $a + b = b$ and b and $b + c = c$, then

$$b + c = (a + b) + c = a + (b + c) = a + c \tag{16.64}$$

Therefore, $\leq$ is transitive. If $a + b = b$ and $b + a = a$, then $a = b$. Thus, $\leq$ is antisymmetric and constitutes a partial ordering of S.

Clearly, $a \leq a + b$ and $b \leq a + b$, so that $a + b$ is an upper bound of $\{a, b\}$. Let u be any upper bound of $\{a, b\}$. Therefore,

$$u = u + a \qquad \text{and} \qquad u = u + b$$

Therefore,

$$u = u + u = (u + a) + (u + b) = u + (a + b)$$

and

$$a + b \leq u \tag{16.65}$$

Similarly, $ab \leq b$ and $ab \leq a$ so that ab is a lower bound of $\{a, b\}$. Also,

$$a = l + a \qquad \text{and} \qquad b = l + b$$

so that

$$ab = (l + a)(l + b) = l + ab$$

where l is any lower bound of $\{a, b\}$. Thus, $l \leq ab$, ab is the greatest lower bound of $\{a, b\}$, and the set S ordered by $\leq$ is a lattice. ∎

16.9   Unate Functions

To permit the definition of a unate function, it is first necessary to define a partial ordering of the vertices of the Boolean hypercube.

Definition 16.7   A vertex $(x_1, x_2, \ldots, x_n)$ is $\leq$ to a vertex $(y_1, y_2, \ldots, y_n)$ if and only if, for every i, $x_i \leq y_i$. We define $x_i \leq y_i$ if $y_i = 1$ whenever $x_i = 1$.

Example 16.14

Each x_i may take on only the values 0 and 1. Thus, $(x_1, x_2, \ldots, x_n)$ is $\leq$ to $(y_1, y_2, \ldots, y_n)$ if and only if y_i is 1 if x_i is 1. For example,

$$(0, 1, 0) \leq (0, 1, 1)$$

whereas $(0, 1, 0)$ and $(1, 0, 1)$ are not ordered by $\leq$. The complete lattice of all vertices of the three-dimensional Boolean cube is shown in Fig. 16.22. A vertex X is $\leq$ to any vertex Y above it in the diagram, to which it is connected by a line.

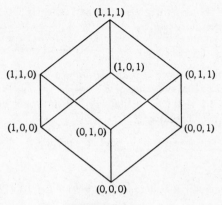

FIGURE 16.22   ■

Definition 16.8 A switching function of n variables is monotone increasing if and only if $f(\mathbf{X}) \le f(\mathbf{Y})$ wherever $\mathbf{X} \le \mathbf{Y}$ where $\mathbf{X}$ represents the vertex $(x_1, x_2, \ldots, x_n)$ and $\mathbf{Y}$ the vertex $(y_1, y_2, \ldots, y_n)$.

The lattice of three-dimensional vertices is repeated in Fig. 16.23a with each vertex, for which the function $x_1 + x_2 x_3 = 1$, indicated by a black circle. The vertices for which $x_1 + \bar{x}_1 \bar{x}_2 \bar{x}_3 = 1$ are similarly indicated in Fig. 16.23b. We see that $x_1 + x_2 x_3$ is monotone increasing. Note however, that $x_1 + \bar{x}_1 \bar{x}_2 \bar{x}_3$ is 1 for $(x_1, x_2, x_3) = (0, 0, 0)$, but is 0 for $(0, 1, 0)$, $(0, 0, 1)$, and $(0, 1, 1)$. As $(0, 0, 0) \le (0, 1, 0)$, $(0, 0, 1)$, the conditions of Definition 16.8 are clearly not satisfied; and $x_1 + \bar{x}_1 \bar{x}_2 \bar{x}_3$ is not monotone increasing.

A monotone increasing function is a special case of what we will now define to be a *unate* function.

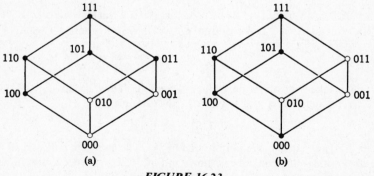

FIGURE 16.23

Definition 16.9   Let $X^J = (x_1^{j_1}, x_2^{j_2}, \ldots, x_n^{j_n})$ where

$$x_i^{j_i} = x_i \quad \text{if} \quad j_i = 0 \qquad \text{and} \qquad x_i^{j_i} = \bar{x}_i \quad \text{if} \quad j_i = 1 \qquad (16.66)$$

A switching function, f, is *unate* if and only if there exists an n-tuple $J = (j_1, j_2, \ldots, j_n)$ such that $f(X) \le f(Y)$, whenever

$$X^J \le Y^J \qquad (16.67)$$

We say that f is unate with respect to the n-tuple J.

Example 16.15

The function $f = x_1 + \bar{x}_2 x_3$ is not monotone increasing as indicated by the functional values superimposed on the lattice of Fig. 16.24a. In Fig. 16.24b, the vertices are ordered in the lattice so as to satisfy Equation 16.67 for $J = (0, 1, 0)$. The functional values are the same. In Fig. 16.24b, we see that wherever $X^J \le Y^J$ that $f(X) \le f(Y)$. Thus,

$$x_1 + \bar{x}_2 x_3 \qquad (16.68)$$

is unate.

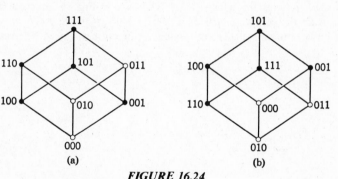

FIGURE 16.24

The truth of the following theorem may seem more evident than its proof. The theorem might be accepted, without proof, as part of the definition of a unate function by the reader interested in obtaining only a minimal background for the consideration of threshold logic.

THEOREM 16.10   A function $f(x_1, x_2, \ldots, x_n)$ is unate with respect to the n-tuple J if and only if it can be expressed as a sum of products of only the literals $x_1^{j_1}, x_2^{j_2}, \ldots, x_n^{j_n}$ where $J = (j_1, j_2, \ldots, j_n)$.

Proof

   A. Let $f(x_1, x_2, \ldots, x_n)$ be a function expressed as a sum of products of the literals $x_1^{j_1}, \ldots, x_b^{j_b}$. Let X^J and Y^J be vertices such that $X^J \le Y^J$. If $f(X) = 0$, then certainly $f(X) \le f(Y)$. Suppose $f(X) = 1$. Then some $P(X) = 1$, where P is one of the products of literals making up the expression for f. Since $\mathbf{X}^{\mathbf{J}} \leq \mathbf{Y}^{\mathbf{J}}$, at least as many of the literals $x_1^{j_1}, x_2^{j_2}, \ldots, x_n^{j_n}$ must be 1 when evaluated at $\mathbf{Y}$ as when evaluated at $\mathbf{X}$. Therefore, $P(\mathbf{Y}) = 1$ and $f(\mathbf{X}) \leq f(\mathbf{Y})$. Therefore, $f(x_1, x_2, \ldots, x_n)$ is unate.

B. Let $f(x_1, x_2, \ldots, x_n)$ be unate with respect to $\mathbf{J} = (j_1, j_2, \ldots, j_n)$. Let $\mathbf{Y}$ be any vertex for which $f(\mathbf{Y}) = 1$. Let $\mathbf{U} = (u_1, u_2, \ldots, u_n)$ be a vertex such that $f(\mathbf{U}^{\mathbf{J}}) = 1$, $\mathbf{U}^{\mathbf{J}} \leq \mathbf{Y}^{\mathbf{J}}$ and such that there exists no other vertex $\mathbf{V}^{\mathbf{J}}$ where $f(\mathbf{V}^{\mathbf{J}}) = 1$ and $\mathbf{V}^{\mathbf{J}} \leq \mathbf{U}^{\mathbf{J}}$. Let $P(\mathbf{X})$ be the product of all literals $x_1^{j_i} = 1$. Therefore, $P(\mathbf{U}^{\mathbf{J}}) = 1$. Since $f(x_1, x_2, \ldots, x_n)$ is unate, $f(\mathbf{Z}) = 1$ for all $\mathbf{Z}$ where $\mathbf{U}^{\mathbf{J}} \leq \mathbf{Z}^{\mathbf{J}}$. However, if $P(\mathbf{Z}) = 1$, then $\mathbf{U}^{\mathbf{J}} \leq \mathbf{Z}^{\mathbf{J}}$. Therefore, if $P(\mathbf{X}) = 1$, then $f(\mathbf{X}) = 1$; and $P(\mathbf{X})$ may be a term in a sum-of-products expression for f.

A similar $P(\mathbf{X})$ may be found corresponding to any vertex $\mathbf{Y}$, for which $f(\mathbf{Y}) = 1$. Thus, f may be expressed as the sum of all such *distinct* products, $P(\mathbf{X})$.

Example 16.16

Determine an SOP expression for the function, tabulated in Fig. 16.25a, which is unate with respect to $\mathbf{J} = (0, 1, 0, 1)$.

Solution The minimal element in the lattice is $(0, \bar{0}, 0, \bar{0}) = (0, 1, 0, 1)$. A partial lattice showing only the vertices for which $f_1 = 1$ is shown in Fig. 16.25b. will be left to the reader to construct the complete lattice and verify that f_1 is unate. According to Theorem 16.10, only the two minimal vertices for which the function is 1 need be considered. As in

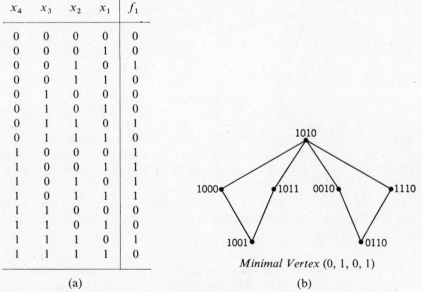

x_4	x_3	x_2	x_1	f_1
0	0	0	0	0
0	0	0	1	0
0	0	1	0	1
0	0	1	1	0
0	1	0	0	0
0	1	0	1	0
0	1	1	0	1
0	1	1	1	0
1	0	0	0	1
1	0	0	1	1
1	0	1	0	1
1	0	1	1	1
1	1	0	0	0
1	1	0	1	0
1	1	1	0	1
1	1	1	1	0

Minimal Vertex $(0, 1, 0, 1)$

(a)   (b)

FIGURE 16.25

the proof of the theorem, we form the functional representation from $x_4^{j_4}$, $x_3^{j_3}$, and $x_2^{j_2}$, $x_1^{j_1}$, since $(1, 0, 0, 1)^{\mathbf{J}} = (1, \bar{0}, 0, \bar{1}) = (1, 1, 0, 0)$ and $(0, 1, 1, 0)^{\mathbf{J}} = (0, \bar{1}, 1, \bar{0}) = (0, 0, 1, 1)$. Therefore,

$$f_1 = x_4 \bar{x}_3 + x_2 \bar{x}_1 \qquad (16.69)$$

$\blacksquare$

The reader will notice that the SOP representation obtained in the previous example was the unique minimal SOP representation. The following theorem is useful in determining whether a given Boolean function is unate. Its proof will be left to the reader.

THEOREM 16.11   A switching function, f, is unate if and only if (1) the set of essential prime implicants of f constitutes a unique minimal cover of f and (2) this unique cover of f results in an SOP representation in which only the literals x_1^1, $x_2^{j_2}$, ..., $x_n^{j_n}$ appear for some $\mathbf{J} = (j_1, j_2, \ldots, j_n)$.

It is evident from Fig. 16.26 that the expression

$$x_1 + \bar{x}_2 \bar{x}_3 + x_2 x_3$$

is a unique minimal cover composed of essential prime subcubes. This unique minimal cover does not satisfy the second condition of Theorem 16.11, so $x_1 + \bar{x}_2 \bar{x}_3 + x_2 x_3$ is not unate.

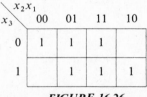

FIGURE 16.26

16.10   Generalized Resistor-Transistor Logic Circuit

One early version of a NOR-gate was a resistor-transistor device much like Fig. 16.27. The reader may wonder about the performance of such a circuit if the input resistors were not all equal. In Fig. 16.27 we see a generalized version of a resistor-transistor logic circuit.

To simplify the analysis of this circuit, we will neglect the very small base-to-emitter input voltage. In this case, we have

$$V x_0 G_0 + V x_1 G_2 + \cdots + V x_{n-1} G_{n-1} - I_t = I_{\text{in}} \qquad (16.70)$$

where the product $V x_i$ indicates that, if the Boolean variable x_i is 1, a voltage V is applied across the conductance G_i. This expression can be written

$$\sum_{i=0}^{n-1} G_i x_i - \frac{I_t}{V} = \frac{I_{\text{in}}}{V} \qquad (16.71)$$

If $\sum_{i=0}^{n-1} G_i x_i$ is sufficiently greater than I_t/V, there will be enough current into the base to switch on the transistor and reduce the output voltage to zero. On the other hand, if $\sum_{i=0}^{n-1} G_i x_i < I_t/V$, the transistor output will be V volts. Thus, by adjusting the conductances $\{G_i\}$ and I_t, the output can be made to take on various functional values for the 2^n different combinations of input variables.

Example 16.17

Suppose that a circuit of the form in Fig. 16.27 is constructed for four input variables with conductances $G_3 = 0.003$, $G_2 = 0.002$, $G_1 = 0.001$, and $G_0 = 0.001$. Let $V = 10$ volts and $I_t = 0.025$ amp. Tabulate the output voltage for the sixteen possible combinations of input values.

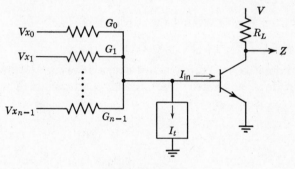

FIGURE 16.27

Solution Let us establish the output for the input values in the order tabulated in Fig. 16.28.

We first calculate $I_t/V = 0.0025$.

Clearly, for the four cases where $x_3 = x_2 = 0$, $G_1 x_1 + G_2 x_2 + G_3 x_3 + G_0 x_0 \leq 0.002$. Therefore, the output is 1 for the first four rows of the table.

For (0100), we have $G_0 \cdot 0 + G_1 \cdot 0 + G_2 \cdot 1 + G_3 \cdot 0 = 0.002$, and again the output is 1. For (0101) and (0110), $\sum_{i=0}^{3} G_i x_i = 0.003$, which is greater than I_t/V. In these cases,

$$I_{in} = 10(0.003 - 0.0025) = 5 \text{ ma}$$

A typical transistor will easily be switched on by this amount of current so the output will be 0 for these two cases. For (0111), the input current will be still greater, so the output will again be 0.

For all eight cases in which $x_3 = 1$, we have

$$\sum_{i=0}^{3} G_i x_i \geq 0.003$$

so the output is similarly 0 for all eight cases.

x_3	x_2	x_1	x_0	f
0	0	0	0	
0	0	0	1	
0	0	1	0	
0	0	1	1	
0	1	0	0	
0	1	0	1	
0	1	1	0	
0	1	1	1	
1	0	0	0	
1	0	0	1	
1	0	1	0	
1	0	1	1	
1	1	0	0	
1	1	0	1	
1	1	1	0	
1	1	1	1	

FIGURE 16.28

The above results are tabulated in Fig. 16.29. The function in Fig, 16.29 may be expressed as

$$f = \overline{x_3 + x_2(x_1 + x_0)} = \bar{x}_3(\bar{x}_2 + \bar{x}_1\bar{x}_0) \tag{16.72}$$

x_3	x_2	x_1	x_0	f
0	0	0	0	1
0	0	0	1	1
0	0	1	0	1
0	0	1	1	1
0	1	0	0	1
0	1	0	1	0
0	1	1	0	0
0	1	1	1	0
1	0	0	0	0
1	0	0	1	0
1	0	1	0	0
1	0	1	1	0
1	1	0	0	0
1	1	0	1	0
1	1	1	0	0
1	1	1	1	0

FIGURE 16.29

Because of the inherent inverting property of transistors, a logic circuit of the form discussed above will be called an inverting threshold element. In the next section, we will turn our attention to identifying the class of functions that may be realized by these devices. It is most convenient, however, to first relate our discussion to elements that do not possess this inverting property.

16.11   Linear Separability*

A Boolean function, $f(x_0, x_1, \ldots, x_{n-1})$, is said to be linearly separable if and only if there exists an n-tuple of ones and zeros $(j_0, j_1, \ldots, j_{n-1})$, a set of weights $N_0, N_1, \ldots, N_{n-1}$, and a threshold value T such that

$$S = \sum_{i=0}^{n-1} N_i x_i^{j_i} \qquad \begin{cases} < T \text{ whenever } f = 0 \\ \geq T \text{ whenever } f = 1 \end{cases} \qquad (16.73)$$

where

$$x_i^{j_i} = \begin{cases} x_i & \text{if } j_i = 0 \\ \bar{x}_i & \text{if } j_i = 1 \end{cases}$$

As was the case in Section 16.10, $N_i x_i^{j_i}$ is to be thought of as the standard algebraic product of a real number and a Boolean literal.

Example 16.18

It is easily seen that $\bar{f} = x_3 + x_2(x_1 + x_0)$ given in Equation 16.72 is linearly separable. If weights and threshold value are chosen proportional to the conductance values and I_t/V given in Example 16.17,

$$\sum_{i=0}^{3} N_i x_i \geq T$$

if and only if

$$\sum_{i=0}^{3} G_i x_i \geq \frac{I_t}{V}.$$

The function $\overline{x_3 + x_2(x_1 + x_0)}$, rather than $x_3 + x_2(x_1 + x_0)$, was realized in Example 16.17 because the transistor served to invert the functional values as defined in Equation 16.73. A convenient set of weights is $N_3 = 3$, $N_2 = 2$, and $N_1 = 1$ with threshold $T = 2.5$.

The function $f = \bar{x}_3(\bar{x}_2 + \bar{x}_1 \bar{x}_0)$ is also linearly separable with n-tuble (1111), the same weights as above, and threshold $T = 5$. Verification of this fact will be left to the reader. ∎

The following theorems, which are useful in the determination of weights for functions that are linearly separable, will be presented without proof.

---

* Linearly separable functions are commonly called *threshold* functions.

THEOREM 16.12   If a function f is linearly separable, it is linearly separable in terms of positive weights and threshold. The signs of negative weights are changed, the corresponding literals are complemented, and the magnitudes of these weights are added to the threshold, T.

Example 16.19

Determine the Boolean function separated by the following summation and threshold, $T = 1$.

$$S = 2x_2 - x_1 + x_0 \tag{16.74}$$

Then find a set of positive weights and a threshold that separates the same function.

Solution   Tabulating the summation for all possible combinations of values results in Fig. 16.30. The function in the right-hand column of Fig. 16.30, which can be expressed as

$$f = x_2 + \bar{x}_1 x_0$$

is the only function separated by the given weights and threshold.

x_2	x_1	x_0	S	f (for $T = 1$)
0	0	0	0	0
0	0	1	1	1
0	1	0	-1	0
0	1	1	0	0
1	0	0	2	1
1	0	1	3	1
1	1	0	1	1
1	1	1	2	1

FIGURE 16.30

Now, applying Theorem 16.12, we have $N'_2 = N_2 = 2$, $N'_1 = |N_1| = 1$, $N'_0 = N_0 = 1$, and

$$T' = T + |N_1| = 1 + 1 = 2 \tag{16.75}$$

It can be easily verified by the reader that

$$S' = 2x_2 + \bar{x}_1 + x_0 \tag{16.76}$$

is greater than or equal to 2 for precisely those combinations of values for which the function in Fig. 16.30 is 1.   ∎

THEOREM 16.13   A linearly separable function is unate.

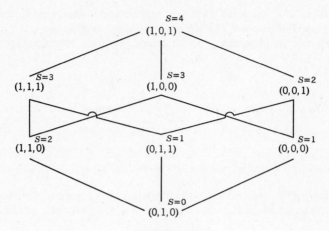

FIGURE 16.31

Example 16.20

Verify that the linearly separable function that is separated by the summation

$$S = 2x_2 + \bar{x}_1 + x_0$$

with threshold, $T = 2$, is unate.

Solution   That the above is true follows from Theorem 16.13. As an illustration of the theorem, let us construct the appropriate lattice for the function. The above summation using positive weights tells us that the least element of the lattice must be $(0, 1, 0)$. The lattice can then be constructed as shown in Fig. 16.31.

The values of S can then be computed for each vertex. These values are also shown in Fig. 16.31. A comparison of the values of S with the threshold, $T = 2$, tells us that only the three vertices for which $S = 1$, or $S = 0$, have functional values 0. As expected, the functional values are nondecreasing along any path from the bottom to the top of the lattice. Definition 16.9 is, therefore, satisfied in this case. Notice that the functional values are those tabulated in Fig. 16.30. The function is, therefore, $f = x_2 + \bar{x}_1 x_0$, which we can easily see is unate.   ∎

Example 16.21

Verify that the unate function

$$f = x_0 x_1 + x_2 x_3 \tag{16.77}$$

is not linearly separable.

Solution   In order that $f(0, 0, 1, 1) = f(1, 1, 0, 0) = 1$, we must have

$$N_0 + N_1 \geq T \tag{16.78}$$

and

$$N_2 + N_3 \geq T \tag{16.79}$$

On the other hand, $f(0, 1, 0, 1)$ and $f(1, 0, 1, 0)$ must be zero. Therefore,

$$N_0 + N_2 < T \tag{16.80}$$

and

$$N_1 + N_3 < T \tag{16.81}$$

Adding expressions 16.78 and 16.79 yields 16.82, and adding 16.80 and 16.81 yields 16.83.

$$N_0 + N_1 + N_2 + N_3 \geq 2T \tag{16.82}$$

$$N_0 + N_1 + N_2 + N_3 < 2T \tag{16.83}$$

Contradiction!   ∎

Once a Boolean function has been found to be unate, the next step is to apply a procedure for finding the weights. If the function is in fact not linearly separable, this process will be unsuccessful. For functions of many variables, the weights can be determined by linear programming. For functions of a small number variables, the following theorem will be helpful.

THEOREM 16.14   Suppose that a function, f, is linearly separable in terms of the positive weights $N_0, N_1, \ldots, N_{n-1}$. Let a_i be the number of vertices for which $f = 1$ and $x_i^{j_i} = 1$. Then the following four conditions hold.

(1) If $N_i \geq N_k$, then $a_i \geq a_k$.
(2) If $N_i = N_k$, then $a_i = a_k$.
(3) If $a_i > a_k$, then $N_i > N_k$.
(4) If $a_i = a_k$, then the weights may be chosen such that $N_i = N_k$.

Example 16.22

Determine whether or not the following two functions are linearly separable and, if so, determine an appropriate set of weights and threshold values.

$$f_1 = x_2 x_1 x_0 + x_4 x_3 + x_4 x_2 + x_3 x_2 + x_3 x_0 + x_4 x_1$$

$$f_2 = x_4 x_3 \bar{x}_0 + x_4 \bar{x}_2 + x_3 \bar{x}_2 + x_3 \bar{x}_1 + x_4 x_3 x_0 + x_4 \bar{x}_1 + x_4 x_0 + \bar{x}_2 \bar{x}_1 x_0$$

Solution   Considering f_1 first, we note that this function is unate. Let us attempt to determine an order for the variables. Toward this end, we first map the function as illustrated in Fig. 16.32.

Here we see that $a_4 = 14$, $a_3 = 14$, $a_2 = 13$, $a_1 = 12$, and $a_0 = 12$. From Theorem 16.14, we conclude that

$$N_4 = N_3 > N_2 > N_1 = N_0 \tag{16.84}$$

Knowledge of the order of the literals suggests the following factorization:

$$x_4(x_3 + x_2 + x_1) + x_3(x_2 + x_0) + x_2 x_1 x_0$$

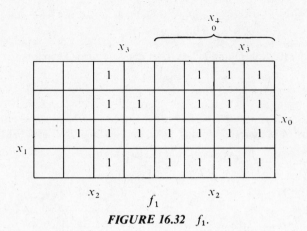

FIGURE 16.32 f_1.

It is now more apparent, perhaps, that $x_4 x_0$ is not included in the function, but that $x_3 x_0$ is. Therefore, we must have

$$N_4 + N_0 < T \quad \text{and} \quad N_3 + N_0 \geq T$$

Consequently,

$$N_4 < N_3 \tag{16.85}$$

The contradiction between Equations 16.84 and 16.85 tells us that f_1 is not linearly separable.

Notice that f_2 appears as a unate function after making the simplification

$$x_4 x_3 \bar{x}_0 + x_4 x_3 x_0 = x_4 x_3$$

The Karnaugh map of this function appears in Fig. 16.33. A count of the vertices reveals that $a_4 = 15$, $a_3 = 14$, $a_2 = 13$, $a_1 = 13$, and $a_0 = 12$. This suggests the factorization

$$f_2 = x_4(x_3 + \bar{x}_2 + \bar{x}_1 + x_0) + x_3(\bar{x}_2 + \bar{x}_1) + \bar{x}_2 \bar{x}_1 x_0 \tag{16.86}$$

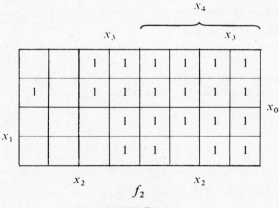

FIGURE 16.33

Since there must be four distinct weights, with only $N_2 = N_1$, we try first the simplest combination,

$$N_4 = 4, \qquad\qquad N_3 = 3$$
$$N_2 = N_1 = 2, \qquad N_0 = 1$$

Note that the smallest summation of weights corresponding to a Boolean product in 16.86 is precisely equal to 5. On the other hand, the summation is less than 5 for any product of the appearing literals, which is not included in f_2. Therefore, this simple set of weights is sufficient to separate f_2 if $T = 5$. ∎

Problems

16.1. Simplify the following symmetric function expressions:

(a) $[S^7_{1, 3, 5, 7}(x_1, \ldots, x_7)] \cdot [S^7_{4, 5, 6, 7}(x_1, \ldots, x_7)]$

(b) $\overline{S^5_{2, 3, 5}(x_1, \ldots, x_5)} + S^5_3(x_1, \ldots, x_5)$

(c) $[S^5_{1, 5, 5}(\bar{x}_1, \bar{x}_2, \ldots, \bar{x}_5)] \cdot [S^5_{2, 3, 4, 5}(x_1, x_2, \ldots, x_5)]$

16.2. How many different functions of the form $f(x_1, x_2, \ldots, x_6)$ are there that are symmetric with respect to the six literals $x_1, x_2, \ldots, x_6$?

16.3. How many of the 16 functions of two variables are symmetric (with respect to any of the four possible pairs of literals)?

16.4. Consider the ten bits $x_1, x_2, \ldots, x_{10}$ arranged in ascending order. There are eight possible ways of selecting three successive bits from this arrangement. The function that is 1 when a majority of some such set of three bits is 1 is, of course, a symmetric function.

(a) Write, as a function of symmetric functions, an expression that will be 1 whenever the majorities of bits in an odd number of the sets of three bits are 1.

(b) Is the resulting expression a symmetric function? Why?

(c) Repeat part (a), this time considering all $_{10}C_3 = 120$ possible combinations of 3 of the 10 bits. Simplify.

(d) Is the function in part (c) symmetric? Why?

16.5. Determine a simple disjoint decomposition for each of the following functions for which such a decomposition exists.

(a) $f(x_1, x_2, x_3, x_4) = \sum m(0, 5, 6, 7, 9, 10, 11, 12)$

(b) $f(x_1, x_2, x_3, x_4) = \sum m(0, 3, 4, 7, 12, 15)$

(c) $f(x_1, x_2, x_3, x_4) = \sum m(0, 2, 4, 9, 6, 13)$

(d) $f(x_1, x_2, x_3, x_4, x_5) = \sum m(3, 10, 14, 17, 18, 22, 23, 24, 27, 28, 31)$

16.6. Repeat Problem 16.5 for the following two functions.

(a) $f(x_1, x_2, x_3, x_4, x_5) = \sum m(1, 3, 5, 7, 9, 10, 13, 14, 21, 23, 26,$
$27, 29, 30)$

(b) $f(x_1, x_2, x_3, x_4, x_5) = \sum m(1, 3, 9, 10, 17, 19, 24, 27)$

16.7. There exists an iterative disjoint decomposition for the following function of five variables. Determine this decomposition and express it in the form of Equation 16.39.

$$f(x_1, x_2, x_3, x_4, x_5) = \sum m(2, 6, 8, 12, 16, 20, 26, 30)$$

16.8. Determine a multiple disjoint decomposition of the 4-variable function.

$$f(x_1, x_2, x_3, x_3, x_4) = \sum m(0, 3, 4, 7, 9, 10, 13, 14)$$

16.9. Develop an original 5-variable example of the multiple disjoint decomposition given by Equation 16.51. Let the variables be partitioned as given by

$$f(x_1, x_2, x_3, x_4, x_5) = H[\lambda(x_1, x_2), \varphi(x_3, x_4), x_5]$$

16.10. Show that a Boolean algebra is a lattice. Which postulates of a Boolean algebra are not satisfied by every lattice?

16.11. Each cell of an iterative network is to have one primary input and one primary output. Let the primary input of cell i be x_i and let the primary output be z_i. Determine the cell table for cell i such that $z_i = 1$ if and only if $x_{i-2} = x_{i-1} = x_1$. Determine a network realization of the typical cell i.

16.12. The n-bit binary adder of Fig. 8.7 may be regarded as an iterative network. A single cell of the network is given in a circuit form as Fig. 8.3. Determine the state table of a single cell of the adder.

16.13. A circuit that will compare the magnitudes of two 18-bit numbers is to be formulated as an iterative network. Each cell is to have two secondary outputs and no primary outputs. The input numbers are given as vectors $X1$ and $X2$. The secondary outputs of cell j should indicate whether the binary number represented by $X1_{0:j-1}$ is greater than, equal to, or less than the binary number represented by $X2_{0:j-1}$. Determine the state table of a single cell and a circuit realization of that cell. Sketch the complete network. What will be the output of the network?

16.14. Which of the following sets are partially ordered? Totally ordered?
(a) Integers.
(b) Real numbers.
(c) Complex numbers (define $\leq$).

(d) A set of all equivalent incompletely specified sequential machines, where a machine S is $\leq$ a machine T if and only if every state in S is compatible with some state in T.

(e) The schools in the Big Ten Conference, where $A \leq B$ if and only if B defeated A in football this fall.

16.15. Which of the sets described in Problem 16.14 form lattices?

16.16. Which of the following switching functions are unate?

(a) $f(x_1, x_2, x_3, x_4) = \sum m(3, 5, 7, 10, 11, 12, 13, 14, 15)$

(b) $f(x_1, x_2, x_3, x_4) = \sum m(4, 8, 10, 12, 13, 14)$

(c) $x_1 \oplus x_2$

(d) $f(x_1, x_2, x_3, x_4) = \sum m(2, 3, 4, 10, 12, 13, 14)$

16.17. Which of the switching functions in Problem 16.5 are unate?

16.18. The weights $N_1 = 4$, $N_2 = 3$, $N_3 = 3$, $N_4 = 2$, $N_5 = 1$, and $T = 6$ can be used to linearly separate $x_1(\bar{x}_2 + x_3 + \bar{x}_4) + \bar{x}_2 x_3 + \bar{x}_2 \bar{x}_4 x_5 + x_3 \bar{x}_4 x_5$. Determine a separating set of weights so that no inverted literals appear in the summation.

16.19. What function is linearly separated by the weights $N_1 = 3$, $N_2 = 1$, $N_3 = -1$, $N_4 = -2$, $T = 1$? Determine a set of nonnegative weights that will separate this function.

16.20. Determine whether or not each of the following functions is linearly separable.

$$f_1 = x_1 x_2 x_4 + x_2 x_3 x_4 + x_1 x_2 x_3 x_5 + x_1 x_4 x_5$$

$$f_2 = \bar{x}_3 x_4 + x_1 \bar{x}_3 \bar{x}_4 + x_1 x_3 x_4 + x_3 x_4 x_5 + x_1 \bar{x}_2 x_3 x_5 + \bar{x}_3 x_5$$
$$+ \bar{x}_1 \bar{x}_2 \bar{x}_3 x_4$$

16.21. Determine weights and threshold values for the functions in Problem 16.20 that are linearly separable.

16.22. List all linearly separable functions of three variables.

16.23. Show that the following function satisfies the orderability criteria in Theorem 16.14, but is not linearly separable.

$$f = x_8[x_7 + x_6 + x_5(x_4 + x_3 + x_2 + x_1) + x_4 x_3 x_2 x_1]$$
$$+ x_7 x_6[x_5(x_4 x_3 x_2 + x_4 x_3 x_1 + x_4 x_2 x_1 + x_3 x_2 x_1) + x_4 x_3 x_2 x_1]$$

16.24. How many functions of five variables are linearly separable?

Bibliography

1. Marcus, M. P. "The Detection and Identification of Symmetric Switching Functions with the use of Tables of Combinations," *IRE Trans. on Electronic Computers*, **EC-5:** 4, 237–239 (Dec. 1956).
2. Ashenhurst R. L. "The Decomposition of Switching Functions," *Proceedings of the Symposium on the Theory of Switching*, April 2–5, 1957, Vol. 29 of *Annals of Computers*, Harvard University, 74–116, 1959.
3. Maley, G. A. and J. Earle, *The Logic Design of Transistor Digital Computers*, Prentice-Hall, Englewood Cliffs, NJ 1963.
4. Curtis, H. A. *Design of Switching Circuits*, Van Nostrand, Princeton, N.J., 1962.
5. Harrison, Micheal. *Introduction to Switching and Automata Theory*, McGraw-Hill, New York, 1965.
6. Birkhoff, G. and S. MacLane. *Survey of Modern Algebra* (3rd ed.), Macmillan, New York, 1965.
7. Shannon, C. E. "A Symbolic Analysis of Relay and Switching Circuits," *Trans. AIEE*, **57,** 713–723 (1938).
8. Chao, S. C. "A Generalized Resistor Transistor Logic Circuit and Some Applications," *IRE Trans. on Electronic Computers*, **PGEC-8:** 1, 8–12 (March 1959).
9. Winder, R. O. "Enumeration of Seven-Argument Threshold Functions," *IEEE Trans. on Electronic Computers*, **EC-14:** 3, 315 (June 1965).
10. Ho, Y. C. and R. L. Kashyap. "An Algorithm for Linear Inequalities and Its Applications," *IEEE Trans. on Electronic Computers*, **EC-14:** 5, 683 (Oct. 1965).
11. Sheng, C. L. and H. R. Hwa. "Testing and Realization of Threshold Functions by Successive Higher Ordering of Incremental Weights," *IEEE Trans. on Electronic Computers*, **EC-15:** 1, 212–219 (April 1966).
12. Haring, D. R. "Multi-Threshold Threshold Elements," *IEEE Trans. on Electronic Computers*, **EC-15:** 1, 45–65 (Feb. 1966).
13. Muroga, S. *Threshold Logic and Its Applications*, John Wiley, New York, 1971.

Appendix A

SELECTION OF MINIMAL CLOSED COVERS

In Chapter 13, after finding the set of all maximal compatibles, the selection of a minimal cover was carried out by an essentially trial-and-error process. This was practical since the number of maximal compatibles was reasonable and there were few implications. In very complex cases, with many compatibles and interacting implications, a more systematic procedure would be highly desirable. Unfortunately, no procedure exists, other than complete enumeration, that guarantees finding the minimal equivalent of a state table in every case. However, a paper published by Grasselli and Luccio[1] provides the basis for a straightforward procedure that will reduce the number of alternatives considerably in most problems.

As a starting point, it is necessary to formalize the procedure of combining sets of compatibility pairs into larger compatibility classes. We have already seen that this can be done, but there is some question as to the effect of this procedure on closure. Theorem A.1 provides the answer to this question.

THEOREM A.1 Let $\{q_1, q_2, \ldots, q_n\} = \{q_i\}$ be a set of states that are pairwise compatible in a single closed collection. Then $\{q_i\}$ is a compatibility class within some closed collection.

Proof Since the initial collection is closed, the next states $\delta(q_1, X_j)$ and $\delta(q_2, X_j)$ for any input X_j are a compatible pair in the same collection. The same holds true for the next states for any other pair in $\{q_i\}$. Thus, all the next states of $\{q_i\}$ for a given input X_j are pairwise compatible and can form a single compatibility class $\{\delta[q_i, X_j]\}$. Therefore, the requirements of Definition 13.7 are satisfied with respect to $\{q_i\}$. Similarly, all the next states of $\{\delta[q_1, X_j]\}$ are pairwise compatible in the original closed collection and can form a single compatibility class, satisfying Definition 13.7 for $\{\delta[q_i, X_j]\}$. This process continues until a complete closed collection is formed. Since only a finite number of distinct classes can be formed from a finite set of states, the process must terminate. Q.E.D.

THEOREM A.2 The union of all distinct maximal compatibles formed from the states of a circuit S is a closed collection of compatibility classes.

Proof Starting with any maximal compatible, generate a closed collection by the process described in the proof of Theorem A.1. Every class in this collection must either be a

593

maximal compatible or else be included in a maximal compatible. Thus, Definition 13.7 is satisfied.   Q.E.D.

The reader can easily verify that the collections of maximal compatibles found in the examples of Chapter 13 are indeed closed collections. These collections of maximal compatibles may be used as a starting point in the process of finding a minimal cover. In these examples, we were able to find minimal covers without great difficulty, simply by inspection. In general, this may not be possible. First, in a problem of any complexity, there will be a large number of maximal compatibles. Second, all possible subsets of the maximal compatibles are also candidates for inclusion in a minimal cover. The first step, then is to reduce the number of candidates.

If C_α and C_β are compatibility classes and P is a collection of classes, we mean by $C_\alpha \subseteq C_\beta$ that all the states in C_α are included in C_β, and by $C_\beta \in P$, that C_β is a member of P.

Definition A.1   The *class set* P_α implied by a compatibility class C_α is the set $\{D_{\alpha_1}, D_{\alpha_2}, \ldots, D_{\alpha_n}\} = \{D_{\alpha_i}\}$, of all classes implied by C_α, such that
1. D_{α_i} has more than one element.
2. $D_{\alpha_i} \nsubseteq C_\alpha$.
3. $D_{\alpha_i} \nsubseteq D_{\alpha_j}$ if $D_{\alpha_j} \in P_\alpha$.

Less formally, the class set is made up of the compatibility classes, other than singletons, implied by a given set and not included in the given set. Figure A.1 shows a small implication table, the determination of the maximal compatibles, and the class sets implied by the maximal compatibles. Note that the class $(abde)$ does not imply a class set since all the pairs implied by this class are subsets of the class. Class (bcd), however, implies three pairs that are not included in the class.

Condition 3 of Definition A.1 may be applicable when some class of three or more states implies all possible pairs of states from another set of three or more states. Consider the two partial state tables shown in Fig. A.2, and assume that

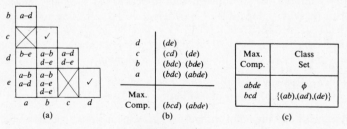

FIGURE A.1

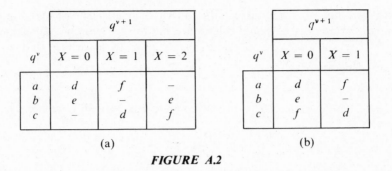

(a) (b)

FIGURE A.2

in both cases (abc) is a compatibility class. In both cases, (abc) implies the classes (de), (df), and (ef). For the table of Fig. A.2, the three classes are implied for different inputs; hence they do not combine into a single implied class. For the table of Fig. A.2b, however, all three classes are implied by a single input, so they combine into a single implied class (def). By Condition 3, only this larger set need be included in the class set. Note that this situation cannot be found with the implication table alone, but requires reference to the state table. In the case of Fig. A.1, it was not necessary to refer to the state table, since the pairs appearing did not make up all possible combinations of some larger set of states.

Definition A.2 A compatibility class C_α *excludes* a compatibility class C_β if and only if
1. $C_\beta \subseteq C_\alpha$ and
2. $P_\alpha \subseteq P_\beta$
where P_α and P_β are the class sets implied by C_α and C_β, respectively.
A class that is not excluded will be called a *prime* compatibility class.

Suppose a minimal set of compatibility classes covering a given state table has been found, including some class C_β that is not prime. Let C_α be a class that excludes C_β. Since C_α includes all the states covered by C_β and does not imply any classes not implied by C_β, both cover and closure will be maintained if C_α replaces C_β. We have thus proved the following theorem.

THEOREM A.3 At least one minimal covering will consist solely of prime compatibility classes.

The concept of "exclusion" is used to eliminate classes from consideration in much the same manner that dominance is used to eliminate prime implicants. An excluded class covers only states that are included in some larger class, as specified by condition 1 of Definition A.2. Condition 2 ensures that closure is not violated by the substitution of the larger class for the smaller one. It is important to note that not all minimal covers consist of prime classes and that not all covers consisting only of prime classes are minimal. Theorem A.3 simply allows us to restrict our attention to prime classes with the assurance that a minimal cover can still be found.

Example A.1

Determine the prime compatibility classes for the circuit partially described by the implication table of Fig. A.3.

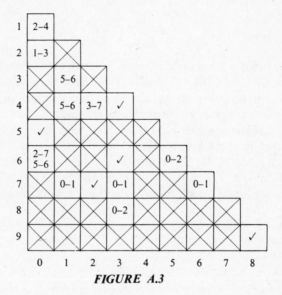

FIGURE A.3

Solution   The first step is the determination of the maximal compatibles, as shown in Fig. A.4.

8	(89)									
7	(89)									
6	(89)	(67)								
5	(89)	(67)	(56)							
4	(89)	(67)	(56)							
3	(89)	(367)	(56)	(38)	(34)					
2	(89)	(367)	(56)	(38)	(34)	(27)	(24)			
1	(89)	(367)	(56)	(38)	(134)	(27)	(24)	(137)		
0	(89)	(367)	(056)	(38)	(134)	(27)	(24)	(137)	(02)	(01)

M.C. | (056) (134) (137) (367) (01) (02) (24) (27) (38) (89)

FIGURE A.4   *Determination of maximal compatibles, Example A.1.*

Max Comp.	056	134	137	367	01	02	24	27	38	89
Class Sets	27 02	56	56 01	01	24	13	37	ϕ	02	ϕ

(a)

Other Prime Classes	05	06	34	17	36	1	56
Class Sets	ϕ	27 56	ϕ	01	ϕ	ϕ	02

(b)

FIGURE A.5   *Determination of prime classes, Example A.1.*

The process of determining the prime compatibles starts with the maximal compatibles, which are themselves prime. These are listed in Fig. A.5a, together with the implied class sets.

The remaining prime classes are found by deleting one state at a time from each of the maximal compatibles. If the implied class set is smaller than that of the maximal compatible or any other prime class containing the subset, then it is prime. After this has been done for all of the maximal compatibles, the procedure is repeated for the resultant prime classes, until no further decomposition takes place.

For example, the maximal compatible (134) calls for a check of (13), (14), and (34). We see that (13) and (14) imply (56), just as does (134), so they are not prime. Class (34), however, implies the empty set and must be included in the list of Fig. 11.15b. Note that a subset must be checked against *all* prime classes that contain it. For example, subset (37), obtained from maximal compatible (137), is not excluded by (137) but is excluded by (367).

Also note that a prime class that implies the empty set excludes all its subsets and thus need not be checked for further prime classes. In this case, only (06), (17), and (56) need be checked further. Class (17) calls for a check of (1) and (7). We see that (7) is excluded by (27), but (1) is included in no prime class implying the empty set, so (1) must be added to the list of prime classes.   ∎

We are now at much the same position as in the Quine-McCluskey method when the determination of the prime implicants has been completed. Just as the prime implicants are the only products that need be considered in realizing a combinational function, so only the prime classes need be considered in finding a cover for a state table. The first step in selecting a cover from the prime classes is to determine any essential prime classes, that is, prime classes that are the only ones including one or more states. Again, the parallel to the Quine-McCluskey method is noted.

Next, we delete the covered states from the unselected classes. If the resultant reduced classes are then excluded, they may be removed from further consideration. This step is roughly comparable to reduction of the prime implicant table. We then check again for classes made essential by this second step. This process may be continued until no further reduction is possible. We will see in the completion of this example how the choice of prime compatibles can often be further narrowed through intuitive treatment of a pictorial model.

Example A.1 (continued)

Find a minimal cover from the set of prime compatibility classes.

We first note that (89) is essential to cover (9). We then delete (8) from (38) leaving (3), which is excluded by (34). So (38) is discarded. This does not make any other sets essential, so we set up a pictorial representation of the implications of the remaining classes (Fig. A.6), which we will refer to as the *implication graph*. Each class that has been selected or is still under consideration (not excluded) is represented by a circle. Selected classes are indicated by an asterisk. Every class implied by a class C_α is indicated by an arrow from C_α to that class. If only one class can satisfy an implication, we indicate a *strict implication* by a solid arrow. If an implication can be satisfied by any of two or more classes, we indicate an *optional implication* by dashed arrows.

A good place to start the reduction process is with classes that imply other classes but

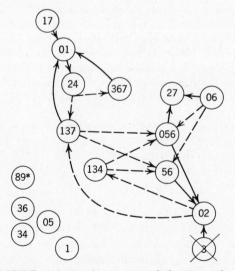

FIGURE A.6   *Implication graph for Example A.1.*

satisfy no implications themselves—for example, class (17). Note that the chain of implications set up by the selection of (17),

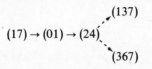

$$(17) \rightarrow (01) \rightarrow (24) \begin{array}{c} (137) \\ \\ (367) \end{array}$$

includes classes that cover both (1) and (7). Therefore, (17) may be discarded. By the same argument, (06) may be discarded. Each time a class is discarded, we should check to see if any of the remaining classes have been made essential as a result. This has not yet occurred, so we continue.

Next, let us compare classes that overlap, such as (056) and (56). Both satisfy the same implications, and both imply (02), thus providing a cover for (0). Thus, (056) can accomplish nothing that (56) cannot, and it implies more classes. Therefore, (056) may be discarded. Even after discarding (056), no classes have become essential. To facilitate the discussion, we redraw the graph, incorporating the above simplifications as shown in Fig. A.7.

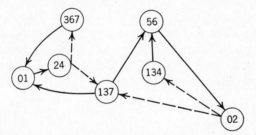

FIGURE A.7 *Simplified implication graph for Example A.1.*

Let us make a tentative selection of (137), to see what sort of cover results. The strict implication of (137) requires that (01), (24), (56), and (02) also be selected. With the addition of the essential class (89), we have a 6-state closed cover,

$$(137)(01)(02)(24)(56)(89)$$

Because of the strict implication, it is evident that this 6-state cover is the smallest that can be found by using (137). Next, let us eliminate (137) and see what possibilities remain. The implication graph with (137) removed is shown in Fig. A.8. Here we see two *cycles*, sets of classes that must be selected as a group because of the cyclic pattern of implication. If we select cycle A, the addition of (05) and (89) will provide a 5-state closed cover,

$$(01)(24)(367)(05)(89) \tag{A.1}$$

If we select cycle B, we also get a five-state closed cover,

$$(02)(134)(56)(27)(89) \tag{A.2}$$

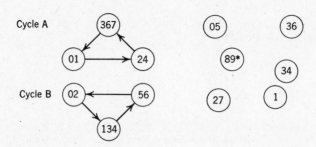

FIGURE A.8   *Final implication graph for Example A.1.*

The reader can quickly verify that any cover not including one of the cycles must have at least six states. Thus, the covers of (A.1) and (A.2) are minimal.   ■

It should be emphasized that the above procedure is not a complete or exact algorithm. Rather, it consists of the application of a few "common-sense" rules in order to reduce the number of candidates for inclusion in a cover to a point where a trial-and-error check of all possibilities is practical. Also, the implication graphs are not an essential feature of the method. The same information is contained in the list of prime classes and class sets. However, the graphs, like Karnaugh maps, make it easier to recognize significant patterns and relationships. The reader who desires practice with this method should rework some of the more difficult exercises of Chapter 13.

Bibliography

1.   Grasselli, A. and F. Lucio. "A Method for Minimizing the Number of Internal States in Incompletely Specified Sequential Networks," *IEEE Transactions on Electronic Computers*, **EC-14:** 3, 330–359 (June 1965).

Appendix B

RELAY CIRCUITS

B.1 Basic Characteristics of Relay Circuits

As was mentioned in Chapter 1, switching theory was originally developed to deal with the problems of designing switch and relay circuits for dial telephone systems. The earliest digital computers were also constructed wholly or in part of relays. Today, the relay has been largely replaced by electronic circuits in the majority of digital systems, so that the emphasis of this book has been on electronic realizations. However, there are still situations where relays are the most practical means of realizing logic circuits. This is particularly true in fields such as industrial control where logic circuits are used to control the application of large amounts of power. For this reason, we present here a brief outline of the means by which the theories and techniques developed in this book can be applied to relay circuit design. The reader interested in more detail is referred to any of several excellent books [1, 3, 4].

A relay consists of an electromagnet and a set of switch contacts. When the relay coil is energized, the switch operates. The three basic types of contacts are shown in Fig. B.1. In each case, the switch contacts, which are generally referred to as *springs*, are shown in the position they take when the relay coil is not energized; this position is known as the *normal* position. Thus, the first relay is normally open (NO), and the contacts close when the coil is energized. The second relay is normally closed (NC), and the contacts open when the relay is

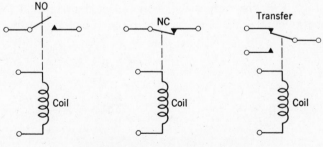

FIGURE B.1 *Basic relay types.*

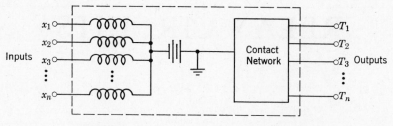

FIGURE B.2   *Standard form of relay circuit.*

energized. The third relay has a *transfer* contact, a combination of NO and NC contacts sharing a common movable spring.

Although we have shown only one set of contacts on each relay in Fig. B.1, practical relays will generally have several sets of contacts, often including all three types.

The standard form of a relay circuit is shown in Fig. B.2. The inputs are signals applied to the coils, which take on the value 1 if the relay is energized, the value 0 if it is not. In this particular form, a coil is energized by connecting the corresponding input terminal to ground, which will probably be accomplished by another switch or relay circuit. But we are not concerned about this any more than we are concerned with the origin of the inputs to an electronic gate network.

The contact network is made up of the contacts of the relays interconnected in some fashion. The outputs are known as the *transmissions*, T_i, and take on the value 1 if the output is connected to ground through the network, the value 0 if it is not.

Consider the trivial relay circuits of Fig. B.3, where the contacts are operated by the coil shown. In Fig. B.3a, the output will be connected to ground $(T = 1)$ only if the relay is operated $(X = 1)$. Therefore, this circuit can be described by

$$T = X$$

Similarly, in Fig. B.3b, the output will be connected to ground $(T = 1)$ if the relay is not operated $(X = 0)$. This circuit can thus be described by

$$T = \bar{X}$$

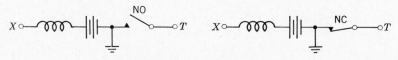

FIGURE B.3   *Basic relay circuits.*

FIGURE B.4 *Schematic representation of circuits of Fig. B.3.*

We thus see that, in writing the equation for transmission, a normally open contact may be represented by the corresponding input variable, a normally closed contact by its complement. Once this is understood, there is nothing to be gained by drawing out all the detail of Fig. B.3. Instead, we draw the circuit as shown in Fig. B.4, where we replace the NO contact with the variable X and the NC contact with $\overline{X}$. It is understood that the transmission is 1 when the terminals are connected by the relay contacts. To clarify this, we show the complete representations for AND and OR circuits, along with the corresponding schematic notation, in Fig. B.5.

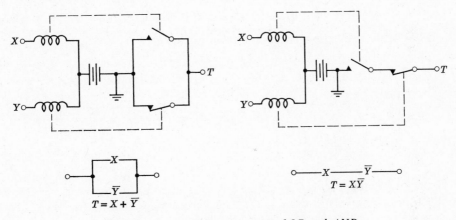

FIGURE B.5 *Relay realizations of OR and AND.*

We have now seen that AND, OR, and NEGATION can all be realized with relays, which tells us that any switching function can be realized with relays. For example, consider the function of Equation 6.6, for which the minimal second-order form is

$$f(A, B, C, D) = AC\overline{D} + A\overline{C}D + BC\overline{D} + B\overline{C}D \tag{B.1}$$

The relay form of this equation can be written directly, as shown in Fig. B.6. The simple and direct translation from the sum-of-products form to a series-parallel relay circuit is most striking. Although the algebraic form is minimal second-order, it is apparent that the relay circuit is more complicated than it need be.

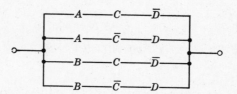

FIGURE B.6   *Relay circuit for Equation B.1.*

First, we see that the two *A* contacts and the two *B* contacts can be combined, as shown in Fig. B.7a. This simplification corresponds to the factoring

$$A\bar{C}D + AC\bar{D} + B\bar{C}D + BC\bar{D} = A(\bar{C}D + C\bar{D}) + B(\bar{C}D + C\bar{D}) \qquad \text{(B.2)}$$

Noting that the terms in parentheses are identical, we can immediately make the further simplification shown in Fig. B.7b. This corresponds to the form

$$f(A, B, C, D) = (A + B)(C\bar{D} + \bar{C}D) \qquad \text{(B.3)}$$

the 3-level gate realization, which is shown in Fig. 6.21.

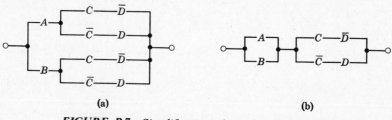

(a)   (b)

FIGURE B.7   *Simplification of circuit of Figure B.6.*

The above example points out that, in relay circuits, unlike electronic gate circuits, there is nothing particularly desirable about the second-order form. The circuit of Fig. B.7b is certainly not slower than that of Fig. B.6. In a contact network, the only delay is in the propagation of a signal down a wire, a delay that is completely negligible compared to the operating speed of the relays. The speed of relay circuits is thus completely independent of the form of the contact networks.

In simplifying relay circuits, the basic objective is to reduce the number of springs. We also would like to eliminate relays if possible, but the number of relays is determined by the number of variables required, a factor not always under our control. One way to reduce the number of springs is to combine NO and NC contacts on the same relay into transfer contacts. The relation between

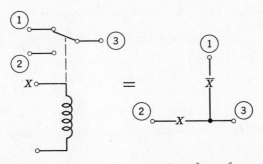

FIGURE B.8 *Schematic representation of transfer contacts.*

a transfer contact and the schematic representation is shown in Fig. B.8. In the circuit of Fig. B.7b, we see immediately that transfer contacts can be used for the C and D variables. Thus, in terms of actual contact configuration, the circuit of Fig. B.7b will appear as shown in Fig. B.9.

Another method of simplification peculiar to relay circuits makes use of the bilateral characters of relay contacts. Contacts conduct current in both directions, which electronic gates do not, and sometimes this fact can be used to advantage. Assume that we wish to realize the function

$$T(A, B, C, D, E) = A\bar{B} + \bar{A}C + \bar{A}\bar{B}E + ACE \tag{B.4}$$

The direct realization of this equation is shown in Fig. B.10a. By inspection, we can simplify to the form of Fig. B.10b. We will leave it to the reader to make sure that the two $\bar{B}$ contacts and the two C contacts cannot be combined into single contacts without introducing undesired transmissions. If E could conduct current in only one direction, nothing further could be done to simplify this circuit. But the bilateral character of the contact makes it possible to simplify to the *bridge* circuit of Fig. B.10c. Path $①\to③\to④$ provides the term $\bar{A}C$, path $①\to②\to④$ the term $A\bar{B}$, path $①\to②\to③\to④$ the term ACE, and path $①\to③\to②\to④$ the term $\bar{A}\bar{B}E$.

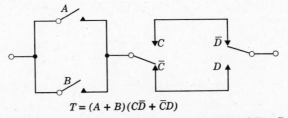

$$T = (A + B)(C\bar{D} + \bar{C}D)$$

FIGURE B.9 *Contact configuration of circuit of Fig. B.7b.*

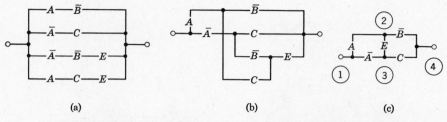

FIGURE B.10   *Simplification of circuit by use of bridge circuit.*

With all these refinements, how do we go about designing relay circuits? First, the techniques presented in this book should be used to find the simplest second-order forms, and the direct realization of these forms should be drawn. Then, by an essentially intuitive combination of visual inspection and algebraic manipulation, contacts are eliminated or combined into transfer contacts wherever possible. Just because there is no single best form, such as second-order, there is no single procedure for finding a minimal circuit or even for determining if a given circuit is minimal. Good relay design is primarily the product of experience.

B.2   Relay Realizations of Symmetric Functions

Symmetric functions are economically realized by using contact networks. Note that the five outputs of the networks in Fig. B11 are the five elementary symmetric functions of four variables. Thus, any symmetric function may be realized by connecting together some combination of outputs. If outputs are connected

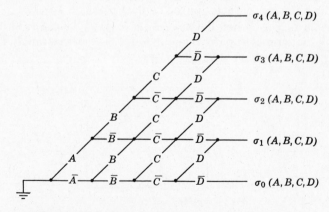

FIGURE B.11   *Relay realization of symmetric functions.*

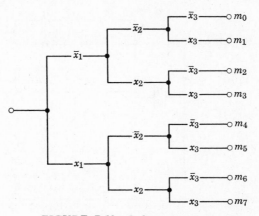

FIGURE B.12   *Relay tree circuit.*

together, some of the contacts become redundant and may be eliminated. For discussion of the procedure involved, the reader is referred to Marcus [1]. Clearly, this realization may be extended to symmetric functions of any number of literals.

Another special network that is conveniently realized by using relays is the standard tree network shown in Fig. B.12. Note that each possible 3-variable minterm is realized at one of the terminals of the tree. Again the extension to n variables is obvious. Networks of this form are useful as selection networks.

B.3   Relays in Sequential Circuits

Relays called latching relays are available; they may be held by a spring in either of two positions when no input is applied. The devices behave in much the same fashion as flip-flops, since when a latching relay is set to one position by the energizing of one of two coils, it remains there until cleared by an energizing of the other coil. Similarly, two ordinary relays can be cross-coupled to form a memory element, as can two NOR or NAND-gates.

Systems featuring latching relays may be considered as synchronous or pulse-mode logic circuits, although this point of view is not particularly common. More often, feedback is added to combinational networks of relays to form fundamental-mode circuits. In Fig. B.13, we see a fundamental-mode relay circuit that will have a 1 output if the most recent input transition has taken place on line A, and a 0 output if the most recent transition has been on line B.

Note that the delay involved in switching a relay is several orders of magnitude longer than the delays in typical electronic switches. The delay in the coils will affect excitation uniformly. Therefore, the switching of contacts due to input

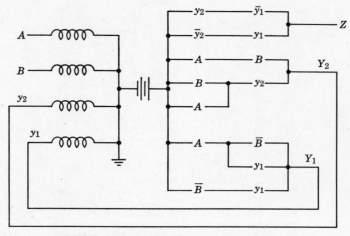

FIGURE B.13 *Relay fundamental mode circuit.*

transitions will always be complete before further transitions are caused by changes in the states of the secondaries. Because of this delay in the feedback loop, essential hazards are not a problem in relay circuits.

The primitive flow table corresponding to the circuit in Fig. B.13 may be found in Fig. B.14a. The implication table, the maximal compatibles, and the minimal state table are found in Figs. B.14b, c, and d, respectively. The excita-

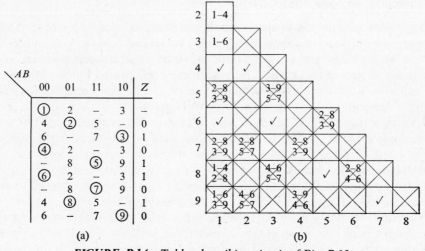

FIGURE B.14 *Tables describing circuit of Fig. B.13.*

(79)(58)(124)(136)

(c)

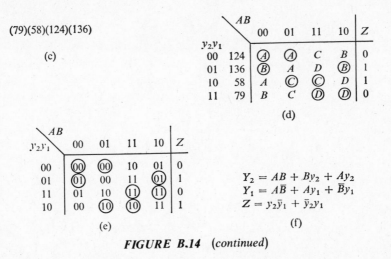

(d)

(e)

$$Y_2 = AB + By_2 + Ay_2$$
$$Y_1 = A\bar{B} + Ay_1 + \bar{B}y_1$$
$$Z = y_2\bar{y}_1 + \bar{y}_2y_1$$

(f)

FIGURE B.14 (continued)

tion map is shown in Fig. B.14e, and the excitation and output expressions are given in Fig. B.14f. Because the outputs are dependent only on the history of input transitions, the minimal state table remains a Moore table, and a hazard-free circuit results without requiring any special precautions.

Bibliography

1. Marcus, M. P. *Switching Circuits for Engineers*, Prentice-Hall, Englewood Cliffs, N.J., 1962.
2. Harrison, Michael. *Introduction to Switching and Automata Theory*, McGraw-Hill, New York, 1965.
3. Caldwell, S. H. *Switching Circuits and Logical Design*, Wiley, New York, 1958.
4. Keister, W., A. Z. Ritchie, and S. H. Washburn. *The Design of Switching Circuits*, Van Nostrand, New York, 1951.

INDEX

613